Genetics and Breeding of Wheat Crops

Genetics and Breeding of Wheat Crops

Editor: Keira Darius

R CALLISTO REFERENCE

www.callistoreference.com

Callisto Reference,
118-35 Queens Blvd., Suite 400,
Forest Hills, NY 11375, USA

Visit us on the World Wide Web at:
www.callistoreference.com

ISBN: 978-1-64116-195-4 (Hardback)

Cataloging-in-Publication Data

Genetics and breeding of wheat crops / edited by Keira Darius.
    p. cm.
Includes bibliographical references and index.
ISBN 978-1-64116-195-4
1. Wheat. 2. Wheat--Genetics. 3. Plant breeding. I. Darius, Keira.
SB191.W5 G46 2019
633.11--dc23

Table of Contents

Preface

In my initial years as a student, I used to run to the library at every possible instance to grab a book and learn something new. Books were my primary source of knowledge and I would not have come such a long way without all that I learnt from them. Thus, when I was approached to edit this book; I became understandably nostalgic. It was an absolute honor to be considered worthy of guiding the current generation as well as those to come. I put all my knowledge and hard work into making this book most beneficial for its readers.

Wheat is a staple food cultivated for its cereal grain. Wheat plants show remarkable genetic variation. Some species of wheat are diploid, but many are polyploids, with four or six pairs of chromosomes. Certain wheat genes have a positive influence on crop yields, such as the dwarfing genes that allow the carbon fixed during photosynthesis to be diverted towards seed production and prevent lodging. The sequencing of the wheat genome and its analysis is valuable for developing an understanding of wheat genome variation, making crosses and alien progression, analyzing evolutionary biology, etc. Identifying the genes responsible for growth, energy production and metabolism also facilitate the breeding of transgenic wheat. Such wheat may be selected for good quality, abiotic stress tolerance and disease resistance, besides high grain yield. This book includes some of the vital pieces of work being conducted across the world, on various topics related to the genetics and breeding of wheat. It unfolds the innovative aspects of wheat breeding and biotechnology, which will be crucial for the progress of this field in the future. It will provide comprehensive knowledge to the readers.

I wish to thank my publisher for supporting me at every step. I would also like to thank all the authors who have contributed their researches in this book. I hope this book will be a valuable contribution to the progress of the field.

Editor

The *in silico* identification and characterization of a bread wheat/*Triticum militinae* introgression line

Michael Abrouk[1], Barbora Balcárková[1], Hana Šimková[1], Eva Komínkova[1], Mihaela M. Martis[2,3], Irena Jakobson[4], Ljudmilla Timofejeva[4], Elodie Rey[1], Jan Vrána[1], Andrzej Kilian[5], Kadri Järve[4], Jaroslav Doležel[1] and Miroslav Valárik[1,*]

[1]Institute of Experimental Botany, Centre of the Region Haná for Biotechnological and Agricultural Research, Olomouc, Czech Republic
[2]Munich Information Center for Protein Sequences/Institute of Bioinformatics and Systems Biology, Institute for Bioinformatics and Systems Biology, Helmholtz Center Munich, Neuherberg, Germany
[3]Division of Cell Biology, Department of Clinical and Experimental Medicine, Bioinformatics Infrastructure for Life Sciences, Linköping University, Linköping, Sweden
[4]Department of Gene Technology, Tallinn University of Technology, Tallinn, Estonia
[5]Diversity Arrays Technology Pty Ltd, Canberra, ACT, Australia

*Correspondence
email
valarik@ueb.cas.cz

Keywords: GenomeZipper, alien introgression, comparative analysis, chromosome rearrangement, chromosome translocation, linkage drag.

Summary

The capacity of the bread wheat (*Triticum aestivum*) genome to tolerate introgression from related genomes can be exploited for wheat improvement. A resistance to powdery mildew expressed by a derivative of the cross-bread wheat cv. Tähti × *T. militinae* (*Tm*) is known to be due to the incorporation of a *Tm* segment into the long arm of chromosome 4A. Here, a newly developed *in silico* method termed rearrangement identification and characterization (RICh) has been applied to characterize the introgression. A virtual gene order, assembled using the GenomeZipper approach, was obtained for the native copy of chromosome 4A; it incorporated 570 4A DArTseq markers to produce a zipper comprising 2132 loci. A comparison between the native and introgressed forms of the 4AL chromosome arm showed that the introgressed region is located at the distal part of the arm. The *Tm* segment, derived from chromosome 7G, harbours 131 homoeologs of the 357 genes present on the corresponding region of Chinese Spring 4AL. The estimated number of *Tm* genes transferred along with the disease resistance gene was 169. Characterizing the introgression's position, gene content and internal gene order should not only facilitate gene isolation, but may also be informative with respect to chromatin structure and behaviour studies.

Introduction

Using interspecific hybridization to widen a crop's gene pool is an attractive strategy for reversing the genetic bottleneck imposed by domestication and for compensating the genetic erosion, which has resulted from intensive selection (Feuillet et al., 2008). Much of the pioneering research in this area has focused on bread wheat (*Triticum aestivum*), in which over 50 related species have been exploited as donors thanks to the plasticity of the recipient's genome (Jiang et al., 1993; Wulff and Moscou, 2014). Typically, introgression events have involved the transfer of a substantially sized donor chromosome segment, which, along with the target, probably bears gene(s), which impact negatively on the host's fitness (a phenomenon also called 'linkage drag') (Gill et al., 2011; Qi et al., 2007; Zamir, 2001). For this reason, very few introgression lines are represented in commercial cultivars (Rey et al., 2015). The prime means of reducing the length of an introgressed segment is to induce recombination with its homoeologous region (Niu et al., 2011). The success of this strategy is highly dependent on the conservation of gene content and order between the donor segment and its wheat equivalent.

The level of resolution with which introgression segments can be characterized has developed over the years along with advances in DNA technology. Large numbers of genetic markers have been identified in many crop species, including wheat (Bellucci et al., 2015; Chapman et al., 2015; Sorrells et al., 2011; Wang et al., 2014). In a recent example, a wheat mapping population has been genotyped with respect to >100 000 markers, but the mapping resolution achieved has only enabled the definition of around 90 mapping bins per chromosome (Chapman et al., 2015). Given that the genomes of most donor species are poorly characterized, marker data at best allow only the position of an introgressed segment to be defined on the basis of the loss of wheat markers; they cannot determine either the size of the introduced segment or analyse its genetic content. The recently developed 'Introgression Browser' (Aflitos et al., 2015) combines genotypic data with phylogenetic inferences to identify the origin of an introgressed segment, but to do so, a high-quality reference sequence of the host genome is needed, along with a large set of donor sequence data. The first of these requirements is being addressed by a concerted effort to acquire a reference sequence for bread wheat (www.wheatgenome.org). So far, only chromosome (3B) has been fully sequenced, and the gene content of each wheat chromosome has been obtained (Choulet et al., 2014; IWGSC, 2014). The so-called GenomeZipper method (Mayer et al., 2011), based on a variety of resources, has been used to predict gene order along each of the 21 bread wheat chromosomes (IWGSC, 2014).

The improved resistance to powdery mildew of an introgressive line 8.1 derived from the cross of bread wheat cv. Tähti (genome formula ABD) and tetraploid *T. militinae* (*Tm*; genome formula A^{t}G) is known to be mainly due to the incorporation of a segment of *Tm* chromatin containing the resistance gene *QPm-tut-4A* into the long arm of chromosome 4A (Jakobson et al., 2006, 2012). Here, a novel *in silico*-based method, termed rearrangement identification and characterization (RICh), has been developed to identify the sequences suitable for generating markers targeting an introgression segment such as the one from *Tm*. The method integrates the GenomeZipper approach with shotgun sequences of chromosome with the introgression. The RICh method was also effective in confirming the identity of the chromosomal rearrangements, which occurred during the evolution of modern wheat.

Results

Chromosome sorting, sequencing and assembly

The flow karyotype derived from the DAPI-stained chromosomes of the DT4AL-TM line included a distinct peak (Figure S1) corresponding to the 4AL telosome (4AL-TM), which enabled it to be sorted to an average purity of 86.2%. The contaminants in the sorted peak comprised a mixture of fragments of various chromosomes and chromatids. DNA of all 45 000 sorted 4AL-TM telosomes was amplified by DNA multiple displacement amplification (MDA). To minimize the risk of representation bias, the products from three independent amplification reactions were pooled. From the resulting 4.5 μg DNA, a total of ~6.2 Gb of sequence was obtained, which was subsequently assembled into 279 077 contigs of individual length >200 bp, with an N50 of 2068 bp (Table 1). When the assembly was aligned with the reference genome sequences of *Brachypodium distachyon* (Vogel *et al.*, 2010), rice (IRGSP, 2005) and sorghum (Paterson *et al.*, 2009), it was apparent that the 4AL-TM telosome shares synteny with segments of *B. distachyon* chromosomes Bd1 and Bd4, rice chromosomes Os3, Os6 and Os11 and sorghum chromosomes Sb1, Sb5 and Sb10 (Figure S2).

Origin of the introgression segment

The chromosomal origin of the *Tm* introgression segment was established by initially flow sorting the *Tm* chromosome complement. This was achieved by pretreating the chromosomes with fluorescence *in situ* hybridization in suspension (FISHIS) (Giorgi *et al.*, 2013) in which GAA microsatellites were fluorescently labelled by FITC. The resulting DAPI *vs* GAA bivariate flow karyotype succeeded in defining 13 distinct clusters (Figure 1). As the haploid chromosome number of *Tm* is 14, one of the clusters was therefore

deemed likely to harbour a mixture of two distinct chromosomes. Two of the clusters (#4 and #8) contained sequences that were amplified by the *Xgwm160* (Roder *et al.*, 1998) and *owm82* primers (these two markers are linked to the *QPm-tut-4A* gene from *Tm* introgression). The dispersed profile of cluster #4 (Figure 1) suggested that it was composed of two different A^t genome chromosomes, because all G chromosomes were identified due to a higher GAA content (Badaeva *et al.*, 2010). The *owm72* marker, also linked to the *QPm-tut-4A* gene, amplified two fragments in *Tm*, one of size 205 bp and the other of size 250 bp; only the former was amplified from 4AL-TM telosome or of cluster #8. The fluorescence *in situ* hybridization (FISH) profile of the chromosomes present in cluster #8 unambiguously identified the introgressed segment as deriving from chromosome 7G.

GenomeZipper improvement

A chromosome 4A zipper was constructed based on Chinese Spring (CS) chromosome specific survey sequences (CSSs) using 1780 specific DArTseq markers ordered in consensus genetic map (Table S1). As DArTseq marker sequences are short (69 nt) and generally nongenic, they were initially anchored to the CSS assembly; this step reduced the number of useful markers to 632 (CSS-DArTseq markers), of which 102 mapped to the short arm and 530 to the long arm. The first version of the zipper comprised a total of 2398 loci. The resulting model for 4AS was collinear with Bd1, Os3 and Sb01, as reported previously (Hernandez *et al.*, 2012). However, the one for 4AL was a mosaic of 15 orthologous blocks (based on the rice genome as the reference), derived from Os11/Bd4/Sb5, Os3/Bd1/Sb1 and Os6/Bd1/Sb10 (Figure S3a). Validation for this complex structure was sought from analysis of the subset of 2638 SNP loci (Wang *et al.*, 2014), which had been assigned a bin locations based on an analysis of a panel of established 4A deletion lines (Endo and Gill, 1996): of these, 750 mapped to five deletion bins on 4AS

Table 1 Assembly statistics of chromosome arms 4AL-TM, 4AS-CS and 4AL-CS

	4AS-CS	4AL-CS	4AL-TM
Sequencing read depth	241x	116x	23x
Total contigs	301 954	362 851	279 077
Total bases (bp)	282 335 959	361 971 522	266 737 930
Assembly coverage*	0.89x	0.67x	0.49x
Min contig length (bp)	200	200	200
Max contig length (bp)	70 057	129 043	28 604
Average contig length (bp)	935	998	956
N50 length (bp)	2782	3053	2068

The data for 4AS-CS and 4AL-CS arms are taken from IWGSC (2014) and data for 4AL-TM were acquired in this study.

*The size of chromosome arms 4AS-CS (318 Mbp) and 4AL-CS (540 Mbp) were taken from Šafář *et al.* (2010). To estimate the assembly coverage of the 4AL-TM arm, the 4AL-CS size was used.

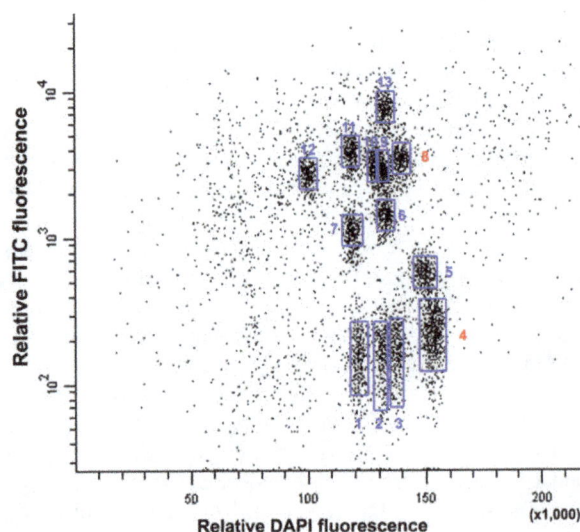

Figure 1 The bivariate flow karyotype of *T. militinae*. Mitotic chromosomes at metaphase were stained with DAPI and GAA microsatellites were labelled with FITC. A set of 13 distinct clusters were obtained (shown boxed). Cluster #8 harbours the *Tm* chromosome (7G) which was the origin of the introgression segment present in line 8.1. Cluster #4 harbours a putative homoeolog of 7G and based on its width and shape most likely comprises a mixture of two distinct chromosomes.

and 1888 to 13 deletion bins on 4AL (Figure S3, Balcárková *et al.*, unpublished). The analysis allowed 329 SNP loci (113 on 4AS, 216 on 4AL) to be integrated into the new 4A zipper. Of the 113 4AS SNP loci, just four mapped to an inconsistent locations, demonstrating the model's accuracy; however, six (#3, #6, #8, #12, #14 and #16) of the 15 4AL blocks were inconsistent with respect to the multiple SNP loci allocations. For example, block #12—positioned in the subtelomeric region according to the zipper—included 18 SNP loci assigned to the pericentromeric region. The GenomeZipper was therefore rerun after first removing the 62 CSS-DArTseq markers associated with the misassignment of the blocks (Table S2); of the 570 CSS-DArTseq markers retained (Table S3), 79 were anchored to at least one of the *B. distachyon*, rice or sorghum scaffolds. The set of 2132 loci (745 on 4AS and 1387 on 4AL) revealed just six (rather than 16) blocks (Figure S3b, Table S2). The final structure resembles that described by Hernandez *et al.* (2012). When the model was retested with SNP markers, no further discrepancies were flagged along distal part of chromosome arm 4AL (Figure S3b).

The *in silico* characterization of the evolutionary chromosome rearrangements on 4AL

The RICh method is based on a stringent identification and density estimation of homoeologs and is validated using a segmentation analysis. To test the approach, the CSS-based scaffolds of chromosome arms 4BS, 4BL, 4DS, 4DL, 5BL, 5DL, 7AS and 7DS (IWGSC, 2014) were compared with that of chromosome 4A, applying as the criteria a 90% level of identity and a minimum alignment length of 100 bp. The numbers of homoeologous loci obtained were, respectively, 719, 762, 636, 877, 850, 673, 602 and 627 (mean 718), but no common distinct blocks allowing for the definition of evolutionary translocations could be identified. A window size of eleven genes was then selected from the 4A zipper for the subsequent segmentation analysis. The ancestral 4AS and 4AL arms had an average density of 0.83, while the remainder of 4AL had a density of only 0.41 (Figure 2a). 4BL and 4DL sequences were homologous to 4AS, and 4BS and 4DS ones to 4AL, confirming the pericentromeric inversion event uncovered before (Devos *et al.*, 1995; Hernandez *et al.*, 2012; Ma *et al.*, 2013; Miftahudin *et al.*, 2004). Immediately following the ancestral 4AL region, the density of homoeologs associated with chromosome group 5 increased (5BL and 5DL: 147 genes, density 0.73), identifying the presence of ancestral 5AL chromatin on this arm (Figure 1b). Finally, the most distal segment of 4AL was associated with an increased density of chromosome group 7 (7AS and 7DS: 557 genes, density 0.45), confirming the ancestral translocation event involving 7BS (Figure 1C).

Characterization of the *Tm* introgression segment

The RICh approach was then used to characterize the 4AL introgressed *Tm* segment. A direct comparison between the 4AL-TM sequence assembly and the 4A-CS zipper (95% identity, 100 bp minimum alignment length) was then made. For the long arm, the segmentation analysis revealed two distinct regions (Figure 3): the more proximal one had a high density of homologous genes (~0.84, 863 loci), so likely corresponds to a region of the 4AL telosome inherited from bread wheat (Figure 3). However, in the distal part of the arm, the homologous gene density fell to ~0.37, suggesting this as the site of the translocation event (Figure 3). Considering the same number of

genes in the homologous regions of CS DT4AL chromosome arm (4AL-CS) and 4AL-TM, the comparison between these proximal segments revealed that 16% of homologous genes (167 of 1030) in the 4AL-TM assembly were not identified and may be accounted to the sequencing and assembly imperfection. If this rate of imperfection is applied to the regions including the introgressed segment (357 CS genes *vs* 131 *Tm* homologous genes), the presence of 169 CS nonhomologous genes in the introgression segment could be estimated. The number of such genes represents the size of linkage drag (neglecting allelic variation of the homologous genes).

Discussion

Introgression from related species provides many opportunities to broaden the genetic base of wheat, but its impact on wheat improvement has been limited by a combination of imperfect homology between donor and recipient chromatin, the loss of key recipient genes, the suppression of recombination and linkage drag effect. Thus, obtaining an accurate understanding of the size, homology, orientation and position of an introgressed segment could help to determine which introgression events are more likely to avoid incurring a performance penalty. Such knowledge would also be informative in the context of isolating a valuable gene introduced via an introgression event. Gaining this information requires saturating the target region with molecular markers. In an effort to clone of *QPm-tut-4A* gene introgressed to the wheat 4A chromosome from *T. militinae*, we developed new method for chromosome rearrangements and introgressions identification and characterization.

The presence of ancient intra- and interchromosomal rearrangements is a known complicating issue in the polyploid wheat genome, and the 4AL chromosome arm, which is one of the site of the introgression event selected in line 8.1, has a particularly complex structure. The composition of the proximal segment of the 4AL telosomes carried by DT4AL-TM and the standard CS DT4AL stock was largely identical, as expected. However, distal part of the telosomes differs in presence of *Tm* introgressive segment (Jakobson *et al.*, 2012), but no difference by synteny blocks could be detected. In hybrids between the tetraploid forms *T. turgidum* and *T. timopheevi*, Gill and Chen (1987) noted that while the latter's G genome chromosomes paired most frequently with those from the B genome, chromosome 4A was occasionally involved in pairing with chromosome 7G, presumably as a result of the presence of the 7BS segment on the *T. turgidum* 4AL arm. The likelihood is therefore that the *Tm* chromosome 7G segment, which has contributed the 4A-based powdery mildew resistance of line 8.1, was introduced via homologous recombination with the segment of 4AL carrying 7BS chromatin.

To increase resolution of the analysis, the GenomeZipper method (Mayer *et al.*, 2011), combining genetic maps, data from chromosome shotgun sequencing, and synteny information with sequenced model genomes has been adopted. The method has been useful for developing virtual gene orders in both wheat and barley chromosomes (IWGSC, 2014; Mayer *et al.*, 2011). The most crucial data set is a reliable genetic map, which serves as backbone to integrate and orient the identified syntenic blocks. Two zippers for chromosome 4A have been published to date. The first was based on relatively low coverage sequencing of the chromosome, employing as its backbone a barley linkage map formed from expressed sequence tags distributed over the

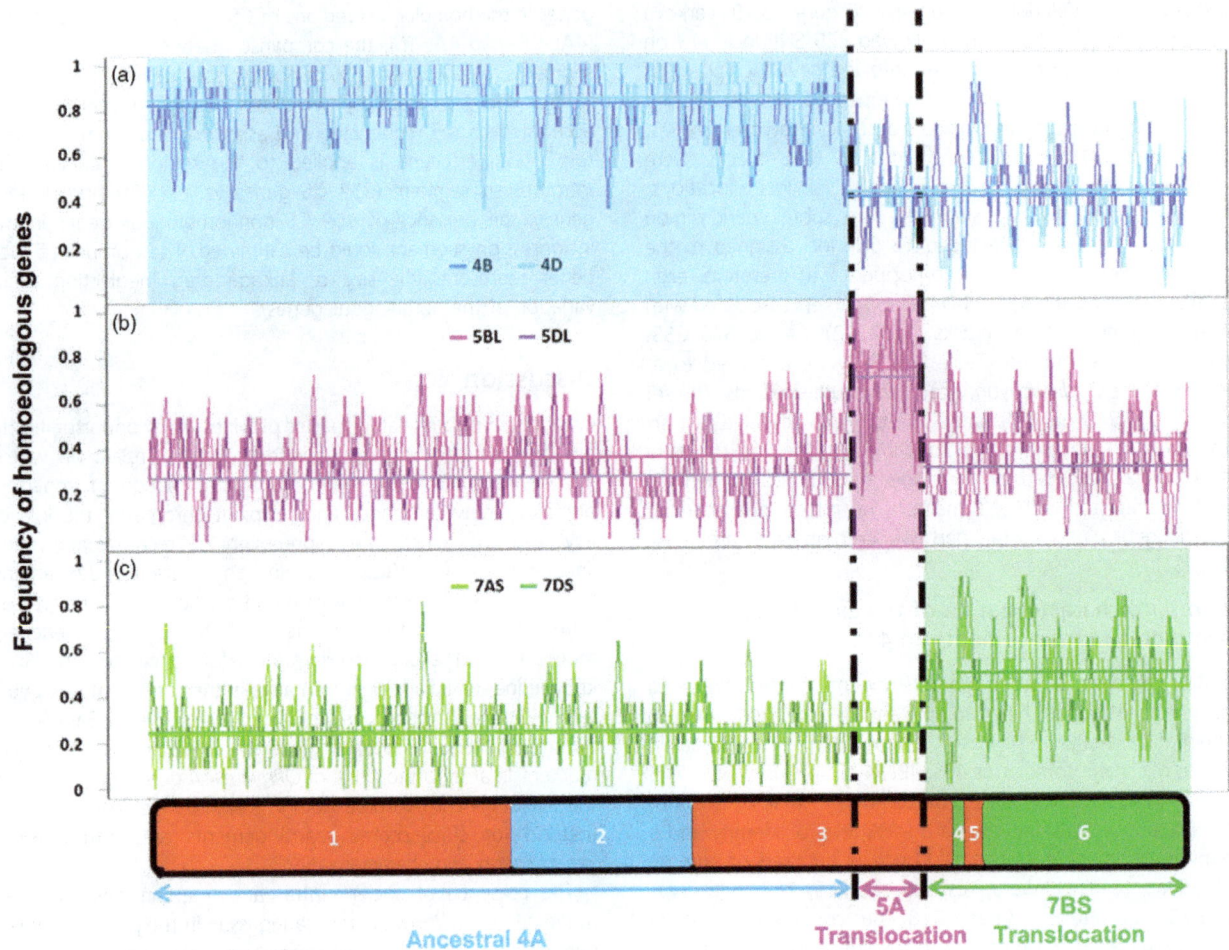

Figure 2 Variation in homoeologous gene density along the various 4A-CS chromosome segments compared to their homoeologous chromosomes. The structure of native 4A-CS chromosome is represented at the bottom with syntenic blocks with rice genome shown in different colours (red = Os3; blue = OS11; green = Os6). (a) The 4A homoeologous gene density compared to 4B and 4D chromosomes, (b) comparison with the 5BL and 5DL chromosome arms and (c) comparison with chromosome arms 7AS and 7DS is shown as homoeologous genes frequency histogram. Homoeologous regions are characterized by a high average frequency (denoted by the horizontal lines). The lower average frequency shown by the group 7 chromosomes reflects a significantly lower sequencing coverage.

chromosome arms 4HS (117 loci), 4HL (16 loci), 5HL (36 loci) and 7HS (36 loci) (Hernandez et al., 2012). The second was based on the 4A CSS and wheat SNP map and consisted of 167 markers on 4AS ordered into 56 mapping bins and 200 (92 mapping bins) markers on 4AL; these were combined with a linkage map developed from a mapping population bred from a cross between bread wheat cv. Opata and a synthetic wheat (Sorrells et al., 2011). Neither of these two zippers was able to provide a sufficient level of resolution to identify the Tm introgression into 4AL chromosome arm. The present new zipper was based on consensus DArT map derived from crosses with CS and comprised 55% more markers and 25% more mapping bins than the latter one, which approximately doubled the number of ordered genes/loci (2132 vs 1004), and was informative with respect to the Tm introgression. When this improved zipper was used in conjunction with the RICh method, it was also possible to recognize the three evolutionary rearrangements, which have long been known to have generated the structure of the modern chromosome arm 4AL (Figure 2) (Devos et al., 1995; Hernandez et al., 2012; Ma et al., 2013; Miftahudin et al., 2004). Similarly, it was able to

identify that a lower density of homologous genes obtained at the distal end of the 4AL-TM telosome (Figure 3) is representing the region harbouring the segment introgressed from Tm. The Tm introgression overlaps with almost the entire chromosome 7BS segment now present on 4AL (Figure 3, Table S2), while the proximal region of the 4AL-CS and 4AL-TM telosomes is essentially of bread wheat origin. The number of wheat loci retained in this latter region did, however, differ by 16% in gene content (4AL-CS—1030 and 4AL-TM—863 genes). This difference may be result of lower sequencing coverage of the 4AL-TM (30x compared to 116x of the 4AL-CS (IWGSC, 2014)) and thus lower representation of the 4AL-TM sequence assembly. If we assume the similar gene density in homologous chromosomes of relative species, as reported before by Tiwari et al. (2015), and if the same rate of missing genes as above due to sequencing and assembly imperfections is assumed, estimated 169 CS nonhomologous genes were carried by the introgression in linkage drag. Knowledgeable selection of parental lines that have relatively high frequency of homologous genes in the region of interest (e.g. QTL for resistance in the Tm introgression, Figure 3) may

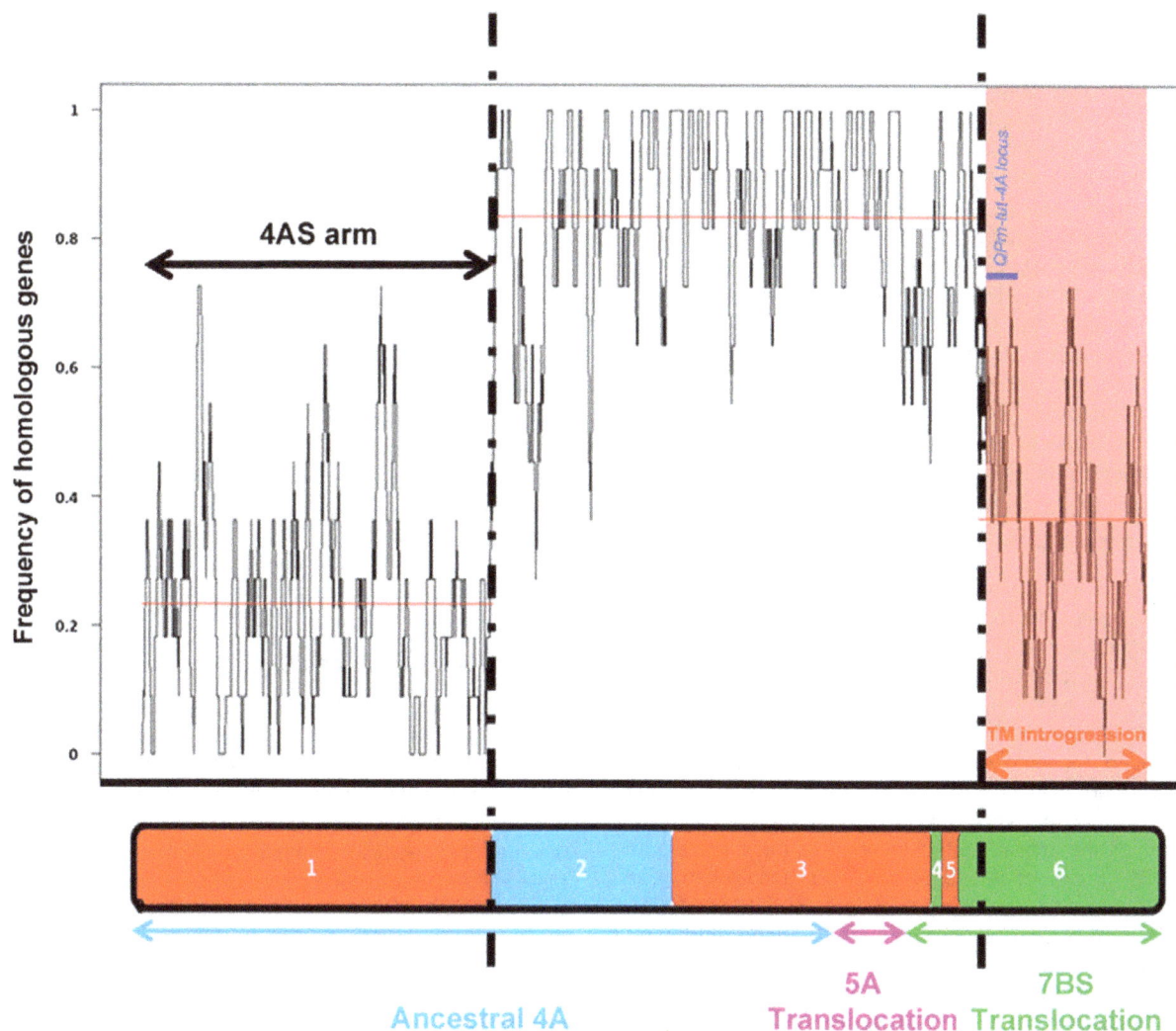

Figure 3 Variation in homologous gene density between 4A-CS chromosome and 4AL-TM telosome. The structure of native 4A-CS chromosome is represented at the bottom with syntenic blocks with rice genome shown in different colours (red = Os3; blue = OS11; green = Os6). The homologous gene density along the 4A-CS zipper compare to 4AL-TM assembly is shown with the black line. The segment of the *Tm* introgression overlaps the 7BS translocation in 4AL (red highlight). The equivalent region on the 4AL-CS telosome harbours 357 genes, only 131 have homologous genes on the *Tm* segment. The dark blue bar represents approximate localization of the *QPm-tut-4A* locus.

increase chances of unobstructed recombination as was observed in the *QPm-tut-4A* locus (Jakobson *et al.*, 2012). So, reducing the length of the introgression segment by inducing further rounds of recombination can lessen (or even eliminate) any negative effects of linkage drag. Application of the RICh approach should prove informative regarding the order or frequency of homologous genes of any such selections. Overall, the RICh method offers a robust means of both characterizing chromosome rearrangements and of predicting the gene content of a specific chromosomal region. Recent advances in high-throughput genotyping permits the elaboration of ever higher density linkage maps (Bellucci *et al.*, 2015; Chapman *et al.*, 2015; Sorrells *et al.*, 2011; Wang *et al.*, 2014). The status of chromosome flow sorting is such that almost any wheat chromosome (Tsõmbalova *et al.*, 2016) and also chromosomes in many crops (Doležel *et al.*, 2014) can now be isolated to a reasonable purity, while the advances in NGS sequencing make RICh widely affordable. These developments should facilitate the preparation of materials needed for

applying the RICh approach, thereby offering novel opportunities for a wide range of prebreeding activities, positional cloning, chromatin hybridization and structural studies.

Experimental procedures

Plant materials

Grains of the bread wheat ditelosomic CS DT4AL line were provided by Dr. Bikram Gill (KSU, Manhattan, KS), those of the two nullisomic–tetrasomic lines N4AT4B and N4AT4D (Sears and Sears, 1978) by the National BioResource Centre (Kyoto, Japan), those of *Tm* ($2n = 4x = 28$, genome formula A^tA^tGG) accession K-46007 by the N.I. Vavilov Institute of Plant Industry (St. Petersburg, Russia). The line denoted DT4AL-TM was generated from the cross CS DT4AL × 8.1: the line carries 40 bread wheat chromosomes and a pair of 4AL telosomes with the *Tm* introgression (4AL-TM) and is resistant to powdery mildew (Jakobson *et al.*, 2012).

Flow sorting and amplification of the 4AL telosome carried by 4AL-TM

Liquid suspensions of mitotic chromosomes were prepared from root tips of 4AL-TM seedlings as described by (Vrána et al., 2000). The telosomes were separated from the rest of the genome by flow sorting, using a FACSAria II SORP flow cytometer and sorter (BD Biosciences, San Jose, CA). The level of contamination within a sorted peak was determined using FISH, based on probes detecting telomeric repeats, the Afa repeat and $(GAA)_n$, following the methods described by Kubaláková et al. (2003). The flow-sorted 4AL-TM telosomes were treated with proteinase, after which DNA was extracted using a Millipore Microcon YM-100 column (www.millipore.com). Chromosomal DNA was MDA amplified using the Illustra GenomiPhi V2 DNA amplification kit (GE Healthcare) as described by Šimková et al. (2008).

Identifying the origin of the introgression segment on the 4AL-TM telosome

Chromosomes of T. militinae were flow sorted as described above. However, prior to flow cytometry, GAA microsatellites on chromosomes were labelled by FITC using FISHIS protocol (Giorgi et al., 2013). Bivariate analysis (DNA content/DAPI vs GAA/FITC) enabled discrimination of 13 of 14 chromosomes of T. militinae. Individual chromosome fractions were sorted into tubes for PCR amplification and onto microscopic slides for identification of sorted chromosomes by FISH. Three markers linked to the Tm powdery mildew resistance gene QPm-tut-4A were used for the selection of the critical cluster: these were the microsatellite Xgwm160 (Roder et al., 1998) and two unpublished, one (owm72) amplified by the primer pair 5'-TGCTTGCTTGTA GATTGTGCA/5'-CCAGTAAGCTTTGCCGTGTG) and the other (owm82) by 5'-GGGAGAGACGAAAGCAGGTA/5'-CTTGCATG CACGCCAGAATA. Each 20 µL PCR contained 0.01% (w/v) o-cresol sulphonephtalein, 1.5% (w/v) sucrose, 0.2 mM dNTP, 0.6 U Taq DNA polymerase and 1 µM of each primer in 10 mM Tris-HCl/50 mM KCl/1.5 mM MgCl₂/0.1% (v/v) Triton X-100. The template comprised about 500 sorted chromosomes. Test reactions were seeded with either 20 ng genomic DNA extracted from CS, Tm, N4AT4B or N4AT4D, or with 50 pg of MDA amplified DNA from 4AL-CS and 4AL-TM telosomes. The reactions were subjected to an initial denaturation (95 °C/5 min), followed by 40 cycles of 95 °C/30 s, 55 °C/30 s and 72 °C/30 s, and completed with an elongation of 72 °C/10 min. The products were electrophoretically separated through 4% nondenaturing polyacrylamide gels and visualized by EtBr staining. The markers were mapped using a F_2 population bred from the cross CS × 8.1 (Jakobson et al., 2012).

Sequencing of the 4AL telosome

A CSS assembly of CS chromosome arms 4AS (4AS-CS) and 4AL (4AL-CS) were acquired from Internation Wheat Genome Sequencing Consortium (IWGSC, 2014). Two sequencing libraries of DNA amplified from the 4AL-TM telosome were constructed using a Nextera kit (Illumina, San Diego, CA) with the insert size adjusted to 500 and 1000 bp. The resulting clones were sequenced as paired-end reads by IGA (Udine, Italy) using a HiSeq 2000 device (Illumina). The 4AL-TM reads were assembled with SOAPdenovo2 software, applying a range of k-mers (54–99, with a step size of 3) to select the assembly with the highest coverage and the largest N50. Assembled scaffolds (k-mer of 69,

minimum length 200 bp) were chosen for further analysis (Table 1).

DArTseq and SNP maps for GenomeZipper construction and validation

A DArTseq consensus map, based on four crosses involving cv. Chinese Spring as a parent has been provided by DArT PL (www.diversityarrays.com). Individual maps were created using DArT PL's OCD MAPPING program (Petroli et al., 2012) to order DArTseq and array-based DArTs. DArT PL's consensus mapping software (Raman et al., 2014) was applied to create a consensus map using similar strategy as described in Li et al. (2015). Version 3.0 of consensus map with approximately 70 000 markers was used in this study.

A SNP deletion map (Balcárková et al., unpublished) was used for validation. Genomic DNAs of a set of 15 chromosome 4A deletion lines (Endo and Gill, 1996) and DNAs amplified from 4AL-CS and 4AS-CS chromosome arms as controls were genotyped at USDA-ARS (Fargo, ND) using a iSelect 90k SNP array (Wang et al., 2014) on Infinium platform (Illumina). The raw genotypic data were manually analysed using GenomeStudio V2011.1 software (Illumina).

Comparative analysis and GenomeZipper analysis

Synteny between related genomic segments was assessed using ChromoWIZ software (Nussbaumer et al., 2014). The number of conserved genes present within a series of 0.5-Mbp genomic windows (window shift 0.1 Mbp) was determined. The consensus chromosome 4A linkage map used as the backbone for the GenomeZipper analysis comprised 1780 DArTseq markers (Table S1). As these sequences are mostly short (69 nt) and few identify coding sequence, they were first aligned to the set of 4A CSS contigs, preserving only those contigs that matched the entire DArTseq marker sequence at a level of at least 98% identity. The retained CSS contigs ('CSS-DArTseq markers') were used for the construction of the zipper, which was subsequently validated against the SNP deletion map (2706 SNPs). Similarly as above, only those 4A CSS contigs that aligned with SNP loci along their entire length (98% identity threshold) were retained. Ordering of the CSS-DArTseq markers was compared with that ordered by SNPs from the deletion bin map and CSS-DArTseq markers which do not follow the SNP order were eliminated, and a second version of the zipper was generated using the remaining markers (Table S3). This version was revalidated against the SNP deletion map.

The RICh approach

To identify introgressed/translocated regions, the final 4A zipper was compared to the complete set of CSS sequences obtained from chromosome arms 4BS, 4BL, 4DS, 4DL, 5BL, 5DL, 7AS and 7DS (IWGSC, 2014). Alignments were performed using the BLAST algorithm (Altschul et al., 1990). The BLAST outputs were filtered by applying the following criteria: a minimum identity of either 90% (translocation analysis) or 95% (introgression analysis) and a minimum alignment length of 100 bp. For each comparison, the density of homologous genes was evaluated using a sliding window of eleven genes (five upstream and five downstream), and a segmentation analysis was performed using the R package changepoint v1.1 (Killick and Eckley, 2014), applying the parameter segment neighbourhoods method with a BIC penalty on the mean change. The method allows a statistical detection of gene density changes along the chromosome, corresponding to an increase or decrease in the level of synteny. For translocation events, an increase in synteny level with one group of homoeologs is required, while for an introgression, a loss of orthology is anticipated.

Acknowledgements

We thank Z. Dubská and R. Šperková for their technical assistance with flow sorting and M. Kubaláková for her FISH-based identification of sorted chromosomes. L. Pingault and E. Paux are acknowledged for their contribution to the statistical analyses, S. Chao for the deletion lines genotyping and J. Bartoš for his critical reading of the manuscript and for his helpful suggestions. This research was supported by grants awarded by the National Program of Sustainability I (No. LO1204), by the Czech Science Foundation (No. 14-07164S) and by the Estonian Ministry of Agriculture.

References

Aflitos, S.A., Sanchez-Perez, G., de Ridder, D., Fransz, P., Schranz, M.E., de Jong, H. and Peters, S.A. (2015) Introgression browser: high-throughput whole-genome SNP visualization. *Plant J.* **82**, 174–182.

Altschul, S.F., Gish, W., Miller, W., Myers, E.W. and Lipman, D.J. (1990) Basic local alignment search tool. *J. Mol. Biol.* **215**, 403–410.

Badaeva, E.D., Budashkina, E.B., Bilinskaya, E.N. and Pukhalskiy, V.A. (2010) Intergenomic chromosome substitutions in wheat interspecific hybrids and their use in the development of a genetic nomenclature of Triticum timopheevii chromosomes. *Russian Journal of Genetics*, **46**, 769–785.

Bellucci, A., Torp, A.M., Bruun, S., Magid, J., Andersen, S.B. and Rasmussen, S.K. (2015) Association mapping in Scandinavian winter wheat for yield, plant height, and traits important for second-generation bioethanol production. *Front. Plant Sci.* **6**, 1046.

Chapman, J.A., Mascher, M., Buluc, A., Barry, K., Georganas, E., Session, A., Strnadova, V. *et al.* (2015) A whole-genome shotgun approach for assembling and anchoring the hexaploid bread wheat genome. *Genome Biol.* **16**, 26.

Choulet, F., Alberti, A., Theil, S., Glover, N., Barbe, V., Daron, J., Pingault, L. *et al.* (2014) Structural and functional partitioning of bread wheat chromosome 3B. *Science*, **345**, 1249721.

Devos, K.M., Dubcovsky, J., Dvořák, J., Chinoy, C.N. and Gale, M.D. (1995) Structural evolution of wheat chromosomes 4A, 5A and 7B and its impact on recombination. *Theor. Appl. Genet.* **91**, 282–288.

Doležel, J., Vrána, J., Cápal, P., Kubaláková, M., Burešová, V. and Simková, H. (2014) Advances in plant chromosome genomics. *Biotechnol. Adv.* **32**, 122–136.

Endo, T. and Gill, B. (1996) The deletion stocks of common wheat. *J. Hered.* **87**, 295–307.

Feuillet, C., Langridge, P. and Waugh, R. (2008) Cereal breeding takes a walk on the wild side. *Trends Genet.* **24**, 24–32.

Gill, B.S. and Chen, P.D. (1987) Role of cytoplasm-specific introgression in the evolution of the polyploid wheats. *Proc. Natl Acad. Sci. USA*, **84**, 6800–6804.

Gill, B.S., Friebe, B.R. and White, F.F. (2011) Alien introgressions represent a rich source of genes for crop improvement. *Proc. Natl Acad. Sci. USA*, **108**, 7657–7658.

Giorgi, D., Farina, A., Grosso, V., Gennaro, A., Ceoloni, C. and Lucretti, S. (2013) FISHIS: fluorescence in situ hybridization in suspension and chromosome flow sorting made easy. *PLoS ONE*, **8**, e57994.

Hernandez, P., Martis, M., Dorado, G., Pfeifer, M., Gálvez, S., Schaaf, S., Jouve, N. *et al.* (2012) Next-generation sequencing and syntenic integration of flow-sorted arms of wheat chromosome 4A exposes the chromosome structure and gene content. *Plant J.* **69**, 377–386.

IRGSP. (2005) The map-based sequence of the rice genome. *Nature*, **436**, 793–800.

IWGSC. (2014) A chromosome-based draft sequence of the hexaploid bread wheat (*Triticum aestivum*) genome. *Science*, **345**, 1251788.

Jakobson, I., Peusha, H., Timofejeva, L. and Järve, K. (2006) Adult plant and seedling resistance to powdery mildew in a *Triticum aestivum* x *Triticum militinae* hybrid line. *Theor. Appl. Genet.* **112**, 760–769.

Jakobson, I., Reis, D., Tiidema, A., Peusha, H., Timofejeva, L., Valárik, M., Kladivová, M. *et al.* (2012) Fine mapping, phenotypic characterization and validation of non-race-specific resistance to powdery mildew in a wheat-*Triticum militinae* introgression line. *Theor. Appl. Genet.* **125**, 609–623.

Jiang, J., Friebe, B. and Gill, B. (1993) Recent advances in alien gene transfer in wheat. *Euphytica*, **73**, 199–212.

Killick, R. and Eckley, I.A. (2014) changepoint: an R package for changepoint analysis. *J. Stat. Softw.* **58**, 1–19.

Kubaláková, M., Valárik, M., Bartoš, J., Vrána, J., Číhalíková, J., Molnár-Láng, M. and Doležel, J. (2003) Analysis and sorting of rye (*Secale cereale* L.) chromosomes using flow cytometry. *Genome*, **46**, 893–905.

Li, H., Vikram, P., Singh, R., Kilian, A., Carling, J., Song, J., Burgueno-Ferreira, J. *et al.* (2015) A high density GBS map of bread wheat and its application for dissecting complex disease resistance traits. *BMC Genom.* **16**, 216.

Ma, J., Stiller, J., Berkman, P.J., Wei, Y., Rogers, J., Feuillet, C., Doležel, J. *et al.* (2013) Sequence-based analysis of translocations and inversions in bread wheat (*Triticum aestivum* L.). *PLoS ONE*, **8**, e79329.

Mayer, K., Martis, M., Hedley, P., Simkova, H., Liu, H. and Morris, J. (2011) Unlocking the barley genome by chromosomal and comparative genomics. *Plant Cell*, **23**, 1249–1263.

Miftahudin, Ross, K., Ma, X.-F., Mahmoud, A.A., Layton, J., Milla, M.A.R., Chikmawati, T. *et al.* (2004) Analysis of expressed sequence tag loci on wheat chromosome group 4. *Genetics*, **168**, 651–663.

Niu, Z., Klindworth, D.L., Friesen, T.L., Chao, S., Jin, Y., Cai, X. and Xu, S.S. (2011) Targeted introgression of a wheat stem rust resistance gene by DNA marker-assisted chromosome engineering. *Genetics*, **187**, 1011–1021.

Nussbaumer, T., Kugler, K.G., Schweiger, W., Bader, K.C., Gundlach, H., Spannagl, M., Poursarebani, N. *et al.* (2014) chromoWIZ: a web tool to query and visualize chromosome-anchored genes from cereal and model genomes. *BMC Plant Biol.* **14**, 348.

Paterson, A.H., Bowers, J.E., Bruggmann, R., Dubchak, I., Grimwood, J., Gundlach, H., Haberer, G. *et al.* (2009) The Sorghum bicolor genome and the diversification of grasses. *Nature*, **457**, 551–556.

Petroli, C., Sansaloni, C., Carling, J., Steane, D., Vaillancourt, R. and Myburg, A. (2012) Genomic characterization of DArT markers based on high-density linkage analysis and physical mapping to the eucalyptus genome. *PLoS ONE*, **7**, e44684.

Qi, L., Friebe, B., Zhang, P. and Gill, B.S. (2007) Homoeologous recombination, chromosome engineering and crop improvement. *Chromosome Res.* **15**, 3–19.

Raman, H., Raman, R., Kilian, A., Detering, F., Carling, J. and Coombes, N. (2014) Genome-wide delineation of natural variation for pod shatter resistance in *Brassica napus*. *PLoS ONE*, **9**, e101673.

Rey, E., Molnár, I. and Doležel, J. (2015) Genomics of wild relatives and alien introgressions. In *Alien Introgression in Wheat: Cytogenetics, Molecular Biology, and Genomics* (Molnár-Láng, M., Ceoloni, C. and Doležel, J., eds), pp. 347–381. Cham: Springer International Publishing.

Roder, M.S., Korzun, V., Wendehake, K., Plaschke, J., Tixier, M.H., Leroy, P. and Ganal, M.W. (1998) A microsatellite map of wheat. *Genetics*, **149**, 2007–2023.

Šafář, J., Šimková, H., Kubaláková, M., Číhalíková, J., Suchánková, P., Bartoš, J. and Doležel, J. (2010) Development of chromosome-specific BAC resources for genomics of bread wheat. *Cytogenet Genome Res.* **129**, 211–223.

Sears, E.R. and Sears, L.M.S. (1978) The telocentric chromosomes of common wheat. In *Proc 5th Int Wheat Genet Symp* (Ramanujam, S., ed), pp. 389–407. New Delhi: Indian Soc of Genet Plant Breed.

Šimková, H., Svensson, J.T., Condamine, P., Hřibová, E., Suchánková, P., Bhat, P.R., Bartoš, J. *et al.* (2008) Coupling amplified DNA from flow-sorted chromosomes to high-density SNP mapping in barley. *BMC Genom.* **9**, 294.

Sorrells, M.E., Gustafson, J.P., Somers, D., Chao, S.M., Benscher, D., Guedira-Brown, G., Huttner, E. *et al.* (2011) Reconstruction of the synthetic W7984 x Opata M85 wheat reference population. *Genome*, **54**, 875–882.

Tiwari, V.K., Wang, S., Danilova, T., Koo, D.H., Vrana, J., Kubalakova, M., Hribova, E. *et al.* (2015) Exploring the tertiary gene pool of bread wheat: sequence assembly and analysis of chromosome 5M(g) of *Aegilops geniculata*. *Plant J.* **84**, 733–746.

Tsõmbalova, J., Karafiátová, M., Vrána, J., Kubaláková, M., Peuša, H., Jakobson, I., Järve, M. *et al.* (2016) A haplotype specific to North European wheat (*Triticum aestivum* L.). *Genet. Resour. Crop Evol.* DOI 10.1007/s10722-016-0389-9

Vogel, J.P., Garvin, D.F., Mockler, T.C., Schmutz, J., Rokhsar, D., Bevan, M.W., Barry, K. *et al.* (2010) Genome sequencing and analysis of the model grass *Brachypodium distachyon*. *Nature,* **463**, 763–768.

Vrána, J., Kubaláková, M., Simková, H., Číhalíková, J., Lysák, M.A. and Doležel, J. (2000) Flow sorting of mitotic chromosomes in common wheat (*Triticum aestivum* L.). *Genetics,* **156**, 2033–2041.

Wang, S., Wong, D., Forrest, K., Allen, A., Chao, S., Huang, B.E., Maccaferri, M. *et al.* (2014) Characterization of polyploid wheat genomic diversity using a high-density 90 000 single nucleotide polymorphism array. *Plant Biotechnol. J.* **12**, 787–796.

Wulff, B.B.H. and Moscou, M.J. (2014) Strategies for transferring resistance into wheat: from wide crosses to GM cassettes. *Front. Plant Sci.* **5**, 692. doi: 10.3389/fpls.2014.00692

Zamir, D. (2001) Improving plant breeding with exotic genetic libraries. *Nat. Rev. Genet.* **2**, 983–989.

2

Characterization of a Wheat Breeders' Array suitable for high-throughput SNP genotyping of global accessions of hexaploid bread wheat (*Triticum aestivum*)

Alexandra M. Allen[1,*], Mark O. Winfield[1], Amanda J. Burridge[1], Rowena C. Downie[1,2], Harriet R. Benbow[1], Gary L. A. Barker[1], Paul A. Wilkinson[1], Jane Coghill[1], Christy Waterfall[1], Alessandro Davassi[3], Geoff Scopes[3], Ali Pirani[3], Teresa Webster[3], Fiona Brew[3], Claire Bloor[3], Simon Griffiths[4], Alison R. Bentley[2], Mark Alda[5], Peter Jack[5], Andrew L. Phillips[6] and Keith J. Edwards[1]

[1]*Life Sciences, University of Bristol, Bristol, UK*

[2]*The John Bingham Laboratory, NIAB, Cambridge, UK*

[3]*Affymetrix UK Ltd, High Wycombe, UK*

[4]*John Innes Centre, Norwich, Norfolk, UK*

[5]*RAGT Seeds, Ickleton, Essex, UK*

[6]*Plant Biology and Crop Science Department, Rothamsted Research, Harpenden, UK*

*Correspondence
email A.Allen@Bristol.ac.uk

Keywords: wheat, genotyping array, single nucleotide polymorphism (SNP).

Summary

Targeted selection and inbreeding have resulted in a lack of genetic diversity in elite hexaploid bread wheat accessions. Reduced diversity can be a limiting factor in the breeding of high yielding varieties and crucially can mean reduced resilience in the face of changing climate and resource pressures. Recent technological advances have enabled the development of molecular markers for use in the assessment and utilization of genetic diversity in hexaploid wheat. Starting with a large collection of 819 571 previously characterized wheat markers, here we describe the identification of 35 143 single nucleotide polymorphism-based markers, which are highly suited to the genotyping of elite hexaploid wheat accessions. To assess their suitability, the markers have been validated using a commercial high-density Affymetrix Axiom® genotyping array (the Wheat Breeders' Array), in a high-throughput 384 microplate configuration, to characterize a diverse global collection of wheat accessions including landraces and elite lines derived from commercial breeding communities. We demonstrate that the Wheat Breeders' Array is also suitable for generating high-density genetic maps of previously uncharacterized populations and for characterizing novel genetic diversity produced by mutagenesis. To facilitate the use of the array by the wheat community, the markers, the associated sequence and the genotype information have been made available through the interactive web site 'CerealsDB'.

Introduction

Increasing wheat yields is a major global priority for feeding the world's growing population. It has been estimated that wheat yields need to increase by 50% by 2050 to meet this demand, yet current trends are exhibiting yield plateaus (Grassini *et al.*, 2013). Hexaploid bread wheat (*Triticum aestivum*) is derived from the hybridization of diploid *Aegilops tauschii* with tetraploid wild emmer, *Triticum turgidum* ssp. *dicoccoides* (Dubcovsky and Dvorak, 2007; Matsuoka, 2011; Shewry, 2009). Hybridization, domestication and strong selection pressure has reduced the level of genetic diversity available to wheat breeders, and this lack of diversity is widely recognized as a limiting factor in the breeding of high yielding varieties, particularly in response to changing biotic and abiotic stresses (Haudry *et al.*, 2007; Tanksley and McCouch, 1997). The ability to assess and fully utilize the genetic diversity present in germplasm collections will inform breeding efforts, enabling potential yield increases to be attained, and it has been recognized in recent years that national efforts should be co-ordinated to maximize progress in wheat breeding (Wheat Initiative, 2011). The ability to assess germplasm on a common genotyping platform will assist exchanges of material between countries for the introduction and mobilization of novel genetic diversity.

High-throughput genotyping in hexaploid wheat has been made possible in recent years through the advent of next-generation sequencing for genotyping-by-sequencing (GbyS; Rife *et al.*, 2015) and SNP discovery (Winfield *et al.*, 2012) and the subsequent development of SNP-based marker technologies. These range from flexible, scalable single PCR-based assays such as KASP (Allen *et al.*, 2011; LGC, Herts, UK) and TaqMan® (Applied Biosystems™, Foster City, CA) assays to high-density fixed-content arrays, for example the Illumina 90k iSelect array (Wang *et al.*, 2014; Illumina, San Diego, CA). We recently reported the generation of an ultra-high-density Affymetrix Axiom® array, containing 820 000 single nucleotide polymorphism (SNP) markers (Winfield *et al.*, 2015). While this array represents a step change in wheat genotyping, the format is not amenable for cost-effective high-throughput genotyping. In addition, the majority of the markers on this array were designed to genotype polymorphisms between wheat and its near relatives and progenitors and hence are of limited direct value to wheat breeders who are specifically interested in comparing hexaploid germplasm. To overcome these limitations, we have utilized the data obtained from using the 820K wheat array in genotyping a range of diverse hexaploid accessions, to identify a set of 35 143 informative markers useful to the breeding community. To confirm the utility of the selected SNP markers, a 384 microplate format Axiom® array (hereafter called the Wheat Breeders' Array) was designed and synthesized to maximize the throughput of sample screening, including algorithms and software to enable rapid automated downstream analysis, therefore reducing required computational load.

Subsequently, we have used the Wheat Breeders' Array to screen a large global collection of hexaploid wheat cultivar and landrace accessions. Additional germplasm screened included lines from five separate genetic mapping populations, which differ in parental material and crossing strategies, novel synthetic hexaploids and accessions subjected to mutagenesis. A diverse range of hexaploid material was included in this initial screen to allow assessment of the performance of the array SNP content in different germplasm across a range of applications of interest to wheat breeders. The design and high-throughput nature of the Wheat Breeders' Array makes it a potentially useful tool for research and breeding applications such as genomewide association studies (GWAS) and genomic selection. By making the array and resulting data available to the global community, we hope to demonstrate the utility of this platform for researchers worldwide. Developing global resources such as these promote rapid germplasm exchanges to boost genetic diversity and facilitate targeted breeding.

Results

SNP selection

SNP markers were selected from a subset of the previously described Axiom® HD 820K wheat array (Table S1). Overall SNP markers were selected as described in methods to include those that were evenly spaced throughout the genome (according to genetic map position) and showed higher levels of polymorphism (measured by minor allele frequency; MAF) in the test range of hexaploid accessions, which included 108 elite hexaploid accessions of which 48 were suggested by a number of commercial wheat breeders (Winfield et al., 2015). Of the 35 143 SNP assays selected, 15 393 (43.8%) were considered to be co-dominant; that is, they were able to discriminate between homozygote and heterozygote states and 19 750 (56.2%) were considered to be dominant. Of the 35 143 SNPs, 24 194 (68.8%) were transitions and 10 949 (31.2%) were transversions, compared with 72% and 28%, respectively, for the larger 820K SNP collection (Winfield et al., 2015).

Genetic mapping

Five mapping populations were genotyped using the Wheat Breeders' Array. Of the 35 143, SNP markers selected 22 001 (62.6%) were placed on one of five genetic maps (Table S3). The five different mapping populations differed in parental accessions, size of population and crossing strategy as detailed in Table 1. Two of the populations (Avalon × Cadenza and Savannah × Rialto) were generated by double haploid production from F_1 plants, two consisted of recombinant inbred lines (RILs) generated from the F_6 generation or F_7 generation (Opata × Synthetic and Chinese Spring × Paragon, respectively), and one was produced by single seed descent to the F_5 generation (Apogee × Paragon). To maximize the number of genetically mapped SNPs, a diverse selection of parental material was used to generate these populations which included spring and winter varieties, a synthetic hexaploid, the model variety Chinese Spring and a 'super-dwarf', 'rapid cycling' cultivar (Apogee, developed for use in controlled environment experiments; Bugbee and Koerner, 1997). The number of SNPs on the array polymorphic between these specific crosses ranged from 6772 to 11 720, suggesting an average of 8793 SNPs on the array (25%) are predicted to be polymorphic between any two varieties.

Markers with greater than 20% missing data were removed prior to map construction. Of the SNPs polymorphic between the parents of the crosses, 86%, 90%, 88% and 80% of markers were able to be assigned to a linkage group on the Avalon × Cadenza, Savannah × Rialto, Opata × Synthetic and Chinese Spring × Paragon maps, respectively (Table 1). The number assigned to the Apogee × Paragon population was considerably lower (2997, 44%) due to the presence of heterozygotes in the population which complicated genotype calling at dominant SNP loci. The number of 'skeleton markers' initially assigned to construct the framework genetic maps was lower (626) in the Savannah × Rialto population compared to the Avalon × Cadenza population (997). This is likely to be due to the smaller number of individuals and therefore recombination events between genomic regions, and also the presence of an identical 1RS translocation on the short arm of chromosome 1B in both varieties. The Opata × Synthetic map contained 1509 skeleton markers, reflecting the greater diversity present between the parents of this cross. The Apogee × Paragon and Chinese Spring × Paragon maps had the highest number of skeleton markers (1537 and 2472) resulting from both the initial diversity present between the parental lines and the large population sizes.

The genotype assignments of SNP markers were tested for deviations from the expected 50:50 parental ratio as such markers can result in distortions in the resulting genetic maps.

Table 1 Mapping populations screened in this study

| | Population | | | | |
	A × C	S × R	O × S	A × P	CS × P
Parent 1	Avalon	Savannah	Opata	Apogee	Chinese Spring
Parent 2	Cadenza	Rialto	Synthetic	Paragon	Paragon
Population type	Double Haploid	Double Haploid	F6-derived RIL	F5 SSD	F7-derived RIL
Number of individuals	128	64	60	349	269
Number of SNPs polymorphic between parents	8498	6997	9978	6772	11 720
Number of markers in genetic map	7328	6303	8820	2997	9434
Number of skeleton markers (unique position)	997	626	1509	1537	2472

The distribution of segregation distortion across the genome was examined for each mapping population (Figure 1, Table S2). The population with the highest number of SNP markers exhibiting significant ($P < 0.005$) distortion of segregation was Chinese Spring × Paragon (317 SNPs), then Avalon × Cadenza (86 SNPs), Apogee × Paragon (54 SNPs) and Savannah × Rialto (38 SNPs). The Opata × Synthetic population had no SNP loci exhibiting significant distortion of segregation. The distorted loci were unevenly distributed across the genome with clusters of SNPs in specific locations (Figure 1). On the Avalon × Cadenza

genetic map, significant SNPs were clustered on 8 chromosomes with the highest number on chromosome 5B (65 SNPs) in the regions 80.9, 116–130 and 157–164 cM (1, 52 and 12 SNPs, respectively). In the Savannah × Rialto genetic map, the significant SNP markers were clustered in four locations on chromosomes 3A and 3B. The clustering of significant SNPs in the Apogee × Paragon genetic map was more widespread with loci mapped to 14 locations on 10 chromosomes. The clusters with the highest number of markers were on chromosomes 2D (17 SNPs) and 3B (15 SNPs). The Chinese Spring × Paragon

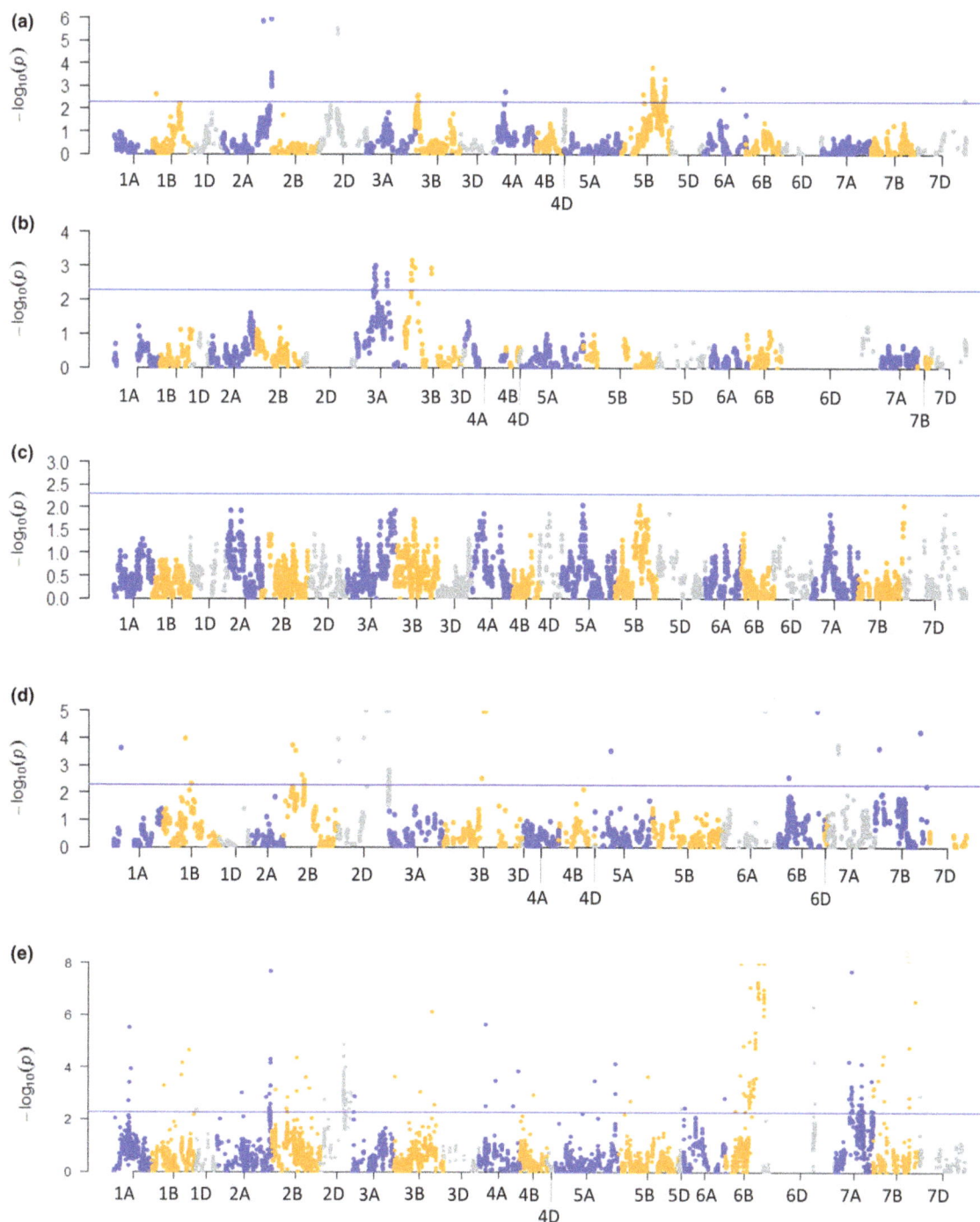

Figure 1 Manhattan plots showing the level of segregation distortion of SNP loci distributed across the wheat genome in four mapping populations: (a) Avalon × Cadenza; (b) Savannah × Rialto; (c) Opata × Synthetic; (d) Apogee × Paragon; (e) Chinese Spring × Paragon. The guideline indicates the significance threshold of the chi-square test at $P = 0.05$.

population had the highest number of distorted SNPs, distributed on almost every chromosome but particularly focussed in regions on chromosomes 2A (23 SNPs), 2D (97 SNPs), 6B (109 SNPs) and 7A (37 SNPs).

The markers with the most significant distortion of segregation in the Avalon × Cadenza population were mapped to chromosomes 2A ($P = 1.12e-6$) and 2D ($P = 3.13e-6$) which equates to parent1 : parent2 ratios of 31 : 83 and 77 : 29, respectively. On the Apogee × Paragon genetic map, the most distorted markers were located on chromosomes 2D ($P = 5.17e-13$), 3B ($P = 5.38e-11$), 6A ($P = 2.0e-10$) and 6B ($P = 1.89e-6$). For the Savannah × Rialto population, the most highly distorted SNP was located on chromosome 3B ($P = 6.7e-4$). The markers exhibiting the highest level of distortion on the Chinese Spring × Paragon map were located in the largest clusters of SNPs on chromosomes 2D and 6B ($P = 2.38e-11$, $P = 7.88e-10$, respectively). The direction of distortion in relation to the parental genotype in the Avalon × Cadenza population appeared biased towards Cadenza with 8 of 11 clusters and 95% SNPs being distorted in favour of the Cadenza genotype. In the Savannah × Rialto map, 24 SNPs in three locations were distorted towards Rialto, while 14 SNPs in one location were distorted towards Savannah. For the Apogee × Paragon population, 57% of the significant SNPs were distorted towards Paragon in 7 of the 14 chromosome locations. A significant bias was seen in the Chinese Spring × Paragon with 81% of SNPs in 23 locations on 16 chromosomes distorted in favour of the Chinese Spring genotype.

Markers exhibiting significant distortion of segregation in any of the populations were removed before creating the consensus genetic map. The five separate genetic maps were merged, and 21 709 markers were placed onto a consensus genetic map of all 21 chromosomes (Table 2, Table S3). The number of markers per chromosome ranged from 157 on chromosome 4D to 2168 on chromosome 2B. Overall, B genome chromosomes had the highest number of mapped polymorphisms (10 745, 48%) and D genome chromosomes had the least (2907, 13%). Individual chromosome map lengths varied from 147.2 cM (1B) to 340.2 cM (3A). The overall map length of the consensus genetic map (4645.8 cM) was higher than the DH population maps (2967.3 and 3284.1 cM) but reduced when compared to RIL-derived population maps (4464.0–6632.3 cM).

Array validation

The Wheat Breeders' Array was used to screen 1843 genomic DNAs derived from 1779 unique hexaploid wheat accessions (listed in Table S4). These unique accessions included an elite collection of 505 breeding lines derived from 17 countries in Africa, Australia, the Americas, the Middle East and Europe; 436 lines from the Gediflux collection (representing Western Europe winter wheat diversity from 1920 to 1990) and 790 accessions from the Watkins global landrace collection assembled from 33 countries in the 1930s (Wingen et al., 2014; Burt et al., 2014; Miller et al., 2001). The unique accessions included eight synthetic hexaploid accessions and forty lines carrying various mutations or deletions in the form of cv. Chinese Spring nullisomics and monosomics (Wheat Genetic and Genomic Resources Centre, Kansas State University, USA), cv. Paragon gamma deletion lines (Wheat Genetic Improvement Network, UK) and cv. Cadenza EMS mutation lines (Table S4). The remaining samples consisted of sixty-four replicates of named accessions sourced from different laboratories.

Genotype calls were generated as described in Experimental Procedures. Across the samples genotyped, the average call rate

Table 2 Distribution of mapped SNP loci on the Wheat Breeders array across the wheat genome

Chromosome	A x C Number of SNPs	A x C Length (cM)	S x R Number of SNPs	S x R Length (cM)	O x S Number of SNPs	O x S Length (cM)	A x P Number of SNPs	A x P Length (cM)	CS x P Number of SNPs	CS x P Length (cM)	Consensus Number of SNPs	Consensus Length (cM)
1A	425	148.1	430	178.5	457	285.1	257	262.1	558	273.9	1245	148.1
1B	956	147.3	323	122.4	759	276.5	239	291.8	795	299.4	1794	148.8
1D	292	124.6	170	71.8	103	275.2	57	171.1	226	278.4	546	238.1
2A	404	178.0	779	173.3	448	244.4	179	163.2	643	376.4	1555	180.0
2B	532	176.9	796	182.5	723	336.7	166	280.2	937	333.7	2107	187.3
2D	216	187.0	60	200.9	304	287.4	138	269.0	219	427.1	612	295.1
3A	339	184.0	375	186.6	445	345.7	168	286.9	479	340.2	1090	340.2
3B	534	179.5	487	217.4	715	313.8	212	412.1	890	344.1	1730	245.6
3D	59	129.2	24	14.7	334	248.8	12	7.7	156	401.1	465	205.0
4A	259	161.6	102	158.7	427	301.4	186	181.2	490	283.6	883	215.5
4B	304	105.0	96	51.1	336	195.7	96	186.9	273	190.2	702	152.1
4D	36	6.3	35	8.19	90	169.0	8	0.1	36	105.4	154	162.1
5A	407	218.0	551	235.4	468	382.7	166	301.6	657	429.8	1300	226.6
5B	559	191.7	305	286.55	673	318.0	194	367.8	847	404.5	1665	325.6
5D	133	126.8	148	208.2	202	347.8	0	0.0	160	349.5	416	219.6
6A	467	164.4	386	141.0	524	269.6	156	292.1	294	287.0	1060	225.3
6B	414	143.0	653	128.4	496	229.6	276	246.4	657	162.3	1509	160.9
6D	58	158.2	53	76.1	145	290.6	35	17.7	122	442.7	244	184.7
7A	395	189.1	310	139.5	477	341.8	176	253.2	546	268.1	1251	201.9
7B	348	181.8	113	56.5	543	336.7	197	281.9	345	324.5	1054	336.7
7D	105	183.6	73	129.6	151	455.8	25	191.0	104	310.4	326	248.2
Total	7242	3284.1	6274	2967.3	8820	6252.3	2943	4464	9434	6632.3	21 708	4647.4

Table 3 Numbers of SNPs unique to and shared between germplasm collections

	Elite cultivars	Gediflux collection	Landraces	Chinese Spring deletion lines	Paragon deletion lines	Cadenza EMS lines	Synthetic hexaploid lines
Elite cultivars	247						
Gediflux collection	31 473	43					
Landraces	32 013	31 388	218				
CS deletion lines	8822	8807	8882	65			
Paragon deletion lines	5932	5906	5913	2778	8		
Cadenza EMS lines	6603	6580	6583	2789	5312	5	
Synthetic hexaploid lines	19 266	18 662	19 035	6350	4342	46 890	144

Table 4 Summary statistics of cultivar collections

	Australia	Central America	Middle East	North America	North Europe	South Africa	South America	South Europe	West Europe	Gediflux
n	146	64	5	40	10	5	6	17	271	436
% P	92.5	85.3	61.2	87.4	69.5	53.8	60.0	81.1	97.3	95.2
H_E	0.229	0.202	0.207	0.232	0.205	0.188	0.200	0.238	0.229	0.214
MAF	0.167	0.146	0.153	0.168	0.150	0.142	0.150	0.173	0.164	0.155
RI	1.029	0.997	1.019	1.029	0.837	0.872	1.155	1.051	0.947	0.826

n, number of samples; % P, percentage of total SNPs on the array which are polymorphic; H_E, expected heterozygosity; MAF, average minor allele frequency; RI, rarity index.

was 97.9%, ranging from 94.1% to 99.2% (Table S4). The accession type with the highest average call rate was mutation lines (98.1%), and the lowest was synthetic hexaploids (96.3%). The relationship between call rate and heterozygous call rate per accession was investigated. A trend was observed where samples with low call rates tended to have a higher than average het rate (a higher percentage of SNPs called AB). The DNA samples for these lines are predicted to be of lower quality as the increase in AB calls and lower call rate represents a higher number of outlier calls from the main clusters. For use as a high-throughput genotyping platform, reproducibility is an important consideration. The call rate among duplicate samples was highly similar (ranging from 99.3% to 99.8%); however, the call rate for replicate samples prepared from the same named accessions, but from different sources, showed more variation (97.5–99.4); this is likely to reflect true within-cultivar variation.

The total number of polymorphic SNPs was 33 326 (94.8%) of the entire array based on the screen of the collection of lines described above. A summary of the numbers of polymorphisms present unique to and shared between germplasm collections is presented in Table 3. The collections with the highest number of unique SNPs were the elite global collection (247 SNPs), the Watkins landrace collection (218 SNPs) and the synthetic hexaploid collection (144 SNPs). The collections sharing the highest number of polymorphisms were the elite cultivars, Gediflux and landrace collections, with up to 32 013 SNPs being transferrable between and useful within different collections. The lower numbers of shared polymorphisms between these and other collections (e.g. deletion and mutation lines) are representative of the narrow genetic base compromising the collections of deletion and mutation lines which are developed in a single genetic background. The effect of collection size on the number of polymorphic SNPs within a collection was also apparent (Table 4). A sharp increase in level of polymorphism was seen between collection sizes of <5 to around 50 individual accessions,

reaching an average of 90% polymorphic SNPs in collections of 100 accessions.

The minor allele frequencies of SNPs within different germplasm collections were calculated as a measure of allelic diversity (Figure 2a). The larger elite cultivars, Gediflux and landrace collections had a higher number of polymorphic SNPs, with a cumulative prevalence of intermediate to high MAF SNP loci, distributions observed previously for similar wheat collections (Wang et al., 2014). The MAF distributions of SNPs within the deletion and mutation lines were more skewed, with a high proportion of polymorphic SNPs showing a MAF of 0 and 0.45–0.5 (57%, Chinese Spring deletion lines; 70%, Paragon deletion lines; 73%, Cadenza EMS mutation lines). This is likely to be due to both the small sample sizes of these collections (16, 9 and 15 samples, respectively) and the limited genetic background of these collections, meaning the samples will be homozygous at most sites represented on the array. The polymorphic SNPs identified in these collections are therefore likely to be associated with deleted or mutated regions. The synthetic collection showed a bimodal MAF distribution; a high proportion of polymorphic SNP loci had either a high or low MAF, although this is partially a reflection of the small sample size of eight individuals. To further study the MAF of the synthetic lines, the average MAF of A, B and D genome mapped SNP loci was calculated (Figure 2b). The average MAF of the cultivar, Gediflux and landrace collections was very similar, with typically higher MAF observed in A and B genome markers compared to D genome markers. The average MAF of A, B and D genome markers in the deletion and mutation line collections was very similar. The synthetic collection had a higher average MAF of D genome markers compared to A and B genome markers highlighting the increase in diversity brought to the D genome by the novel A. tauschii accessions used in the creation of these lines.

The relationship between accessions was visualized by calculating a pairwise similarity matrix that was used to

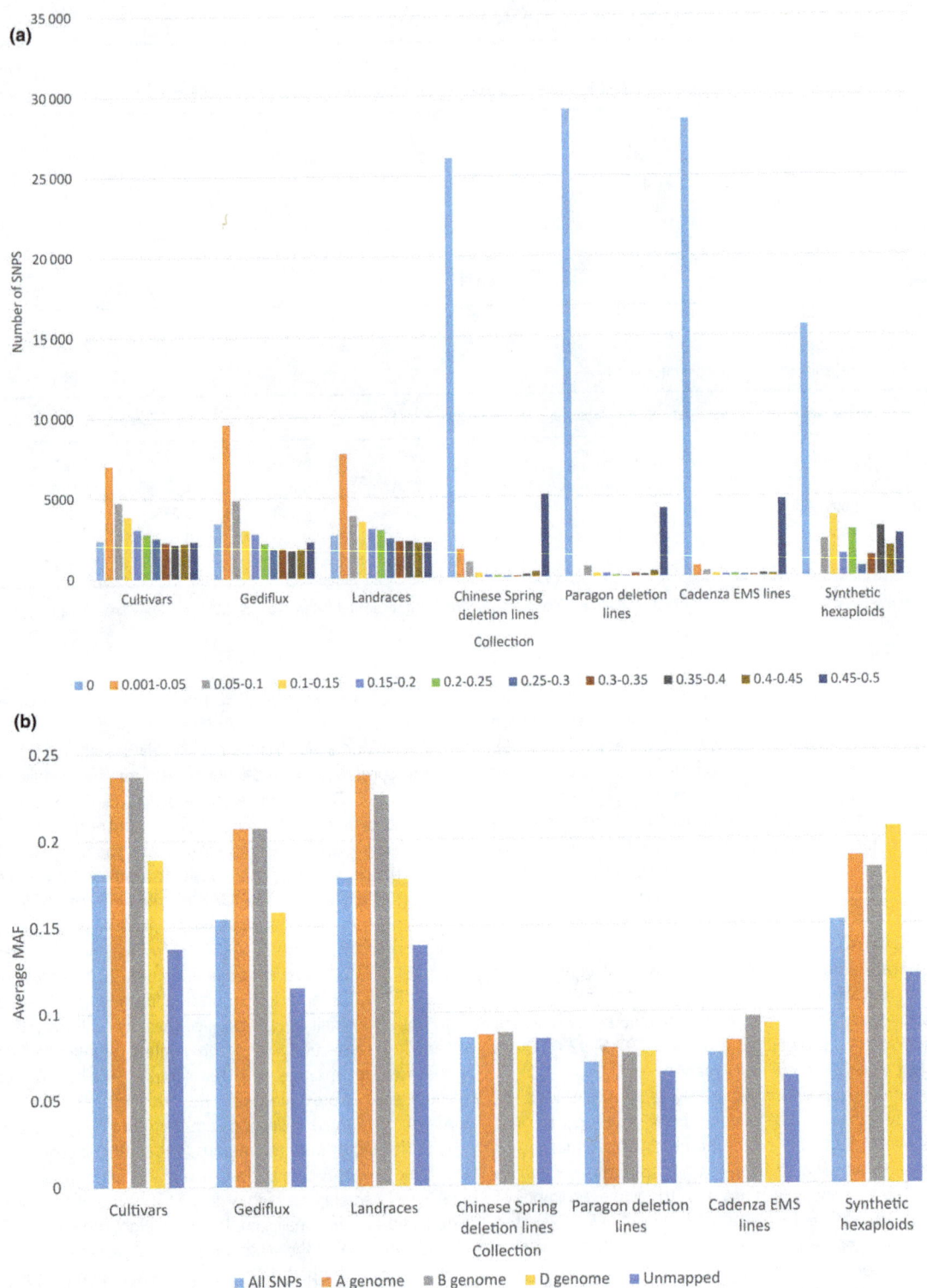

Figure 2 (a) Distribution of minor allele frequencies (MAFs) of SNP loci within germplasm collections. (b) Average MAF of A, B and D genome mapped SNPs in the different germplasm collections.

perform multidimensional scaling (MDS) and create principal coordinate (PCoA) plots (Figure 3). With the different germplasm collections, it was clear that accession type contributes to the structure of the PCoA plot (Figure 3a). Elite cultivars were split primarily into two clusters representing spring (negative PCo2 values) and winter (positive PCo2 values) accessions, with the winter accessions interspersed with the Gediflux collection. The Watkins landrace collection slightly overlapped in distribution with the elite cultivars but formed a distinct cluster extending towards negative PCo2 values. The deletion/mutation lines formed tight clusters representing the common genetic backgrounds to these lines. The eight novel synthetic hexaploid accessions clustered within spring cultivars and landrace accessions.

Relationship between global elite germplasm collections

To further examine the diversity present in different elite germplasm collections, they were defined by geographical region of origin. The separate components of the collections are summarized in Table 4 including number of accessions, % of polymorphic markers, measures of diversity (HE and MAF) and rarity index (a measurement of the number of rare alleles present in each subcollection). The subcollections with the highest numbers of polymorphic markers were both from Western Europe; these were also the two largest subpopulations; the smallest subpopulations had the lowest numbers of polymorphic markers, reinforcing the relationship seen between collections size and number of polymorphic markers seen in Table 4. The subcollections with the highest genetic diversity measures were from Southern Europe and North America. The subcollection with the highest rarity index was South America (despite being one of the smallest subcollections), and the lowest was the Gediflux

collection, suggesting that polymorphic alleles are widespread in this collection.

The relationship between different subpopulations was further examined by calculating the number of shared polymorphisms and genetic differentiation between subcollections (Table 5). The number of shared polymorphisms was highest between the Gediflux and Western Europe elite accessions; these also had the lowest F_{ST} value, which is not surprising given that they are both of Western European origin. The Western Europe and Australian collections appeared to have a high degree of similarity with considerable overlaps in polymorphic markers, although this is may also be attributed to both collections having the highest number of polymorphic SNPs overall. Overall, the majority of polymorphic SNPs were shared among populations, suggesting that there is a high transferability of SNP markers across global elite germplasm collections. High F_{ST} measures between populations of different geographical origin is likely to be caused by the usage of different

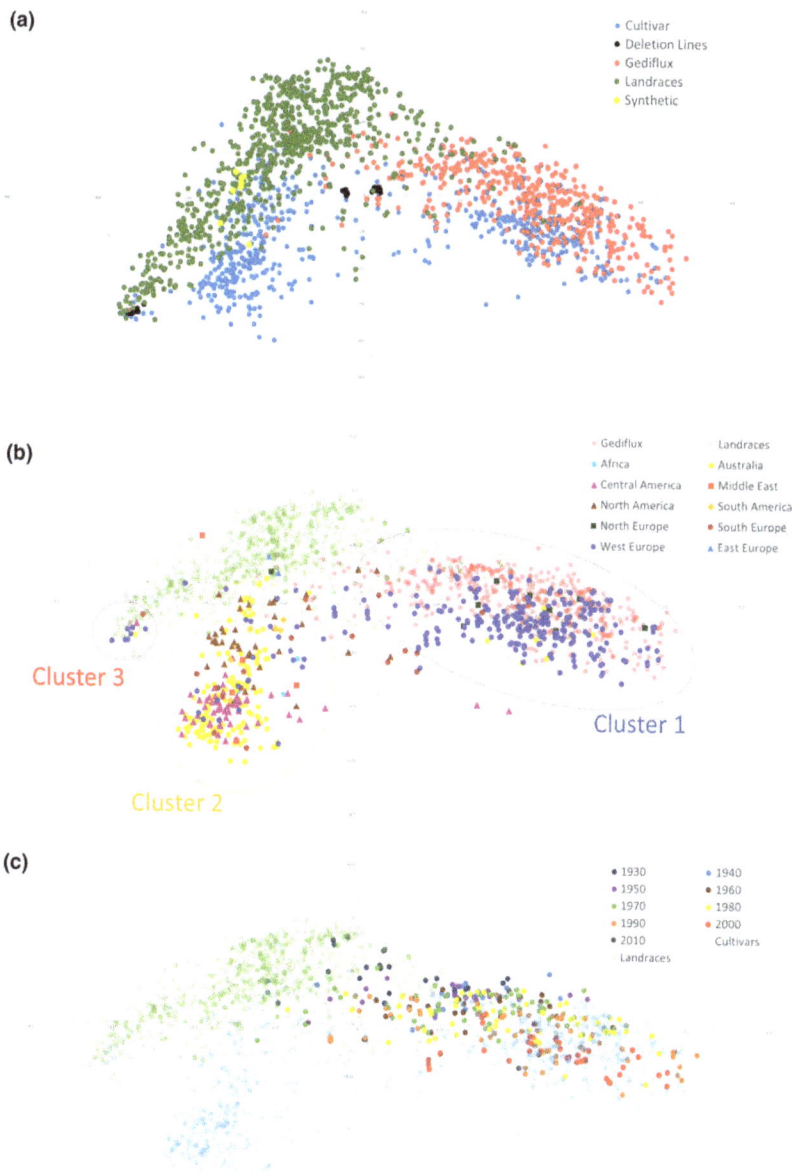

Figure 3 Principal coordinate analysis (PCoA) plots coloured by (a) collection, (b) country of origin (c) date of line release. Coordinate 1 is plotted along the x-axis, coordinate 2 is plotted along the y-axis.

founders or by allele frequency divergence during the development of locally adapted populations. High F_{ST} measures were seen in particular between Western European and Central American subpopulations. Conversely, the lowest F_{ST} measures were seen between Middle Eastern, Southern American and Southern Europe subpopulations suggesting use of similar founders or overlap in these breeding programmes.

To further examine the relationship between cultivars of different geographical origins, samples plotted by the first two principal co-ordinates were coloured by region of origin (Figure 3b). Three main clusters were observed for elite cultivars. One cluster defined by positive PCo1 values consists of mainly Western European accessions and also includes a small number of Australian and Northern European accessions. The second largest cluster consists primarily of Australian and American accessions. Further structure is observed for this cluster with lower PCo2 values for Central American and Australian accessions and higher PCo2 values for Middle Eastern, Southern and Northern American accessions. A third small cluster consisted of accessions from Western and Southern Europe, Australia and Central America. On further investigation, the samples supplied from Western Europe are actually of Asian origin. To investigate the impact of cultivar age on genetic diversity, the date of cultivar release was used to colour accessions from Western Europe (Figure 3c). Accessions were grouped per decade, and a trend was observed with pre-1960s accessions locating closer to the central landrace cluster and later accessions extending along the PCo1 axis, particularly in the 1980s and 1990s, suggesting an expansion of diversity during this period.

Using the Wheat Breeders' Array to characterize novel genetic diversity including deletions, introgressions and genomic rearrangements

The collection of lines screened on the Wheat Breeders' Array included 40 accessions with known deletions of various types; monosomic (missing one chromosome of a pair), nullisomic (both chromosomes of a pair are deleted but substituted with those from a homoeologous genome), ditelosomic (missing part of an end of a chromosome) and gamma-irradiated deletions (smaller deletions within a chromosome). In the case of the gamma-irradiated lines, we were able to identify between 299 and 796 polymorphic markers compared to the Paragon control representing between 1.86% and 1.95% of the markers scored. For the five lineages examined, the majority of variable markers mapped to specific regions of which most were associated with

the known deleted regions. For the Cadenza EMS mutagenized lines, for the 15 lines examined, there were between 424 and 709 polymorphic markers compared to the Cadenza control representing between 2.64% and 4.41% of the markers scored. A number of these polymorphic markers appeared to be common to all or most of the mutagenized lines, suggesting that some polymorphisms might have existed in the original stock used for mutagenesis. The breeding schedule for cv. Cadenza shows that cultivar is based on a single individual from the F_6 generation resulting in ~3% residual heterogeneity in the final cultivar. Unique SNPs to each accession were mostly randomly distributed throughout the genome and are likely to reside in mutated regions specific to each line.

We used copy number variation (CNV) analysis to characterize the various deletion lines; using this procedure, we were able to highlight SNP loci associated with each of the deleted regions where a reduced hybridization signal was observed (Figure 4). Interestingly, evidence was also seen for an increase in signal in certain chromosome regions of some of the gamma deletion lines (4e). This apparent over-representation implies duplication of large chromosomal segments, presumably as a result of the gamma irradiation. The same approach was applied to analyse the genotyping data from the elite cultivars. For certain accessions, regions of reduced signal were observed, associated with varieties carrying known introgressions, that is the 1RS introgression from rye (e.g. cv. Savannah, 4f). A number of additional cultivars (Keilder, Gulliver, Mercato) showed significantly low signal strength for one or more chromosomes potentially representing deletions or ancestral introgressions (Table S5).

Discussion

We designed the Wheat Breeders' Array to be a high-throughput platform for the cost-efficient generation of genetic maps between a range of parental lines and for genotyping hexaploid wheat derived from a variety of sources. To confirm the utility of the platform, we screened several mapping populations, generated from a range of hexaploid lines via different crossing strategies, to produce five high-density genetic maps. The relatively high level of codominant SNP assays on the array enabled a map to be produced from an F_5 population containing heterozygotes in addition to the more traditional DH populations. This may be a more cost-effective mapping population development strategy than DHs for some purposes.

Table 5 Number of shared polymorphisms (above diagonal) and genetic differentiation, F_{ST}, (below diagonal) between cultivar subcollections

	Australia	Central America	Middle East	North America	North Europe	South Africa	South America	South Europe	West Europe	Gediflux
Australia		27 615	20 172	28 446	22 693	17 813	19 746	26 594	30 488	30 115
Central America	0.077		19 676	26 603	21 252	17 300	19 181	25 237	28 130	27 732
Middle East	−0.047	−0.022		19 938	16 177	14 497	16 010	19 305	20 312	20 195
North America	0.046	0.079	−0.0594		22 005	17 575	19 429	25 963	28 927	28 731
North Europe	0.141	0.185	0.023	0.101		13 757	11 009	21 085	23 057	23 027
South Africa	−0.063	0.005	−0.247	−0.051	0.043		13 536	16 970	17 879	17 718
South America	0.039	0.063	−0.134	−0.004	0.082	−0.086		18 690	19 931	19 699
South Europe	0.032	0.030	−0.116	0.009	0.044	−0.085	−0.018		26 919	26 728
West Europe	0.139	0.166	0.063	0.116	−0.033	0.069	0.111	0.068		31 371
Gediflux	0.180	0.210	0.114	0.152	−0.030	0.121	0.156	0.103	0.013	

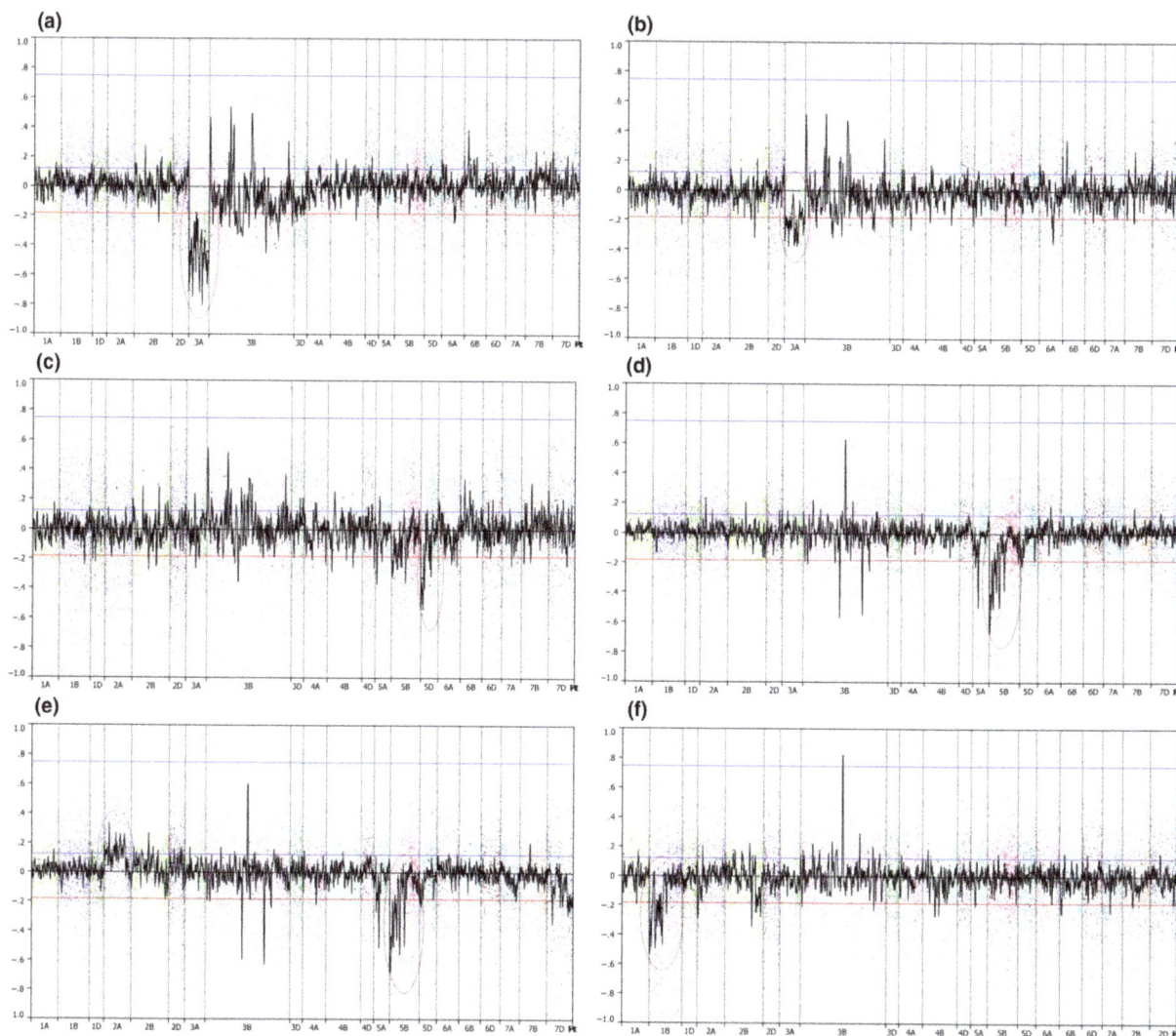

Figure 4 Signal intensity (Log$_2$R ratio) plots of copy number variation (CNV) across the genome for different hexaploid wheat accessions. The accessions displayed are as follows: (a) Chinese Spring nullisomic 3A deletion; (b) Chinese Spring monosomic 3A deletion; (c) Chinese Spring ditelosomic 5DS deletion; (d) Paragon gamma-irradiated 5B deletion; (e) Paragon gamma-irradiated line exhibiting CNV loss and gain; (f) cv. Savannah, carrying the 1RS translocation from rye. Blue circles highlight copy number-gained CNV regions, red circles highlight copy number-loss CNV regions.

The genetic map produced using the Apogee × Paragon population revealed a high number of recombinants within the population, indicated by the high number of skeleton markers, and inclusion of this in the consensus map helped to resolve marker order in regions consigned to large 'bins' of markers in the other maps. Similarly, the Opata × Synthetic map increased the overall number of mapped SNPs, particularly in the D genome due to the increased diversity incorporated through use of a synthetic hexaploid as a parent of the mapping population. The lack of genetic diversity in the D genome of hexaploid wheat cultivars is a well-documented phenomenon attributed to the genetic bottle-neck experienced during the initial hybridization to create the hexaploid and the subsequent limited gene flow into bread wheat from *A. tauschii* compared to that from tetraploids into the A and B genomes (Dvorak, 2006, Halloran et al., 2008). This was also reflected in the analysis of the average MAF of A, B and D genome markers, where the synthetic hexaploid lines (bred to specifically increase D genome diversity) screened showed higher MAF in the D genome compared to the A and B genomes, a trend

opposite to that observed in conventional cultivars and landraces. Using the five mapping populations, we were able to generate a consensus map consisting of 22 001 SNP markers or 63% of the total SNP markers on the array. This compares favourably to maps generated previously for similar wheat SNP arrays such as the Illumina iSelect 90k wheat array (Wang et al., 2014; 46 977 mapped markers, 58% of total) and the Affymetrix Axiom® HD Wheat Genotyping Array (Winfield et al., 2015; 56 505 mapped markers, 7% of total).

The high marker density of the constructed maps highlighted features of the genome such as regions of distorted segregation which were unequally distributed across the genome. Segregation distortion of genetic loci is a potentially powerful evolutionary force that allows the enhanced transmission of a specific genetic locus (Taylor and Ingvarsson, 2003). A number of examples of significant segregation distortion were observed in the mapping populations analysed in this study. The Avalon × Cadenza population had several peaks of highly significant segregation distortion, in particular on chromosomes

2A, 2D and 5B. The bimodal peak on 5B is likely to represent the effect of the 5B–7B reciprocal translocation present in the population and donated from Avalon. This translocation is a relatively widespread chromosomal rearrangement in Western European cultivars and is thought to be of adaptive value in controlling plant growth and development. In some populations, the translocated chromosomes have been reported to be preferentially transmitted (Schlegel, 1996; Friebe and Gill, 1994). Overall, the Avalon × Cadenza population showed significant bias (95% of loci) towards inheriting the Cadenza genotype.

The Savannah × Rialto population exhibited distortion of segregation of 38 SNPs representing four loci on two homoeologous chromosomes, 3A and 3B. It is interesting to note that very similar patterns are observed on both of these homoeologous group 3 chromosomes, both having bimodal peaks which may represent genomic rearrangements (as described above) or genes of large effect. In contrast, the Apogee × Paragon population had numerous regions of distorted segregation, typically consisting of relatively small numbers of SNPs with a high level of significance, with no significant bias towards either parent. The difference in pattern of distorted loci may partially reflect how each population was produced. The DH populations were in effect 'fixed' at the F_1 cross, and any regions of segregation distortion present were transferred into the DH and maintained in the population; these may have been of large effect and size. The Apogee × Paragon population has undergone further inbreeding to the F_5 generation and multiple distorted loci of small size are observed.

Screening the Wheat Breeders' Array with a range of hexaploid lines demonstrated its utility on a wide range of germplasm from different geographical areas and ages. Overall, a high number of polymorphisms were shared between collections, with an average of 23% of SNPs on the array predicted to be polymorphic between two random accessions. A relationship between polymorphism level and collection size was observed, with an indication that at least 30 accessions are needed to maximize the chances of fully utilizing the polymorphism content of the array. The genotyping data were further explored to examine the relationships between diverse collections of global breeding lines. In general, a high number of shared polymorphisms and low F_{ST} was observed between populations of different geographical origin, suggesting that there has been an overlap of germplasm used within these breeding programmes. The principal co-ordinate plots reflect low F_{ST} measures with overlaps in particular between (i) Western Europe, Northern Europe and Gediflux accessions; (ii) Australian and Central American accessions; (iii) Northern American, Southern American, Southern Europe, Middle Eastern and Southern African accessions. Cluster 1 is unsurprising given the overlap between the geographical origins of these collections. Cluster 2 reflects the significant impact the CIMMYT developed lines have had on Australian breeding programmes since 1965 (Brennan and Quade, 2004). The relationships between the populations overlapping in cluster 3 are less clear, although climatic conditions within these countries are similar, making the exchange of adapted germplasm conceivable. It has been observed that during the 20th century, the global community of wheat breeders freely shared genetic materials (Kronstad, 1997), particularly in efforts led by the International Maize and Wheat Improvement Center (CIMMYT) and the International Center for Agricultural Research in the Dry Areas (ICARDA).

The hexaploid nature of the bread wheat genome means that it is amenable for both crossing with a range of wheat relatives and large-scale mutagenesis such as gamma irradiation. As both of these procedures can increase the diversity of the hexaploid gene pool, they are becoming more widely employed by breeders and academics alike. Hence, we investigated the ability of the Wheat Breeders' Array to characterize such material via the use of the CNV tool developed by Affymetrix (Axiom™ CNV Summary Tools Software v 1.1, part #600 733). By first using a collection of lines containing known deletions of different sizes and locations, we were able to characterize a range of deletions in terms of both their size and nature, that is monosomic or nullisomic. Examination of a number of lines of the variety Paragon, which had undergone gamma irradiation, allowed us to identify previous uncharacterized deletions and in addition show that a number of these irradiated lines also potentially carry duplicated regions. Finally, we used the fact that the SNP markers on the Wheat Breeders' Array were specific for hexaploid wheat, to screen a range of hexaploid lines for the evidence of either introgressions, such as the 1RS introgression from rye, or deletions. This screen generated evidence that numerous lines probably carry deletions and introgressions, and hence, our analysis suggests that further work is needed to characterize the extent of copy number variation within the hexaploid gene pool.

The Wheat Breeders' Array has been demonstrated to be useful for screening germplasm collections from across the globe and for characterizing sources of novel variation in a hexaploid background. As such, and given the design and high-throughput nature of the Wheat Breeders' Array, this tool may be applied to research and breeding approaches such as genomewide association studies (GWAS) and genomic selection. To further increase the utility of the array, we have screened five mapping populations and constructed a consensus genetic map to allocate a position to over 63% of the markers on the array. Further analysis has indicated that the markers on the array may be successfully used to identify regions of CNV and distorted segregation in the wheat genome, which in turn point towards chromosomal rearrangements and the presence of introgressions. To facilitate the use of the array by the global wheat community, the markers, the associated sequence and the genotype information have been made available through the interactive web site 'CerealsDB' (Wilkinson et al., 2012, 2016).

Experimental procedures

SNP selection

The original SNP collection consisted of 819 571 SNPs obtained from genic sequences derived via targeted capture re-sequencing of numerous wheat lines and validated on the Axiom® HD Wheat Genotyping Array (Winfield et al., 2015; Affymetrix UK Ltd, High Wycombe, UK). To select the most informative ~35 000 SNPs (the maximum permissible on the 384 Axiom® genotyping platform) for inclusion on the Wheat Breeders' Array, each SNP was assigned to an IWGSC scaffold via BLAST (Winfield et al., 2015). Once assigned, SNPs unique to a particular contig were selected. In cases where there was more than one SNP per contig, SNPs which had been genetically mapped on one or more of the three mapping populations used in the original analysis were selected. In cases where more than one SNP had been mapped, co-dominant SNP markers were preferentially selected and of these the SNP marker with the highest Polymorphic Information Content (PIC) score was selected. Where no SNPs in an IWGSC contig had been mapped, one SNP was selected with co-dominant SNPs being selected in preference to dominant SNPs and SNPs with high PIC scores being selected in preference to those with lower scores.

Plant material

The accessions grown for DNA extraction (listed in Table S4) were grown in peat-based soil in pots and maintained in a glasshouse at 15–25 °C with 16-h light, 8-h dark. Leaf tissue was harvested from 6-week-old plants, immediately frozen on liquid nitrogen and then stored at −20 °C prior to nucleic acid extraction. Genomic DNA was prepared from leaf tissue using a phenol–chloroform extraction method (Sambrook II., 1989). Genomic DNA samples were treated with RNase-A (New England Biolabs UK Ltd. Hitchin, UK), according to the manufacturer's instructions and purified using the QiaQuick PCR purification kit (QIAGEN Ltd., Manchester, UK).

Genotyping

The Axiom® Wheat Breeders' Array was used to genotype 2713 samples (Table S4) using the Affymetrix GeneTitan® system according to the procedure described by Affymetrix (Axiom® 2.0 Assay for 384 samples P/N 703154 Rev. 2). Allele calling was carried out using the Affymetrix proprietary software package Axiom Analysis Suite, following the Axiom® Best Practices Genotyping Workflow (http://media.affymetrix.com/support/downloads/man uals/axiom_genotyping_solution_analysis_guide.pdf).

Genetic map construction

Individuals from five mapping populations were genotyped with the Axiom® Wheat Breeders' Array (Table 1). For each popula-tion, markers with more than 20% missing data were removed and markers were binned based on their pattern of segregation in each respective population using the 'bound' function in Multi-point ULD (MultiQTL Ltd., Haifa, Israel). Markers were placed into the same bin if the correlation coefficient between them was 1, and therefore, the recombination frequency between them was estimated as 0. Following binning, linkage groups were ordered and then all markers which displayed a unique pattern of segregation and did not previously fall into a bin were iteratively added into each linkage group. During this process, the inflation coefficient was set to 1.2 to ensuring that markers which caused map inflations (likely to be due to genotyping errors) were not retained.

Markers were tested for significant segregation distortion using a chi-square test. The log10 value of the chi-square test statistic for each marker was plotted against marker position using the R package qqman. SNP loci exhibiting significant distortion of segregation and ambiguous markers mapping to different chromosomes in different populations were removed from individual maps before creating the consensus map. The consensus map was constructed using the R package LPmerge (Endelman and Plomion, 2014). No weighting was given to the component maps.

Dimensionality reduction

The relationship between the lines was determined by calculating a similarity matrix for all the lines. This was calculated as number of markers shared by any two lines divided by total number of markers for the two lines; markers that had missing calls for either

of the lines were not used to estimate similarity. The matrices were imported into R and used to create principal coordinate plots using the classic multidimensional scaling (MDS) method, cmdscale.

Summary statistics of germplasm collections

Summary statistics were calculated using StAMPP v1.0 (Pemble-ton et al., 2013) and the following formulae:

$$\text{Expected heterozygosity} = He = 1 - \Sigma p_i^2$$

$$\text{Rarity index (RI)} = RI_j = \frac{1}{I} \sum_{i=1}^{I} \frac{p_{ij}}{p_i}$$

where I is the number of markers, p_{ij} is the frequency of ith marker in a group of cultivars j, and P_i is the frequency of ith marker in the total dataset.

CNV analysis

CEL files from the Wheat Breeders' Array were processed using the Axiom Analysis Suite, with option set to Polyploid and Inbred, with the inbred het penalty set to 4. The annotation file was generated using the Affymetrix Annotation Converter, using chromosomal locations for SNPs downloaded from the IWGSCv1 assembly on Ensembl Plants. CNV analyses were visualized in Biodiscovery Nexus Copy Number (El Segundo, CA).

Acknowledgements

We are grateful to the Biotechnology and Biological Sciences Research Council, UK, for funding this work (award BB/1002278/ 1). AP is funded under the 2020 Wheat Institute Strategic Programme of the BBSRC, award BBS/E/C/00005202. ARB is supported by BB/I002561/1. We are grateful to the Wheat Genetic Improvement Network for making public the mapping data relating to the Avalon × Cadenza population. This popula-tion of doubled-haploid (DH) individuals was developed by Clare Ellerbrook, Liz Sayers and the late Tony Worland (John Innes Centre), as part of a Defra funded project led by ADAS. The parents, having contrasting canopy architectures, were originally chosen by Steve Parker (CSL), Tony Worland and Darren Lovell (Rothamsted Research). We thank Simon Berry of Limagrain UK limited for supplying the Savannah × Rialto mapping population and related marker data. We are grateful to all international collaborators who sent material for screening, as detailed in Table S4.

References

Allen, A.M., Barker, G.L., Berry, S.T., Coghill, J.A., Gwilliam, R., Kirby, S., Robinson, P., et al. (2011) Transcript-specific, single-nucleotide polymorphism discovery and linkage analysis in hexaploid bread wheat (Triticum aestivum L.). Plant Biotechnol. J. **9**, 1086–1099.

Allen, A.M., Barker, G.L., Wilkinson, P., Burridge, A., Winfield, M., Coghill, J., Uauy, C., et al. (2013) Discovery and development of exome-based, codominant single nucleotide polymorphism markers in hexaploid wheat (Triticum aestivum L.). Plant Biotechnol. J. **11**, 279–295.

Brennan, J.P. and Quade, K.J. (2004) Analysis of the Impact of CIMMYT Research on the Australian Wheat Industry. Economic Research Report no.

25, NSW Department of Primary Industries, Wagga Wagga. Available from http://www.agric.nsw.gov.au/reader/10550.

Burt, C., Griffe, L.L., Ridolfini, A.P., Orford, S., Griffiths, S. and Nicholson, P. (2014) Mining the Watkins collection of wheat landraces for novel sources of eyespot resistance. *Plant. Pathol.* **63**, 1241–1250.

Dubcovsky, J. and Dvorak, J. (2007) Genome plasticity a key factor in the success of polyploid wheat under domestication. *Science*, **316**, 1862–1866.

Dvorak, J., Akhunov, E.D., Akhunov, A.R., Deal, K.R. and Luo, M.C. (2006) Molecular characterization of a diagnostic DNA marker for domesticated tetraploid wheat provides evidence for gene flow from wild tetraploid wheat to hexaploid wheat. *Mol Biol Evol.* **23**, 1386–1396.

Endelman, J.B. and Plomion, C. (2014) LPmerge: an R package for merging genetic maps by linear programming. *Bioinformatics*, **30**, 1623–1624.

Friebe, B. and Gill, B.S. (1994) C-band poly-morphism and structural rearrangements detected in common wheat (Triticum aestivum). *Euphytica*, **78**, 1–5.

Grassini, P., Eskridge, K.M. and Cassman, K.G. (2013) Distinguishing between yield advances and yield plateaus in historical crop production trends. *Nat. Commun.* **4**, 2918–000.

Halloran, G.M., Ogbonnaya, F.C. and Lagudah, E.S. (2008) *Triticum (Aegilops) tauschii* in the natural and artificial synthesis of hexaploid wheat. *Aust. J. Agric. Res.* **59**, 475–490.

Haudry, A., Cenci, A., Ravel, C., Bataillon, T., Brunel, D., Poncet, C., Hochu, I., et al. (2007) Grinding up wheat: a massive loss of nucleotide diversity since domestication. *Mol. Biol. Evol.* **24**, 1506–1517.

Kronstad, W.E. (1997) *Agricultural development and wheat breeding in the 20th Century.* In: Wheat: Prospects for Global Improvement. Proceedings of the 5th International Wheat Conference, 10–14 June, 1996, Ankara, Turkey.

Matsuoka, Y. (2011) Evolution of polyploidy triticum wheats under cultivation: the role of domestication, natural hybridization and allopolyploid speciation in their diversification. *Plant Cell Physiol.* **52**, 750–764.

Miller, T., Ambrose, M. and Reader, S. (2001) The Watkins collection of landrace derived wheats. In *Wheat Taxonomy: The Legacy of John Percival. The Linnean Special Issue* (Caligari, P.D. and Brandham, P.E., eds), pp. 113–120. London, UK: Academic Press.

Pembleton, L.W., Cogan, N.O.I. and Forster, J.W. (2013) StAMPP: an R package for calculation of genetic differentiation and structure of mixed-ploidy level populations. *Mol. Ecol. Resour.* **13**, 946–952.

Rife, T., Wu, S., Bowden, R. and Poland, J. (2015) Spiked GBS: a unified, open platform for single marker genotyping and whole-genome profiling. *BMC Genom.* **16**, 248–000.

Schlegel, R. (1996) A compendium of reciprocal translocations in wheat. 2nd Edition. *Wheat Inf Serv.* **83**, 35–46.

Shewry, P.R. (2009) Wheat. *J. Exp. Bot.* **60**, 1537–1553.

Tanksley, S.D. and McCouch, S.R. (1997) Seed banks and molecular maps: unlocking genetic potential from the wild. *Science*, **277**, 1063–1066.

Taylor, D.R. and Ingvarsson, P.K. (2003) Common features of segregation distortion in plants and animals. *Genetica*, **117**, 27–35.

Wang, S., Wong, D., Forrest, K., Allen, A., Chao, S., Huang, B.E., Maccaferri, M., et al. (2014) Characterization of polyploid wheat genome diversity using a high-density 90,000 single nucleotide polymorphism array. *Plant Biotechnol. J.* **12**, 787–796.

Wheat Initiative (2011) http://www.wheatinitiative.org/sites/default/files/wysiwyg/g20-proposal.pdf.

Wilkinson, P.A., Winfield, M.O., Barker, G.L.A., Allen, A.M., Burridge, A., Coghill, J.A., Burridge, A., et al. (2012) CerealsDB 2.0: an integrated resource for plant breeders and scientists. *BMC Bioinform.* **13**, 219–000.

Wilkinson, P.A., Winfield, M.O., Barker, G.L.A., Tyrrell, S., Bian, X., Allen, A.M., Burridge, A., et al. (2016) CerealsDB 3.0: expansion of resources and data integration. *BMC Bioinform.* **17**, 256–000.

Winfield, M.O., Wilkinson, P.A., Allen, A.M., Barker, G.L.A., Coghill, J.A., Burridge, A., Hall, A., et al. (2012) Targeted re-sequencing of the allohexaploid wheat exome. *Plant Biotechnol. J.* **10**, 733–742.

Winfield, M.O., Allen, A.M., Burridge, A.J., Barker, G.L., Benbow, H.R., Wilkinson, P.A., Coghill, J., et al. (2015) High-density SNP genotyping array for hexaploid wheat and its secondary and tertiary gene pool. *Plant Biotechnol. J.* **000**, 000–000.

Wingen, L.U., Orford, S., Goram, R., Leverington-Waite, M., Bilham, L., Patsiou, T.S., Ambrose, M., et al. (2014) Establishing the A.E. Watkins landrace cultivar collection as a resource for systematic gene discovery in bread wheat. *Theor. Appl. Genet.* **127**, 1831–1842.

3

Targeting of prolamins by RNAi in bread wheat: effectiveness of seven silencing-fragment combinations for obtaining lines devoid of coeliac disease epitopes from highly immunogenic gliadins

Francisco Barro[1,*], Julio C. M. Iehisa[1,†], María J. Giménez[1], María D. García-Molina[1], Carmen V. Ozuna[1], Isabel Comino[2], Carolina Sousa[2] and Javier Gil-Humanes[1,‡]

[1]Departamento de Mejora Genética, Instituto de Agricultura Sostenible (IAS), Consejo Superior de Investigaciones Científicas (CSIC), Córdoba, Spain
[2]Departamento de Microbiología y Parasitología, Facultad de Farmacia, Universidad de Sevilla, Sevilla, Spain

---

*Correspondence
email fbarro@ias.csic.

[†]Present Address: Departamento de Biotecnología, Facultad de Ciencias Químicas, Universidad Nacional de Asunción, San Lorenzo, Paraguay.
[‡]Present Address: Department of Genetics, Cell Biology and Development, University of Minnesota, Cargill Building, 1500 Gortner Avenue, Saint Paul, MN, 55108, USA

Keywords: CD epitopes, gluten-free diet, gluten intolerance, immunogenic peptides, RNAi fragments, wheat breeding.

Summary

Gluten proteins are responsible for the viscoelastic properties of wheat flour but also for triggering pathologies in susceptible individuals, of which coeliac disease (CD) and noncoeliac gluten sensitivity may affect up to 8% of the population. The only effective treatment for affected persons is a strict gluten-free diet. Here, we report the effectiveness of seven plasmid combinations, encompassing RNAi fragments from α-, γ-, ω-gliadins, and LMW glutenin subunits, for silencing the expression of different prolamin fractions. Silencing patterns of transgenic lines were analysed by gel electrophoresis, RP-HPLC and mass spectrometry (LC-MS/MS), whereas gluten immunogenicity was assayed by an anti-gliadin 33-mer monoclonal antibody (moAb). Plasmid combinations 1 and 2 downregulated only γ- and α-gliadins, respectively. Four plasmid combinations were highly effective in the silencing of ω-gliadins and γ-gliadins, and three of these also silenced α-gliadins. HMW glutenins were upregulated in all but one plasmid combination, while LMW glutenins were downregulated in three plasmid combinations. Total protein and starch contents were unaffected regardless of the plasmid combination used. Six plasmid combinations provided strong reduction in the gluten content as measured by moAb and for two combinations, this reduction was higher than 90% in comparison with the wild type. CD epitope analysis in peptides identified in LC-MS/MS showed that lines from three plasmid combinations were totally devoid of CD epitopes from the highly immunogenic α- and ω-gliadins. Our findings raise the prospect of breeding wheat species with low levels of harmful gluten, and of achieving the important goal of developing nontoxic wheat cultivars.

Introduction

Cereal grains contain about 10–15% (dry weight) of protein, from which gluten is the most important fraction as it is major determinant of the technological properties of baking cereals. However, gluten is not a single protein but a complex mix of proteins, which are deposited in the starchy endosperm during grain development. Gluten proteins are divided into two major fractions: the gliadins and the glutenins, which are different in terms of structure and functionality. In turn, gliadins are formed by three different fractions; ω-, γ-, and α-gliadins, while glutenins comprise two fractions; the high molecular weight (HMW) and the low molecular weight (LMW) subunits. The gliadins are generally present as monomers and contribute extensibility to wheat flour dough. The glutenins contribute elasticity to dough and form large polymers linked by disulphide bonds.

These proteins make up a complex mixture that in a typical bread wheat cultivar may be comprised of up to 45 different gliadins, 7–16 LMW glutenin subunits and 3–6 HMW glutenin subunits. Gliadins and glutenins are not present at the same amount in the grain of cereals, and their proportions can vary within a broad range depending on both genotype (variety) and growing conditions (soil, climate, fertilization, etc.). The ratio of gliadins to glutenins was examined in a range of cereals (Wieser and Koehler, 2009), and hexaploid common wheat showed the lowest ratio (1.5 : 1–3.1 : 1), followed by durum wheat (3.1 : 1–3.4 : 1), emmer wheat (3.5 : 1–7.6 : 1) and einkorn wheat (4.0 : 1–13.9 : 1).

In addition to their unique viscoelastic properties, gluten proteins are responsible for triggering certain pathologies in susceptible individuals: (i) coeliac disease (CD), which affects both children and adults throughout the world at various frequencies (from 0.1% to >2.8%) (Abadie et al., 2011; Mustalahti et al., 2010), and (ii) noncoeliac gluten sensitivity, a newly recognized pathology of intolerance to gluten (Sapone et al., 2011) with an estimated prevalence of 6% for the USA population. However, gliadins and glutenins do not contribute equally to CD, and gliadins are indubitably the main toxic component of gluten as most (DQ2 or DQ8-specific) CD4+ T-lymphocytes obtained from small intestinal biopsies from coeliac patients seem to recognize this fraction (Arentz Hansen et al., 2002;). In the immune epitope database (IEDB) (http://

www.iedb.org/), 190 T-lymphocytes stimulating epitopes related to CD can be found. Of these, 180 (95%) map to gliadins, while only 10 (5%) map to glutenins.

However, not all gliadin epitopes are equally important in triggering CD. The α-2-gliadin family contains the 33-mer peptide, present in the N-terminal repetitive region, with six overlapping copies of three different DQ2-restricted T-cell epitopes with high stimulatory properties and highly resistant to human intestinal proteases (Shan et al., 2002; Tye-Din et al., 2010). The α-gliadins also contain the peptide p31-43, which has been reported to induce mucosal damage via a non-T-cell-dependent pathway (innate response) (Di Sabatino and Corazza, 2009; Maiuri et al., 2003). Moreover, an additional DQ2-restricted epitope (DQ2.5-glia-α3) which partially overlaps with 33-mer peptide (Vader et al., 2002) is present in α-2-gliadins.

Tye-Din et al. (2010) comprehensively assessed the potentially toxic peptides contained within wheat, but also barley, and rye, and identified which ones stimulate T-cells from patients with coeliac disease. They found that the 33-mer peptide from wheat α-gliadin was highly stimulatory, and another peptide (QPFPQPE QPFPW) from ω-gliadin/C-hordein was immunodominant after eating wheat, barley and rye. These two peptides present in wheat, plus another from barley, can elicit 90% of the immunogenic response induced by wheat, barley and rye (Tye-Din et al., 2010). These findings showed that the immunotoxicity of gluten could be reduced to three highly immunogenic peptides, which make the development of varieties with low-toxic epitopes more affordable.

One promising approach for reducing gluten toxicity and, therefore the incidence of gluten-related intolerances in cereals, is the downregulation of immunodominant peptides by RNAi. This technology was applied to downregulate the expression of γ-gliadins (Gil-Humanes et al., 2008), ω-5 gliadins (Altenbach and Allen, 2011), α-gliadins (Becker et al., 2012) and all gliadins (Gil-Humanes et al., 2010) in bread wheat. In all these reports, the silencing of gliadin fractions was accompanied by an increase in other storage proteins or nongluten proteins (Rosell et al., 2014). Protein extracts from transgenic lines with all three gliadin fractions downregulated were tested for stimulation of DQ2- and DQ8-restricted T-cell clones of patients with coeliac disease, and a pronounced reduction in proliferative responses was seen in some transgenic lines (Gil-Humanes et al., 2010).

In the present work, we have tested the effectiveness of seven plasmid combinations encompassing RNAi fragments from α-, γ-, ω-gliadins, and LMW glutenin subunits for selectively silencing different fractions of prolamins. The results from different fragments are comparable as all seven constructs were placed under the control of the same endosperm-specific promoter. Wheat grain protein composition was substantially modified by some of the plasmid combinations, which provided wheat lines devoid of CD epitopes from α- and ω-gliadins, the most immunogenic gluten fractions in wheat. The plasmid combinations reported here may be useful for the development of new wheat cultivars with low levels of harmful gluten.

Results

RNAi sequence combinations and plasmid design

Figure 1a shows the structure of the α-, γ-, and ω-gliadins, and LMW glutenins indicating the gene regions where seven RNAi

sequences were designed. These RNAi sequences varied in length from 109 to 377 (nucleotide sequences provided in Table S1) and were cloned from different domains of the gliadin sequences. The alpha/beta ZR RNAi fragment covered part of the repetitive domain of α-gliadins plus the poly-Q domain, and a small region of the nonrepetitive domain I, whereas the alpha RNAi fragment was PCR-amplified from the nonrepetitive domain I. The γ-gliadin RNAi (g8.1) sequence encompassed part of the nonrepetitive domain I of the γ-gliadins, plus the Q-rich domain and part of the nonrepetitive domain II. Among the ω-gliadins, the omega RNAi sequence was designed containing part of the signal peptide, the N-terminal domain, and part of the repetitive domain; whereas both the omega4 and omega8 were cloned from the 3′ end of the repetitive domain of the ω-gliadins. The LMW fragment contained part of the signal peptide, the N-terminal domain, and part of the repetitive domain of the LMW glutenins. The identity observed between the IR fragments and the nontargeted prolamin fractions was in all cases between 40% and 60% (Table S2). The RNAi sequences were in some cases combined by fusion PCR to provide five different RNAi fragments, which were inserted downstream a D-hordein promoter to produce five hairpin plasmids (Figure 1b). Plasmids were further used alone or combined with others providing seven plasmid combinations (Figure 1c). Plasmids pghpg8.1 and pDhp_ω/α were described previously (Gil-Humanes et al., 2008, 2010), whereas all other plasmids and plasmid combinations are compared for the first time in this work. In total, 21 transgenic wheat lines (Table S3) containing any of the seven combination of plasmids (1–6 different lines for each combination) were assayed for the downregulation of gliadins and effects on other protein fractions, agronomic parameters, and the content of coeliac disease-related gliadins as determined using the G12 monoclonal antibody and CD immunogenic peptides identified by LC-MS/MS. Parental line and null segregants for most of the transgenic lines were generated and assayed. However, no significant differences were found between these null segregants and the parental line for the above parameters. Therefore, only parental line was included in all comparisons with transgenic lines.

Plasmid combinations and patterns of silencing

Silencing patterns in all 21 transgenic lines were analysed by A-PAGE and SDS-PAGE, and the gliadin and glutenin fractions quantified by RP-HPLC and compared with the untransformed wild-type BW208 (denoted as Plasmid Combination 0). Examples of gliadin and glutenin patterns obtained with all seven plasmid combinations are in Figure S1. Detailed results for each line are in Supplementary Table S4. Figure 2 shows the transgenic lines and the effects of each plasmid combinations on the gliadin and glutenin fractions. Plasmid Combination 1 clearly targeted the γ-gliadin fraction while ω-, and α-gliadins were upregulated. This Plasmid Combination did not significantly affect the total amount of the glutenin fractions. However, not all LMW glutenins were equally affected: while the amount of LMW glutenins appearing between 45 and 60 min of retention time decreased, LMW peaks between 35 and 45 min of retention time were slightly increased compared with the wild type (Figure S1). Plasmid Combination 2 achieved a greater reduction in the α-gliadin fraction, with no major effects on other gliadin fractions, and an important increment in

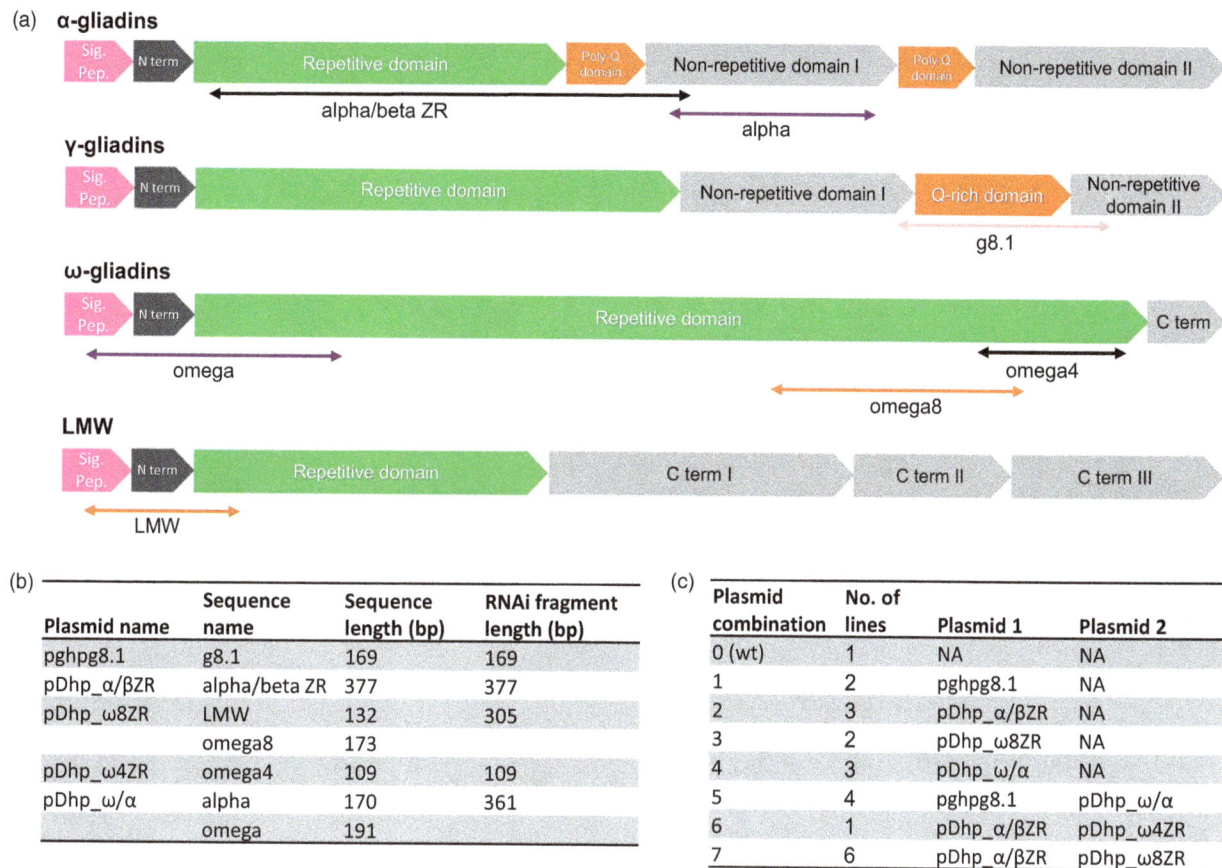

Figure 1 Plasmid design. (a) Structure of α-, γ-, and ω-gliadins, and LMW glutenins as reviewed in (Qi *et al.*, 2006), indicating the domains for each family where sequences for RNAi fragments were designated. (b) Sequences were or were not combined into a unique RNAi fragment and inserted downstream a D-hordein promoter providing five different hairpin plasmids. (c) Plasmids were used alone or combined into seven combinations where number 0 corresponds to BW208 wild type (wt).

the HMW glutenin fraction compared with the wild type. Plasmid Combination 3 provided a significant reduction in all three ω-, γ-, and α-gliadin fractions, and also a strong reduction in the LMW glutenin content, but an important increment in the HMW glutenin fraction (Figure 2b). In contrast to Plasmid Combination 1, Combination 3 affected mainly the LMW in the region between 35 and 45 min of retention time (Figure S1). Plasmid Combination 4 also provided a strong reduction in all three gliadin fractions, and a significant increment in the HMW glutenin fraction. However, in contrast to Plasmid Combination 3, the LMW glutenins were not reduced. Plasmid Combinations 5, 6 and 7 were equally effective in the downregulation of ω-gliadins (Figure 2a), although Combination 5 was more effective in the downregulation of γ-gliadins, and Combinations 6 and 7 in the downregulation of α-gliadins (Figure 2a). These Plasmid Combinations (5, 6, and 7) also had an important impact in the glutenin fraction; while all three Plasmid Combinations decreased the content of LMW glutenins (with wide differences within Combination 5), only Combination 7 showed a strong increment in the HMW glutenin fraction (Figure 2b).

Protein fraction redistributions kernel composition and agronomic traits

Figure S2 shows the fold change in kernel composition of transgenic lines with the different plasmid combinations. Gliadins,

Gli/Glu ratio and total prolamin content were significantly reduced in all the transgenic combinations compared with the control (Combination 0). Plasmid Combinations 5, 6 and 7, all three combinations of two different plasmids, were the most effective in the downregulation of total gliadin content in transgenic wheat (Table 1). Only three plasmid combinations, 2, 4 and 6, had significant effect on total glutenin content; Plasmid Combinations 2 and 4 increased the total glutenin content in transgenic wheat, whereas Combination 6 decreased it (Table 1 and Figure S2). The Gli/Glu ratio decreased significantly in all seven plasmid combinations compared with the wild type, with the largest differences found for Combinations 4, 5 and 7. As consequence of these changes, the prolamin fraction (comprising gliadins plus glutenins) was strongly decreased when Plasmid Combinations 5, 6 and 7 were used. In contrast, nongluten proteins, which represented about 4.3% DW in the wild type (Plasmid Combination 0), increased up to 9.9%, 10.6% and 9.4% DW in Plasmid Combinations 5, 6 and 7, respectively (Table 1). Nongluten proteins compensated the downregulation of prolamins, and total protein was not affected by any of the seven plasmid combinations, all of them showing protein contents around 13% DW. The major component of wheat grain, starch, was not significantly affected by the changes in the protein fraction observed with the seven combinations of plasmids. Hence, the lowest value for starch content was

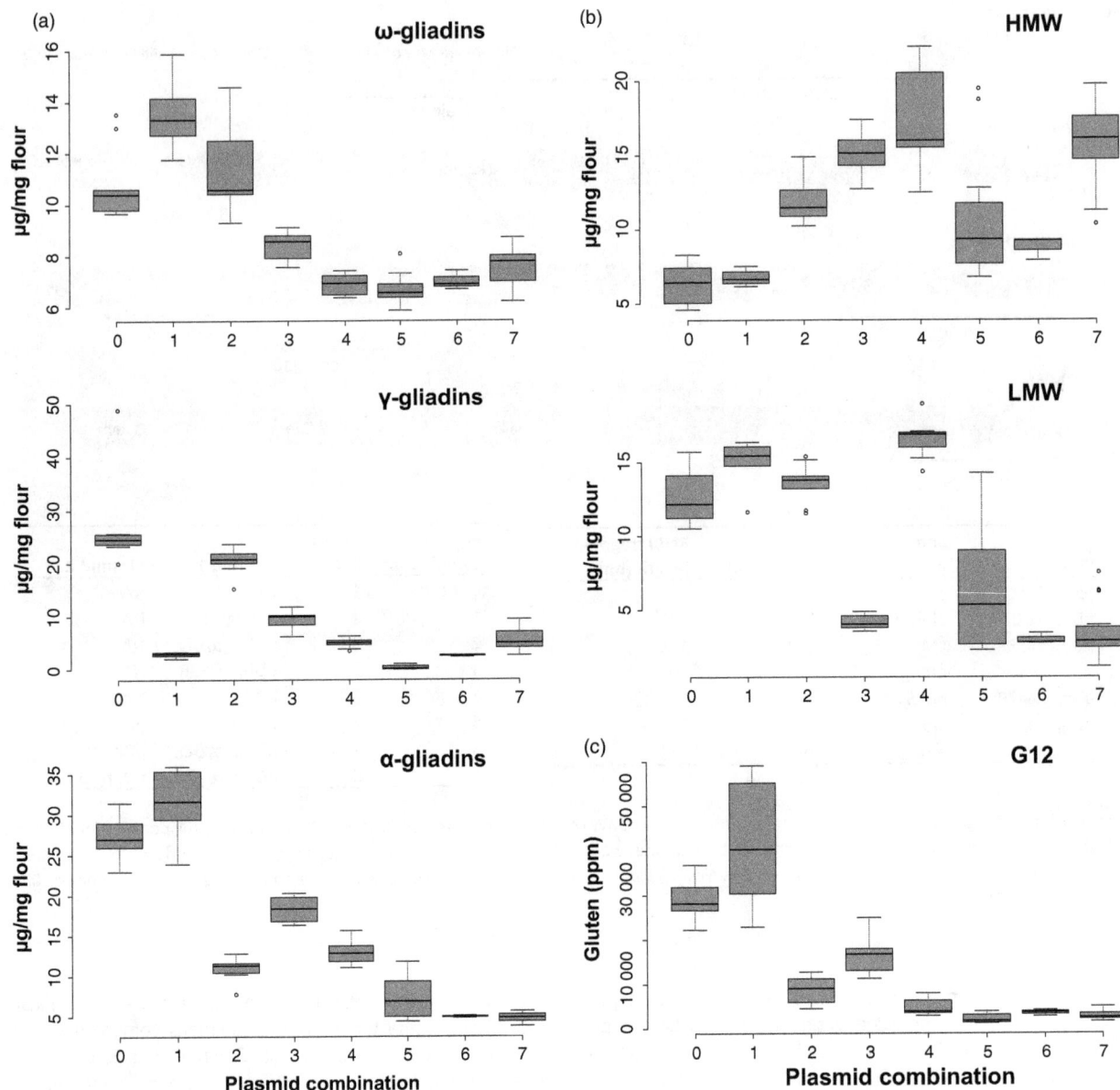

Figure 2 Box plots of gliadins (a) and glutenins (b) distribution in the grain of transgenic lines with seven plasmid combinations expressed as μg protein/ mg flour. HMW and LMW denote the high molecular weight and low molecular weight subunits of glutenins, respectively. (c) Box plot of gluten content with seven plasmid combinations calculated by G12 competitive ELISA and expressed in parts per million (ppm). Plasmid combinations are as indicated in Figure 1c. Plasmid Combination 0 corresponds to BW208 wild type.

62.8% DW for Plasmid Combinations 4 and 6, which was not significantly different to the 66.0% DW of the wild type (Table 1).

Agronomic traits such as days to anthesis, kernel weight and test weight were determined in transgenic plants from all seven plasmid combinations and compared with that of the wild type. As showed in Table 1, the number of days from sowing to anthesis was around the 149 days in the wild type, with no significant differences for transgenic plants from all seven plasmid combinations. Kernel weights showed higher values than the wild type in Plasmid Combination 1, whereas Combinations 5, 6 and 7 had lower weights (Table 1 and Figure S2). With respect to kernel weight, these differences were significant for Combinations 1, 5, 6 and 7, and in the case of test weight for Combinations 5 and 6.

LC-MS/MS analysis of the protein fraction in plasmid combinations

Total protein extracted from all plasmid combinations was also analysed by LC-MS/MS after PT-digestion. The total fragments detected by MS ranged from 46,287 (Plasmid Combination 1) to 52,129 (Plasmid Combination 7) (Table 2). Approximately 3% to 10% of these fragments were identified using the NCBI protein database restricted to *Triticum* species, comprising 146–198 proteins. More than 70% and 38% of the identified fragments and proteins, respectively, were seed storage proteins (SSPs). Although Plasmid Combination 1 targets γ-gliadins, the number of identified γ-gliadin fragments was not as low as would be expected (Table 2). In addition, the number of LMW glutenins was higher and that of ω-gliadins was lower. In Plasmid

Table 1 Average values for protein fractions, and quality and agronomic parameters for each plasmid combination

Parameter	Plasmid combination							
	0 (wt)	1	2	3	4	5	6	7
Total gliadins (µg/mg flour)	65	47.8**	43.0**	36.2**	24.9**	14.4**	14.6**	18.1**
Total glutenins (µg/mg flour)	18.9	21.6	25.4**	19.2	33.7**	16.7	11.8*	19.2
Gli/Glu ratio	3.5	2.2**	1.7**	1.9**	0.7**	0.9**	1.3**	1.0**
Total prolamins (µg/mg flour)	84.0	69.4**	68.4**	55.4**	58.6**	31.1**	26.4**	37.3**
Nongluten proteins (% DW)	4.3	5.9*	5.9*	7.2**	7.4**	9.9**	10.6**	9.4**
Grain protein (% DW)	13.4	13.3	13.2	13.1	13.7	13.3	13.5	13.4
Gluten content (ppm)	29494	41332**	8511**	16721**	4796**	2146**	3479**	2720**
Starch (% DW)	66.0	64.5	64.0	66.8	62.8	63.5	62.8	65.4
SDSS (mL)	9.3	8.6	10.4**	6.6**	9.6	7.0**	6.2**	6.6**
Anthesis (days)	149	147	147	152	148	148	152	152
Kernel weight (g)	36.3	39.9*	37.5	36.9	32.6	30.2**	31.4*	32.0*
Test weight (g/L)	829.2	845.8	834.7	827.8	812.5	797.4*	791.7*	808.2

Plasmid combinations are as indicated in Figure 1C; DW, Dried Weight; ppm, parts per million; SDSS, sodium dodecyl sulphate sedimentation.

Significant differences were identified at the 5% (*) and 1% (**) probability levels by the two-sided Dunnett's multiple comparisons with control indicated as (0) in plasmid combination.

Combination 2, the number of α-gliadin peptides was lower but that of γ- and ω-gliadins and glutenins was similar to the wild type. The profile of gliadin fraction in Plasmid Combination 3 was similar to the control, but the proportion of fragments derived from LMW glutenin was lower and that of nongluten proteins was higher. In contrast to Plasmid Combination 3, in Plasmid Combination 4 the number of gliadin fragments was lower and that of HMW glutenins was higher, while LMW glutenins were similar to the wild type. In this combination, the proportion of nongluten fragments was also higher. In Plasmid Combinations 5–7, the gliadin fraction was greatly decreased, affecting mainly the α-gliadin in Combinations 6 and 7 and all gliadins in the Combination 5. The proportion of LMW glutenin peptides was also decreased with higher effect in Plasmid Combination 7, and with a strong increase in HMW glutenins in Plasmid Combination 5. There was also an increase in the proportion of nongluten protein fragments in these three combinations. In general, the higher proportion of nongluten fragments was due to the increase in serpins, globulins and triticins.

CD immunotoxic peptides in fragments identified by LC-MS/MS

In peptides identified by LC-MS/MS analysis against *Triticum* proteins, CD immunogenic epitopes (Sollid et al., 2012) and the p31-43 fragment of α-gliadins were searched allowing up to one mismatch (Tables 3 and S5). In the wild type, a total of 645 epitopes with perfect match were found in 3,240 identified peptides, being most of them epitopes reported in γ-gliadins (482 epitopes) followed by α-gliadin (92 epitopes) and ω-gliadin (24 epitopes) (Table 3). HMW glutenin epitopes were not found in these peptides. In transgenic lines, total perfect match epitopes ranged from 55 in Plasmid Combination 5 to 849 in Plasmid Combination 3 (Table 3), γ-gliadin epitopes being the most abundant in all combinations. α-gliadin epitopes were not found in the identified peptides in Plasmid Combinations 2, 5 and 6, and were much lower than wild type in Combinations 4 and 7. Furthermore, the number of ω-gliadin epitopes found was lower in Plasmid Combinations 4, 5 and 7. Plasmid Combination 5 also exhibited the lowest number of γ-gliadin epitopes and an increase in HMW glutenin epitopes. The p31-43 fragment was lower in all

transgenic lines except in Plasmid Combination 1, and it was not found in Plasmid Combinations 2, 6 and 7. Of the α-gliadin epitopes, the most abundant was DQ2.5-glia-α2, followed by DQ2.5-glia-α1b and DQ2.5-glia-α1a (Table 3). Among the γ-gliadin epitopes, DQ2.5-glia-γ4c/DQ8-glia-γ1a and DQ2.5-glia-γ5 were the majority. LMW glutenin epitopes were not found in Combinations 3 and 4, were lower in Combinations 2 and 7 and were similar to wild-type levels in 5 and 6, and higher than wild type in Plasmid Combination 1. Epitopes matched to that reported in hordeins were lower in all transgenics.

The epitopes with one mismatch also showed a similar pattern to those with perfect match. A total of 842 epitopes were found in the wild type, where 603 were γ-gliadin epitopes, 95 ω-gliadin and 82 α-gliadin epitopes (Table S5). Similarly, a drastic decrease in all gliadin epitopes and an increase in HMW glutenin epitopes were observed in Plasmid Combination 5. Epitopes with one mismatch of all gliadins were also decreased in Combinations 4 and 7 compared with the wild type, with a decrease in LMW glutenin epitopes in the latter. Plasmid Combinations 2, 3 and 6 presented lower number of α-gliadin epitopes, but differing in γ-gliadin (slightly lower in combination 2) and ω-gliadin (slightly increased in Combinations 3 and 6) epitopes. In Plasmid Combination 1, γ- and ω-gliadin epitopes were slightly decreased, but LMW glutenin was increased compared with wild type. A drastic increase in Hordein epitopes was observed in Plasmid Combination 2. These results suggest that the Plasmid Combinations 4 and 5 have lower CD toxicity than wild type and other transgenics.

Transgenic lines and gluten immunogenicity by anti-gliadin 33-mer moAb

The 33-mer peptide derived from α-2 gliadin (residues 57–89) is one of the most highly antigenic peptides identified to date (Shan et al., 2002; Tye-Din et al., 2010). This peptide was identified as the primary initiator of the inflammatory response to gluten in patients with CD. This 33-mer fragment is naturally formed by digestion with gastric and intestinal proteases, binds to DQ2 after deamidation by tissue transglutaminases (tTG), contains a cluster of 6 T-cell epitopes and is the most immunodominant peptide in

Table 2 Fragments and proteins identified by LC-MS/MS analysis in all seven plasmid combinations

Plasmid Combination	0 BW208		1 D623		2 I17		3 H754		4 D783		5 E82		6 H320		7 H811	
Line	Fragments	Proteins	Fragments	Proteins	Fragments	Proteins	Fragments	Proteins	Fragments	Proteins	Fragments	Proteins	Fragments	Proteins	Fragments	Proteins
Total	48405	ND	46287	ND	51660	ND	48480	ND	50980	ND	51221	ND	50365	ND	52129	ND
Triticum	3240	184	4832	198	1639	149	3436	197	1639	187	1464	162	2141	181	1616	146
α-gliadins	1554	32	3107	44	21	2	1323	29	262	11	37	3	0	0	7	1
γ-gliadins	404	14	341	15	409	14	641	18	189	13	33	5	555	20	361	13
ω-gliadins	347	9	244	5	264	6	279	9	22	2	10	1	226	6	99	4
Total gliadins	2305	55	3692	64	694	22	2243	56	473	26	80	9	781	26	467	18
LMW	104	12	317	22	83	12	50	7	150	15	91	14	82	11	0	0
HMW	11	1	13	2	11	2	13	2	42	6	172	7	64	6	35	3
Total glutenins	115	13	330	24	94	14	63	9	192	21	263	21	146	17	35	3
Total prolamins	2420	68	4022	88	788	36	2306	65	665	47	343	30	927	43	502	21
γ-secalin	72	3	39	2	28	1	123	5	12	1	0	0	109	3	20	1
ω-secalin	93	3	36	1	82	3	51	3	7	1	0	0	54	2	0	0
γ-hordein-like	2	1	16	2	2	1	2	1	7	2	5	2	2	1	0	0
γ-gliadin-like	73	3	61	4	12	1	83	4	22	2	7	1	55	2	85	4
Gliadin/avenin-like	38	3	44	5	50	5	33	4	41	5	29	4	24	4	46	4
AAI	48	9	53	8	67	10	72	10	69	11	78	13	71	11	77	10
Globulins	106	5	105	4	132	5	178	6	157	5	210	6	211	6	228	8
Triticins	8	1	10	1	17	1	18	1	25	1	60	2	27	1	28	1
Serpin	70	6	114	6	137	7	182	8	231	9	314	9	247	8	255	8
Nonglutem Proteins	270	24	326	24	403	28	483	29	523	31	691	34	580	30	634	31
LTPs	3	1	3	2	4	1	4	1	3	1	4	2	3	1	3	1
β-amylase	44	2	47	2	41	2	62	2	54	2	65	2	57	2	64	2
Total SSPs	2737	95	4398	116	1236	67	2855	97	1245	81	1103	68	1567	76	1203	55

Table 3 Number of CD immunogenic epitopes with perfect match and found in peptides identified by MS

Prolamin	Epitope	Deaminated sequence	Original sequence	0 BW208	1 D623	2 I17	3 H754	4 D783	5 E82	6 H320	7 H811
γ-gliadin	DQ2.5-glia-g4c/DQ8-glia-g1a	QQPEQPFPQ/EQPQQPFPQ	QQPQQPFPQ	262	203	225	383	72	18	272	148
γ-gliadin	DQ2.5-glia-g5	QQPFPEQPQ	QQPFPQQPQ	181	185	122	241	15	12	210	67
γ-gliadin	DQ2.5-glia-g1/DQ8.5-glia-g1	PQQSFPEQQ/PQQSFPEQE	PQQSFPQQQ	17	7	10	36	16	1	25	17
γ-gliadin	DQ2.5-glia-g3/DQ8-glia-g1b	QQPEQPYPQ/EQPQQPYPE	QQPQQPYPQ	7	5	3	15	8	0	14	8
γ-gliadin	DQ2.5-glia-g4a	SQPEQEFPQ	SQPQQQFPQ	5	0	5	8	1	0	3	5
γ-gliadin	DQ2.5-glia-g2	IQPEQPAQL	IQPQQPAQL	4	4	7	7	4	2	4	7
γ-gliadin	DQ2.5-glia-g4b	PQPEQEFPQ	PQPQQQFPQ	3	5	0	20	1	0	11	3
γ-gliadin	DQ2.5-glia-g4d	PQPEQPFCQ	PQPQQPFCQ	3	0	0	3	0	0	1	0
Total γ-gliadin				482	409	372	713	117	33	540	255
α-gliadin	DQ2.5-glia-a2	PQPELPYPQ	PQPQLPYPQ	43	63	0	49	4	0	0	2
α-gliadin	DQ2.5-glia-a1b	PYPQPELPY	PYPQPQLPY	30	40	0	30	4	0	0	1
α-gliadin	DQ2.5-glia-a1a	PFPQPELPY	PFPQPQLPY	15	29	0	17	1	0	0	0
α-gliadin	DQ2.5-glia-a3	FRPEQPYPQ	FRPQQPYPQ	0	3	0	2	0	0	0	0
α-gliadin	DQ8-glia-a1	EGSFQPSQE	QGSFQPSQQ	4	13	0	7	0	0	0	0
Total α-gliadin				92	148	0	105	9	0	0	3
ω-gliadin	DQ2.5-glia-w1	PFPQPEQPF	PFPQPQQPF	17	8	11	14	2	0	13	3
ω-gliadin	DQ2.5-glia-w2	PQPEQPFPW	PQPQQPFPW	7	4	4	1	0	0	3	1
Total ω-gliadin				24	12	15	15	2	0	16	4
HMW-GS	DQ8-glut-1	QGYYPTSPQ	QGYYPTSPQ	0	0	1	2	2	9	4	3
α-gliadin	p31-43	PGQQQPFPPQQPY	PGQQQPFPPQQPY	13	14	0	3	3	0	0	0
α-gliadin	p31-43	LGQQQPFPPQQPY	LGQQQPFPPQQPY	8	6	0	6	3	1	0	0
Total α-gliadin p31-43				21	20	0	9	6	1	0	0
LMW-GS	DQ2.5-glut-L1	PFSEQEQPV	PFSQQQQPV	16	24	4	0	0	11	15	2
LMW-GS	DQ2.5-glut-L2	FSQQQESPF	FSQQQQSPF	0	1	0	0	0	1	0	0
Total LMW-GS				16	25	4	0	0	12	15	2
Hordein	DQ2.5-hor-2	PQPEQPFPQ	PQPQQPFPQ	10	1	4	5	1	0	3	2
Hordein	DQ2.5-hor-3	PIPEQPQPY	PIPQQPQPY	0	0	0	0	0	0	0	0
Total Hordein				10	1	4	5	1	0	3	2

patients with CD after eating wheat-based products (Shan et al., 2002; Tye-Din et al., 2010).

The G12 monoclonal antibody (moAb) is able to recognize with great sensitivity peptides (besides the 33-mer) inmunotoxic for patients with CD. The sensitivity and epitope preferences of this antibody were found to be useful for detecting gluten-relevant peptides to infer the potential toxicity of cereals for patients with CD (Comino et al., 2011; Morón et al., 2008b). Immunotoxic properties of the seven lines were studied with G12 to study the reduced toxicity profile.

As shown in Figure 2c, the reactive gluten identified by G12 moAb strongly decreased in transgenic lines for Plasmid Combinations 3, 4, 5, 6 and 7, but increased in wheat lines for Combination 2. We observed reductions of 84%, 93%, 88% and 92% gluten content by competitive G12 moAb in transgenic lines for Plasmid Combinations 4, 5, 6, and 7, respectively, in comparison with the wild type. G12 values for individual lines are shown in Table S4.

Discussion

IR sequences were designed using different regions of the gliadin and LMW coding sequences and combined to form seven different RNAi constructs (Figure 1). These combinations promoted differential changes in the prolamin (gliadins and glutenins) composition of the transgenic plants generated (Fig-

ure 2). Interestingly, pleiotropic effects (i.e. stronger downregulation of the targeted gliadin groups) were observed when two RNAi constructs with IR fragments targeting different groups were combined in the same transgenic plant. For example, downregulation of α- and γ-gliadins was stronger in Combination 5 (pDhp_ω/α and the pghpg8.1 vectors) than when they were used alone (Combinations 4 and 1, respectively). Similarly, the downregulation of ω-gliadins was significantly stronger in the Combination 7, in which the pDhp_ω8ZR was cotransformed with the pDhp_αβ/ZR, than when only the pDhp_ω8ZR was used (Combination 3). The same was observed in the α-gliadins, with a stronger downregulation in the Combinations 6 and 7 (both with the pDhp_αβ/ZR plasmid plus a second plasmid targeting different prolamins) than in the Combination 2 (the pDhp_αβ/ZR plasmid alone). In summary, the on-target silencing was normally higher when a second IR sequence (targeting a different gliadin group) was cotransformed, which might be a consequence of (i) off-target activity of the second IR fragment, (ii) reduction in the cells of potential off-target mRNA sequences that compete with the on-target gliadin mRNAs for recognition and degradation by the post-transcriptional silencing machinery, or (iii) a combination of both effects. When short interfering RNAs (siRNAs) from a certain double-stranded RNA (dsRNA) sequence are formed, they lead the RNA-induced silencing complex (RISC) to recognize homologous mRNA sequences for their cleavage and degradation (Hammond et al., 2000). It has been previously reported that nonspecific downregulation can occur during RNAi as 100% identity between the silencing siRNA and the target gene is not absolutely required to promote gene silencing (Senthil-Kumar et al., 2007; Xu et al., 2006). In fact, only 14 nucleotides (or even less) of sequence complementarity between siRNA and mRNA can lead to gene silencing in plants (Jackson and Linsley, 2004; Jackson et al., 2006). Therefore, by reducing the number of possible off-target sequences in the cell, the probability of the siRNA to find the on-target sequences is enhanced, and so is the efficiency of the gene silencing.

The identity observed between the IR fragments and the nontargeted prolamin fractions (40–60%) would make it possible for there to be, among the pool of siRNAs formed from each IR fragment, some with total (or almost total) identity with nontargeted prolamins. This fact could explain the off-target silencing observed with some plasmid combinations. For example, α- and γ-gliadins were downregulated with Combination 3 that only contains the LMW and omega8 IR fragments (Figure 2). Similarly, Combination 6 showed off-target reduction in γ-gliadins and LMW glutenins (Figure 2). These results fit with the off-target downregulation of γ-gliadins and LMW glutenins reported in previous studies when using the IR ω/α (omega and alpha RNAi fragments) (Gil-Humanes et al., 2010, 2014) (Gil-Humanes et al., 2011). By contrast, Plasmid Combinations 1 and 2 were highly specific to their on-target gliadin group (γ- and α-gliadins, respectively) and did not silence any of the other prolamin groups. Moreover, the high number of γ- gliadin fragments found by LS-MS/MS for Combination 1 could be explained as a consequence of the high specificity of the pgphp8.1 fragment in silencing just some of the proteins in this family. The analysis of the 11 active genes reported for the cultivar Chinese Spring (Anderson et al., 2013) shows that the identity between those sequences and the pgphp8.1 fragment varies from less than 60% to nearly 100%. Two of the proteins with lower identity (γ-gliadin 3, 60.4%, and γ-gliadin 4, 61.6%) are those containing an odd number of cysteine residues, supporting the hypothesis that the amount of gamma gliadin as measured by RP-HPLC in our lines may be underestimated because they could be part of the polymer fraction.

The silencing of gliadins and LMW glutenins did not affect total protein and starch content or kernel weight. In contrast, quality parameters like SDSS and the gliadins to glutenins ratio were affected by the silencing of gliadins. The SDS sedimentation test provides information on the protein quantity and the quality of ground wheat and flour samples (Carter et al., 1999). Positive correlations were observed between sedimentation volume and gluten strength, and hence, SDSS test should be an effective small-scale test for quality assessment. As plasmid combinations 3, 5, 6 and 7 had lower SDSS values than wild type, they might exhibit also lower gluten strength. The fact that Plasmid Combination 4 showed a SDSS value similar to the wild type, with lower gliadin content, but significantly higher contents of glutenins may indicate that glutenins are major determinants of SDSS as previously reported (Piston et al., 2011).

Because we did not perform quantitative LC-MS/MS analysis, the number of protein fragments and immunotoxic epitopes identified do not necessarily reflect their abundance in the total gluten protein. In addition, not all epitopes are equally immunogenic for CD (Tye-Din et al., 2010). The number of α-gliadin peptides and epitopes were well correlated with α-gliadin quantified by RP-HPLC. The DQ2.5-glia-α3 epitope, downstream of the 33-mer fragment, was the least abundant probably due to

its digestibility with Trypsin. Because γ-gliadins contain 5–10 epitopes per sequence (Salentijn et al., 2012), a high number of these epitopes was identified in almost all lines analysed, most prevalent being the epitopes DQ2.5-glia-γ4c/DQ8-glia-γ1a and DQ2.5-glia-γ5. In general, a lower number of CD epitopes was observed in transgenic lines generated using two plasmids, with the exception of Plasmid Combination 6.

As expected, the changes observed in the prolamin fractions led to changes in the immunotoxicity of lines generated from the different combinations of plasmids, as determined by the G12 moAb assay (Figure 2c). The 33-mer peptide is highly resistant to digestion and one of the main immunodominant toxic peptide in patients with coeliac disease (Tye-Din et al., 2010). The sensitivity and epitope preferences of the G12 antibody has been found to be useful for detecting gluten-relevant peptides to predict the potential toxicity of cereal foods for patients with CD (Morón et al., 2008a). Moreover, the reactivity of G12 moAb with cereal storage proteins of different varieties of cereals has been correlated with the known dietary immunotoxicity of the different grains (Comino et al., 2011). All the combinations of plasmids, except Combination 1, resulted in a reduction in the reactivity of the G12 moAb, and consequently a decrease in the immunotoxicity of the gluten proteins. The level of reduction in the different combinations was strongly correlated (92.38%, $P < 0.001$) with the level of reduction in the α-gliadin fraction, and was also well correlated with the number of α-gliadin epitopes found in MS analysis. This observation may be explained by the fact that the G12 moAb mainly recognizes the hexapeptide QPQLPY found in the 33-mer peptide of α2-gliadin, although it also binds to other related peptide variants of immunotoxic gluten proteins (Morón et al., 2008a).

Our present results indicate that RNAi can be a very effective approach for obtaining wheat lines without, or with very low levels of gliadins and LMW glutenins, the major gluten proteins containing epitopes triggering coeliac disease. Six of the plasmid combinations tested showed strong reduction in the gluten content as measured by competitive anti-gliadin 33-mer moAb and in two combinations, this reduction was higher than 90% by comparison with the wild type. Transgenic lines from three plasmid combinations were found to be totally devoid of CD epitopes from the highly immunogenic α- and ω-gliadins. These lines have potential to be used directly in breeding programs for obtaining wheat cultivars suitable for coeliac or other gluten intolerant patients, and the effective plasmid combinations identified may be used to downregulate toxic epitopes in other wheat backgrounds.

Experimental procedures

Plasmid design and construction

For the synthesis of the RNAi vectors, conserved regions of the α-, γ-, and ω-gliadins were PCR-amplified from genomic DNA isolated from the bread wheat (Triticum aestivum) cv. Bobwhite 208 (BW208) and cloned in sense and antisense orientation, separated by the Ubi1 intron to form the inverted repeat (IR) sequences. The relative position of each IR on their corresponding gliadin sequences is shown in Figure 1a. In total, we synthesized five different plasmids (Figure 1b). The synthesis of plasmids pghpg8.1 and pDhp_ω/α was described in (Gil-Humanes et al., 2008, 2011), respectively. The vectors pDhp_α/βZR, pDhp_ω4ZR and pDhp_ω8ZR were synthesized for this work using the GATEWAY (Invitrogen) recombination technology. Three different pairs of primers were used for the amplification of the IR fragments of these vectors: primers alpha/betaF (AATTGCAGCCA

CAAAATCCATCTCAG) and alpha/betaR (CATCCMTGCATGGAA TCAGTTGTTG) for the alpha/betaZR fragment; primers omega8F (CCTATCTTTGTCCTCCTTGCC) and omega8R (CATCGTTACATT-GAACGCTCA) for the omega8 and LMW fragments; and primers omega4F (CAACAATCCCCTGAACAACA) and omega4R (GCTG GGGTGGGTATGGTATT) for the omega4 fragment. The expression of the IRs was driven by an endosperm-specific D-hordein promoter (Piston et al., 2008), with the nopaline synthase (nos) as terminator sequence.

Plant material and genetic transformation

All lines described in this work are transgenic lines derived from bread wheat cv. BW208. Lines C655, D623, 28A, 28B, D783, E33, E42, E82 and E83 were reported previously (Piston et al., 2013). The rest of transgenic lines were produced in the present work using immature scutella as explants for genetic transformation as described in León et al., 2009;. Plasmids carrying the RNAi fragments were used in combination with plasmid pAHC25 containing the selectable bar gene (Christensen and Quail, 1996). For bombardment, plasmids were precipitated onto 0.6-µm gold particles at 0.5 pmol/mg gold for pAHC25 and 0.75 pmol/mg gold for the plasmids containing the RNAi fragments. Putative transgenic plants were then transferred to soil and grown to maturity in the greenhouse. Homozygous progeny of plants containing the RNAi plasmids were first identified by PCR using the forward primer prHorD*3 (GGGGTACCCATTAATTGAACTCATTCGGGAA GC) and one of the reverse primers specific for each RNAi construct: SUbiR (GCGTACCTTGAAGCGGAGGTGGTCGACTCTAGATTGCA ACACCAATGATCTGATCG) for the pghpg8.1 plasmid, alpha/betaR for the pDhp_α/βZR plasmid, omega8R for the pDhp_ω8ZR plasmid, omega4R for the pDhp_ω4ZR plasmid and omega_III_R_overlapping (CAGTTGTTGTTGAAATGGTTGTTGCGATGG) for the pDhp_ω/α plasmid. Then, PCR-positive transgenic plants were analysed by A-PAGE of endosperm proteins in lines produced by single half-seed descent. Homozygous lines were self-pollinated for three generations and assayed as described below.

Experimental design and statistical analysis

The homozygous transgenic lines were assayed using randomized complete block designs with three replicates of five plants each. All five plants per line and block were bulked, and each block was treated as biological replications and analysed separately. For each line and block, at least two technical replications were used. Data were analysed with the statistical software R version 3.0.1 (Ihaka and Gentleman, 1996). Figures were drawn using the user interface GrapheR (Hervé, 2013). The differences in the data were assessed using analysis of the variance (ANOVA) (function aov, package agricolae), followed by the two-tailed Dunnett's post hoc test for median multiple comparisons. P values lower than 0.05 were considered significant.

Polyacrylamide gel electrophoresis analysis

Ten mature wheat grains per line were crushed into a fine powder and used to extract the endosperm storage proteins. Gliadins and glutenins were sequentially extracted and separated in A-PAGE and SDS-PAGE gels, respectively, as described in (Gil-Humanes et al., 2012).

Reversed-phase high-performance liquid chromatography (RP-HPLC)

Gliadins and glutenins were extracted and quantified by RP-HPLC following the protocol reported by Piston et al. (2011). Three

biological replications (one per block) and two technical repetitions per each block were carried out for each transgenic line and control.

Mass spectrometry analysis

Total protein was extracted from 1 g of flour with 10 ml of SDS buffer (0.5% SDS, 0.1 M sodium phosphate pH 6.9) shaking (120 r.p.m.) at 60 °C during 80 min. Sonication (Ultrasonic cleaner USLU-5.7; Fungilab SA, Barcelona, Spain) was carried out at medium intensity during 1 min at 24 °C, and the supernatant was collected centrifuging at 16 000 **g** during 15 min. The extraction was performed twice using the pellet to finally obtain 30 mL of extract. Proteins were precipitated adding four volumes of cold acetone and kept at −20 °C during 30 min. After centrifuging at 16 000 **g** during 15 min, proteins were dissolved in 5 mL of 0.01 M acetic acid.

Pepsin digestion was carried out adjusting the pH to 1.8 with hydrochloric acid, adding pepsin (Sigma, St. Louis, MO) at a 1 : 100 ratio (w/w) and stirring during 4 h at 37 °C. The pepsin digestion was finished by bringing the pH to 7.8 with sodium hydroxide solution. A second digestion with trypsin [Sigma, 1 : 100 ratio (w/w)] was performed at the same incubation conditions. Trypsin was deactivated by heating at 85 °C for 45 min. The pH of the peptic-tryptic (PT) digest was adjusted to 4.5 with hydrochloric acid, dialysed (1 kDa cut-off) against 0.01 M ammonium bicarbonate and lyophilized.

The PT-digest was cleaned using a SEP-PAK C18 cartridge (Waters, Milford, MA), and 1.5 µg of total peptide was injected into Eksigent NanoLC-1D Plus (AB SCIEX, Madrid, Spain, CA) coupled to TripleTOF® 5600 System (AB SCIEX) (LC-MS/MS). The HPLC precolumn was Acclaim® PepMap 100, 100 µm × 2 cm (Thermo Fisher Scientific, Waltham, MA) and HPLC column NanoACQUITY UPLC® 1.7 µm BEH130 C18, 75 µm × 150 mm (Waters). All liquid chromatography–tandem mass spectrometry (LC-MS/MS) data sets were searched using MASCOT version 2.4 (http://www.matrixscience.com/) against NCBI protein database of the species Triticum without any enzyme restriction, peptide error tolerance of 25 ppm and MS/MS fragment error tolerance of 0.05 Da. Only peptides with scores higher than 20 were extracted for further analyses.

CD epitope analysis

The peptides identified in LC-MS/MS analysis were BlastP-searched against CD epitopes described by Sollid and collaborators (Sollid et al., 2012) and the α-gliadin peptide 31–43 [p31-43; PGQQQPFPPQQPY (Maiuri et al., 1996) and LGQQQPFPPQQPY (Maiuri et al., 2003)], able to induce innate immune response, setting the parameters: -task blastp-short, -ungapped, -seg no, -max_target_seqs 5 and allowing one mismatch. CD epitopes were searched in peptides longer than eight amino acids as described above.

Seed quality determinations

Kernel weight (g) was determined using 1000 seeds from each sample. Test weight (g/L) was calculated by weighing 100 mL of cleaned grains from each sample. Two measurements were carried out for each sample.

The protein content of whole flour was calculated from the Kjeldahl nitrogen content (%N × 5.7) according to the standard ICC method no. 105/2 (ICC, 1994), and the starch content was determined according to the standard ICC method no. 123/1

(ICC, 1994). Both parameters were expressed on a 14% moisture basis.

The nongluten proteins, expressed in percentage of dried weight (% DW), were calculated as follow: [Total protein in % − (Prolamin content in μg/mg*10)/(100 − moisture in %)]. The sodium dodecyl sulphate sedimentation (SDSS) volume was determined as described by (Williams et al., 1988). Three technical replicates were carried out for each biological sample.

Gluten content determination by competitive ELISA

Gluten proteins were extracted according to manufacturer's instructions using Universal Gluten Extraction Solution UGES (Biomedal SL, Seville, Spain). Maxisorp microtiter plates (Nunc, Roskilde, Denmark) were coated with Prolamin Working Group (PWG) gliadin solution and incubated overnight at 4 °C. The plates were washed with PBS-Tween 20 buffer and blocked with blocking solution (phosphate-buffered saline (PBS)-5% nonfat dry milk) for 1 h at RT. Different dilutions of each sample as well as standard solution of PWG gliadin were made in PBS-bovine serum albumin 3%, to each of which was added horseradish peroxidase–conjugated G12 moAb solution. The samples were pre-incubated for 2 h at RT with gentle stirring, and then added to the wells. After 30 min of incubation at RT, the plates were washed, and 3,3′,5,5′-tetramethylbenzidine (TMB) substrate solution (Sigma) was added. After 30 min of incubation at RT in the dark, the reaction was stopped with 1 m sulphuric acid, and the absorbance at 450 nm was measured (microplate reader UVM340; Asys Hitech GmbH, Eugendorf, Austria). Results were expressed in parts per million (ppm) in dry matter.

Acknowledgements

The Spanish Ministry of Economy and Competitiveness (Project AGL2013-48946-C3-1-R), the European Regional Development Fund (FEDER) and Junta de Andalucía (Project P11-AGR-7920) supported this work. The technical assistance of Ana García is also acknowledged.

References

Abadie, V., Sollid, L.M. and Barreiro, L.B. (2011) Integration of genetic and immunological insights into a model of celiac disease pathogenesis. *Annu. Rev. Immunol.* **29**, 493–525.

Altenbach, S.B. and Allen, P.V. (2011) Transformation of the US bread wheat 'Butte 86' and silencing of omega-5 gliadin genes. *GM Crops*, **2**, 66–73.

Anderson, O.D., Huo, N. and Gu, Y.Q. (2013) The gene space in wheat: the complete γ-gliadin gene family from the wheat cultivar Chinese Spring. *Funct. Integr. Genomics*, **13**, 261–273.

Arentz Hansen, H., Mcadam, S.N., Molberg, Ø., Fleckenstein, B., Lundin, K.E.A., Jørgensen, T.J.D., Jung, G., Roepstorff, P. and Sollid, L.M. (2002) Celiac lesion T cells recognize epitopes that cluster in regions of gliadins rich in proline residues. *Gastroenterology*, **123**, 803–809.

Becker, D., Wieser, H., Koehler, P., Folck, A., Mühling, K.H. and Zörb, C. (2012) Protein composition and techno-functional properties of transgenic wheat with reduced α-gliadin content obtained by RNA interference. *J. Appl. Bot. Food Qual.* **85**, 23.

Carter, B.P., Morris, C.F. and Anderson, J.A. (1999) Optimizing the SDS sedimentation test for end-use quality selection in a soft white and club wheat breeding program. *Cereal Chem.* **76**, 907–911.

Christensen, A.H. and Quail, P.H. (1996) Ubiquitin promoter-based vectors for high-level expression of selectable and/or screenable marker genes in monocotyledonous plants. *Transgenic Res.* **5**, 213–218.

Comino, I., Real, A., de Lorenzo, L., Cornell, H., López-Casado, A., Barro, F., Lorite, P., Torres, M.I., Cebolla, Á. and Sousa, C. (2011) Diversity in oat potential immunogenicity: basis for the selection of oat varieties with no toxicity in coeliac disease. *Gut*, **60**, 915–922.

Di Sabatino, A. and Corazza, G.R. (2009) Coeliac disease. *Lancet*, **373**, 1480–1493.

Gil-Humanes, J., Piston, F., Hernando, A., Alvarez, J.B., Shewry, P.R. and Barro, F. (2008) Silencing of γ-gliadins by RNA interference (RNAi) in bread wheat. *J. Cereal Sci.* **48**, 565–568.

Gil-Humanes, J., Pistón, F., Tollefsen, S., Sollid, L.M. and Barro, F. (2010) Effective shutdown in the expression of celiac disease-related wheat gliadin T-cell epitopes by RNA interference. *Proc. Natl Acad. Sci. USA*, **107**, 17023–17028.

Gil-Humanes, J., Pistón, F., Shewry, P.R., Tosi, P. and Barro, F. (2011) Suppression of gliadins results in altered protein body morphology in wheat. *J. Exp. Bot.* **62**, 4203–4213.

Gil-Humanes, J., Piston, F., Giménez, M.J., Martin, A. and Barro, F. (2012) The introgression of RNAi silencing of gamma-gliadins into commercial lines of bread wheat changes the mixing and technological properties of the dough. *PLoS ONE*, **7**, e45937.

Gil-Humanes, J., Pistón, F., Barro, F. and Rosell, C.M. (2014) The shutdown of celiac disease-related gliadin epitopes in bread Wheat by RNAi provides flours with increased stability and better tolerance to over-mixing. *PLoS ONE*, **9**, e91931.

ICC. (1994) Determination of crude protein in cereals and cereal products for food and for feed. International Association for Cereal Science and Technology. Method No. 105/2.

ICC. (1994) Determination of starch content by hydrochloric acid dissolution. International Association for Cereal Science and Technology. Method No. 123/1.

Hammond, S.M., Bernstein, E., Beach, D. and Hannon, G.J. (2000) An RNA-directed nuclease mediates post-transcriptional gene silencing in Drosophila cells. *Nature*, **404**, 293–296.

Hervé, M. (2013) GrapheR: a multiplatform GUI for drawing customizable graphs in R. *R J.* **3**, 45–53.

Ihaka, R. and Gentleman, R. (1996) R: a language for data analysis and graphics. *J. Comp. Graph. Stat.* **5**, 299–314.

Jackson, A.L. and Linsley, P.S. (2004) Noise amidst the silence: off-target effects of siRNAs? *Trends Genet.* **20**, 521–524.

Jackson, A.L., Burchard, J., Leake, D., Reynolds, A., Schelter, J., Guo, J., Johnson, J.M., Lim, L., Karpilow, J., Nichols, K., Marshall, W., Khvorova, A. and Linsley, P.S. (2006) Position-specific chemical modification of siRNAs reduces 'off-target' transcript silencing. *RNA*, **12**, 1197–1205.

León, E., Marín, S., Giménez, M.J., Piston, F., Rodríguez-Quijano, M., Shewry, P.R. and Barro, F. (2009) Mixing properties and dough functionality of transgenic lines of a commercial wheat cultivar expressing the 1Ax1, 1Dx5 and 1Dy10 HMW glutenin subunit genes. *J. Cereal Sci.* **49**, 148–156.

Maiuri, L., Picarelli, A., Boirivant, M., Coletta, S., Mazzilli, M.C., De Vincenzi, M., Londei, M. and Auricchio, S. (1996) Definition of the initial immunologic modifications upon in vitro gliadin challenge in the small intestine of celiac patients. *Gastroenterology*, **110**, 1368–1378.

Maiuri, L., Ciacci, C., Ricciardelli, I., Vacca, L., Raia, V., Auricchio, S., Picard, J., Osman, M., Quaratino, S. and Londei, M. (2003) Association between innate response to gliadin and activation of pathogenic T cells in coeliac disease. *Lancet*, **362**, 30–37.

Morón, B., Bethune, M.T., Comino, I., Manyani, H., Ferragud, M., López, M.C., Cebolla, Á., Khosla, C. and Sousa, C. (2008a) Toward the assessment of food toxicity for celiac patients: characterization of monoclonal antibodies to a main immunogenic gluten peptide. *PLoS ONE*, **3**, e2294.

Morón, B., Cebolla, Á., Manyani, H., Álvarez-Maqueda, M., Megías, M., Thomas, M.C., López, M.C. and Sousa, C. (2008b) Sensitive detection of cereal fractions that are toxic to celiac disease patients by using monoclonal antibodies to a main immunogenic wheat peptide. *Am. J. Clin. Nutr.* **87**, 405–414.

Mustalahti, K., Catassi, C., Reunanen, A., Fabiani, E., Heier, M., McMillan, S., Murray, L., Metzger, M.H., Gasparin, M., Bravi, E. and Mäki, M. (2010) The prevalence of celiac disease in Europe: results of a centralized, international mass screening project. *Ann. Med.* **42**, 587–595.

Piston, F., León, E., Lazzeri, P.A. and Barro, F. (2008) Isolation of two storage protein promoters from *Hordeum chilense* and characterization of their expression patterns in transgenic wheat. *Euphytica*, **162**, 371–379.

Piston, F., Gil-Humanes, J., Rodríguez-Quijano, M. and Barro, F. (2011) Down-regulating γ-gliadins in bread wheat leads to non-specific increases in other gluten proteins and has no major effect on dough gluten strength. *PLoS ONE*, **6**, e24754.

Piston, F., Gil-Humanes, J. and Barro, F. (2013) Integration of promoters, inverted repeat sequences and proteomic data into a model for high silencing efficiency of coeliac disease related gliadins in bread wheat. *BMC Plant Biol.* **13**, 136.

Qi, P.F., Wei, Y.M., Yue, Y.W., Yan, Z.H. and Zheng, Y.L. (2006) Biochemical and molecular characterization of gliadins. *Mol. Biol.* **40**, 796–807.

Rosell, C.M., Barro, F., Sousa, C. and Mena, M.C. (2014) Cereals for developing gluten-free products and analytical tools for gluten detection. *J. Cereal Sci.* **59**, 354–364.

Salentijn, E.M., Mitea, D.C., Goryunova, S.V., van der Meer, I.M., Padioleau, I., Gilissen, L.J.W.J., Koning, F. and Smulders, M.J.M. (2012) Celiac disease T-cell epitopes from gamma-gliadins: immunoreactivity depends on the genome of origin, transcript frequency, and flanking protein variation. *BMC Genom.* **13**, 277.

Sapone, A., Lammers, K.M., Casolaro, V., Cammarota, M., Giuliano, M.T., De Rosa, M., Stefanile, R., Mazzarella, G., Tolone, C. and Russo, M.I. (2011) Divergence of gut permeability and mucosal immune gene expression in two gluten-associated conditions: celiac disease and gluten sensitivity. *BMC Med.* **9**, 23.

Senthil-Kumar, M., Hema, R., Anand, A., Kang, L., Udayakumar, M. and Mysore, K.S. (2007) A systematic study to determine the extent of gene silencing in *Nicotiana benthamiana* and other Solanaceae species when heterologous gene sequences are used for virus-induced gene silencing. *New Phytol.* **176**, 782–791.

Shan, L., Molberg, Ø., Parrot, I., Hausch, F., Filiz, F., Gray, G.M., Sollid, L.M. and Khosla, C. (2002) Structural basis for gluten intolerance in celiac sprue. *Science*, **297**, 2275–2279.

Sollid, L.M., Qiao, S.-W., Anderson, R.P., Gianfrani, C. and Koning, F. (2012) Nomenclature and listing of celiac disease relevant gluten T-cell epitopes restricted by HLA-DQ molecules. *Immunogenetics*, **64**, 455–460.

Tye-Din, J.A., Stewart, J.A., Dromey, J.A., Beissbarth, T., van Heel, D.A., Tatham, A., Henderson, K., Mannering, S.I., Gianfrani, C., Jewell, D.P., Hill,

A.V., McCluskey, J., Rossjohn, J. and Anderson, R.P. (2010) Comprehensive, quantitative mapping of T cell epitopes in gluten in celiac disease. *Sci. Transl. Med.* **2**, 41ra51.

Vader, W., Kooy, Y., van Veelen, P., de Ru, A., Harris, D., Benckhuijsen, W., Peña, S., Mearin, L., Drijfhout, J.W. and Koning, F. (2002) The gluten response in children with celiac disease is directed toward multiple gliadin and glutenin peptides. *Gastroenterology*, **122**, 1729–1737.

Wieser, H. and Koehler, P. (2009) Is the calculation of the gluten content by multiplying the prolamin content by a factor of 2 valid? *Eur. Food Res. Technol.* **229**, 9–13.

Williams, P., Jaby el-Haramein, F., Nakkoul, H. and Rihawi, S. (1988) *Crop Quality Evaluation Methods and guidelines, Cereals - Food legumes - Forages.* Technical Manuel No 14, ICARDA, Aleppo, Syria. Second edition May, 1988. p. 145.

Xu, P., Zhang, Y., Kang, L., Roossinck, M.J. and Mysore, K.S. (2006) Computational estimation and experimental verification of off-target silencing during posttranscriptional gene silencing in plants. *Plant Physiol.* **142**, 429–440.

4

Gene expression in the developing aleurone and starchy endosperm of wheat

Susan A. Gillies[1,]*, Agnelo Futardo[1,2] and Robert J. Henry[2]

[1]Southern Cross Plant Science, Southern Cross University, Lismore, NSW, Australia
[2]Queensland Alliance for Agriculture and Food Innovation, The University of Queensland, St Lucia, Qld, Australia

*Correspondence
email susan.gillies@scu.edu.au

Keywords: *Triticum aestivum*, aleurone, endosperm, transcriptome, next generation sequencing, grain.

Summary

Wheat is a critical food source globally. Food security is an increasing concern; current production levels are not expected to keep pace with global demand. New technologies have provided a vast array of wheat genetic data; however, best use of this data requires placing it within a framework in which the various genes, pathways and interactions can be examined. Here we present the first systematic comparison of the global transcriptomes of the aleurone and starchy endosperm of the developing wheat seed (*Triticum aestivum*), at time points critical to the development of the aleurone layer; 6-, 9- and 14-day post-anthesis. Illumina sequencing gave 25—55 million sequence reads per tissue, of the trimmed reads, 70%—81% mapped to reference expressed sequence transcripts. Transcript abundance was analysed by performing RNA-Seq normalization to generate reads per kilobase of exon model per million mapped reads values, and these were used in comparative analyses between the tissues at each time point using Kal's Z-test. This identified 9414—13 202 highly differentially expressed transcripts that were categorized on the basis of tissue and time point expression and functionally analysed revealing two very distinct tissues. The results demonstrate the fundamental biological reprogramming of the two major biologically and economically significant tissues of the wheat seed over this time course. Understanding these changes in gene expression profiles is essential to mining the potential these tissues hold for human nutrition and contributing to the systems biology of this important crop plant.

Introduction

The most recent Food and Agriculture Organization and International Grain Council estimates give the global estimates for harvested wheat grain (*Triticum aestivum*) in 2010/11 at over 650 million tonnes (http://faostat.fao.org/; http://www.igc.int/). Although principally grown for human nutrition, wheat is also a source of food for livestock and is easily processed to provide the raw materials for a myriad of other foodstuffs and industrial commodities.

The economic and nutritional importance of wheat grain and the need for a better understanding of grain development to underpin any future potential manipulation of the seed have led to an abundance of studies addressing aspects of the grain transcriptome. These, however, examine the total grain or one tissue in isolation, or have followed the progress of specific genes and use a variety of sources for functional annotation (Laudencia- Chingcuanco et al., 2007; Stamova et al., 2009; Wan et al., 2008, 2009). The work presented here follows the parallel development of the transcriptomes of two of the most important tissues of the wheat seed. Although our primary focus is in understanding the genetic mechanisms behind development of the aleurone layer, its derivation from endosperm makes interpretation of the data difficult if removed from the context of the endosperm. Consequently, both tissues have been included at each time point in our analysis to provide a direct comparison. Additionally, the controlled gene ontology vocabulary provided by the Gene Ontology project (http://www.geneontology.org/) was selected to provide a standardized vocabulary for functional analysis.

The starchy endosperm of the mature wheat seed is presently a primary source of nutrition, particularly in the western diet. It consists principally of carbohydrates in the form of simple starches, typically 55%—75% of total dry grain weight, with a storage protein content of 10%—20%. The aleurone layer develops at the perimeter of this tissue, where its primary biological role is in the digestion of the starchy endosperm to release nutrients in the form of free sugars and amino acids to the germinating embryo. A secondary role is in protection of the endosperm by storage of proteins which protect against stress and pathogens (Jerkovic et al., 2010). In the wheat seed and in maize this layer is a single cell thick, whereas in other major cereals there are often multiple layers; in barley, there are typically three, and in rice one to several.

The aleurone layer is the most concentrated source of vitamins and minerals in the wheat seed and is additionally rich in proteins and lipids (Geisler-Lee and Gallie, 2005; Liu, 2011; Regvar et al., 2011). Recent clinical trials have demonstrated its potential for beneficial health outcomes. Aleurone flour is a rich source of natural folate, a B-group vitamin critical in nucleotide synthesis, DNA methylation and gene expression, essential for the prevention of neurological deficits in the newborn. Increased consumption of aleurone has been shown to increase plasma folate (Fenech et al., 2005) or increase plasma betaine and lower plasma homocysteine and low density lipoproteins (Price et al., 2010), the latter three of which are associated with a lower risk of heart disease. Additionally, wheat aleurone is abundant in dietary fibre that is fermented in the gut, releasing short-chain fatty acids, which have potential roles in cancer prevention. Recent *in vitro* and animal studies indicate a possible role for aleurone intake in intestinal cancer prevention and treatment (McIntosh *et al.*, 2001; Stein *et al.*, 2010).

Aleurone development and maintenance is a dynamic process. This layer develops at the latter stages of endosperm specification. In a mechanism common amongst angiosperms, the endosperm is produced from a double fertilization event when a single sperm nuclei fuses with the two polar nuclei in the central cell. This is followed by a series of free nuclear divisions and directed migration of this nuclear material to the periphery of the central cell. Cell walls are formed from microtubular systems arising from the nuclear membranes. An initial anticlinal plane of cell division is followed by a periclinal plane of division, accompanied by cytokinesis which completes the first peripheral layer (Brown *et al.*, 1994). In wheat seed, this layer is complete by 6—7 days (S. Gillies, unpublished data, http://www.wheatbp.net).

Starchy endosperm cellularization is finalized by repetition of this process extending into the centre of the endosperm cavity, with progressive loss of the strict control of the planes of cell division (Becraft, 2001; Brown *et al.*, 1994; Olsen, 2001). Inner cells grow in size as they accumulate starch and protein bodies. The initial periclinal division signals the initiation of aleurone specification and gives rise to daughter cells which have distinct fates. As cellularization nears completion, the inner cells become starchy endosperm, the perimeter cells become aleurone (Becraft, 2001; Becraft and Yi, 2011; Olsen, 2001, 2004) which becomes resistant to desiccation, developing the capacity to remain living in the mature seed until germination of the embryo (Young and Gallie, 2000; Young *et al.*, 2004). In wheat, maturation of the aleurone is complete by 12 days after anthesis (personal observation, http://www.wheatbp.net).

The current understanding of the molecular mechanisms involved in aleurone development in cereals arises primarily from studies in maize. Aleurone fate is plastic and appears determined by positional cues, as it develops at the periphery of starchy endosperm masses of both a maize endosperm mutant and *in vitro* endosperm cultures (Geisler-Lee and Gallie, 2005; Gruis *et al.*, 2006; Olsen, 2004). Starchy endosperm, however, appears to be the default cell type as aleurone fails to develop at the periphery in the absence of appropriate cues (Becraft and Asuncion-Crabb, 2000; Becraft *et al.*, 1996; Gruis *et al.*, 2006; Lid *et al.*, 2002; Shen *et al.*, 2003).

Despite the importance of the wheat seed, the mechanism of development is far from established. Here we report the comparison of the transcriptomes of the developing aleurone and starchy endosperm of the wheat grain at time points critical in aleurone development using next generation sequencing. This work demonstrates two distinct tissues which undergo dynamic shifts over the time course examined and provides a genetic resource which may be used to promote further crop improvement.

Results

Experimental design

Six-, 9- and 14-day post-anthesis (DPA) tissues were selected as representative of pre/initial, mid and completed aleurone layer development. Six DPA was determined in preliminary experiments to be the first time point at which the initial peripheral layer of the endosperm could be isolated (Figures 1 and 2,

http://www.wheatbp.net), which is enriched for the developing aleurone layer (Drea *et al.*, 2005). At 14 DPA, the aleurone layer in wheat is expected to be and was consistently found experimentally to be in its mature form 9 DPA represents a midpoint in development. Visibly the layer at 9 DPA resembled more the 6 DPA tissue, however, with a clearly more ordered and robust cell structure (Figure 2). During the time course studied starchy endosperm almost completes cellularization, it is complete by 16 DPA (Drea *et al.*, 2005; Wegel *et al.*, 2005). Seed was collected from appropriately aged wheat plants; the tissues were surgically separated, isolated and snap frozen; and cDNA was prepared and sequenced using an Illumina system. The level of cross contamination of tissues was low as demonstrated by the ability to detect genes with highly differential expression between the two tissues. A highly expressed gene originally detected in SAGE analysis (McIntosh *et al.*, 2007) has been the subject on ongoing analysis (Furtado unpublished) and was found to be expressed only in starchy endosperm at 6 and 9 DPA and at more than ten times the level in starchy endosperm at 14 DPA. Sequences were optimized by stringent removal of poor-quality reads and contaminating sequences.

RNA-Seq analysis using CLC Genomics Workbench (Version 4.0) (Katrinebjerg, Aarhus N, Denmark) was performed on the sequences after trimming. The Dana Farber Cancer Institute (DFCI) (Boston, MA) Wheat Gene Index was used as the reference database (Release 12.0, http://compbio.dfci.harvard.edu). This DFCI EST database is comprised of 93 508 tentative consensus sequences, 128 166 singleton ESTs and 251 mature transcripts for a total of 221 925 unique EST sequences. Of the trimmed sequences, 69.9%—81.4% mapped back to the DFCI ESTs. Of the mapped sequences, 42.1%—52.9% mapped specifically to one reference sequence, and 41.2%—57.9% nonspecifically (Table 1). Nonunique reads were discarded if matching more than ten distinct sequences, and if <10, reads were proportionately assigned. Sequenced transcripts were analysed after normalization of the counts to give reads per kilobase of exon model per million mapped reads (RPKM) (Mortazavi *et al.*, 2008).

Hierarchical cluster of sample analysis gives an overview of the total data. Cluster analysis of all RPKM values displays the close relationship between 6 and 9 DPA aleurone tissue (Figure 3). In contrast, in endosperm the 9 DPA and 14 DPA tissues are far more closely aligned. Of particular interest, 14 DPA aleurone tissue, after it has achieved its mature form, is

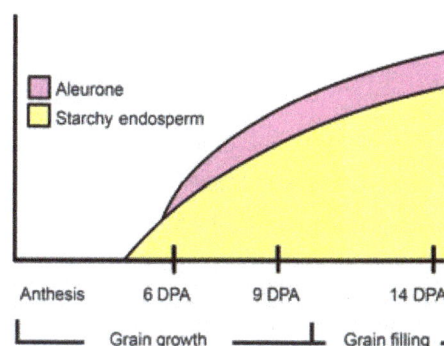

Figure 1 Schematic of development of the aleurone and endosperm layers. (Adapted from: Wheat: The big picture http://www.wheatbp.net).

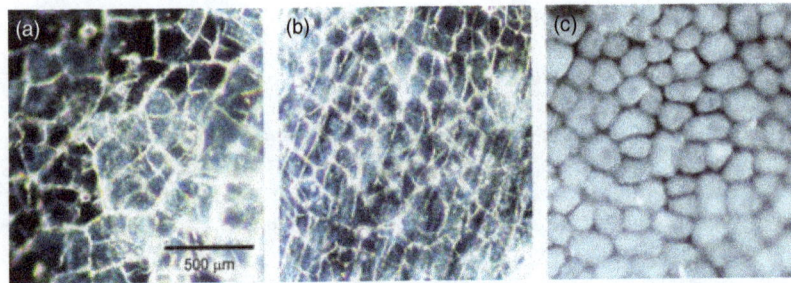

Figure 2 Development of the aleurone layer. Light microscopy at 10 × magnification demonstrates the increasing order and strength within this layer over time. (a) 6 day post-anthesis (DPA). (b) 9 DPA. (c) 14 DPA.

Table 1 Summary statistics for Illumina reads

	6 DPA		9 DPA		14 DPA	
	Al	En	Al	En	Al	En
Total reads	53 580 004	48 126 050	25 897 678	25 523 822	56 859 592	55 417 756
Mapped sequences (%)	27 150 859 (78.4)	30 581 013 (78.3)	4 549 034 (69.9)	4 104 682 (71.9)	38 911 327 (80.3)	37 502 901 (81.4)
Uniquely (%)	14 161 063 (52.1)	16 154 259 (52.9)	2 051 587 (45.1)	1 726 574 (42.1)	18 965 616 (50.0)	16 034 744 (42.8)
Non specifically (%)	12 989 796 (47.8)	14 426 759 (47.2)	2 497 447 (54.9)	2 378 108 (57.9)	18 945 711 (50.0)	21 466 157 (57.2)

The total reads produced by Illumina sequencing at each time point were trimmed and mapped against the Dana Farber Cancer Institute Wheat EST reference data base (Mapped sequences). Of the trimmed sequences 69.9–81.4% mapped back to the DFCI ESTs. Of these 42.1–52.9% mapped specifically to one reference sequence (Uniquely), and 41.2–57.9% non-specifically. Al, aleurone; En, starchy endosperm.

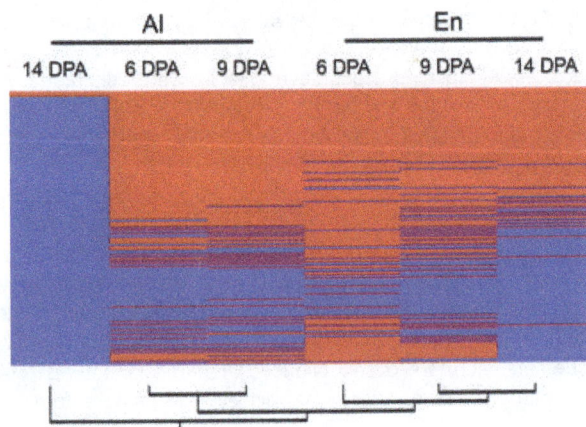

Figure 3 Hierarchical clustering of samples using all reads per kilobase of exon model values displays the relationship between each tissue and each time point. Clustering is calculated using the 1-Pearson correlation and single linkage. Al, aleurone; En, starchy endosperm.

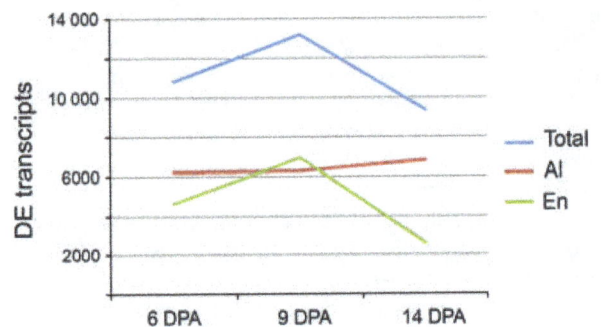

Figure 4 Differentially expressed transcripts over time. RNA-Seq reads per kilobase of exon model values were used in paired expression analysis between tissues at each time point using Kal's test. Multiple testing was corrected for by false discovery rate analysis of p-values and those with values less than or equal to 0.05 retained. Total, All differentially expressed transcripts at each time point. Al, up-regulated in aleurone; En, up-regulated in endosperm.

determined to be a very distinct tissue, as closely related to the endosperm tissues as aleurone tissues at earlier time points.

Differentially expressed (DE) transcripts

Aleurone (Al) was compared to starchy endosperm (En) at each time point. RPKM values were used in two-group comparisons using Kal's Z-test for statistical analysis, and false discovery rate (FDR) corrected p-values were generated to correct for multiple testing. ESTs that were expressed significantly differently (FDR-corrected P-value of 0.05 or less) were separated into those more highly expressed in the aleurone and those more highly

expressed in the endosperm. As shown in Figure 4, the number of ESTs significantly up-regulated in the aleurone was relatively constant across the time points, in contrast the numbers of DE ESTs in the endosperm was highly variable.

These significantly differentially expressed ESTs were further divided into those expressed at one time point and those expressed at overlapping time points (Figure 5). Throughout the major stages of development of the aleurone layer, the numbers of these transcripts in aleurone remain relatively consistent (Figure 5a). Only an increase in the number of genes differentially expressed at 14 DPA exclusively is notable when the tissue has completed development, an observation

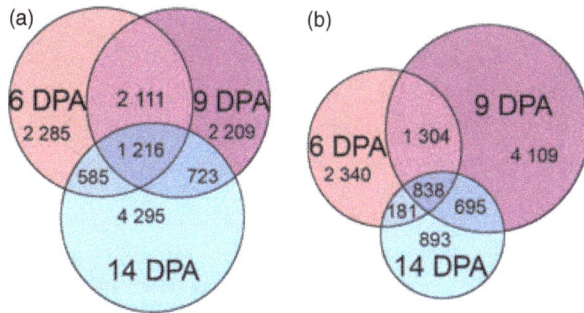

Figure 5 Differentially Expressed transcripts over time. Transcripts were categorized on the basis of those detected at one time point alone and those found at overlapping time points. (a) aleurone, (b) endosperm.

supported by Figure 3, where 14 DPA aleurone is displayed as a very distinct tissue from 6 to 9 DPA aluerone. As expected based on visual inspection of the tissues (Figure 2) and hierarchical clustering (Figure 3), the overlap between 6 and 9 DPA aleurone is far more extensive than the overlap between other time points. Over half (53%) of the DE transcripts identified in 6 DPA aleurone are also detected at 9 DPA. Endosperm, however, (Figure 5b) exhibits a greater fluctuation in numbers of highly differentially expressed genes as the seed moves from grain growth to grain filling. Interestingly, in both aleurone and endosperm the proportion of DE transcripts common to all time points is relatively low, comprising just 9.1% of the total in aleurone and 8.1% in starchy endosperm.

Functional annotation of differentially expressed transcripts identified all time points in aleurone or endosperm

The sequences represented in each sector of Figure 4 were subject to functional analysis. Annotation relied upon homology with distant species in many cases because of our relatively poor knowledge of the wheat genome. Repeated analysis of this data with more species-specific annotation will be worthwhile in the future. Analysis using the Mapman metabolic overview pathway (Usadel et al., 2005, 2009) (Figure 6a) shows the expected results of high concentrations of transcripts binned into lipid metabolism (exotics, arrow 1; fatty acid synthesis and degradation, arrow 2) (Geisler-Lee and Gallie, 2005; Regvar et al., 2011) and a high proportion of transcripts identified as associated with cell wall synthesis (arrow 3) in this tissue undergoing rapid cellularization. Less expected may be the significant concentration of transcripts identified as involved in the tricarboxylic acid cycle and mitochondrial electron transport (arrows 4, 5). These are not identified in DE expressed transcripts in endosperm and are indicative of the energy-intensive nature of aleurone development in comparison with endosperm.

Transcripts identified at all time points in endosperm (Figure 6b) are grouped primarily into the major carbohydrate bins of starch synthesis and degradation (arrows 1 and 2) (Morrison et al., 1975a,b). Examination of secondary metabolites also displays distinct differences between aleurone and endosperm.

Analysis using level 2 GO terms gives an alternate overview of this DE data. Graphing each term as a proportion of the total demonstrates those DE transcripts common to all time points in either tissue (central region of Figure 5a,b) are, as expected for two closely related tissues at identical time points, broadly

similar at level 2 (Figure 7a,c,e). However, although the GO terms identified are the same, the individual sequences within each category are unique to each tissue.

Biological process terms describe the operations or sets of molecular events with a defined beginning and end (http://www.geneontology.org/). Figure 6a demonstrates one of the few differences in terms is that aleurone has a greater proportion of transcripts identified as involved in localization (GO:0051179). These are the processes involved in transport to and maintenance of substances in a specific location. In endosperm, transcripts involved in response to stimulus (GO:0050896) are more frequently identified. These are described as transcripts involved in a change in state or activity of a cell as a result of a stimulus (http://www.geneontology.org/).

Analysis of biological process terms using the more detailed level 4 GO vocabulary, however, shows the emergence of clear differences between the tissues (Figure 7b). In aleurone, a higher proportion of DE transcripts are principally involved in processes which could be expected to be required for the development of specialized cells at a specific site. These processes include transport (transport (GO:0006810)) and definition of cell structure (regulation of anatomical structure size (GO:0090066); anatomical structure morphogenesis (GO:0009653); cell differentiation (GO:0030154); signal transmission (GO:0023060)). Additionally, as expected, transcripts identified as involved in lipid and protein metabolism (Geisler-Lee and Gallie, 2005; Regvar et al., 2011) (lipid metabolic process (GO:0046493); cellular amino acid and derivative metabolic process (GO:0006519); cellular nitrogen compound metabolic process (GO:0034641)), are more prevalent in aleurone.

Conversely, transcripts that promote the rapid accumulation of carbohydrate and other storage macromolecules are more frequently detected in endosperm as predicted from previous analysis (Emes et al., 2003; Morrison et al., 1975b), including those involved in carbohydrate metabolic processes (GO:0005975); generation of precursor metabolites and energy (GO:0006091); cellular biosynthetic processes (GO:0044249); cellular macromolecule metabolic process (GO:0044260); and macromolecule biosynthetic process (GO:0009059).

Examining molecular function terms, which describe the elemental activities of a gene product, again at level 2 GO analysis the differences between the tissues are not marked (Figure 7c). However, consistent with above (Figure 7b), there are a higher proportion of sequences identified as being involved in transport (transporter activity (GO:0005215)) in aleurone. Endosperm has a higher proportion of transcripts having enzyme regulator activity (GO:0030234). These transcripts could be expected to assist in control of the rate of cellularization.

Level 4 molecular function analysis shows pronounced differences in activities between the tissues (Figure 7d). Aleurone has a higher ratio of sequences identified as involved in receptor activity (GO:0004872). This is defined as combining with an extracellular or intracellular messenger to initiate a change in cell activity and is consistent with this tissue also identified as having a higher proportion of sequences involved in signal transmission (Figure 7b). There is also a small proportion of transcripts involved in hydrolase activity, acting on acid anhydrides (GO:0016817), a category not found in starchy endosperm. This corresponds to a cluster of enzymes that hydrolyse the diphosphate bonds in nucleosidic phosphates and sulfonyl containing anhydrides (Enzyme Nomenclature, 1992, EC 3.6., http://enzyme.expasy.org/) such as those involved in hydrolysis

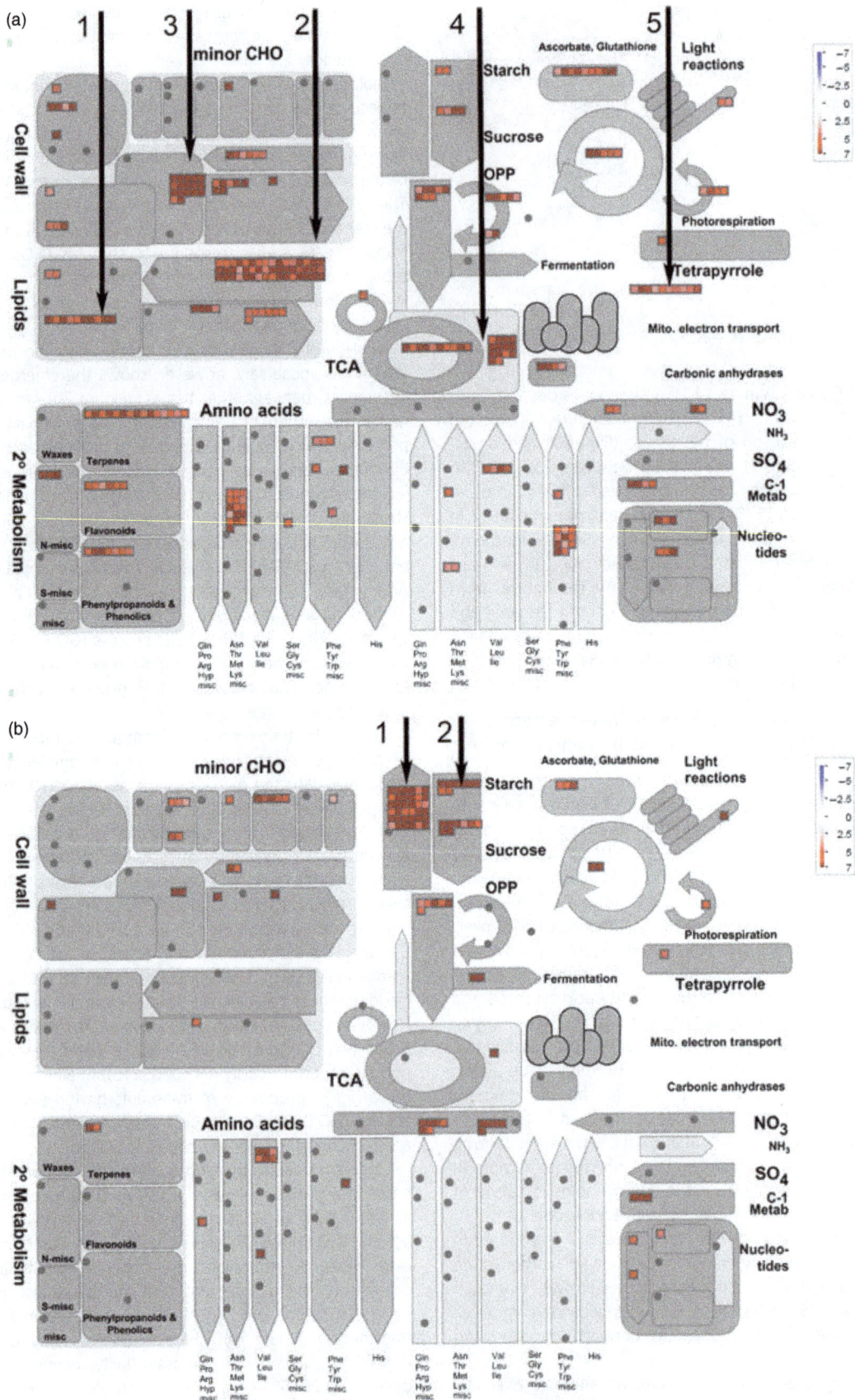

Figure 6 Mapman metabolic overview pathway of transcripts differentially expressed at all time points in either tissue. (a) aleurone, (b) endosperm.

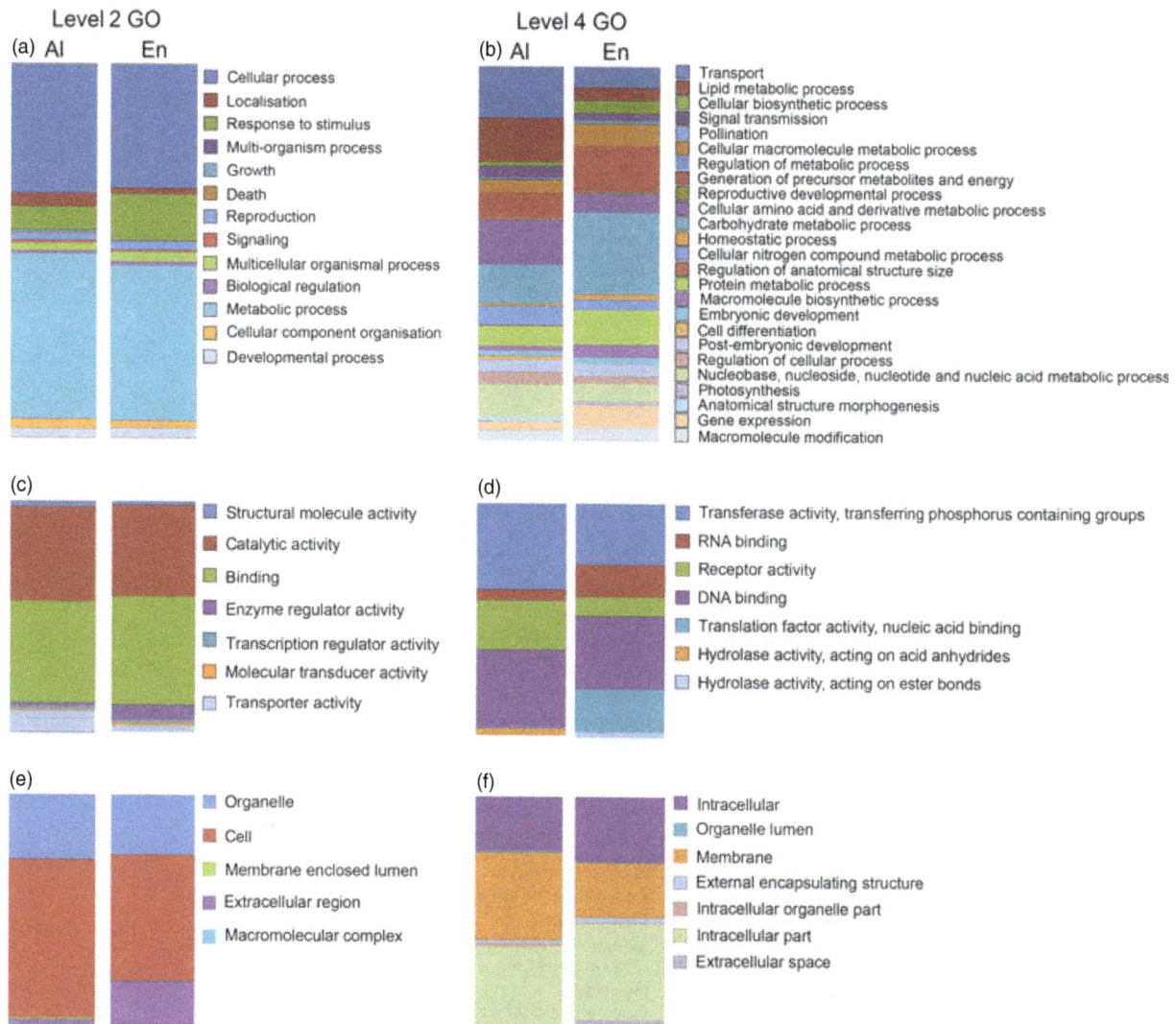

Figure 7 Gene Ontology (GO) classification of differentially expressed (DE) transcripts. The Level 4 and Level 2 GO vocabulary provided by the Gene Ontology project was used for functional annotation of DE transcripts detected at all time points in either tissue. (a, b) biological process terms. (c, d) molecular function terms. (e, f) cellular component terms. Al, aleurone; En, starchy endosperm.

of ATP and GTP to provide energy, which in agreement with Figure 6 indicates aleurone differentiation is a highly energy-intensive process.

In contrast, endosperm produces a small proportion of sequences identified as involved in hydrolase activity, acting on ester bonds (GO:0016788), a term not detected in DE transcripts up-regulated in aleurone. This group of hydrolases includes the deoxyribonucleases, ribonucleases, endonucleases, exonucleases and the phosphoric ester hydrolases (Enzyme Nomenclature, 1992, EC 3.1., http://enzyme.expasy.org/). This indicates rapid DNA/RNA turnover as phosphodiester bonds form the backbone of DNA/RNA molecules. Starchy endosperm additionally produces a far higher proportion of transcripts involved with RNA binding (GO:0003723) and translation factor activity, nucleic acid binding (GO:0008135).

Level 2 cellular component analysis (the parts of a cell or its extracellular environment) (Figure 7e) reflects the differing states of each tissue. Aleurone as a single layer tissue in wheat displays far fewer sequences identified as located in the extracellular region (GO:0005576) and has a small proportion of

sequences identified as being targeted to the membrane enclosed lumen (GO:0031974), a category not detected in endosperm. Level 4 analysis (Figure 7f) shows only endosperm identifies sequences as being present in the extracellular space (GO:0005615), consistent with a tissue in the process of completing cellularization (Becraft, 2001; Drea et al., 2005; Wegel et al., 2005). In aleurone, a higher proportion of sequences are predicted to be found in membrane enclosed spaces identified by such terms as membrane (GO:0016020), organelle lumen (GO:0043233) and intracellular organelle part (GO:0044446).

Shifts in Level 4 GO categories in differentially expressed transcripts of aleurone and endosperm during aleurone development

Further analysis demonstrates the changes in each tissue over the time period critical to aleurone development. GO analyses use Level 4 terms. This level was selected as it was found to provide sufficiently detailed terms to clearly reveal the changes over time in each of the tissues. Those sequences that were

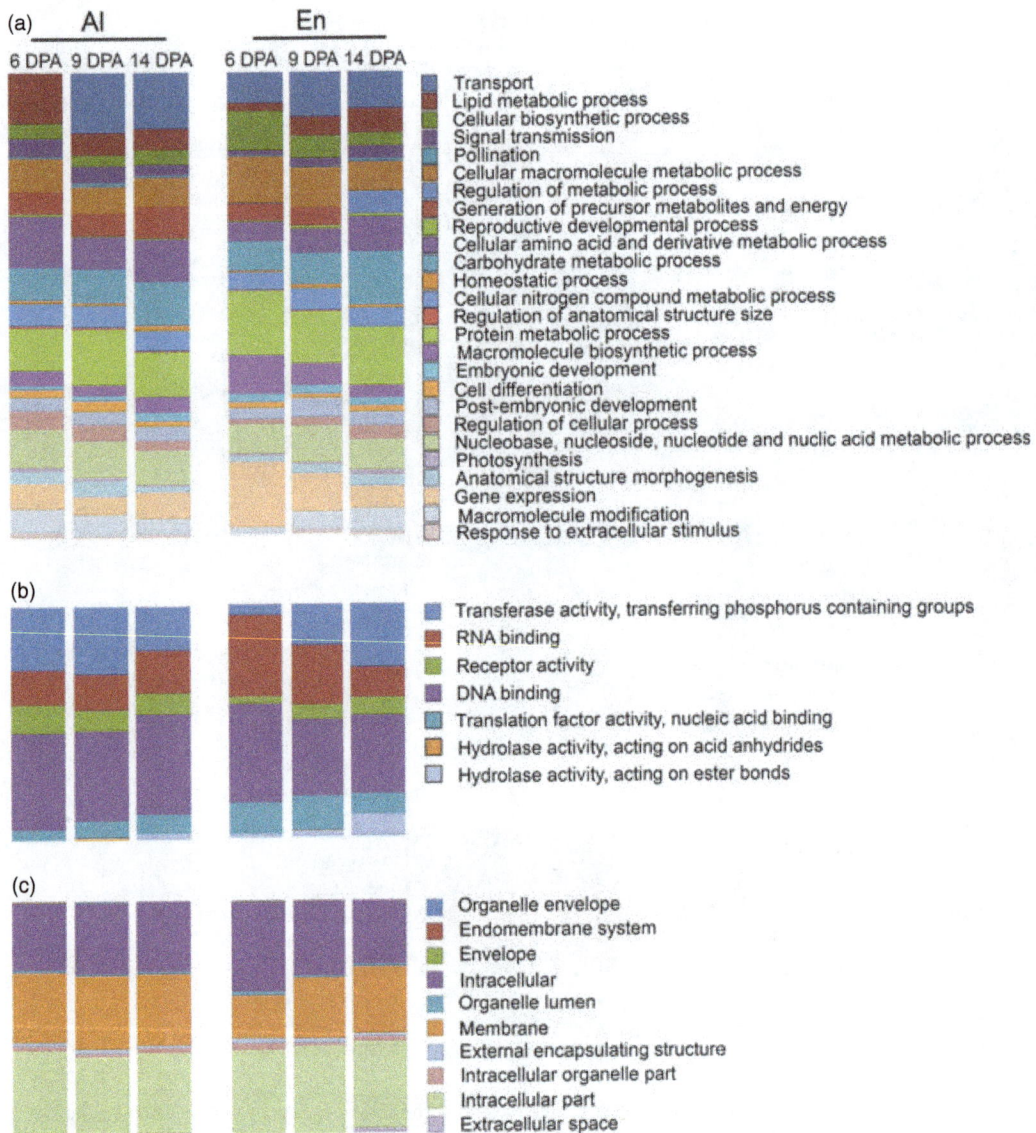

Figure 8 Functional Annotation of differentially expressed (DE) transcripts over time. Level 4 GO terms provided by the Gene Ontology project were used to annotate DE transcripts found exclusively at one time point in each tissue. (a) biological process terms; (b), molecular function terms; (c), cellular component terms. Al, aleurone; En, starchy endosperm.

found to be differentially expressed exclusively at one time point only (see Figure 5) were isolated and sent through the blast2GO program. Almost identical terms are retrieved for each tissue at each time point to those found for sequences common to all time points (Figure 6) and the total complement of sequences DE at any one time point (Figure S1). There are, however, clear changes in the proportion of transcripts in particular categories over time. Figure 8 shows the level 4 gene ontology terms retrieved at each time point, demonstrating the proportion of transcripts associated with that process/function/component over and above those produced at all time points. Figure S1 graphs the total data. This is the functional annotation of the total complement of all DE transcripts identified at any one time point. This is shown in comparison with DE transcripts identified at one time point exclusively (Only) and those transcripts differentially expressed at all time points (All). For almost every GO term, the total data produced proportions intermediate between 'Only' and 'All', demonstrating the 'Only'

data on which the majority of the analysis rests represents a meaningful subset of the total data.

The biological process data (Figure 8a) show a sudden dramatic rise in aleurone in the proportion of transcripts associated with transport from 9 DPA. This is maintained at 14 DPA (transport (GO:0006810)), one of the few terms to display a proportional increase over time. This indicates that although by 14 DPA the tissue is in its final form, it remains very much an active tissue. Other increasing categories include generation of precursor metabolites and energy (GO:0006091), with a slight increase in carbohydrate metabolic process (GO:0005975) as the tissue matures and moves towards accumulation of storage molecules (Becraft, 2001; Becraft and Yi, 2011). Consistent with this completion of differentiation is the down-regulation of transcripts identified as involved in cell specification including cell differentiation (GO:0030154), anatomical structure morphogenesis (GO:0006810), signal transmission (GO:0023060) and regulation of cellular process (GO:0050794). Macromolecule modification (GO:0043412) and lipid metabolic process (GO:0046493) are also reduced, indicating as expected, these constituents are nearing their final form.

Figure 8a demonstrates the endosperm is clearly in the process of reprogramming from rapid biosynthesis of a restricted range of storage macromolecules to a tissue requiring a broader range of molecules, with the requisite regulation required for the control of this diversified requirement. Cellular biosynthetic processes are down-regulated (GO:0044249) as are macromolecule biosynthetic process (GO:0009059) and cellular macromolecular metabolic process (GO:0044260), with an abrupt drop in transcripts involved in generation of precursor metabolites and energy (GO:0006139) after 9 DPA. There is a concurrent sharp rise in transcripts involved in regulation of metabolic process (GO:0019222) and macromolecule modification (GO:0043412), with an increase in regulation of cellular process (GO:0019222), signal transmission (GO:0023060), transport (GO:0006810) and anatomical structure morphogenesis (GO: 0009653). Additionally, there is an increase in lipid metabolism (lipid metabolic process GO:0046493) and cellular amino acid and derivative metabolic process (GO:0006519). However, the number of transcripts involved in carbohydrate metabolic processes (GO:0005975) continues to increase over time reflecting that endosperm remains principally a carbohydrate storage tissue. Interestingly, there is also a noticeable decline in the number of transcripts identified as involved in gene expression. This broad term describes the process by which a gene's sequence is converted into a mature gene product or products (proteins or RNA) (http://www.geneontology.org/). The decline in the proportion of these transcripts may be associated with a reduction in the rate of molecule accumulation.

Analysis of the molecular function data (Figure 8b) shows aleurone, as seen in previous analyses, changes little from 6 to 9 DPA. By 14 DPA, there is a drop in the number of sequences identified as involved in transferase activity, transferring phosphorus-containing groups (GO:0016772). This term covers a broad range of phosphotransferases and kinases. There is additionally a decline in receptor activity (GO:0004872). However, there is an increase in translation factor activity, nucleic acid binding (GO:0008135), again indicating at 14 DPA this tissue is still involved with the manufacture of new compounds.

Changes in endosperm are more marked. In contrast to aleurone, there is a large increase in the proportion of transcripts involved in the transfer of phosphorus-containing groups (GO:0016772), and additionally in transcripts associated with the hydrolysis of ester bonds (GO:0016788), again indicating rapid nucleotide turnover is necessary during starchy endosperm cellularization. Consistent with this is a large drop in transcripts involved in RNA binding (GO:0003723). There is also an increase in the proportion of terms identified as involved in receptor activity (GO:0004872).

Cellular component analysis (Figure 8c) shows the positional association of DE aleurone transcripts changing little over the time course examined, as expected as this is a single layer tissue. Again as endosperm becomes increasingly cellularized, this tissue exhibits more dynamic changes with transcripts moving from a simple intracellular location (intracellular GO:0005622) to being more commonly found at the membrane (GO:0016020) in proportions similar to that of aleurone. This finding is consistent with the finding of an increase in signal transmission over the time course examined (Figure 8a).

The Mapman cellular response overview visualization tool (Figure 9) displays the quite distinct stress responses of the two tissues. At 6 DPA, the majority of aleurone transcripts associated with abiotic stress are binned as touch/wounding transcripts (Figure 9a). At 9 DPA (Figure 9c), a greater proportion is identified as associated with drought/salt tolerance, while at 14 DPA (Figure 9e), heat and cold transcripts are commonly identified. In contrast, in endosperm the proportion of DE transcripts associated with heat reduces over the time course examined (Figure 9b,d,f), and touch/wounding transcripts increase over time. Both tissues display a dynamic stress response which alters rapidly over this short time course.

Of the antioxidant defence pathways, thioredoxin and ascorbate/glutathione become increasingly important in aleurone. Interestingly, the peroxiredoxins become suddenly apparent at 14 DPA, the importance of this group in redox regulatory reactions has recently been revealed (Dietz et al., 2006; Tovar-Mendez et al., 2011). In endosperm, few DE transcripts are identified as involved in redox reactions by 14 DPA, suggesting this stress response pathway has a narrow window for activation for wheat endosperm.

Aleurone displays a decreasing proportion of transcripts associated with cell division as expected as the tissue matures, whereas the proportion of transcripts identified as associated with cell cycling and cell development increase. Endosperm shows the proportion of DE transcripts binned as associated with cell division also reduces over time; however, in this tissue cell cycling also reduces, while development associated transcripts remains fairly static in proportions.

Discussion

Transcriptome sequencing provides a vast array of valuable data, in this analysis generating a total of over 250 million reads of approximately 75 base pairs. One of the challenges to be overcome is to reduce the complexity inherent in data sets of this size so as to provide an overview from which the tissue as a whole can be viewed and a framework in which the complex interplay of processes, functions and components can be integrated. To create this framework, the strategy of directly comparing each tissue at critical time points was selected, extracting the highly differentially expressed transcripts and annotating these. Here the GO vocabulary of the Gene Ontology project was utilized so that these standardized terms can be directly applied to other analyses. Further overviews of the data were examined using the Mapman tool. These data will provide a basis from which more detailed analysis can be begun in any number of experimental settings, particularly as wild-type wheat plants were examined. These can be used in comparisons with other wild-type varieties or treated/mutant plants. Here we provide a direct comparison of the most important biological and economic tissues of the wheat seed.

Endosperm is principally a carbohydrate storage tissue, and aleurone additionally accumulates lipids and proteins

These data validate findings from previous analyses. Figure 5b shows that of the DE transcripts detected at all time points, in starchy endosperm almost half are in the categories of carbohydrate metabolic process and generation of precursor metabolites and energy, as expected of what is principally a carbohydrate storage tissue (Emes et al., 2003; Morrison et al.,

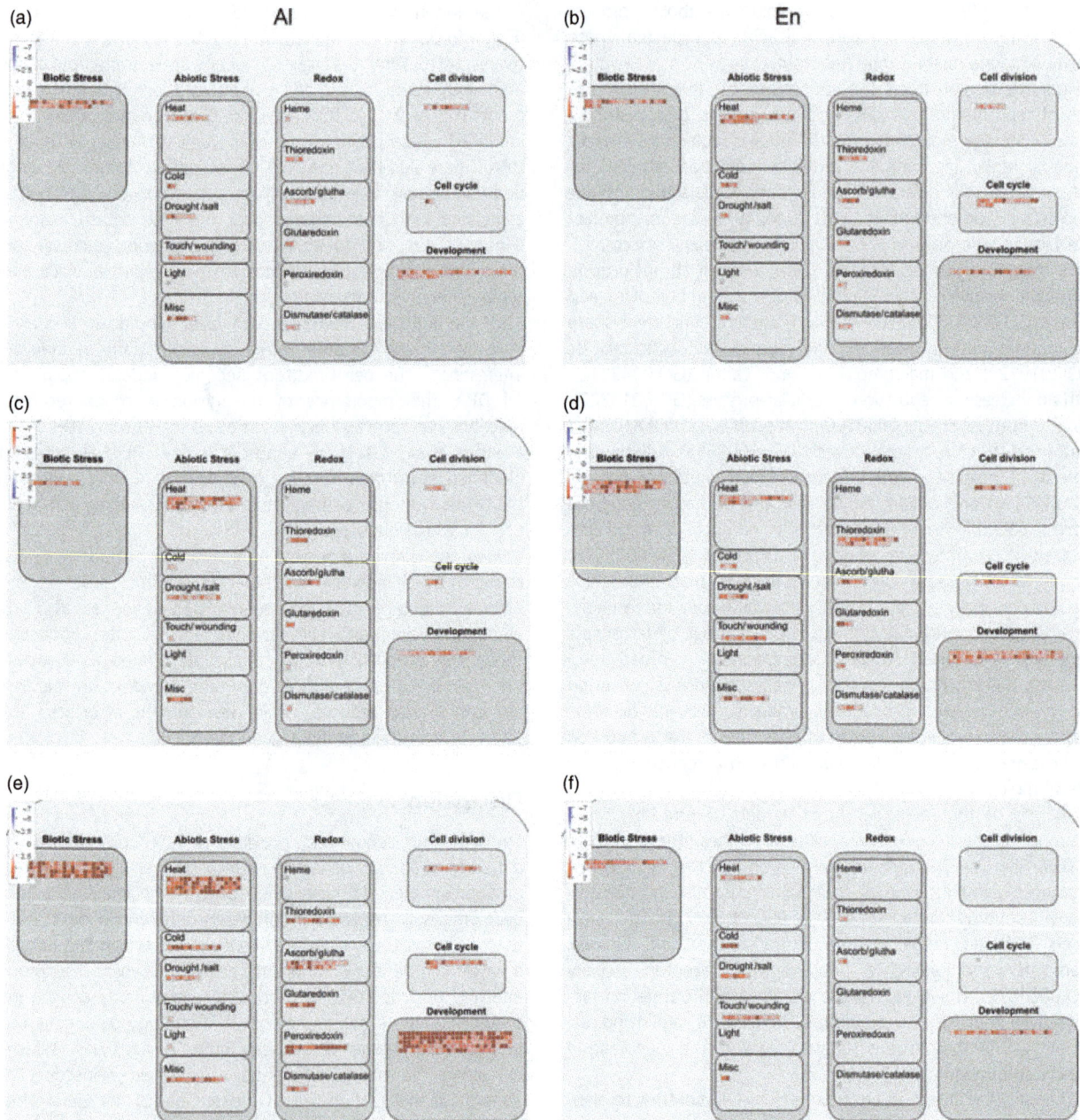

Figure 9 Mapman cellular response overview showing transcripts differentially expressed at each time point. 6 day post-anthesis (DPA) a, b, 9 DPA c, d, 14 DPA e, f. Al, aleurone; En, endosperm.

1975b; Stamova *et al.*, 2009). Additionally, as expected, aleurone shows a comparable proportion of DE transcripts for lipid metabolic process, carbohydrate metabolic process and cellular amino acid metabolic process (Liu, 2011; Morrison *et al.*, 1975a; Regvar *et al.*, 2011). Figure 7 demonstrates the consistent gradual changes in the proportions of the GO terms over the time course examined in each of the tissues, which when combined with the expected relationships shown by the hierarchical structure analysis demonstrates the validity of the approach taken. Additionally, this investigation demonstrates that although wheat currently does not have a completed annotated genomic sequence, an EST database can be used as a valid reference source.

It is essential to note, however, this analysis required the use of the more detailed Level 4 GO terms to clearly reveal the differences between the tissues, reinforcing the view that the selection of appropriate and consistent terms is critical in the meaningful interpretation of the data.

Aleurone and endosperm develop at significantly different rates

A combination of hierarchical sample clustering, DE transcript number and GO term analysis demonstrates the close yet distinct relationship between aleurone and starchy endosperm. Cluster analysis (Figure 3) demonstrates by 6 DPA, when aleurone is still being defined, it is already a very distinct tissue from the endosperm from which it has been derived, far more closely related to 9 DPA aleurone than starchy endosperm. Clearly a large proportion of genes involved in aleurone development have already been activated by 6 DPA. Additionally, these data in combination with visual inspection (Figure 2) demonstrate 14 DPA aleurone is a very distinct tissue to aleurone at earlier time points. This observation is consistent with Figure 5, which shows the number of DE transcripts up-regulated in aleurone at 14 DPA alone is double that of 6 and 9 DPA. This finding, however, would not be detected by examining total numbers of DE transcripts alone (Figure 4) or by examining the changes in proportions of transcripts identified by GO terms over time (Figure 7).

Aleurone and endosperm reprogram over the time course examined

Examination of the functional annotation of DE transcripts emphasizes the marked differences between these tissues and reveals the reprogramming of each of these tissues over time. Starchy endosperm shifts from a tissue dominated by the rapid accumulation of storage macromolecules as discussed above (Figures 6 and 7) to one requiring a broader array of molecules, evidenced by the increasing proportion of transcripts associated with lipid metabolic process (GO:0005975), cellular amino acid and derivative metabolic process (GO:0006575) together with transcripts identified as involved in regulation of metabolic process (GO:0019222) and regulation of cellular process (GO:0050794). Aleurone, in contrast, as it finalizes cell specification, shows a slight shift towards accumulation of storage macromolecules as evidenced by the accumulation of the proportion of transcripts associated with generation of precursor metabolites and energy and carbohydrate metabolic process (Figure 7). Interestingly, however, the high and increasing proportion of transcripts identified as involved in transport indicate aleurone remains a very active tissue even after differentiation is complete.

Aleurone and endosperm display different molecular functions

An important finding is the fundamental differences in molecular function between the two tissues, which may point the way for manipulation of these tissues in the future. Although the proportion of sequences predicted to be involved in DNA binding is consistent between the tissues and across the time course (Figures 7b and 8b), there is significant differences and quite profound changes in all other categories, particularly in the starchy endosperm which has a far higher proportion of sequences involved with RNA binding and translation factor activity. This indicates that endosperm utilizes extensive post-transcriptional control of the cellular machinery, allowing the cell to rapidly manipulate protein production without new mRNA synthesis, processing or export (Dever, 2002; Mata et al., 2005). Interestingly, the proportion of each of these transcripts drops significantly as the tissue approaches complete cellularization, suggesting post-transcriptional control is a particularly important process to facilitate rapid cellularization.

Aleurone and endosperm have distinct stress responses

Both aleurone and endosperm displayed rapidly changing stress response particularly in association with abiotic stress and redox reactions. Endosperm, however, identified few DE transcripts as involved in any redox pathway by 14 DPA. These data may help identify particular windows of vulnerability to particular stressors in early grain development.

Conclusion

By 6 DPA, aleurone and endosperm are distinct tissues which develop at different rates. In aleurone, highly differentially expressed transcripts are fairly evenly distributed between lipid, protein and carbohydrate biogenesis. In endosperm, carbohydrate accumulation is by far the dominant activity. However, the proportions of transcripts associated with this activity reduce over time, perhaps surprisingly as cellularization of the endosperm is not yet complete. Rapid development of the starchy endosperm appears facilitated by post-transcriptional factors, and these tissues have disparate and fast altering responses to stress.

This work provides a framework for future analysis of the developing cereal grain. It will be of particular value when the annotated wheat sequence becomes available so that the analysis of transcription start sites, polyadenylation signals, alternative splice sites and post-transcriptional modifications can be applied to these data. While this overview reinforces much current experimental data on the development of the aleurone and starchy endosperm layers, it additionally throws up new possible avenues for investigation. These data contribute to the understanding of the development of the aleurone and starchy endosperm layers in cereal grains and contribute in particular to the systems biology of the wheat seed, an essential food source.

Experimental procedures

Plant tissue collection

The *Triticum aestivum* cultivar Banks (Australian Spring wheat) was planted at four seeds per pot in 25-cm pots in soils containing equal parts Lithuanian peatmoss, perlite (size P500) and vermiculite (size 2) with 5 kg/m^3 long-term release fertilizer, 1 kg/m^3 trace element mix and 1 kg/m^3 dolomite and grown to maturity in a greenhouse during spring and early summer under natural light. Greenhouses were kept below 37°C by evaporative cooling. Plants were tagged at anthesis and seed was collected from appropriately aged (6, 9 and 14 DPA) plants from the central third of the spike. The stages of development in this study correspond to those analysed in an earlier SAGE study of whole seed transcripts (McIntosh et al., 2007) with the 14 DPa point being at a midpoint of seed development prior to seed desiccation and with a wide diversity of genes being transcribed. Generally, aleurone was separated from endosperm by microdissection over ice by incision of the whole seed, prising off the pericarp and dissecting away the embryo/presumptive embryo. The testa was then gently scraped from the endosperm layers. The starchy endosperm was rolled from the aleurone/presumptive aleurone layer and the desired tissue quickly rinsed in ice-cold phosphate buffer saline, then snap frozen in liquid nitrogen. At 6 and 9 DPA, the endosperm is not completely cellularized, and therefore the contents of the central cavity were scraped into a cooled tube and snap frozen. At 6 DPA, these contents are principally liquid and at 9 DPA principally cellular. A minimum of five individual seeds from three separate spikes from three individual plants were pooled for each tissue preparation.

RNA isolation

Total RNA was prepared using a Trizol (Invitrogen, Carlsbad, CA) protocol and purified using an RNeasy Plant Kit (Qiagen, Hilden, Germany). Total RNA quality and concentration were determined using the RNA 6000 Pico kit (Agilent, Santa Clara, CA) on a 2100 Bioanalyzer (Agilent).

cDNA construction

Approximately 1.5 µg of purified cDNA was prepared from each tissue. Total cDNA was prepared using the Ambion Message-Amp II aRNA Amplification Kit (Invitrogen) essentially as per the manufacturer's instructions. However, the first-strand synthesis primer was replaced by one designed to contain a BpuE1 site $(CTTGAG(N)_{16/14})$ 19 bp upstream from a VN clamp adjacent to the polyA tail. This site was used to remove the polyA tail after cDNA synthesis following BpuE1 digestion (NEB, Ipswich, MA). The polyA tails were removed by spinning through Qiaquick PCR purification (Qiagen) size exclusion columns after enzymatic digestion.

Sequencing

Sequencing was performed in-house on the Illumina GAIIx system with a paired end 75-bp system.

Data analysis

Approximately 25—55 million reads were obtained per tissue. Sequences were initially trimmed using default quality parameters of the CLC Genomics Workbench Version 4.0. (CLC bio, Aarhus, Denmark). Remaining contaminating sequences were removed using an editing package (010 Editor, http://www.sweetscape.com/010editor/) and the reads returned to CLC for a further round of trimming based on quality scores to optimize the quality of the sequences used in subsequent analyses. The trimmed sequences were sent through the RNA-Seq analysis program of the CLC platform, mapping against the unannotated reference sequences derived from the DFCI Wheat Gene Index, Release 12.0 (The Computational Biology and Functional Genomics Laboratory, Dana Farber Cancer Institute and Harvard School of Public Health) using default parameters. The minimum length fraction was 0.9 and the minimum similarity fraction was 0.8. RPKM was selected as the normalized expression value. Dataset S1 provides the complete set of raw sequences and RNA-Seq data for each tissue.

Tissues that had undergone RNA-Seq analysis were used in paired expression analysis experiments between aleurone and starchy endosperm at 6, 9 and 14 DPA using a two-group comparison (Kal's test) and existing RPKM expression values. Multiple testing was corrected for by obtaining FDR analysis of p-values. Sequences with a FDR-corrected P-value of less than or equal to 0.05 were selected. These were separated into those differentially expressed highly in the aleurone and those expressed in the starchy endosperm. These were then categorized into those DE at one time point or at common time points.

Each category of ESTs was submitted to the BLAST2GO program (http://www.blast2go.org/, (Bioinformatics and Genomics Department, Centro de Investigación Príncipe Valencia, Valencia, Spain), (Conesa and Gotz, 2008; Conesa et al., 2005) for functional annotation. This retrieves the gene ontology (GO) terms using the structured vocabulary provided by the Gene Ontology project (http://www.geneontology.org/)

which separates the annotations into the three vocabularies of biological processes, molecular function and cellular components. To expand and refine the annotation data, the InterPro Scan, ANNEX and GOSlim functions were applied to the GO annotations from within the BLAST2GO program (BioBam Bioinformatics, Valencia, Spain). InterPro combines a number of protein signature searches (Zdobnov and Apweiler, 2001), the results of which were attached to the GO terms. ANNEX deduces new relationships between the GO terms (Botton et al., 2008; Myhre et al., 2006), and these results were further incorporated adding on average a further 18% of annotations. To refine the annotations to allow adequate visualization, GOslim was applied using the GOPlant database. GOSlim reduces the annotations used to those of the selected level of categories or nodes. This allows a higher-level simplified visualization of the data. The GOSlim function was applied in a species-specific context (plant.obo), which has been observed to increase retrieval of plant-specific terminology and reduce category fragmentation (Botton et al., 2008).

Alternately, DE ESTs were visualized using the Mapman tools (Usadel et al., 2005, 2009). Mapping files were created using the Mercator tool (http://mapman.gabipd.org/web/guest/mercator) which bins all DE transcripts according to hierarchical ontologies. Default parameters were retained and ORYZA TIGR5 rice protein and IPR Interpro scans selected. Experimental files were created using the test statistic calculated from the Kal's test described above.

Acknowledgement

This work was supported by the Grain Foods Cooperative Research Centre.

References

Becraft, P.W. (2001) Cell fate specification in the cereal endosperm. *Semin. Cell Dev. Biol.* **12**, 387–394.

Becraft, P.W. and Asuncion-Crabb, Y. (2000) Positional cues specify and maintain aleurone cell fate in maize endosperm development. *Development*, **127**, 4039–4048.

Becraft, P.W. and Yi, G. (2011) Regulation of aleurone development in cereal grains. *J. Exp. Bot.*, **62**, 1669–1675.

Becraft, P.W., Stinard, P.S. and McCarty, D.R. (1996) CRINKLY4: a TNFR-like receptor kinase involved in maize epidermal differentiation. *Science*, **273**, 1406–1409.

Botton, A., Galla, G., Conesa, A., Bachem, C., Ramina, A. and Barcaccia, G. (2008) Large-scale Gene Ontology analysis of plant transcriptome-derived sequences retrieved by AFLP technology. *BMC Genomics*, **9**, 347.

Brown, R.C., Lemmon, B.E. and Olsen, O.A. (1994) Endosperm development in barley: microtubule involvement in the morphogenetic pathway. *Plant Cell*, **6**, 1241–1252.

Conesa, A. and Gotz, S. (2008) Blast2GO: a comprehensive suite for functional analysis in plant genomics. *Int J Plant Genomics*, **2008**, 619832.

Conesa, A., Gotz, S., Garcia- Gomez, J.M., Terol, J., Talon, M. and Robles, M. (2005) Blast2GO: a universal tool for annotation, visualization and analysis in functional genomics research. *Bioinformatics*, **21**, 3674–3676.

Dever, T.E. (2002) Gene-specific regulation by general translation factors. *Cell*, **108**, 545–556.

Dietz, K.J., Jacob, S., Oelze, M.L., Laxa, M., Tognetti, V., de Miranda, S.M., Baier, M. and Finkemeier, I. (2006) The function of peroxiredoxins in plant organelle redox metabolism. *J. Exp. Bot.* **57**, 1697–1709.

Drea, S., Leader, D.J., Arnold, B.C., Shaw, P., Dolan, L. and Doonan, J.H. (2005) Systematic spatial analysis of gene expression during wheat caryopsis development. *Plant Cell*, **17**, 2172–2185.

Emes, M.J., Bowsher, C.G., Hedley, C., Burrell, M.M., Scrase- Field, E.S. and Tetlow, I.J. (2003) Starch synthesis and carbon partitioning in developing endosperm. *J. Exp. Bot.* **54**, 569–575.

Fenech, M., Noakes, M., Clifton, P. and Topping, D. (2005) Aleurone flour increases red-cell folate and lowers plasma homocyst(e)ine substantially in man. *Br. J. Nutr.* **93**, 353–360.

Geisler-Lee, J. and Gallie, D.R. (2005) Aleurone cell identity is suppressed following connation in maize kernels. *Plant Physiol.* **139**, 204–212.

Gruis, D.F., Guo, H., Selinger, D., Tian, Q. and Olsen, O.A. (2006) Surface position, not signaling from surrounding maternal tissues, specifies aleurone epidermal cell fate in maize. *Plant Physiol.* **141**, 898–909.

Jerkovic, A., Kriegel, A.M., Bradner, J.R., Atwell, B.J., Roberts, T.H. and Willows, R.D. (2010) Strategic distribution of protective proteins within bran layers of wheat protects the nutrient-rich endosperm. *Plant Physiol.* **152**, 1459–1470.

Laudencia- Chingcuanco, D.L., Stamova, B.S., You, F.M., Lazo, G.R., Beckles, D.M. and Anderson, O.D. (2007) Transcriptional profiling of wheat caryopsis development using cDNA microarrays. *Plant Mol. Biol.*, **63**, 651–668.

Lid, S.E., Gruis, D., Jung, R., Lorentzen, J.A., Ananiev, E., Chamberlin, M., Niu, X., Meeley, R., Nichols, S. and Olsen, O.A. (2002) The defective kernel 1 (dek1) gene required for aleurone cell development in the endosperm of maize grains encodes a membrane protein of the calpain gene superfamily. *Proc. Natl. Acad. Sci. USA*, **99**, 5460–5465.

Liu, K. (2011) Comparison of lipid content and fatty acid composition and their distribution within seeds of 5 small grain species. *J. Food Sci.* **76**, C334–C342.

Mata, J., Marguerat, S. and Bahler, J. (2005) Post-transcriptional control of gene expression: a genome-wide perspective. *Trends Biochem. Sci.* **30**, 506–514.

McIntosh, G.H., Royle, P.J. and Pointing, G. (2001) Wheat aleurone flour increases cecal beta-glucuronidase activity and butyrate concentration and reduces colon adenoma burden in azoxymethane-treated rats. *J. Nutr.* **131**, 127–131.

McIntosh, S., Watson, L., Bundock, P.C., Crawford, A.C., White, J., Cordeiro, G.M., Barbary, D., Rooke, L. and Henry, R.J. (2007) SAGE of the developing wheat Caryopsis. *Plant Biotechnol. J.* **5**, 69–83.

Morrison, I.N., Kuo, J. and O' Brien, T.P. (1975a) Histochemistry and fine structure of developing wheat aleurone cells. *Planta*, **123**, 105–116.

Morrison, W.R., Mann, D.L., Soon, W. and Coventry, A.M. (1975b) Selective extraction and quantitative analysis of non-starch and starch lipids from wheat flour. *J. Sci. Food Agric.* **26**, 507–521.

Mortazavi, A., Williams, B.A., McCue, K., Schaeffer, L. and Wold, B. (2008) Mapping and quantifying mammalian transcriptomes by RNA-Seq. *Nat. Methods*, **5**, 621–628.

Myhre, S., Tveit, H., Mollestad, T. and Laegreid, A. (2006) Additional gene ontology structure for improved biological reasoning. *Bioinformatics*, **22**, 2020–2027.

Olsen, O.A. (2001) ENDOSPERM DEVELOPMENT: cellularization and Cell Fate Specification. *Annu. Rev. Plant Physiol. Plant Mol. Biol.* **52**, 233–267.

Olsen, O.A. (2004) Nuclear endosperm development in cereals and Arabidopsis thaliana. *Plant Cell*, **16**(Suppl), S214–S227.

Price, R.K., Keaveney, E.M., Hamill, L.L., Wallace, J.M., Ward, M., Ueland, P.M., McNulty, H., Strain, J.J., Parker, M.J. and Welch, R.W. (2010) Consumption of wheat aleurone-rich foods increases fasting plasma betaine and modestly decreases fasting homocysteine and LDL-cholesterol in adults. *J. Nutr.* **140**, 2153–2157.

Regvar, M., Eichert, D., Kaulich, B., Gianoncelli, A., Pongrac, P., Vogel-Mikus, K. and Kreft, I. (2011) New insights into globoids of protein storage vacuoles in wheat aleurone using synchrotron soft X-ray microscopy. *J. Exp. Bot.* **62**, 3929–3939.

Shen, B., Li, C., Min, Z., Meeley, R.B., Tarczynski, M.C. and Olsen, O.A. (2003) sal1 determines the number of aleurone cell layers in maize endosperm and encodes a class E vacuolar sorting protein. *Proc. Natl. Acad. Sci. USA*, **100**, 6552–6557.

Stamova, B.S., Laudencia-Chingcuanco, D. and Beckles, D.M. (2009) Transcriptomic analysis of starch biosynthesis in the developing grain of hexaploid wheat. *Int. J. Plant Genomics*, **2009**, 407426.

Stein, K., Borowicki, A., Scharlau, D. and Glei, M. (2010) Fermented wheat aleurone induces enzymes involved in detoxification of carcinogens and in antioxidative defence in human colon cells. *Br. J. Nutr.* **104**, 1101–1111.

Tovar-Mendez, A., Matamoros, M.A., Bustos-Sanmamed, P., Dietz, K.J., Cejudo, F.J., Rouhier, N., Sato, S., Tabata, S. and Becana, M. (2011) Peroxiredoxins and NADPH-dependent thioredoxin systems in the model legume Lotus japonicus. *Plant Physiol.* **156**, 1535–1547.

Usadel, B., Nagel, A., Thimm, O., Redestig, H., Blaesing, O.E., Palacios-Rojas, N., Selbig, J., Hannemann, J., Piques, M.C., Steinhauser, D., Scheible, W.R., Gibon, Y., Morcuende, R., Weicht, D., Meyer, S. and Stitt, M. (2005) Extension of the visualization tool MapMan to allow statistical analysis of arrays, display of corresponding genes, and comparison with known responses. *Plant Physiol.* **138**, 1195–1204.

Usadel, B., Poree, F., Nagel, A., Lohse, M., Czedik-Eysenberg, A. and Stitt, M. (2009) A guide to using MapMan to visualize and compare Omics data in plants: a case study in the crop species, Maize. *Plant Cell Environ.* **32**, 1211–1229.

Wan, Y., Poole, R.L., Huttly, A.K., Toscano-Underwood, C., Feeney, K., Welham, S., Gooding, M.J., Mills, C., Edwards, K.J., Shewry, P.R. and Mitchell, R.A. (2008) Transcriptome analysis of grain development in hexaploid wheat. *BMC Genomics*, **9**, 121.

Wan, Y., Underwood, C., Toole, G., Skeggs, P., Zhu, T., Leverington, M., Griffiths, S., Wheeler, T., Gooding, M., Poole, R., Edwards, K.J., Gezan, S., Welham, S., Snape, J., Mills, E.N., Mitchell, R.A. and Shewry, P.R. (2009) A novel transcriptomic approach to identify candidate genes for grain quality traits in wheat. *Plant Biotechnol. J.* **7**, 401–410.

Wegel, E., Pilling, E., Calder, G., Drea, S., Doonan, J., Dolan, L. and Shaw, P. (2005) Three-dimensional modelling of wheat endosperm development. *New Phytol.* **168**, 253–262.

Young, T.E. and Gallie, D.R. (2000) Programmed cell death during endosperm development. *Plant Mol. Biol.* **44**, 283–301.

Young, T.E., Meeley, R.B. and Gallie, D.R. (2004) ACC synthase expression regulates leaf performance and drought tolerance in maize. *Plant J.* **40**, 813–825.

Zdobnov, E.M. and Apweiler, R. (2001) InterProScan – an integration platform for the signature-recognition methods in InterPro. *Bioinformatics*, **17**, 847–848.

5

Dispersion and domestication shaped the genome of bread wheat

Paul J. Berkman[1,2,3], Paul Visendi[1,2], Hong C. Lee[1], Jiri Stiller[1,3], Sahana Manoli[1], Michał T. Lorenc[1,2], Kaitao Lai[1,2], Jacqueline Batley[1], Delphine Fleury[4], Hana Šimková[5], Marie Kubaláková[6], Song Weining[6], Jaroslav Doležel[5] and David Edwards[1,2,*]

[1]School of Agriculture and Food Sciences, University of Queensland, Brisbane, QLD, Australia

[2]Australian Centre for Plant Functional Genomics, University of Queensland, Brisbane, QLD, Australia

[3]CSIRO Plant Industry, St Lucia, QLD, Australia

[4]Australian Centre for Plant Functional Genomics, University of Adelaide, Glen Osmond, SA, Australia

[5]Centre of the Region Haná for Biotechnological and Agricultural Research, Institute of Experimental Botany, Olomouc, Czech, Republic

[6]State Key Laboratory of Crop Stress Biology in Arid Areas, College of Agronomy and Yangling Branch of China Wheat Improvement Center, Northwest A&F University, Yangling, Shaanxi, China

*Correspondence
email Dave.
Edwards@uq.edu.au

Summary

Despite the international significance of wheat, its large and complex genome hinders genome sequencing efforts. To assess the impact of selection on this genome, we have assembled genomic regions representing genes for chromosomes 7A, 7B and 7D. We demonstrate that the dispersion of wheat to new environments has shaped the modern wheat genome. Most genes are conserved between the three homoeologous chromosomes. We found differential gene loss that supports current theories on the evolution of wheat, with greater loss observed in the A and B genomes compared with the D. Analysis of intervarietal polymorphisms identified fewer polymorphisms in the D genome, supporting the hypothesis of early gene flow between the tetraploid and hexaploid. The enrichment for genes on the D genome that confer environmental adaptation may be associated with dispersion following wheat domestication. Our results demonstrate the value of applying next-generation sequencing technologies to assemble gene-rich regions of complex genomes and investigate polyploid genome evolution. We anticipate the genome-wide application of this reduced-complexity syntenic assembly approach will accelerate crop improvement efforts not only in wheat, but also in other polyploid crops of significance.

Keywords: *Triticum aestivum*, genome sequencing, evolution.

Introduction

Wheat is a major food crop and is used widely for making breads, pastries, noodles and dumplings. It was domesticated around 10 000 years ago in both tetraploid and hexaploid forms (Dubcovsky and Dvorak, 2007), and today, hexaploid bread wheat (*Triticum aestivum*) provides roughly a fifth of world's food (Food and Agriculture Organisation of the United Nations, 2012). Genome analysis in bread wheat is a challenge as it has a genome nearly six times larger than the human genome and consists of between 80 and 90% repetitive sequence (Wanjugi *et al.*, 2009). Bread wheat is also hexaploid with 21 pairs of chromosomes, being derived from a combination of three diploid donor species each with seven pairs of chromosomes. The donor species are proposed to have diverged from an ancestral diploid species between 2.5 and 6 MYA (Chantret *et al.*, 2005; Huang *et al.*, 2002). Hexaploid bread wheat evolved through two interspecific hybridization events, each accompanied by polyploidization. The first, occurring between 0.5 and 3 MYA, combined the genomes of *Triticum urartu* (AuAu) and an unidentified species (BB) highly similar to *Aegilops speltoides* to produce the allotetraploid genome of wild emmer wheat or *Triticum turgidum* (AuAuBB) (Chantret *et al.*, 2005; Eckardt, 2001; Huang *et al.*, 2002). The second event combined the genomes of *T. turgidum* (AuAuBB) and *Aegilops tauschii* (DD) to produce the allohexaploid genome of *T. aestivum* (AuAuBBDD) (McFadden and Sears, 1946). Each diploid progenitor genome is around 5500 million base pairs, almost twice the size of the human genome, and consists of between 80% and 90% repetitive elements (Dvořák, 2009). A greater number of genes for domestication traits are found on the A and B genomes (Gegas *et al.*, 2010), suggesting that the tetraploid was domesticated prior to the emergence of the hexaploid. No wild hexaploid wheats are known, and it is accepted that *T. aestivum* originated from a cross between *Ae. tauschii* and domesticated tetraploid emmer, probably South or West of the Caspian Sea around 8000 years ago (Giles and Brown, 2006; Nesbitt and Samuel, 1995; Salamini *et al.*, 2002).

The evolution of the homoeologous gene space in polyploid genomes can be investigated by comparative genomics approaches, including differential gene loss and single-nucleotide polymorphisms (Bekaert *et al.*, 2011; Thomas *et al.*, 2006). In addition, the group 7 chromosomes are known to contain QTL associated with boron tolerance, drought tolerance and pathogen resistance (Dolores Vazquez *et al.*, 2012; Genc *et al.*, 2010; Schnurbusch *et al.*, 2007). Consequently, while the group 7 has been sequenced as study of genome evolution, sequencing this chromosome group also represents an opportunity to pursue agriculturally important traits.

Results

Using second-generation sequencing (2GS) of isolated group 7 chromosome arms, as per our previously described methodology (Berkman *et al.*, 2011, 2012), we have assembled genomic regions containing all or nearly all genes for these chromosomes and ordered and aligned the majority of these genes based on synteny with *B. distachyon*, *O. sativa*, and *S. bicolor* (Table 1, Table S1). We identified 9258 genes in total to be present on the three chromosomes, 5532 (59.8%) of which were placed into the syntenic builds (Table S1).

We obtained from the GrainGenes database cDNA sequences representing 18 785 loci (Carollo *et al.*, 2005; Matthews *et al.*, 2003) that were located to defined chromosomal regions by hybridization to DNA from wheat deletion lines (Hossain *et al.*, 2004). These cDNAs were reciprocal best BLAST aligned with each of the group 7 assemblies (Table S4). The results are consistent with our previous results for 7DS and the expected error rate for bin mapping cDNAs in wheat (Berkman *et al.*, 2011). The results showed group 7 mapped ESTs preferentially aligning with the group 7 chromosome assemblies. Some sequences within the 7A and 7D assemblies are also aligned with 4A, which is consistent with the previously characterized ancestral translocation between 7BS and 4AL (Berkman *et al.*, 2012; Hernandez *et al.*, 2012) (Figure S2). Genetic markers from *Ae. tauschii* chromosome 7 (Luo *et al.*, 2009) demonstrated a high degree of concordance with the order and orientation of the 7D syntenic build (Figure S3), as per our previous results (Berkman *et al.*, 2011).

Comparison of assemblies demonstrated that the majority of genes remain conserved, with a copy present on each of the 3 homoeologous chromosomes (Figure 1). Within the syntenic builds, we identified 1291 genes to be present on all three chromosomes and were orthologous to predicted syntenic *B. distachyon* genes. In addition, we identified 550 *B. distachyon* genes with orthologs present on only 2 wheat genomes and 545 *B. distachyon* genes with orthologs present on only a single wheat genome. Greater gene loss was observed in the A and B genomes compared with the D genome, with 2988, 2905 and 3351 genes identified in the A, B and D genomes, respectively (Figure 1, Table S2).

We predicted the gene ontology (GO) terms for genes retained in the wheat group 7 chromosomes compared with all genes present in syntenic regions of *O. sativa*, *S. bicolor* and *B. distachyon*. GO terms with a STRING (Szklarczyk *et al.*, 2011) enrichment score >1 are listed in Table 2. The genes retained in

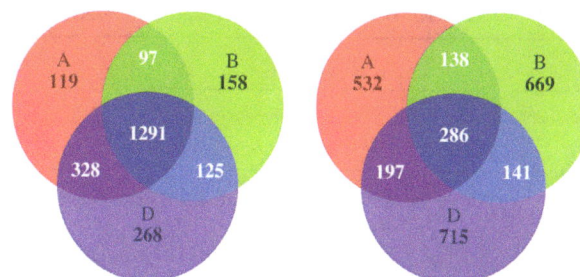

Figure 1 Venn diagram representing genes present on 7A (red), 7B (green) and 7D (blue). Genes within the syntenic build are represented on the left. Genes not found within syntenic regions of related grass species are represented on the right.

the three subgenomes present differing GO profiles in preferentially retained genes when ranked by enrichment index (Table 2, Table S3). The A and B subgenomes have preferentially retained genes relating to basic cellular processes, while the D subgenome has maintained a greater proportion of genes related to specialized processes.

Protein interaction network analysis was undertaken based on the protein interaction network of *Arabidopsis thaliana*. This revealed an abundance of genes from both 7A and 7B involved in gene networks of ribosomal proteins, transcription factors and RNA polymerases and a reduced representation of these large networks observed for 7D (Figure S1). Network analysis also revealed an increased proportion of 7D genes not associated with a network, as well as the presence of a proteolysis gene network not observed in 7A or 7B (Table 2, Figure S1).

By comparing whole-genome sequence data for four Australian wheat varieties with the group 7 assemblies, we predicted more than 900 000 intervarietal SNPs on the assembled group 7 contigs with an accuracy of 93% (Lorenc *et al.*, 2012). Syntenic builds for chromosomes 7A and 7B showed significantly greater intervarietal SNP polymorphism than 7D (Figure 2, Table 3), with 14 059, 9396 and 3137 SNPs identified on syntenic builds 7A, 7B and 7D, respectively.

Discussion

Wheat has a complex evolutionary history, with the tetraploid forming in the wild, and later combining with the D genome

Table 1 Syntenic build summary table

Chromosome arm			Velvet assembly statistics			Syntenic build statistics	
Name	Size (Mbp)[†]	2GS data coverage (×)	N50	Total (Mbp)	Longest (bp)	Length (Mbp)	Gene density (per Mbp on chromosome)
7AS	407	27.96	2503	210.57	49,550	6.85	3.55
7AL	407	46.04	2893	256.46	51,272	7.75	3.80
7BS	360	44.81	3596	220.93	56,148	6.62	3.29
7BL	540	27.96	1929	253.91	36,580	5.97	3.19
7DS	381	72.62	2098	210.62	49,276	7.47	4.45
7DL	346	76.68	5194	238.07	59,194	13.48	4.79

[†]Šafář *et al.* (2010).

Table 2 Enriched GO terms of subgenomes against hexaploid gene content

7A		7B		7A & 7B (tetraploid)		7D	
Enrichment score	GO Term	Enrichment score	GO Term	Enrichment score	GO Term	Enrichment score	GO Term
2.72	Primary metabolism	1.64	Cytoplasmic protein	1.84	Cytoplasmic protein	1.44	DNA replication and transposition
1.8	Cytoplasmic protein	1.28	DNA metabolism	1.36	Primary metabolism	1.29	Reproductive structure development
1.47	Chloroplast protein	1.09	Nucleolar lumen protein	1.18	Chloroplast envelope protein	1.26	Metal ion binding
1.18	Proteolysis						
1.06	DNA metabolism						

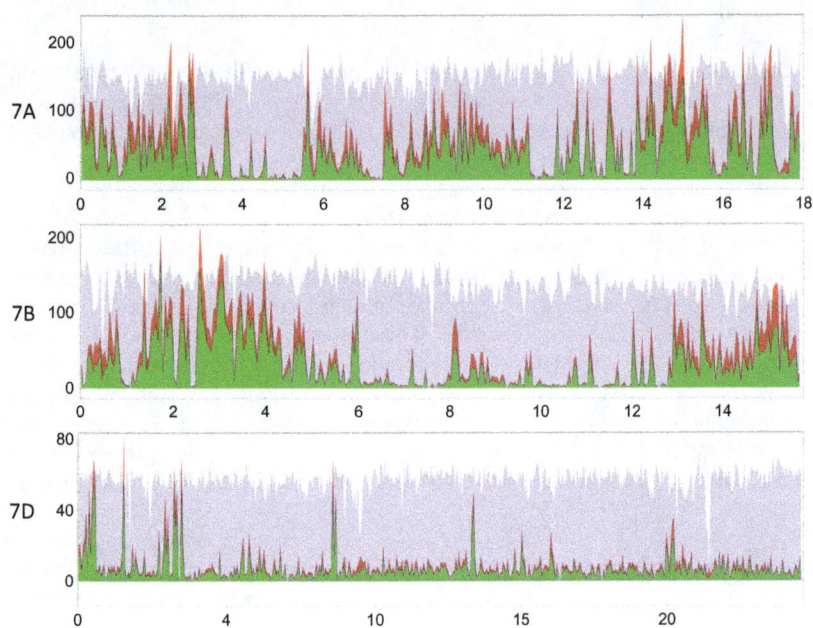

Figure 2 Single-nucleotide polymorphism (SNP) distribution across the syntenic builds of 7A, 7B and 7D. Raw SNP density numbers (SNPs/50 kbp) are represented in green, SNP densities normalized for coverage are represented in red, and read coverage is represented in blue. The X-axis represents the position on the syntenic build (Mbp) from the short-arm telomere (left) to the long-arm telomere (right).

species to produce hexaploid bread wheat. By sequencing and assembling the low copy and unique gene regions of isolated chromosome arms representing homoeologous chromosomes, we can observe the impact of this history on genome structure.

Gene loss is common following whole-genome duplication (WGD) (Paterson et al., 2009; Schnable et al., 2009; Thomas et al., 2006; Woodhouse et al., 2010). WGD also provides the opportunity for the neo- or subfunctionalization of duplicated genes, and this has been suggested as an explanation of the prevalence of polyploidy in plants (Freeling and Thomas, 2006). In *Brassica rapa*, ancient WGD events have been identified as the basis for the increased morphological plasticity observed in the *Brassica* family (Wang et al., 2011). The dosage balance hypothesis suggests that gene networks in plants significantly influence the process of gene loss following genome duplication (Thomas et al., 2006). While the dosage balance hypothesis holds true in a constant selective environment, recent evidence suggests that a change in the selective environment at the time of WGD can result in the emergence of new genes or functions that provide a specific selective advantage (Bekaert et al., 2011). By using the same method to generate each of the syntenic builds, the observed differential retention of genes between chromosomes represents true differences in gene content between the genomes.

Greater gene loss was observed in the A and B genomes compared with the D genome, (Figure 1, Table S2) which, when combined with the smaller chromosome size of 7D, appears to have resulted in a higher overall gene density on 7D (Table 1). This reflects two rounds of fractionation. Initial gene loss occurred in the A and B genomes following tetraploidy between 0.5 and 3 MYA, with subsequent loss from all three genomes following the formation of the hexaploid around 8000 years ago. The genes retained in the three subgenomes present differing GO profiles in preferentially retained genes when ranked by enrichment index (Table 2, Table S3). This suggests differential selection pressure between the first and second polyploidization events. Chromosome 7D possesses enriched GO terms related to metal ion

Table 3 Subgenomic varietal SNP profiles from four Australian cultivars

Chromosome	# SNPs	SNPs/Mb
7A	14059	963.3
7B	9396	746.2
7D	3137	149.7

processing and reproductive development, which is consistent with a recent suggestion of the D genome's contribution to adaptability and dispersion of hexaploid bread wheat (Dubcovsky and Dvorak, 2007).

Changes in the selective environment may cause the retention of genes that provide a selective advantage, with the absolute dosage of advantageous genes taking precedence over the dosage balance of highly networked genes which applies under 'normal' conditions (Bekaert et al., 2011). Protein interaction network analysis revealed a reduced representation of large networks for 7D and an increased proportion of 7D genes not associated with a network (Table 2, Figure S1). Following from the above-described work of Bekaert et al. (2011), the differences we observed in the gene network profile of the D genome, in which minimally networked genes are preserved, suggest changes in the selective pressure of wheat following hexaploidization. This is consistent with the shift from natural to human selection following T. turgidum/A. tauschii hybridization associated with domestication and the expansion of environments in which hexaploid wheat was grown. The differential gene loss between the A, B and D genomes further supports the recently proposed theories for selective gene retention and loss following polyploidization (Bekaert et al., 2011). Our results therefore suggest that following the emergence of T. aestivum, the expansion of hexaploid wheat to new environments has driven genome fractionation and evolution in this species.

The abundance and type of single-nucleotide polymorphisms (SNPs) in a genome reflect the history of selection and evolution in the species. By comparing whole-genome sequence data for four Australian wheat varieties with the group 7 assemblies, we identified more than 900,000 SNPs on the assembled group 7 contigs. Syntenic builds for chromosomes 7A and 7B showed significantly greater intervarietal SNP polymorphism than 7D (Figure 2, Table 3). This is consistent with previous results (Chao et al., 2009). Following the origin of T. aestivum, the tetraploid and hexaploid existed in sympatry for a period (Dvorak et al., 1998), and it has been suggested that gene flow between the two species resulted in greater sequence diversity within the A and B subgenomes when compared with the D subgenome (Caldwell et al., 2004; Dvorak et al., 2006; Talbert et al., 1998). The greater genetic diversity we observe on the A and B genomes is most likely due to early gene flow occurring between hexaploid T. aestivum (AuAuBBDD) and its tetraploid progenitor T. turgidum (AuAuBB), without a similar flow occurring between the hexaploid and Ae. tauschii (DD) (Dvorak et al., 2006).

Summary

The greater levels of gene conservation, together with the relative abundance of non-networked genes, suggests that that the D genome has played an important role in bread wheat evolution and that the hexaploid wheat genome has been shaped by domestication and dispersal. This analysis also demonstrates the utility of second-generation sequencing technologies in the analysis of complex polyploid genomes without a reference genome sequence. While the absence of a reference genome sequence hampers investigation of issues of biological and economic importance in polyploid crops, this study demonstrates that new sequencing technologies can yield important insight into polyploid genome evolution and crop physiology, prior to the availability of a complete genome sequence.

Experimental procedures

Data generation, assembly and validation

Professor Bikram Gill (Kansas State University, Manhattan, USA) provided seeds of double ditelosomic lines for the group 7 chromosomes of Triticum aestivum cv. Chinese Spring. The seeds were germinated and root tips of young seedlings were used for the preparation of liquid suspensions of intact chromosomes as previously described (Vrána et al., 2000). Group 7 chromosome arms were flow-sorted as telocentric chromosome in batches of 19–39 000 chromosomes representing 20–30 ng DNA. To estimate contamination with other chromosomes, 1000 chromosomes were sorted onto a microscope slide in three replicates and used for fluorescence in situ hybridization (FISH) with probes for Afa family and telomeric repeats (Kubaláková et al., 2005). Purity in sorted fractions varied from 84% (7AS, 7AL, 7DS) to 94% (7BS, 7BL, 7DL). Chromosomal DNA was purified and subsequently amplified using the Illustra GenomiPhi V2 DNA Amplification Kit (GE Healthcare, Chalfont St. Giles, UK). A total of 200 ng of pooled, amplified DNA from each chromosome arm was used to prepare Illumina paired-end libraries, which were sequenced on the Illumina GAIIx and Illumina HiSeq2000 platforms as previously described (Berkman et al., 2011, 2012). All B. distachyon, O. sativa and S. bicolor gene sequences excluding intron regions were downloaded from Phytozome version 7 (Goodstein et al., 2012).

Details of the paired-end data generated on Illumina GAIIx and HiSeq 2000 for each chromosome arm is presented in Table S1. All data have been submitted to the NCBI short-read archive, references SRA025100.1, SRA028115.1, SRA049415.1, SRA049416.1, SRA049417.1 and SRA049418.1. All sequence data were filtered and trimmed using an in-house script, trimConverter.py, to produce reads with a Phred quality score of at least 20 at each nucleotide position and a minimum length of 63 bp. The trimmed and filtered read set was assembled using Velvet version 1.1.04 (Zerbino and Birney, 2008) on a DELL R905 server with 128 GB RAM. Each Velvet assembly used a kmer size of 63 bp and an expected coverage representing the read depth after filtering (Table S1). The assembly statistics and predicted size of each chromosome arm are displayed in Table S1 with more detailed assembly statistics presented in Table S1.

Producing syntenic builds and assembly annotation

A comparative genomics approach was applied to order and orientate the wheat contigs into a draft syntenic build, as previously described (Berkman et al., 2011, 2012; Mayer et al., 2009), with results presented in Table S2A. In this approach, B. distachyon, O. sativa and S. bicolor were used as the genomes for syntenic comparison, with this order of evolutionary closeness to wheat determining the decision hierarchy in the algorithm. In the case of infrequent tandem gene duplications, the duplications were assembled individually and placed adjacent to each other in the final syntenic build. The process of producing the syntenic build also identified wheat homologs to B. distachyon genes outside of the predicted syntenic regions and reflect previously described gene movement (Berkman et al., 2012; Wicker et al., 2010, 2011) (Table S2B). Syntenic builds for each chromosome arm were combined to form a single syntenic build for each chromosome, with the remaining assembled contigs placed into a chromosome-specific 'extra_contigs' files. The full syntenic chromosome sequences of B. distachyon version 192, O. sativa version 193 and *S. bicolor* version 79 were all downloaded from Phytozome (Goodstein *et al.*, 2012; International Rice Genome Sequencing Project, 2005; Paterson *et al.*, 2009; Vogel *et al.*, 2010). Following the generation of the syntenic builds, the number of genes on each of the group 7 chromosomes was calculated as per the methodology previously described (Berkman *et al.*, 2011, 2012). The majority of contigs from each short-read assembly did not match any *B. distachyon* gene, and reviewing the annotation of these contigs suggests that they are predominantly made up of nested transposable element insertions (data not shown), consistent with our previous results (Berkman *et al.*, 2011, 2012).

We annotated each syntenic build with predicted genes, gene functional annotations, as well as homoeologous and varietal SNPs. Predicted genes were identified by comparing all *B. distachyon* genes with the syntenic builds using WU-BLASTN (http://hg.wustl.edu/info/README.html) with an E-value cut-off of 1e-5 and a minimum distance between HSPs of 4 kbp. Gene functional annotations were predicted by comparing each syntenic build and its extra contigs with the Uniref90 database using BLASTX (Altschul *et al.*, 1990) with an E-value cut-off of 1e-10. Homoeologous SNPs were detected by comparison of chromosome arm syntenic builds to raw short-read sequences using custom in-house software SGSautoSNP, while varietal SNPs were detected by aligning varietal WGS data from four Australian cultivars (Gladius, Drysdale, Excalibur, and RAC875) using the same tool. All annotations and sequences are available for public access at www.wheatgenome.info (Lai *et al.*, 2012).

Comparison of homoeolog gene content

Wheat gene sequences predicted by comparison of the chromosome assemblies with *B. distachyon* genes were compared with the Swissprot database (downloaded 16th August 2011) using BLASTX with an E-value cut-off of 1e-5. The same comparison with Swissprot was performed with all predicted genes from *B. distachyon* genome as well as a subset of *B. distachyon* genes within regions syntenic with wheat chromosome 7. The resulting lists of UNIPROT accession numbers were analysed using the DAVID GO functional annotation clustering tool (Huang *et al.*, 2008, 2009) with the background of the complete *B. distachyon* genome as a control measure of enrichment for this particular genomic region in *B. distachyon*. UNIPROT accession numbers identified from 7A, 7B and 7D, as well as a combined list of all three were compared, with the background set of all genes from the *B. distachyon* syntenic region. Annotations of COG_ONTOLOGY in the Functional_Categories database as well as GOTERM_BP_ALL, GOTERM_BP_FAT, GOTERM_CC_ALL, GOTERM_CC_FAT, GOTERM_MF_ALL, and GOTERM_MF_FAT in the Gene_Ontology database were considered in characterizing enriched function.

Brassica distachyon gene sequences with an ortholog on each of the 7A, 7B and 7D syntenic builds were identified as per our previously described methodology (Berkman *et al.*, 2011, 2012). The full set of *Arabidopsis thaliana* coding regions was downloaded from Phytozome version 7 (Goodstein *et al.*, 2012) and compared with the *B. distachyon* gene sequences using BLASTN (Altschul *et al.*, 1990) with an E-value cut-off of 10e-05. The list of names for the *A. thaliana* genes with top hits against the three sets of *B. distachyon* genes was then loaded into STRING (Szklarczyk *et al.*, 2011) to identify the representation of networked genes on each group 7 chromo-

some. 120 connecting nodes were provided to the network produced by each chromosome gene list, resulting in minimum confidence scores of 0.933, 0.916 and 0.930 for 7A, 7B and 7D, respectively. The confidence score is a measure of the reliability of evidence for network interactions; therefore, the number of nodes applied and consistent confidence scores represent the optimal network generation for accurate comparison. Figure S1 displays the full gene networks representing each chromosome arm.

SNP density analysis and variation

Between 8.8× and 10.8× coverage of Illumina whole-genome paired read sequence data were generated for four Australian wheat cultivars Drysdale (173.11 Gbp), Gladius (185.60 Gbp), Excalibur (161.22 Gbp) and RAC875 (149.59 Gbp). The data were mapped to the three group 7 wheat chromosome assemblies as well as an assembly of chromosome arm 4AL (Hernandez *et al.*, 2012) to prevent the erroneous mapping of reads associated with 7BS/4AL translocation. Mapping was conducted with SOAP v2.21 (Li *et al.*, 2008) using the default parameters, allowing up to 2 mismatches per read. Only uniquely mapped paired reads were retained. The number of reads mapped from each dataset is presented in Table S5 and is consistent with the proportion of reads that could be expected to uniquely map with high stringency to the assembled portion of the genome.

SNPs were predicted using our in-house SGSautoSNP pipeline (Lorenc et al. 2012), which is available on request and based on the autoSNP algorithm (Barker *et al.*, 2003; Duran *et al.*, 2009). All predicted SNPs are presented on a GBrowse database at www.wheatgenome.info and summarized in Table 3 (full details in Table S6). SNP density was measured across each syntenic build using in-house scripts that count the number of SNPs occurring in a window of 50 kbp moving in a frame of 10 kbp (Figure 2). SNP density varied across each of the chromosomes, with a greater number of SNPs identified on chromosome 7A, and significantly fewer SNPs identified on chromosome 7D. Calculating the proportion of SNP transitions indicates the historic methylation profile of each subgenome (Table S6).

Acknowledgements

We thank Dr. Jarmila Čihalíková, Romana Šperková, Bc. and Ms. Zdeňka Dubská for their assistance with chromosome sorting. We also thank J. Perry Gustafson for valuable discussions on the evolutionary history of wheat. The authors would like to acknowledge funding support from Bioplatforms Australia, the Australian Research Council (Projects LP0882095, LP0883462 and DP0985953), the Department of Industry, Innovation, Science, Research and Tertiary Education (project CG120174), the Czech Science Foundation (grant No. P501/12/2554), and from the European Union (grant No. ED0007/01/01 Centre of the Region Haná for Biotechnological and Agricultural Research). Support also came from the Australian Genome Research Facility (AGRF), the Queensland Cyber Infrastructure Foundation (QCIF) and the Australian Partnership for Advanced Computing (APAC).

References

Altschul, S.F., Gish, W., Miller, W., Myers, E.W. and Lipman, D.J. (1990) Basic Local Alignment Search Tool. *J. Mol. Biol.* **215**, 403–410.

Barker, G., Batley, J., O' Sullivan, H., Edwards, K.J. and Edwards, D. (2003) Redundancy based detection of sequence polymorphisms in expressed sequence tag data using autoSNP. *Bioinformatics*, **19**, 421–422.

Bekaert, M., Edger, P.P., Pires, J.C. and Conant, G.C. (2011) Two-phase resolution of polyploidy in the Arabidopsis metabolic network gives rise to relative and absolute dosage constraints. *Plant Cell*, **23**, 1719–1728.

Berkman, P.J., Skarshewski, A., Lorenc, M., Lai, K., Duran, C., Ling, E.Y.S., Stiller, J., Smits, L., Imelfort, M., Manoli, S., McKenzie, M., Kubaláková, M., Šimkovž, H., Batley, J., Fleury, D., Doleel, J. and Edwards, D. (2011) Sequencing and assembly of low copy and genic regions of isolated *Triticum aestivum* chromosome arm 7DS. *Plant Biotechnol. J.* **9**, 768–775.

Berkman, P.J., Skarshewski, A., Manoli, S., Lorenc, M.T., Stiller, J., Smits, L., Lai, K., Campbell, E., Kubaláková, M., Šimková, H., Batley, J., Doleel, J., Hernandez, P. and Edwards, D. (2012) Sequencing wheat chromosome arm 7BS delimits the 7BS/4AL translocation and reveals homoeologous gene conservation. *Theor. Appl. Genet.* **124**, 423–432.

Caldwell, K.S., Dvorak, J., Lagudah, E.S., Akhunov, E., Luo, M.-C., Wolters, P. and Powell, W. (2004) Sequence Polymorphism in Polyploid Wheat and Their D-Genome Diploid Ancestor. *Genetics*, **167**, 941–947.

Carollo, V., Matthews, D.E., Lazo, G.R., Blake, T.K., Hummel, D.D., Lui, N., Hane, D.L. and Anderson, O.D. (2005) GrainGenes 2.0. an improved resource for the small-grains community. *Plant Physiol.* **139**, 643–651.

Chantret, N., Salse, J., Sabot, F., Rahman, S., Bellec, A., Laubin, B., Dubois, I., Dossat, C., Sourdille, P., Joudrier, P., Gautier, M.F., Cattolico, L., Beckert, M., Aubourg, S., Weissenbach, J., Caboche, M., Bernard, M., Leroy, P. and Chalhoub, B. (2005) Molecular basis of evolutionary events that shaped the hardness locus in diploid and polyploid wheat species (*Triticum* and *Aegilops*). *Plant Cell*, **17**, 1033–1045.

Chao, S., Zhang, W., Akhunov, E., Sherman, J., Ma, Y., Luo, M.-C. and Dubcovsky, J. (2009) Analysis of gene-derived SNP marker polymorphism in US wheat (*Triticum aestivum* L.) cultivars. *Mol Breed*, **23**, 23–33.

Dolores Vazquez, M., James Peterson, C., Riera-Lizarazu, O., Chen, X., Heesacker, A., Ammar, K., Crossa, J. and Mundt, C. (2012) Genetic analysis of adult plant, quantitative resistance to stripe rust in wheat cultivar 'Stephens' in multi-environment trials. *Theor. Appl. Genet.* **124**, 1–11.

Dubcovsky, J. and Dvorak, J. (2007) Genome plasticity a key factor in the success of polyploid wheat under domestication. *Science*, **316**, 1862–1866.

Duran, C., Appleby, N., Clark, T., Wood, D., Imelfort, M., Batley, J. and Edwards, D. (2009) AutoSNPdb: an annotated single nucleotide polymorphism database for crop plants. *Nucleic Acids Res.* **37**, D951–953.

Dvořák, J. (2009) Crops and Models. In:*Plant Genetics and Genomics* (Muehlbauer, G.M. and Feuillet, C., eds), pp. 685–711. New York: Springer.

Dvorak, J., Luo, M.C., Yang, Z.L. and Zhang, H.B. (1998) The structure of the *Aegilops tauschii* genepool and the evolution of hexaploid wheat. *Theor. Appl. Genet.* **97**, 657–670.

Dvorak, J., Akhunov, E.D., Akhunov, A.R., Deal, K.R. and Luo, M.C. (2006) Molecular characterization of a diagnostic DNA marker for domesticated tetraploid wheat provides evidence for gene flow from wild tetraploid wheat to hexaploid wheat. *Mol. Biol. Evol.* **23**, 1386–1396.

Eckardt, N.A. (2001) A Sense of Self: the Role of DNA Sequence Elimination in Allopolyploidization. *Plant Cell*, **13**, 1699–1704.

Food and Agriculture Organisation of the United Nations (2012) *FAOSTAT Food Supply – Crops Primary Equivalent.*

Freeling, M. and Thomas, B.C. (2006) Gene-balanced duplications, like tetraploidy, provide predictable drive to increase morphological complexity. *Genome Res.* **16**, 805–814.

Gegas, V.C., Nazari, A., Griffiths, S., Simmonds, J., Fish, L., Orford, S., Sayers, L., Doonan, J.H. and Snape, J.W. (2010) A genetic framework for grain size and shape variation in wheat. *Plant Cell*, **22**, 1046–1056.

Genc, Y., Oldach, K., Verbyla, A., Lott, G., Hassan, M., Tester, M., Wallwork, H. and McDonald, G. (2010) Sodium exclusion QTL associated with improved seedling growth in bread wheat under salinity stress. *Theor. Appl. Genet.* **121**, 877–894.

Giles, R.J. and Brown, T.A. (2006) GluDy allele variations in *Aegilops tauschii* and *Triticum aestivum*: implications for the origins of hexaploid wheats. *Theor. Appl. Genet.* **112**, 1563–1572.

Goodstein, D.M., Shu, S., Howson, R., Neupane, R., Hayes, R.D., Fazo, J., Mitros, T., Dirks, W., Hellsten, U., Putnam, N. and Rokhsar, D.S. (2012) Phytozome: a comparative platform for green plant genomics. *Nucleic Acids Res.* **40**, D1178–1186.

Hernandez, P., Martis, M., Dorado, G., Pfeifer, M., Galvez, S., Schaaf, S., Jouve, N., Šimková, H., Valárik, M., Doležel, J. and Mayer, K.F. (2012) Next-generation sequencing and syntenic integration of flow-sorted arms of wheat chromosome 4A exposes the chromosome structure and gene content. *Plant J.* **69**, 377–386.

Hossain, K.G., Kalavacharla, V., Lazo, G.R., Hegstad, J., Wentz, M.J., Kianian, P.M., Simons, K., Gehlhar, S., Rust, J.L., Syamala, R.R., Obeori, K., Bhamidimarri, S., Karunadharma, P., Chao, S., Anderson, O.D., Qi, L.L., Echalier, B., Gill, B.S., Linkiewicz, A.M., Ratnasiri, A., Dubcovsky, J., Akhunov, E.D., Dvorak, J., Miftahudin, Ross, K., Gustafson, J.P., Radhawa, H.S., Dilbirligi, M., Gill, K.S., Peng, J.H., Lapitan, N.L., Greene, R.A., Bermudez-Kandianis, C.E., Sorrells, M.E., Feril, O., Pathan, M.S., Nguyen, H.T., Gonzalez-Hernandez, J.L., Conley, E.J., Anderson, J.A., Choi, D.W., Fenton, D., Close, T.J., McGuire, P.E., Qualset, C.O. and Kianian, S.F. (2004) A chromosome bin map of 2148 expressed sequence tag loci of wheat homoeologous group 7. *Genetics*, **168**, 687–699.

Huang, S., Sirikhachornkit, A., Su, X., Faris, J., Gill, B., Haselkorn, R. and Gornicki, P. (2002) Genes encoding plastid acetyl-CoA carboxylase and 3-phosphoglycerate kinase of the *Triticum/Aegilops* complex and the evolutionary history of polyploid wheat. *Proc. Natl. Acad. Sci. USA*, **99**, 8133–8138.

Huang, D.W., Sherman, B.T. and Lempicki, R.A. (2008) Systematic and integrative analysis of large gene lists using DAVID bioinformatics resources. *Nat. Protoc.* **4**, 44–57.

Huang, D.W., Sherman, B.T. and Lempicki, R.A. (2009) Bioinformatics enrichment tools: paths toward the comprehensive functional analysis of large gene lists. *Nucleic Acids Res.* **37**, 1–13.

International Rice Genome Sequencing Project (2005) The map-based sequence of the rice genome. *Nature*, **436**, 793–800.

Kubaláková, M., Kovářová, P., Suchánková, P., Číhalíková, J., Bartoš, J., Lucretti, S., Watanabe, N., Kianian, S.F. and Doležel, J. (2005) Chromosome sorting in tetraploid wheat and its potential for genome analysis. *Genetics*, **170**, 823–829.

Lai, K., Berkman, P.J., Lorenc, M.T., Duran, C., Smits, L., Manoli, S., Stiller, J. and Edwards, D. (2012) WheatGenome.info: an integrated database and portal for wheat genome information. *Plant Cell Physiol.* **52**, e2. (1-7).

Li, R., Li, Y., Kristiansen, K. and Wang, J. (2008) SOAP: short oligonucleotide alignment program. *Bioinformatics*, **24**, 713–714.

Lorenc, M.T., Hayashi, S., Stiller, J., Lee, H., Manoli, S., Ruperao, P., Visendi, P., Berkman, P.J., Lai, K., Batley, J. and Edwards, D. (2012) Discovery of Single Nucleotide Polymorphisms in Complex Genomes Using SGSautoSNP. *Biology*, **1**, 370–382.

Luo, M.C., Deal, K.R., Akhunov, E.D., Akhunova, A.R., Anderson, O.D., Anderson, J.A., Blake, N., Clegg, M.T., Coleman-Derr, D., Conley, E.J., Crossman, C.C., Dubcovsky, J., Gill, B.S., Gu, Y.Q., Hadam, J., Heo, H.Y., Huo, N., Lazo, G., Ma, Y., Matthews, D.E., McGuire, P.E., Morrell, P.L., Qualset, C.O., Renfro, J., Tabanao, D., Talbert, L.E., Tian, C., Toleno, D.M., Warburton, M.L., You, F.M., Zhang, W. and Dvorak, J. (2009) Genome comparisons reveal a dominant mechanism of chromosome number reduction in grasses and accelerated genome evolution in *Triticeae*. *Proc. Natl Acad. Sci. USA* **106**, 15780–15785.

Matthews, D.E., Carollo, V.L., Lazo, G.R. and Anderson, O.D. (2003) GrainGenes, the genome database for small-grain crops. *Nucleic Acids Res.* **31**, 183–186.

Mayer, K.F., Taudien, S., Martis, M., Šimkova, H., Suchánková, P., Gundlach, H., Wicker, T., Petzold, A., Felder, M., Steuernagel, B., Scholz, U., Graner, A., Platzer, M., Doležel, J. and Stein, N. (2009) Gene content and virtual gene order of barley chromosome 1H. *Plant Physiol.* **151**, 496–505.

McFadden, E.S. and Sears, E.R. (1946) The origin of *Triticum spelta* and its free-threshing hexaploid relatives. *J. Hered.* **37**, 81–107.

Nesbitt, M. and Samuel, D. (1995) From staple crop to extinction? The archaeology and history of the hulled wheats. In: *First International Workshop on Hulled Wheats* (Padulosi, S., Hammer, K. and Heller, J., eds), pp. 40–99. Castelvecchio Pascoli, Tuscany, Italy: Bioversity International.

Paterson, A.H., Bowers, J.E., Bruggmann, R., Dubchak, I., Grimwood, J., Gundlach, H., Haberer, G., Hellsten, U., Mitros, T., Poliakov, A., Schmutz, J., Spannagl, M., Tang, H., Wang, X., Wicker, T., Bharti, A.K., Chapman, J., Feltus, F.A., Gowik, U., Grigoriev, I.V., Lyons, E., Maher, C.A., Martis, M., Narechania, A., Otillar, R.P., Penning, B.W., Salamov, A.A., Wang, Y., Zhang, L., Carpita, N.C., Freeling, M., Gingle, A.R., Hash, C.T., Keller, B., Klein, P., Kresovich, S., McCann, M.C., Ming, R., Peterson, D.G., Mehboob ur, R., Ware, D., Westhoff, P., Mayer, K.F., Messing, J. and Rokhsar, D.S. (2009) The *Sorghum bicolor* genome and the diversification of grasses. *Nature*, **457**, 551–556.

Šafář, J., Šimková, H., Kubaláková, M., Číhalíková, J., Suchánková, P., Bartoš, J. and Doležel, J. (2010) Development of chromosome-specific BAC resources for genomics of bread wheat. *Cytogenet. Genome Res.* **129**, 211–223.

Salamini, F., Ozkan, H., Brandolini, A., Schafer-Pregl, R. and Martin, W. (2002) Genetics and geography of wild cereal domestication in the near east. *Nat. Rev. Genet.* **3**, 429–441.

Schnable, P.S., Ware, D., Fulton, R.S., Stein, J.C., Wei, F., Pasternak, S., Liang, C., Zhang, J., Fulton, L., Graves, T.A., Minx, P., Reily, A.D., Courtney, L., Kruchowski, S.S., Tomlinson, C., Strong, C., Delehaunty, K., Fronick, C., Courtney, B., Rock, S.M., Belter, E., Du, F., Kim, K., Abbott, R.M., Cotton, M., Levy, A., Marchetto, P., Ochoa, K., Jackson, S.M., Gillam, B., Chen, W., Yan, L., Higginbotham, J., Cardenas, M., Waligorski, J., Applebaum, E., Phelps, L., Falcone, J., Kanchi, K., Thane, T., Scimone, A., Thane, N., Henke, J., Wang, T., Ruppert, J., Shah, N., Rotter, K., Hodges, J., Ingenthron, E., Cordes, M., Kohlberg, S., Sgro, J., Delgado, B., Mead, K., Chinwalla, A., Leonard, S., Crouse, K., Collura, K., Kudrna, D., Currie, J., He, R., Angelova, A., Rajasekar, S., Mueller, T., Lomeli, R., Scara, G., Ko, A., Delaney, K., Wissotski, M., Lopez, G., Campos, D., Braidotti, M., Ashley, E., Golser, W., Kim, H., Lee, S., Lin, J. and Dujmic, Z., Kim, W., Talag, J., Zuccolo, A., Fan, C., Sebastian, A., Kramer, M., Spiegel, L., Nascimento, L., Zutavern, T., Miller, B., Ambroise, C., Muller, S., Spooner, W., Narechania, A., Ren, L., Wei, S., Kumari, S., Faga, B., Levy, M.J., McMahan, L., Van Buren, P., Vaughn, M.W., Ying, K., Yeh, C.T., Emrich, S.J., Jia, Y., Kalyanaraman, A., Hsia, A.P., Barbazuk, W.B., Baucom, R.S., Brutnell, T.P., Carpita, N.C., Chaparro, C., Chia, J.M., Deragon, J.M., Estill, J.C., Fu, Y., Jeddeloh, J.A., Han, Y., Lee, H., Li, P., Lisch, D.R., Liu, S., Liu, Z., Nagel, D.H., McCann, M.C., SanMiguel, P., Myers, A.M., Nettleton, D., Nguyen, J., Penning, B.W., Ponnala, L., Schneider, K.L., Schwartz, D.C., Sharma, A., Soderlund, C., Springer, N.M., Sun, Q., Wang, H., Waterman, M., Westerman, R., Wolfgruber, T.K., Yang, L., Yu, Y., Zhang, L., Zhou, S., Zhu, Q., Bennetzen, J.L., Dawe, R.K., Jiang, J., Jiang, N., Presting, G.G., Wessler, S.R., Aluru, S., Martienssen, R.A., Clifton, S.W., McCombie, W.R., Wing, R.A. and Wilson, R.K. (2009) The B73 maize genome: complexity, diversity, and dynamics. *Science* **326**, 1112–1115.

Schnurbusch, T., Collins, N.C., Eastwood, R.F., Sutton, T., Jefferies, S.P. and Langridge, P. (2007) Fine mapping and targeted SNP survey using rice-wheat gene colinearity in the region of the Bo1 boron toxicity tolerance locus of bread wheat. *Theor. Appl. Genet.* **115**, 451–461.

Szklarczyk, D., Franceschini, A., Kuhn, M., Simonovic, M., Roth, A., Minguez, P., Doerks, T., Stark, M., Muller, J., Bork, P., Jensen, L.J. and von Mering, C. (2011) The STRING database in 2011: functional interaction networks of proteins, globally integrated and scored. *Nucleic Acids Res.* **39**, D561–D568.

Talbert, L.E., Smith, L.Y. and Blake, N.K. (1998) More than one origin of hexaploid wheat is indicated by sequence comparison of low-copy DNA. *Genome*, **41**, 402–407.

Thomas, B.C., Pedersen, B. and Freeling, M. (2006) Following tetraploidy in an Arabidopsis ancestor, genes were removed preferentially from one homeolog leaving clusters enriched in dose-sensitive genes. *Genome Res.* **16**, 934–946.

Vogel, J.P., Garvin, D.F., Mockler, T.C., Schmutz, J., Rokhsar, D., Bevan, M.W., Barry, K., Lucas, S., Harmon-Smith, M., Lail, K., Tice, H., Schmutz Leader, J., Grimwood, J., McKenzie, N., Huo, N., Gu, Y.Q., Lazo, G.R., Anderson, O.D., Vogel Leader, J.P., You, F.M., Luo, M.C., Dvorak, J., Wright, J., Febrer, M., Idziak, D., Hasterok, R., Lindquist, E., Wang, M., Fox, S.E., Priest, H.D., Filichkin, S.A., Givan, S.A., Bryant, D.W., Chang, J.H., Mockler Leader, T.C., Wu, H., Wu, W., Hsia, A.P., Schnable, P.S., Kalyanaraman, A., Barbazuk, B., Michael, T.P., Hazen, S.P., Bragg, J.N., Laudencia-Chingcuanco, D., Weng, Y., Haberer, G., Spannagl, M., Mayer Leader, K., Rattei, T., Mitros, T., Lee, S.J., Rose, J.K., Mueller, L.A., York, T.L., Wicker Leader, T., Buchmann, J.P., Tanskanen, J., Schulman Leader, A.H., Gundlach, H., Bevan, M., Costa de

Oliveira, A., da, C.M.L., Belknap, W., Jiang, N., Lai, J., Zhu, L., Ma, J., Sun, C., Pritham, E., Salse Leader, J., Murat, F., Abrouk, M., Mayer, K., Bruggmann, R., Messing, J., Fahlgren, N., Sullivan, C.M., Carrington, J.C., Chapman, E.J., May, G.D. and Zhai, J., Ganssmann, M., Guna Ranjan Gurazada, S., German, M., Meyers, B.C., Green Leader, P.J., Tyler, L., Wu, J., Thomson, J., Chen, S., Scheller, H.V., Harholt, J., Ulvskov, P., Kimbrel, J.A., Bartley, L.E., Cao, P., Jung, K.H., Sharma, M.K., Vega-Sanchez, M., Ronald, P., Dardick, C.D., De Bodt, S., Verelst, W., Inze, D., Heese, M., Schnittger, A., Yang, X., Kalluri, U.C., Tuskan, G.A., Hua, Z., Vierstra, R.D., Cui, Y., Ouyang, S., Sun, Q., Liu, Z., Yilmaz, A., Grotewold, E., Sibout, R., Hematy, K., Mouille, G., Hofte, H., Michael, T., Pelloux, J., O'Connor, D., Schnable, J., Rowe, S., Harmon, F., Cass, C.L., Sedbrook, J.C., Byrne, M.E., Walsh, S., Higgins, J., Li, P., Brutnell, T., Unver, T., Budak, H., Belcram, H., Charles, M., Chalhoub, B. and Baxter, I. (2010) Genome sequencing and analysis of the model grass *Brachypodium distachyon*. *Nature*, **463**, 763–768.

Vrána, J., Kubaláková, M., Šimková, H., Číhalíková, J., Lysák, M.A. and Doležel, J. (2000) Flow sorting of mitotic chromosomes in common wheat (*Triticum aestivum* L.). *Genetics*, **156**, 2033–2041.

Wang, X., Wang, H., Wang, J., Sun, R., Wu, J., Liu, S., Bai, Y., Mun, J.H., Bancroft, I., Cheng, F., Huang, S., Li, X., Hua, W., Freeling, M., Pires, J.C., Paterson, A.H., Chalhoub, B., Wang, B., Hayward, A., Sharpe, A.G., Park, B.S., Weisshaar, B., Liu, B., Li, B., Tong, C., Song, C., Duran, C., Peng, C., Geng, C., Koh, C., Lin, C., Edwards, D., Mu, D., Shen, D., Soumpourou, E., Li, F., Fraser, F., Conant, G., Lassalle, G., King, G.J., Bonnema, G., Tang, H., Belcram, H., Zhou, H., Hirakawa, H., Abe, H., Guo, H., Jin, H., Parkin, I.A., Batley, J., Kim, J.S., Just, J., Li, J., Xu, J., Deng, J., Kim, J.A., Yu, J., Meng, J., Min, J., Poulain, J., Hatakeyama, K., Wu, K., Wang, L., Fang, L., Trick, M., Links, M.G., Zhao, M., Jin, M., Ramchiary, N., Drou, N., Berkman, P.J., Cai, Q., Huang, Q., Li, R., Tabata, S., Cheng, S., Zhang, S., Sato, S., Sun, S., Kwon, S.J., Choi, S.R., Lee, T.H., Fan, W., Zhao, X., Tan, X., Xu, X., Wang, Y., Qiu, Y., Yin, Y., Li, Y., Du, Y., Liao, Y., Lim, Y., Narusaka, Y., Wang, Z., Li, Z., Xiong, Z. and Zhang, Z. (2011) The genome of the mesopolyploid crop species *Brassica rapa*. *Nat. Genet.* **43**, 1035–1039.

Wanjugi, H., Coleman-Derr, D., Huo, N., Kianian, S.F., Luo, M.-C., Wu, J., Anderson, O. and Gu, Y.Q. (2009) Rapid development of PCR-based genome-specific repetitive DNA junction markers in wheat. *Genome*, **52**, 576–587.

Wicker, T., Buchmann, J.P. and Keller, B. (2010) Patching gaps in plant genomes results in gene movement and erosion of colinearity. *Genome Res.* **20**, 1229–1237.

Wicker, T., Mayer, K.F.X., Gundlach, H., Martis, M., Steuernagel, B., Scholz, U., Šimková, H., Kubaláková, M., Choulet, F., Taudien, S., Platzer, M., Feuillet, C., Fahima, T., Budak, H., Doležel, J., Keller, B. and Stein, N. (2011) Frequent Gene Movement and Pseudogene Evolution Is Common to the Large and Complex Genomes of Wheat, Barley, and Their Relatives. *Plant Cell*, **23**, 1706–1718.

Woodhouse, M.R., Schnable, J.C., Pedersen, B.S., Lyons, E., Lisch, D., Subramaniam, S. and Freeling, M. (2010) Following tetraploidy in maize, a short deletion mechanism removed genes preferentially from one of the two homologs. *PLoS Biol.* **8**, e1000409.

Zerbino, D.R. and Birney, E. (2008) Velvet: algorithms for de novo short read assembly using de Bruijn graphs. *Genome Res.* **18**, 821–829.

6

Identification and molecular characterization of the nicotianamine synthase gene family in bread wheat

Julien Bonneau[1], Ute Baumann[2], Jesse Beasley[1], Yuan Li[2] and Alexander A. T. Johnson[1,*]

[1]School of BioSciences, The University of Melbourne, Melbourne, Vic., Australia
[2]Australian Centre for Plant Functional Genomics, The University of Adelaide, Adelaide, SA, Australia

*Correspondence

email johnsa@unimelb.edu.au
Accession numbers: The *Triticum aestivum* GenBank accession numbers for the NAS genes described in this article are as follows: *TaNAS3-A*, KU529947; *TaNAS1-A*, KU529948; *TaNAS9-A*, KU529949; *TaNAS3-B*, KU529950; *TaNAS1-B*, KU529951; *TaNAS9-B*, KU529952; *TaNAS9-D*, KU529953; *TaNAS5-B*, KU529954; *TaNAS6-A*, KU529955; *TaNAS6-B*, KU529956; *TaNAS6-D*, KU529957; *TaNAS4-A*, KU529958; *TaNAS7-A2*, KU529959; *TaNAS7-A1*, KU529960; *TaNAS2-A*, KU529961; *TaNAS7-D*, KU529962; *TaNAS2-D1*, KU529963; *TaNAS2-D2*, KU529964; *TaNAS4-D*, KU529965; *TaNAS4-U*, KU529966.

Keywords: phytosiderophore, iron, *Triticum aestivum*, gene expression, nutrient and metal transport, ion transport.

Summary

Nicotianamine (NA) is a non-protein amino acid involved in fundamental aspects of metal uptake, transport and homeostasis in all plants and constitutes the biosynthetic precursor of mugineic acid family phytosiderophores (MAs) in graminaceous plant species. Nicotianamine synthase (NAS) genes, which encode enzymes that synthesize NA from S-adenosyl-L-methionine (SAM), are differentially regulated by iron (Fe) status in most plant species and plant genomes have been found to contain anywhere from 1 to 9 NAS genes. This study describes the identification of 21 NAS genes in the hexaploid bread wheat (*Triticum aestivum* L.) genome and their phylogenetic classification into two distinct clades. The *TaNAS* genes are highly expressed during germination, seedling growth and reproductive development. Fourteen of the clade I NAS genes were up-regulated in root tissues under conditions of Fe deficiency. Protein sequence analyses revealed the presence of endocytosis motifs in all of the wheat NAS proteins as well as chloroplast, mitochondrial and secretory transit peptide signals in four proteins. These results greatly expand our knowledge of NAS gene families in graminaceous plant species as well as the genetics underlying Fe nutrition in bread wheat.

Introduction

Nicotianamine (NA) is a non-protein amino acid found in all higher plants (Noma and Noguchi, 1976; Noma et al., 1971). Nicotianamine was originally described as the 'normalizing factor' for tomato (*Lycopersicon esculentum* L) mutant *chloronerva* (Budesinsky et al., 1980). The *chloronerva* plants displayed retarded growth and signs of iron (Fe) deficiency despite containing high concentrations of Fe and other heavy metals in vegetative tissues, symptoms which were reversed upon exogenous application of NA. These findings, combined with three-dimensional studies of NA structure, indicated that NA was crucial for *in planta* chelation, transport and homeostasis of Fe and other heavy metals (Budesinsky et al., 1980; Scholz et al., 1985). Later studies confirmed important roles for NA in short- and long-distance transport of ferrous Fe (Fe^{2+}) and other divalent metal cations such as zinc (Zn^{2+}), manganese (Mn^{2+}), copper (Cu^{2+}) and nickel (Ni^{2+}) as well as ferric iron (Fe^{3+}) (Scholz et al., 1985; Stephan and Scholz, 1993; von Wiren et al., 1999). Synthesis of NA occurs via trimerization of S-adenosyl-L-methionine (SAM) in a process catalysed by nicotianamine synthase (NAS) enzymes (Higuchi et al., 1994). Nicotianamine is

further converted to mugineic acid family phytosiderophores (MAs), such as 2'-deoxymugineic acid (DMA), specifically in graminaceous plant species that utilize Strategy II Fe uptake (Marschner and Romheld, 1994; Marschner et al., 1986). Strategy II Fe uptake has been characterized in a number of graminaceous crops including rice (*Oryza sativa* L.), barley (*Hordeum vulgare* L.), maize (*Zea mays* L.) and wheat (*Triticum aestivum* L.) and involves the release of MAs into the rhizosphere, particularly under plant Fe deficiency, for chelation of Fe^{3+} and other essential transition metals (Marschner et al., 1986). Following chelation of Fe^{3+} with DMA or other MAs, the MAs-Fe^{3+} complexes are reabsorbed through yellow stripe-like (YSL) transporters at the root surface (Curie et al., 2001). The release of MAs facilitates the growth of Strategy II plant species on calcareous soils where high pH impedes the Fe reductive approach utilized by Strategy I plants (Guerinot and Yi, 1994; Romheld et al., 1982; Scholz et al., 1992).

Studies of the NAS genes and proteins of rice, barley and maize have revealed significant differences in NAS gene number and expression level as well as NAS protein structure, length and function across the three species. Three NAS genes are present in the rice genome (hereafter referred to as *OsNAS* genes), consisting of the highly homologous *OsNAS1* and *OsNAS2* genes on chromosome Os3 and the *OsNAS3* gene on chromosome Os7 (Inoue *et al.*, 2003). The presence of 9-10 NAS genes in the maize genome (hereafter referred to as *ZmNAS* genes) located on chromosomes 1, 7 and 9 represents a gene family three times larger than that of rice (Mizuno *et al.*, 2003; Zhou *et al.*, 2013a, b). The ZmNAS proteins vary between 327 and 601 amino acid (aa) in length and phylogenetic comparison to the NAS proteins of rice (three proteins), barley (nine proteins), *Arabidopsis thaliana* (three proteins) and *Solanum lycopersicum* (one protein) identified two clades or classes of NAS proteins within the Gramineae; these are referred to as clade I and II NAS proteins in this paper (Inoue *et al.*, 2003; Mizuno *et al.*, 2003; Perovic *et al.*, 2007; Zhou *et al.*, 2013a,b). The pairs of clade I NAS genes in maize are believed to result from chromosomal block duplication (Wei *et al.*, 2007). Such duplication of NAS genes in rice and barley has also been described (Higuchi *et al.*, 2001; Perovic *et al.*, 2007). Expression of the clade I NAS genes in rice and maize is root specific, induced ubiquitously in root and shoot tissues under Fe deficiency and decreased under conditions of Fe excess; these findings suggest that clade I NAS genes function primarily in the regulation of NA and MAs biosynthesis, Fe uptake and long-distance Fe translocation. Uniquely to maize, a pair of expressed clade I NAS genes (*ZmNAS2;1/2;2*) encodes NAS proteins approximately twice as long (601 aa) as the clade I NAS proteins found in rice and barley; however, enzymatic activity of these maize NAS proteins has not been detected (Mizuno *et al.*, 2003; Zhou *et al.*, 2013b). Expression of the clade II NAS genes in rice and maize is shoot specific, decreased and/or relocalized to root tissues under Fe deficiency and induced ubiquitously in root and shoot under conditions of Fe excess. These expression patterns suggest that clade II NAS proteins do not contribute to MAs biosynthesis under Fe deficiency and are instead primarily involved in NA biosynthesis for Fe loading of vascular tissues and maintenance of cellular Fe homeostasis.

Evidence suggests that clade I and II NAS proteins are targeted to various compartments within the plant cell; a recent bioinformatics study of the ZmNAS protein sequences revealed signal peptides localizing ZmNAS5 (clade II) and ZmNAS6;1 (clade I) to the mitochondria and chloroplasts, respectively (Zhou *et al.*, 2013a). In rice, the OsNAS2 (clade I) protein contains two motifs (YXXΦ and LL) involved in vesicular transport and OsNAS2 enzyme activity was found to be localized to rice root vesicles under Fe deficiency (Nozoye *et al.*, 2014).

The presence of 9-10 NAS genes in the barley genome (hereafter referred to as *HvNAS* genes) represents a gene family similar in size to that of maize (Herbik *et al.*, 1999; Higuchi *et al.*, 1999a,b, 2001; Mizuno *et al.*, 2003; Perovic *et al.*, 2007). The *HvNAS* genes have been classified into three synteny-based groups through comparison with the rice genome (Perovic *et al.*, 2007). Group 1 consists of the *HvNAS2*, *HvNAS3* and *NASHOR1a* genes and is located on barley chromosome 4H. The *NASHOR1a* gene is orthologous and syntenic to *OsNAS1* and *OsNAS2* on rice chromosome Os3. Group 2 contains the *NASHOR2* gene which is orthologous and syntenic to *OsNAS3* on rice chromosome Os7. Group 3 consists of five *HvNAS* genes that are nonsyntenic with rice. Four of the group 3 genes are located on chromosome 6H (*HvNAS1*, *HvNAS4*, *HvNAS7*, *NASHOR1b*) while *HvNAS5* is located on chromosome 2HS; these *HvNAS* genes are considered unique to the barley genome and are thought to have arisen from ectopic *HvNAS* gene duplication events (Perovic *et al.*, 2007). In contrast to rice and maize, all *HvNAS* genes investigated to date show Fe deficiency inducible expression specifically in root tissues

with no expression under Fe sufficiency nor in shoot tissues under any Fe status (Higuchi *et al.*, 2001). The HvNAS proteins vary between 267 and 340 aa in length and little is known about their cellular localization.

While the NAS gene families of rice, maize and barley have been described in publications for over 10 years, there is only one publication to date (Pearce *et al.*, 2014) identifying two genes belonging to the NAS gene family of bread wheat (*Triticum aestivum*), hereafter referred to as the *TaNAS* gene family. The paucity of information about *TaNAS* genes is surprising in the light of recent publications identifying NA and/or DMA as the predominant chelators of Fe in wheat white flour (Eagling *et al.*, 2014a,b). In this paper, we describe the identification and characterization of a large *TaNAS* gene family in bread wheat using cv. Chinese Spring—the reference genome of the International Wheat Genome Sequence Consortium—and high-yielding Australian cv. Gladius as our plant materials.

Results

Identification of 21 *TaNAS* genes located on multiple bread wheat chromosomes and subgenomes

We identified a total of 21 unique *TaNAS* genes and determined the genomic location of 19 of these genes on chromosome groups 2, 3, 4, 5 and 6 (Figure 1, Table 1). Several sets of homeologous *TaNAS* genes were identified on chromosome groups 2, 4 and 6 (Figure 1, Table S5). Chromosome group 2 contains three homeologous *TaNAS* genes located on the A, B and D subgenomes (*TaNAS9-A/TaNAS9-B/TaNAS9-D*) and two pairs of homeologous *TaNAS* genes located only on the A and B subgenomes (*TaNAS3-A/TaNAS3-B* and *TaNAS1-A/TaNAS1-B*). Chromosome group 4 contains three homeologous *TaNAS* genes located on the A, B and D subgenomes (*TaNAS6-A/TaNAS6-B/ TaNAS6-D*). The homeologous *TaNAS4-D* and *TaNAS4-A* genes on chromosome groups 4 and 5 were likely separated by the known 4AL-5AL wheat chromosome translocation (Hernandez *et al.*, 2012). Chromosome group 6 contains three homeologous *TaNAS* genes located on the A and D subgenomes (*TaNAS2-A/ TaNAS2-D1/TaNAS2-D2*) and the two 6DS genes share 98.5% genomic sequence identity. Chromosome group 6 also contains an additional three homeologous *TaNAS* genes located on the A and D subgenomes (*TaNAS7-A1/TaNAS7-A2/TaNAS7-D*) and the 6AS and 6DL genes share 96.2% genomic sequence identity. We identified one *TaNAS* gene on chromosome group 3 without any homeologues (*TaNAS5-B*).

The genomic location of two *TaNAS* genes (*TaNAS4-U* and *TaNAS8-U*) could not be determined based on currently available wheat genome sequence. The genomic sequence of *TaNAS4-U* shares over 95% nucleic acid identity with the *TaNAS4-A* and *TaNAS4-D* genes and most likely represents the homeologous gene located on subgenome B. The *TaNAS4-U* gene could only be amplified from cv. Gladius DNA, suggesting that the gene is present in only one of the investigated wheat cultivars or has significant polymorphisms between the two wheat cultivars. The partial (168 bp) sequence obtained for the *TaNAS8-U* gene shares 94.6% nucleic acid identity with the two homeologous *TaNAS3-A* and *TaNAS3-B* genes located on wheat chromosome group 2 and likely represents the homeologous gene *TaNAS3-D* located on subgenome D. Overall, the 20 full-length *TaNAS* genes identified in this study share medium to high identity in terms of nucleic acid sequence (67.8%–98.5%) as well as medium to high similarity in protein sequence

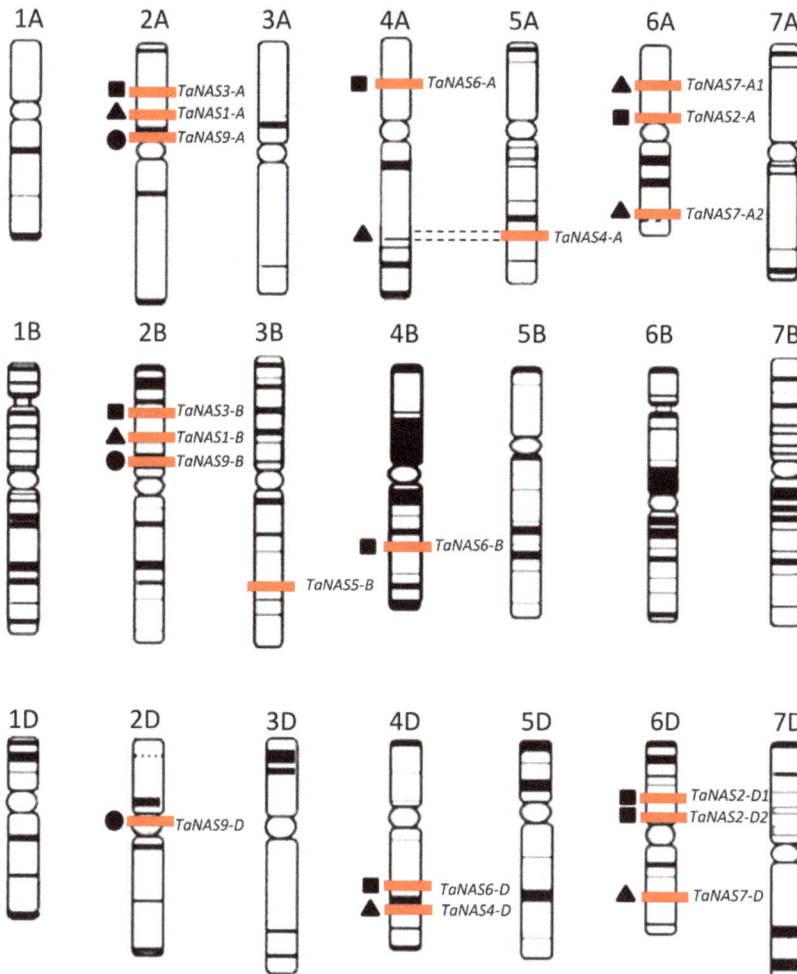

Figure 1 Distribution of 19 *TaNAS* genes across the seven chromosome groups and A, B and D subgenomes of bread wheat. Genes are represented by red bars and homeologous genes within chromosome groups, when present, are indicated by common symbol (■▲●). Dotted lines represent the 4AL-5AL chromosomal translocation. Figure adapted from the IWGSC website (http://www.wheatgenome.org/).

(57.8%–99.4%) based on chemical reactivity of amino acid side chains (Figure 2).

The *TaNAS* genes separate into two distinct clades and show high similarity to *HvNAS* genes

Our phylogenetic analyses identified two distinct clades of *TaNAS* genes that share most similarity with barley *HvNAS* genes (Figure 3). Clade I contains 17 of the *TaNAS* genes as well as the previously described clade I NAS genes of rice, maize and barley; clade I further separates into three subgroups with bootstrap proportions of <75%. Phylogenetic analyses revealed that the *TaNAS* and *HvNAS* genes within each clade I subgroup are most closely related with bootstrap proportions of 99.6%–100%. Clade I subgroup 1 contains two *TaNAS* genes from wheat chromosome groups 2A and 2B (*TaNAS1-A* and *TaNAS1-B*) and the orthologous barley gene *HvNAS5-1/2* located on chromosome 2HS. Clade I subgroup 2 contains three *TaNAS* genes from wheat chromosome groups 6A and 6D (*TaNAS2-A, TaNAS2-D1* and *TaNAS2-D2*) and the orthologous barley gene *HvNAS1* located on chromosome 6HS. Clade I subgroup 3 is the largest of the subgroups and comprises 12 *TaNAS* genes located on various wheat chromosome groups and six orthologous *HvNAS* genes. Clade II is a much smaller NAS gene clade with no subgroups and contains three homeologous *TaNAS* genes located on wheat chromosome group 2 (*TaNAS9-A/TaNAS9-B/TaNAS9-D*) as well as the previously described clade II NAS genes of rice, maize and barley.

The TaNAS proteins show variation in the length of N- and C-terminal regions and are targeted to several subcellular locations

Alignment of the protein sequences encoded by the 20 full-length *TaNAS* genes showed that TaNAS proteins vary in length from 287 to 385 amino acids and that several regions of the proteins are highly conserved (Figure 2). Two of the proteins have uncharacteristically long N-terminal regions that contain either mitochondrial targeting or chloroplast transit peptides (TaNAS5-B and TaNAS6-D), two others have uncharacteristically short C-terminal regions (TaNAS1-A and TaNAS1-B), while one protein has both an uncharacteristically long N-terminal containing a chloroplast transit peptide and a short C-terminal region (TaNAS6-A). The TaNAS2-A protein has an N-terminal secretory pathway signal peptide (Table S4). All of the TaNAS proteins contain YXXΦ (Φ = amino acids with bulky hydrophobic residues) and di-leucine (LL) motifs in the N-terminal region that have previously been described as characteristic features of NAS superfamily proteins in plants and indicate that they are localized to cytoplasmic vesicles (Figure 2).

Clade I and II *TaNAS* genes are expressed during germination, seeding growth and reproductive development

Eighteen of the 21 *TaNAS* genes were expressed in young root tissues, seedling leaf and/or developing reproductive organs of cv.

Table 1 Identification of 21 *TaNAS* genes in bread wheat. Each *TaNAS* gene was assigned a unique name (first column) based on homeologous grouping and subgenome A, B or D (U indicates unknown). An additional number was added to the gene name when more than one NAS gene belonging to the same homeologous group was located on the same chromosome. The third and fourth columns provide the IWGSC contig (https://urgi.versailles.inra.fr/blast/blast.php) on which the gene was identified and the MIPS database gene annotation (http://pgsb.helmholtz-muenchen.de/plant/index.jsp) for each of the *TaNAS* genes. The sequences of PCR primers used to amplify each gene from genomic DNA, as well as the expected PCR product sizes (bp), are provided in columns 5–7.

Gene name	Chromosome location	IWGSC contig	MIPS database gene annotation	Forward primer sequence 5′ to 3′	Reverse primer sequence 5′ to 3′	PCR product length
TaNAS1-A	2AS	ctg5260464	Traes_2AS_1FB8F6760.1	TCAAGCATCAGCTCCATCAT	AGGCTCCCCCTTTTCTTCA	973
		ctg5186953	Traes_2AS_D50EEDA84.1	–		
TaNAS1-B	2BS	ctg5223571	Traes_2BS_A984C71A6.1	TCAAGCATCAGCTCCATCAA	ATGTGCGGGGAATCATCTAC	1026
TaNAS2-A	6AS	ctg4374229	Traes_6AS_15ABF3B3E1.1	ATCAGCACGCACATTTTCAA	GGTCACCAAGAAGCAACGAT	1025
		ctg4418337	Traes_6AS_89FA598FB.1	–		–
TaNAS2-D1	6DS	ctg2123013	Traes_6DS_1965FC73E.1	CACATTTTCTCCTGCTTCCT	ATGATCGGCACGCACATTTCG	1067
TaNAS2-D2	6DS	ctg2092960	Traes_6DS_92693EC0A.1	GAGATGATCAGCACGCACAT	CGGTCACCAAGAAGCAATTA	1041
TaNAS3-A	2AS	ctg5220125	Traes_2AS_F55243E0C.1	AGCTCGATCAACCACTCTCC	GCACATGCACAACCACTACC	1089
TaNAS3-B	2BS	ctg5244458	Traes_2BS_98815CE06.1	AGCTCCATCAACCACTCTCT	GCACATGCACAACCACTACT	1091
TaNAS4-A	5AL	ctg624885	Traes_5AL_8715BE19D.1	GCCTGCACTGAGGTACCAAC	GATCAAGCAGCGAGGTAGGA	1099
TaNAS4-U	U	–	–	CATGGATGCACACCTTTTGT	CTTGGCAACGATCGATCAGAAGA	1056
TaNAS4-D	4DL	ctg14334745	Traes_4DL_4E10A6DFB.1	CATGGATGCACACCTTTTGT	CAATTACCACGTGTGGTTGC	1220
TaNAS5-B	3BU	ctg10414150	Traes_3B_3AC6B469E.1	CGAGCCTTATTGGGAGTGTC	GACCACACATGCACACGTTC	1205
		ctg10751094	Traes_3B_A52E3C298.1	–	–	–
TaNAS6-A	4AS	ctg6009291	Traes_4AS_34072BFC9.1	GGCATGAAACCAGCCATATT	AGCAGCCACGTATGGATGAG	2392
TaNAS6-B	4BL	ctg6980195	Traes_4BL_BB0FC3BD3.1	TGGTGCACGCTCTTTATTCG	ACGTACGGACGATGACCAC	1489
TaNAS6-D	4DL	ctg14432849	Traes_4DL_AF0869DDB.1	TGGTGCAGGCTCTTTATTCA	ACGTACTGATGATGACCGC	1361
TaNAS7-A1	6AS	ctg4338157	Traes_6AS_5D0B8CF891.1	GGTACCAAGGCAAGAACACAC	GATGATGACCTCGCAGTCG	1043
TaNAS7-A2	6AL	ctg5810064	Traes_6AL_F4AEC0314.1	TTCCATAGCTCATCAAGCAA	AACTCCTCTCTCTTCTGGGTCA	1070
TaNAS7-D	6DL	ctg3268810	Traes_6DL_1470C4162.1	CATTAAAATGGACGCCCAGA	GCACGGATGATGACCTCAC	1024
TaNAS8-U	U	–	Traes_XX_432D1C804.1	–	–	
TaNAS9-A	2AS	ctg5303755	Traes_2AS_DEDC612AE.1	AGCTAGCTAGTGCCCTCTGC	CATGTATGTATCGGTCGGTGA	1221
		ctg5195617	Traes_2AS_452FED53F.1	–		
TaNAS9-B	2BS	ctg5246738	Traes_2BS_CB79BAFB1.1	CTCACTCTCAGAGCCCCTCA	TCCATGCATGAGGACAAACG	1263
TaNAS9-D	2DS	ctg5343018	Traes_2DS_FE40FC64C.1	CAATTAGCAGCTGATCTGTCG	TCCATGCATGAGGACAAGC	1182
		ctg2341167	–	–	–	

Chinese Spring (Figure 4). In most cases, homeologous *TaNAS* genes displayed similar expression profiles; however, in some instances specific homeologous genes were either more tissue specific or higher in expression relative to the corresponding homeologues. The homeologous *TaNAS3-A*/*TaNAS3-B* genes as well as the *TaNAS7-A1*, *TaNAS7-D* and *TaNAS2-D2* genes were most highly expressed in the embryonic radicle 2 days after sowing (DAS) and seedling root 10-12 DAS (Figures 4a-e). The *TaNAS2-A* gene (Figure 4f), a homeologous gene to *TaNAS2-D2*, was not only highly expressed in young root tissues but also in seedling leaf and reproductive organs with highest expression in the caryopsis tissue 3–5 days after pollination (DAP). The *TaNAS7-A2* gene, a homeologous gene to *TaNAS7-D*, displayed highly anther-specific expression with approximately 100-fold higher expression in anther tissue relative to all other organs (Figure 4g). The homeologous *TaNAS4-D*/*TaNAS4-A* genes and the homeologous clade II *TaNAS9-A*/*TaNAS9-B*/*TaNAS9-D* genes were all most highly expressed in leaf tissues (Figure 4h-l). Expression of the subgenome A *TaNAS9-A* gene was 10-fold higher than the corresponding subgenome B and D homeologues (*TaNAS9-B* and *TaNAS9-D*). The *TaNAS5-B* gene as well as the homeologous *TaNAS6-A*/*TaNAS6-B*/*TaNAS6-D* genes had generally low expression levels with most expression detected in the caryopsis tissue 3–5 DAP (Figure 4l-p). Expression of the subgenome D *TaNAS6-D* gene was threefold higher than the corresponding subgenome A and B homeologues (*TaNAS6-A* and *TaNAS6-B*). The *TaNAS1-A* and *TaNAS8-U* genes had the most ubiquitous expression profiles of the *TaNAS* genes with expression detected in most tissue types and developmental stages (Figure 4q-r). No expression of the *TaNAS1-B*, *TaNAS2-D1* and *TaNAS4-U* genes was detected in cv. Chinese Spring.

Clade I *TaNAS* genes are expressed in root tissues and up-regulated under iron deficiency during vegetative growth

The analysis of *TaNAS* gene expression in the root and shoot of 4-week-old cv. Gladius plants found that nearly all of the clade I *TaNAS* genes were expressed in root tissues of the vegetatively growing bread wheat plants under conditions of Fe sufficiency and deficiency (Figure 5). Under Fe-sufficient growth conditions (control treatment), 14 of the clade I *TaNAS* genes were expressed at low levels in root tissues until day 5 of the treatment period and then moderately expressed from days 5–7 (dashed lines in Figure 5a-n). Increased levels of *TaNAS* gene expression towards the end of the control treatment period may reflect increased demands for NA- and/or DMA-mediated Fe transport as

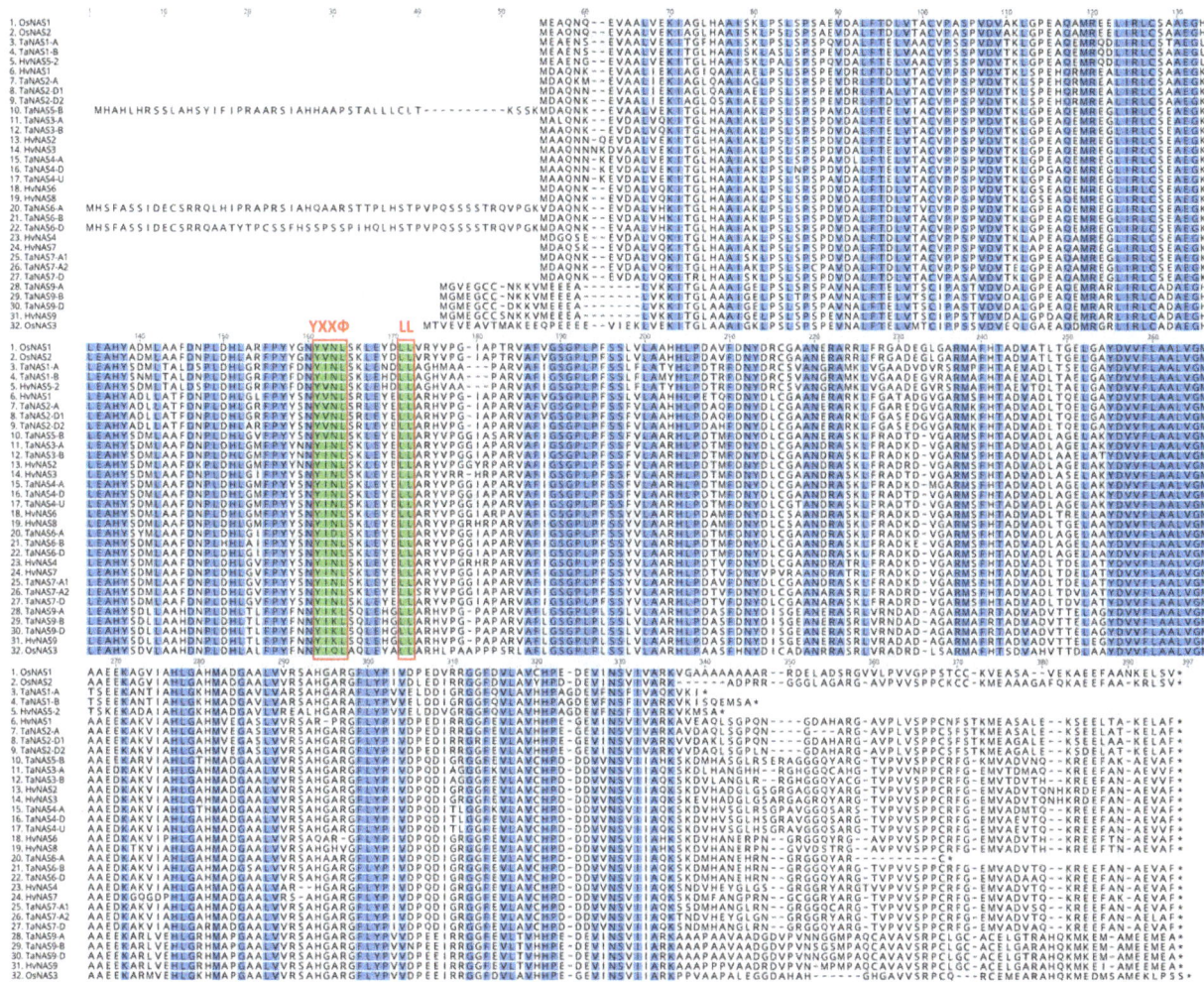

Figure 2 Amino acid sequence alignment of the NAS proteins from wheat (TaNAS), rice (OsNAS) and barley (HvNAS) identifies conserved regions and motifs. Blue shading corresponds to 100% conservation of amino acids between species. The YXXΦ and LL motifs of the NAS superfamily protein are shaded in green and outlined in red (Y = tyrosine; X = any amino acid residue; Φ = amino acids with bulky hydrophobic residues; LL = di-leucine).

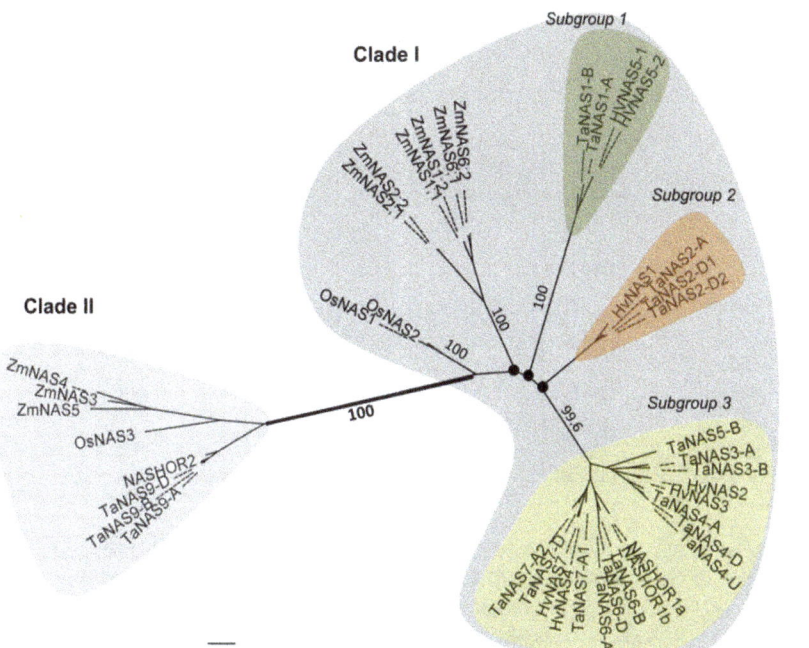

Figure 3 Phylogenetic relationship of the wheat, rice, barley and maize NAS genes. An unrooted tree of NAS gene coding sequences from wheat (*TaNAS*), rice (*OsNAS*), barley (*HvNAS*) and maize (*ZmNAS*) identifies the clade I and II NAS genes of these species. Clade I separates into three subgroups comprising only wheat and barley genes. Black nodes (●) represent weak bootstrap values (<75%). The scale bar corresponds to branch length and longer branches correspond to greater numbers of nucleic acid polymorphisms along the sequence.

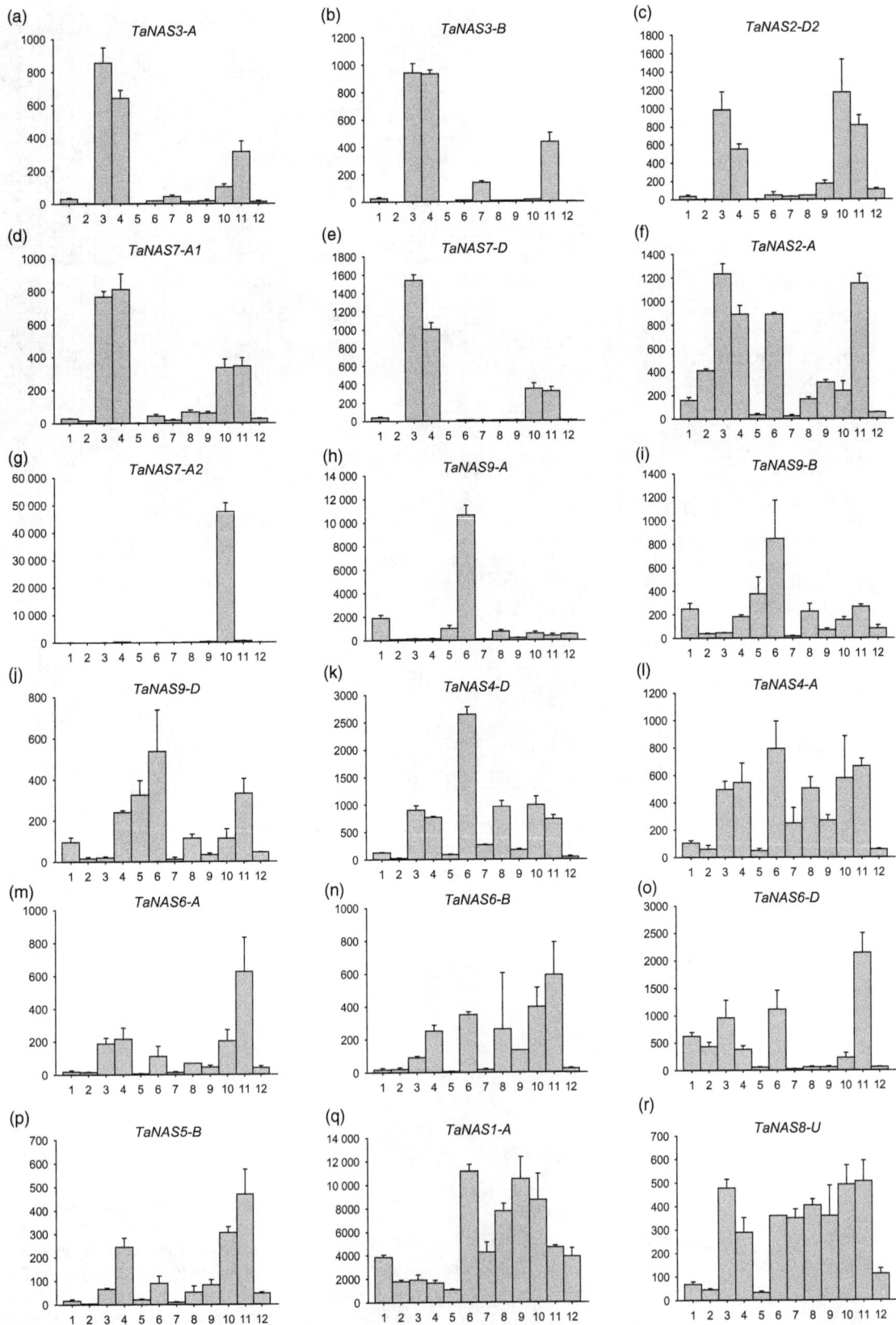

Figure 4 Relative expression of 18 *TaNAS* genes in 12 different tissues and developmental stages of bread wheat cv. Chinese Spring. Relative expression of *TaNAS* genes is provided for the following: (1) embryo 1 DAS; (2) coleoptile 2 DAS; (3) radicle 2 DAS; (4) seedling root 10-12 DAS; (5) seedling crown 10–12 DAS; (6) seedling leaf 10–12 DAS; (7) immature inflorescence; (8) bracts; (9) pistil; (10) anthers; (11) caryopsis 3-5 DAP; (12) embryo 22 DAP. Units on the y-axis indicate normalized mRNA copies per µg of total RNA. Error bars indicate SEM of three technical replicates derived from one bulked biological replicate.

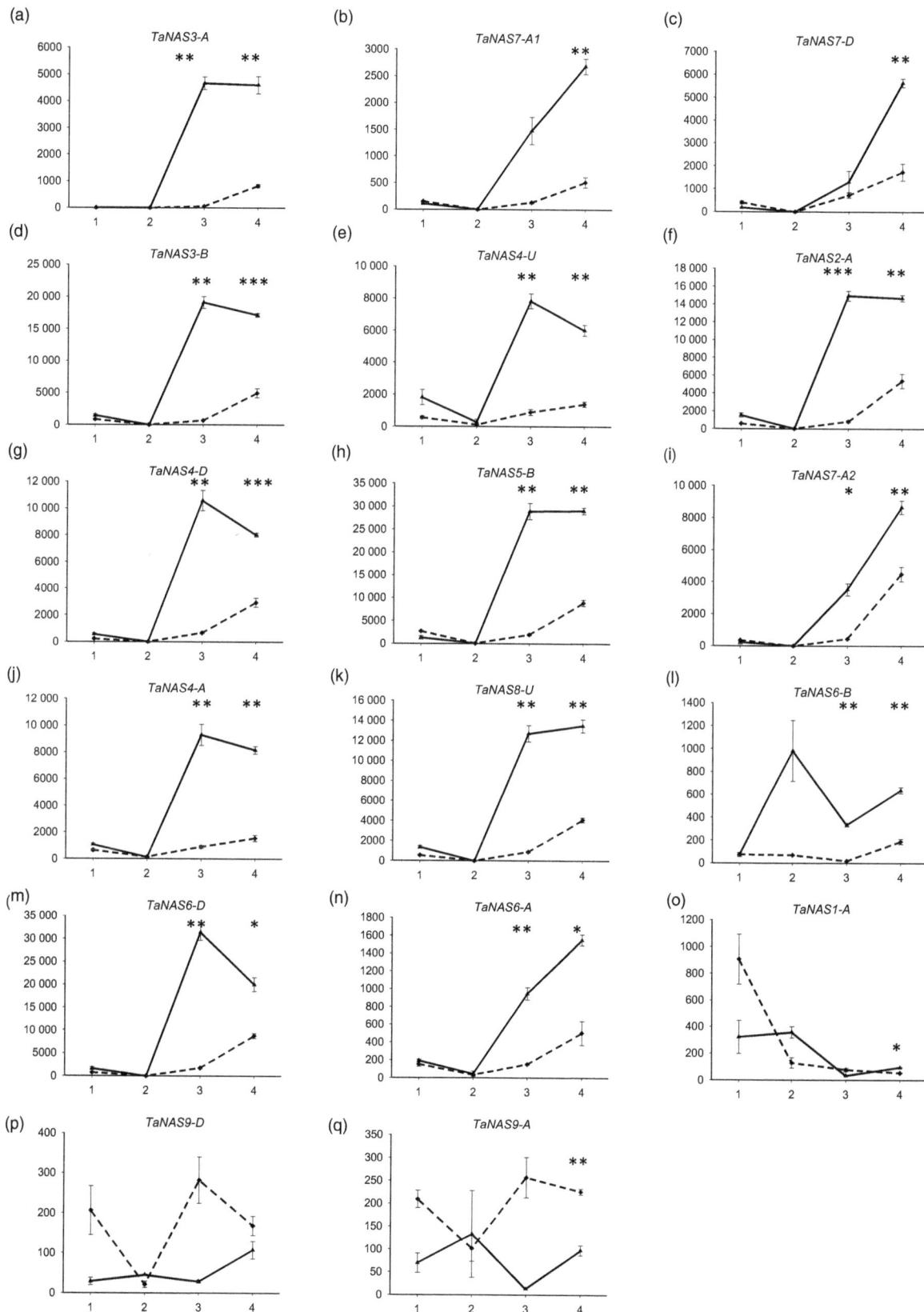

Figure 5 Relative expression of 17 *TaNAS* genes in root tissues of bread wheat cv. Gladius under iron-sufficient/deficient conditions. Gene expression in plants grown under Fe-sufficient (dashed line) or Fe-deficient (solid line) conditions is presented at four time points of the 7 day treatment: day 0 (1), day 1 (2), day 5 (3) and day 7 (4). Units on the *y*-axis indicate normalized mRNA copies per μg of total RNA. Error bars indicate SEM of three biological replicates for time points 1, 3 and 4 and two biological replicates for time point 2. Asterisks indicate statistically significant differences for the effect of condition (+Fe and −Fe) at each time point (one-way ANOVA, Tukey's test,*P value ≤0.05; **P value ≤0.01; ***P value ≤0.001).

wheat plants begin to rapidly grow and tiller at the 4-week growth stage. Under Fe-deficient growth conditions, expression of the 14 clade I *TaNAS* genes was significantly up-regulated in root tissues with approximately 15-fold higher expression at day 5 and threefold higher expression at day 7 of the treatment relative to controls (solid lines in Figure 5a-n). Three additional *TaNAS* genes, including the clade II *TaNAS9-A* and *TaNAS9-D* genes, were differentially expressed under Fe deficiency in the root; however, their expression levels were significantly lower than that of the previously mentioned 14 genes and no clear expression patterns were identified (Figure 5o-q).

Only seven *TaNAS* genes, including the three homeologous clade II *TaNAS* genes located on wheat chromosome group 2 (*TaNAS9-A/TaNAS9-B/TaNAS9-D*), were expressed at low levels in cv. Gladius shoot tissues (Figure S3). The average level of *TaNAS* gene expression under both Fe-sufficient and Fe-deficient growth conditions in shoots was 30-fold lower than that of roots and bordered on the detection limit for qRT-PCR. Unlike the expression patterns observed in roots, Fe-deficient growth conditions tended to decrease the already low *TaNAS* gene shoot expression levels. No expression of the *TaNAS1-B*, *TaNAS2-D1* and *TaNAS2-D2* genes was detected in cv. Gladius.

Discussion

The 21 *TaNAS* genes located on the A, B and D subgenomes of bread wheat represent the largest plant NAS gene family identified to date. Most members of this large gene family are transcriptionally active at multiple stages of bread wheat growth and development as well as under conditions of Fe deficiency. The results obtained from our molecular characterization of the *TaNAS* gene family allow several conclusions to be made about the origin of *TaNAS* genes as well as their function within the wheat plant.

Many clade I and II *TaNAS* genes share a common origin with *HvNAS* genes

Our study identified two distinct clades of *TaNAS* genes, a result that is consistent with previous studies of maize, barley and rice NAS genes (Perovic et al., 2007; Zhou et al., 2013b). The separation into the two clades is supported by high bootstrap values and branch lengths. Seventeen of the 20 full-length *TaNAS* genes belong to clade I, and in our phylogenetic analyses, these genes clustered as three subgroups comprising only wheat and barley NAS genes. This suggests that each *Triticeae* subgroup is derived from distinct ancestors.

Clade I subgroups 1 and 2 contain two groups of homeologous *TaNAS* genes—the *TaNAS1-A/TaNAS1-B* genes located on chromosome group 2 and the *TaNAS2-A/TaNAS2-D1/TaNAS2-D2* genes located on chromosome group 6—which are syntenic with the duplicated *HvNAS5-1/2* and *HvNAS1* genes located on barley chromosomes 2 and 6, respectively (Perovic et al., 2007). It was originally hypothesized that a barley-specific NAS gene duplication event gave rise to the *HvNAS5-1* and *HvNAS1* genes; however, our results indicate that this gene duplication event likely occurred in a diploid wheat–barley ancestor.

Clade I subgroup 3 contains 12 *TaNAS* and 6 *HvNAS* genes and our results suggest that there is macro-synteny of NAS genes between wheat and barley in this subgroup. In addition, our results indicate that intra- and interchromosomal NAS gene duplication events gave rise to this large subgroup of NAS genes in both wheat and barley. The homeologous *TaNAS4-D/TaNAS4-*

A genes located on wheat chromosome groups 4 and 5 and the homeologous *TaNAS7-A2/TaNAS7-D* genes located on wheat chromosome group 6 are orthologous and syntenic to the *HvNAS2/HvNAS3* and *HvNAS4/HvNAS7* intrachromosomal gene duplication pairs located on barley chromosomes 4HL and 6HL, respectively (Perovic et al., 2007). The homeologous *TaNAS6-A/TaNAS6-B/TaNAS6-D* genes located on wheat chromosome group 4 are orthologous and syntenic to the *NASHOR1a* gene located on barley chromosome 4HL. The putatively homeologous *TaNAS3-A/TaNAS3-B/TaNAS8-U* genes are not orthologous nor syntenic to any of the barley NAS genes, indicating that wheat-specific NAS gene duplication events most likely occurred in a diploid wheat ancestor to give rise to these genes. Similar wheat-specific gene duplication events and/or chromosomal translocations are also likely responsible for origin of the *TaNAS7-A1*, *TaNAS2-D1*, *TaNAS2-D2*, *TaNAS4-U* and *TaNAS5-B* genes. Interestingly, the *TaNAS4-U* and *TaNAS5-B* genes of clade I subgroup 3 contain significant differences between cvs. Gladius and Chinese Spring, suggesting that some genes within this subgroup are evolving rapidly. The *TaNAS4-U* gene may be specific to cv. Gladius or have significant polymorphisms in cv. Chinese Spring as it could not be amplified from genomic DNA nor cDNA of cv. Chinese Spring. The *TaNAS5-B* gene, which has no homeologues on the A and D subgenomes, is functional in cv. Gladius yet contains a SNP (G→C) in the cv. Chinese Spring coding sequence that creates a premature stop codon (Figure S4).

Clade II contains the homeologous *TaNAS9-A/TaNAS9-B/TaNAS9-D* genes that are orthologous and syntenic to the barley *NASHOR2* gene, orthologous to the rice *OsNAS3* gene and orthologous to the maize *ZmNAS3*, *ZmNAS4* and *ZmNAS5* genes (Figure 3). In contrast to the clade I *TaNAS* genes, there is no evidence for gene duplication, translocation or deletion in the origin of the clade II *TaNAS* genes.

The *TaNAS* gene expression profiles indicate key roles for nicotianamine in bread wheat germination, seedling growth, reproductive development and iron deficiency response

The finding that several *TaNAS* genes are expressed during seed germination and seedling growth (namely in the embryonic radicle, seedling root and seedling shoot) indicates a requirement for NA biosynthesis at early stages of bread wheat growth and development. The rice *OsNAS* genes are also expressed at early stages of rice seed germination (Nozoye et al., 2007). We can hypothesize that NA functions as an important chelator of divalent metal cations such as Fe^{2+} and Zn^{2+} during wheat germination and seedling growth and helps to redistribute these metals at the tissue/cellular level and maintain metal homeostasis. Nicotianamine is also the biosynthetic precursor to mugineic acid family phytosiderophores such as DMA that wheat plants secrete from root tissues in order to absorb rhizospheric Fe^{3+} and it is likely that a significant proportion of wheat seedling root NA is diverted towards DMA biosynthesis. In general, the clade I *TaNAS* genes appear to play a major role in root-specific NA production at the germination and seedling stages while the clade II *TaNAS* genes contribute more towards shoot-specific NA production at these stages. Similar expression trends have been observed for the clade I and II NAS genes of rice and maize (Inoue et al., 2003; Zhou et al., 2013b).

Expression of several *TaNAS* genes in tissues of the bread wheat inflorescence, particularly the anther, and in developing grain indicates that NA is essential for bread wheat reproduction and complements the results of biochemical studies that have identified NA and/or DMA as predominant chelators of Fe in bread wheat white flour (Eagling et al., 2014a,b; Walker and Waters, 2011). Anthers are known to be major sink tissues for Fe in Arabidopsis and high expression of several *TaNAS* genes in the wheat anther, in particular the anther-specific *TaNAS7-A2* gene, suggests that wheat anthers are major Fe sinks and that NA is required to chelate Fe and protect against oxidative damage in this reproductive organ (Chu et al., 2010; Roschzttardtz et al., 2013; Takahashi et al., 2003). The finding that several *TaNAS* genes are expressed in the caryopsis 3-5 DAP and the embryo 22 DAP likely reflects a requirement for NA to mobilize Fe and other divalent metals to the developing embryo. Expression of the rice *OsNAS1* and *OsNAS2* genes has also been detected in developing rice seed and the embryo (Dash et al., 2012; Fujita et al., 2010; Jain et al., 2007).

Nearly all of the clade I *TaNAS* genes were expressed at low to moderate levels in the roots of 4-week-old cv. Gladius plants grown under Fe sufficiency. These low levels of root-specific *TaNAS* expression presumably provide sufficient NA to enable DMA biosynthesis as well as root to shoot translocation of Fe under sufficient conditions. Expression of the clade I *TaNAS* genes was significantly up-regulated in root tissues under Fe deficiency and this result is consistent with the Fe deficiency response of barley where all of the *HvNAS* genes are significantly up-regulated in root tissues in response to Fe deficiency (Higuchi et al., 2001). Increased expression of the clade I *TaNAS* genes in root tissues should result in greater quantities of NA (and DMA) to increase Fe uptake from the rhizosphere as well as Fe mobilization within the plant. The Fe deficiency response of rice differs somewhat from that of barley in that the rice clade I *OsNAS1* and *OsNAS2* genes are up-regulated in both root and leaf tissues under Fe deficiency (Inoue et al., 2003). The lack of significant *TaNAS* gene expression in the shoot of vegetatively growing wheat plants under Fe deficiency indicates that the wheat Fe deficiency response is more similar to that of barley than rice and involves root-specific induction of clade I *TaNAS* genes.

Our finding that, in some cases, *TaNAS* genes located on one subgenome were more highly expressed than corresponding homeologous genes on other subgenomes is a common feature of gene expression in hexaploid bread wheat. Genetic studies have found that approximately 55% of bread wheat genes in the three subgenomes display 'unbalanced' expression where at least one homeologous gene dominates. Silencing or loss of homeologous genes also frequently occurs in the bread wheat genome (Borg et al., 2012; Leach et al., 2014; Nussbaumer et al., 2015).

The TaNAS proteins localize to cytoplasmic vesicles, chloroplasts and mitochondria within bread wheat cells and show variation in the length of N- and C-terminal regions

The 20 full-length TaNAS proteins identified in this study contain di-leucine (LL) and YXXΦ endocytosis motifs in the N-terminal region as previously found in rice and barley proteins which suggests that the TaNAS proteins are enzymatically active and localized to cytoplasmic vesicles where NA biosynthesis is likely to take place (Figure 2, Geldner and Robatzek, 2008; Robert et al., 2005; Zuo et al., 2000). Rice OsNAS proteins also contain these motifs and substitution mutations in either motif of the OsNAS2 protein disrupt MAs vesicle formation and movement in rice root cells (Nozoye et al., 2014).

Three of the clade I TaNAS proteins have uncharacteristically long N-terminal regions that were determined by *in silico* analyses to contain peptides for mitochondrial targeting (TaNAS5-B) or chloroplast transit (TaNAS6-A and TaNAS6-D). We identified an EST corresponding to the long N-terminal region of the *TaNAS6-D* gene (GenBank identifier CA637385), indicating that these long N-terminal regions are transcribed and that specific clade I TaNAS proteins are targeted to mitochondria and chloroplasts to facilitate NA biosynthesis and maintenance of Fe homeostasis within these organelles. Furthermore, the presence of a secretory signal peptide in the N-terminal region of the TaNAS2-A protein suggests that certain TaNAS proteins may be secreted from the cell to enable NA biosynthesis in other parts of the plant. Similar subcellular targeting of NAS proteins has been suggested in maize, where four of the nine ZmNAS proteins contain peptides for secretion or targeting to chloroplasts and mitochondria (Zhou et al., 2013a). The function of the short C-terminal regions of the TaNAS1-A and TaNAS1-B proteins is currently unknown; however, a similar short C-terminal region is present in the orthologous HvNAS5-1 protein of barley (Figure 2).

Applications towards plant breeding

The 21 *TaNAS* genes described in this study represent a valuable resource for plant breeders working to improve the growth, yield and nutritional value of bread wheat, a major food staple that provides 20% of the calories and protein consumed by humans worldwide. Iron toxicity often occurs when wheat plants are cultivated under waterlogged conditions, while Fe deficiency can occur in plants grown on calcareous soil. These abiotic stresses frequently lead to early leaf senescence and negatively impact on grain yield and quality. Altered expression of specific *TaNAS* genes could either increase or suppress the Fe deficiency response and thereby facilitate the development of bread wheat cultivars adapted to Fe-rich or Fe-poor growth conditions. Furthermore, increased activity of *TaNAS* genes with expression within the developing grain could lead to the production of more nutritious wheat varieties containing enhanced concentrations of bioavailable grain Fe.

Experimental procedures

Bread wheat cv. Gladius germination and hydroponic growth conditions

Seeds of cv. Gladius were germinated on moist filter paper for 5 days. Seedlings were transferred to 20-L tubs (30 seedlings per tub) containing hydroponic growth solution as described in Genc et al. (2007) and grown in a University of Melbourne glasshouse (Parkville, Vic., Australia) for 4 weeks over January 2014 with average daily temperatures of 24.6 °C and 58.8% humidity. Hydroponic solutions were replaced every 2 days and pH was maintained between 6.5 and 7. Iron deficiency treatments commenced when plants were 3 weeks old; one tub of seedlings was grown for seven additional days in hydroponic solution lacking Fe, whereas a control tub of seedlings was grown for seven additional days in hydroponic solution with Fe. Three plants were individually harvested from the Fe-deficient and control tubs at four time points—days 0 (experiment start), 1, 5 and 7—and the root and shoot tissues were snap-frozen in liquid nitrogen. Prior to harvest, shoot weight and length and leaf SPAD value (Chlorophyll Meter SPAD-502Plus, Konica Minolta, Europe) were recorded for each plant. Plants grown under Fe deficiency developed leaf Fe deficiency symptoms as indicated by significantly reduced chlorophyll content of the fourth fully expanded leaf on day 5 ($P < 0.05$) and the fifth fully expanded leaf on day 7 ($P < 0.001$) of the treatment (Figure S1).

Identification and naming of the *TaNAS* genes

The *HvNAS* coding sequences (Table S1) were used as queries in BLAST searches against the 41 genomic databases of the International Wheat Genome Sequencing Consortium (IWGSC, 2014—http://www.wheatgenome.org) to identify sequences encoding putative *TaNAS* genes. Matching sequences from the IWGSC portal were annotated using the TriAnnot v1.4 (http://wheat-urgi.versailles.inra.fr) and FGENESH (http://www.soft-berry.com/) pipelines (Solovyev *et al.*, 2006). In cases where a full gene sequence was not available from IWGSC, we further searched the *Triticum aestivum* cv. Chinese Spring 5× genome database (http://www.cerealsdb.uk.net). Gene sequences were merged and assembled using Geneious Pro 5.6.6 (Biomatters, http://www.geneious.com/). Each *TaNAS* gene was assigned a unique name based on homeologous grouping and subgenome and followed the recommended rules for gene symbolization in wheat (http://wheat.pw.usda.gov/ggpages/wgc/98/Intro.htm, Table 1). Gene names *TaNAS1-A* and *TaNAS9-A/TaNAS9-B* correspond to the *TaNAS1* and *TaNAS3* genes, respectively, described in Pearce *et al.* (2014). Homeologous *TaNAS* genes were identified on the basis of their chromosome group, arm location and percentage of nucleic acid identity with a 94% identity threshold level for homeologous genes. The classification of homeologous genes also took into account the known 4AL-5AL wheat chromosome translocation (Hernandez *et al.*, 2012) and the pericentromeric inversion of wheat chromosome 4AS-4AL (Liu *et al.*, 1992).

Gene prediction models including predicted CDS and protein sequence were obtained for each of the identified *TaNAS* genes using FGENESH software (http://www.softberry.com). CDS

alignments (CLUSTALW) and protein alignments based on the degree of amino acid conservation (CLUSTALW—cost matrix BLOSUM62, threshold 1) were performed in Geneious Pro 5.6.6. The TaNAS protein sequences were analysed for the presence of target peptides using the recommended protocol for TargetP 1.1 as described by Emanuelsson *et al.* (2007). Validation of the *TaNAS* genes was obtained by BLASTN searches of the 19 *TaNAS* coding sequences to the recently released TGACv1 genome assembly (http://pre.plants.ensembl.org/Triticum_aestivum/Info/Index), all of which returned a single TGAC scaffold with identity equal to or higher than 99%.

Gene sequencing

Subgenome-specific PCR primers for each *TaNAS* gene were designed using the respective IWGSC contig sequences and Primer3 software (Table 1; Untergasser *et al.*, 2012); the primer pairs were used to amplify fragments from cvs. Gladius and Chinese Spring genomic DNA. The subgenome specificity of each primer pair was verified using cv. Chinese Spring nulli-tetrasomic DNA (Sears, 1954). PCR amplification cycles consisted of 1 cycle = 1 min 95 °C; 35 cycles = 20 s 95 °C, 30 s 61–62 °C, 1 min 72 °C; 1 cycle = 5 min 72 °C. PCRs were performed in a final volume of 30 μL according to manufacturer instructions for MyTaq™ HS DNA polymerase (Bioline, Boston, MA, USA). Amplification products were sequenced by Sanger sequencing at the Australian Genome Research Facility Ltd (http://www.agr-f.org.au/). Most primer pairs amplified expected PCR products from both cvs. Gladius and Chinese Spring except for those amplifying the *TaNAS3-A*, *TaNAS1-A*, *TaNAS7-A2*, *TaNAS7-D* and *TaNAS4-D* genes; in these cases, the primer pairs amplified the PCR products from cv. Chinese Spring only. In the case of the *TaNAS4-U* gene, the primer pairs amplified a PCR product from cv. Gladius only.

Table 2 Quantitative reverse transcription PCR (qRT-PCR) analysis of the 21 *TaNAS* genes. The table provides gene name, forward and reverse primer sequences, PCR product length (bp) and the qRT-PCR efficiency

Gene name	Forward primer sequence (5′-3′)	Reverse primer sequence (5′-3′)	PCR product length	qRT-PCR efficiency
TaNAS1-A	AAGTCGACGCGCTCTTCACC	GCTCAGGTTGATGTAGTTGTCA	230	0.98
TaNAS1-B	AAGTCGACGCGCTCTTCACG	GCTCAGGTTGATGTAGTTGTCG	230	–
TaNAS2-A	GTGCGTACGACGTGGTCTTC	GCTCTTTGGTGGCCAACTCT	388	0.92
TaNAS2-D1	CTCTTTGGCGGCCAGCTCT	CACCCCGAAGGTGAGGTG	189	–
TaNAS2-D2	CTCTTTGGTGGCCAACTCGT	CACCCCGAAGGTGAGGTC	189	0.99
TaNAS3-A	GGTTCCTGTACCCGATCGTA	TGTCACCATCTCACCAAACC	221	0.94
TaNAS3-B	GCTGCTGAGCTCGCTACATA	CCTCTCTCTTGTGGGTCACG	391	0.97
TaNAS4-A	CCTTCGACAACCCTCTGGAC	ACATGGTGTCGGGCAGGT	198	1.03
TaNAS4-U	ACTACGACCTGTGTGGCACA	ATCACCTTCGCCTTGTCCT	187	0.97
TaNAS4-D	ACGTCGACGCGCTCTTCACT	ATGCCGAGGTGATCCAGA	197	1.01
TaNAS5-B	GCGGGTTCCTATACCCGAT	TGCATGTCCTTCGACTTGTG	130	0.96
TaNAS6-A	CTGGAGGCGCACTACTCCTA	GTCGGCGGTGTGGAAAGA	330	0.97
TaNAS6-B	CACCACCTCGCCATCTTC	ACGTCAGCGGTGTGGAAC	283	0.95
TaNAS6-D	GACGACGACGTGGTGAACT	GATCAGAAGGCCACTTCA	203	0.92
TaNAS7-A1	GTGCGAGCAAGCTGTTCC	AAACTCCTCTCTCTTCTGGCTCA	473	0.91
TaNAS7-A2	TGCCCTCGCTCAGCCCGTG	AGTAGTGCGCCTCCAGCTTC	174	0.98
TaNAS7-D	TCAGCAAGCTGGAGTACGAG	CATGCATGTCGTTGGACTTT	528	0.88
TaNAS8-U	GTCCAAAAGATCACCGGACT	GAGTCCCTCCCGCATTTTCT	168	0.9
TaNAS9-A	GCCTCCTTCGACAACTACGA	GGAACACCACGTCGTAGC	154	1.07
TaNAS9-B	CCTGTACCCGGTGGTGAA	CATGGCCATCTCCTTCATCT	265	0.92
TaNAS9-D	GCCTCCTTCGACAACTACGA	ACCACCGGGTACAGGAAG	287	0.92

Phylogenetic analysis

Phylogenetic analysis of the 20 full-length *TaNAS* coding sequences was performed in Geneious Pro 5.6.6 using the Geneious plugin PhyML (Guindon *et al.*, 2010) and a Kimura 2-parameter substitution model with bootstrap value of 1000. Published coding sequences of 3 rice *OsNAS* (*OsNAS1*, LOC_Os03 g19427.1; *OsNAS2*, LOC_Os03 g19420.1; *OsNAS3*, LOC_Os07 g48980.1), 10 barley *HvNAS* (Table S1) and nine maize *ZmNAS* (Table S2) genes were included in the phylogenetic analysis.

Quantitative reverse transcription PCR (qRT-PCR) analysis of *TaNAS* genes

Preparation of a wheat cv. Chinese Spring cDNA library representing a range of tissues and developmental stages has been described previously, where each cDNA was generated from pooled RNA representing 7–10 individual plants (Schreiber *et al.*, 2009). We prepared a cv. Gladius cDNA library; the snap-frozen root and shoot tissues harvested from individual 4-week-old hydroponically grown plants were pulverized and total RNA was extracted from 90 to 100 mg of ground plant material using TRIzol® Reagent (Life Technologies, Carlsbad, CA, USA) according to manufacturer instructions with the following modifications: 250 µL of a 0.8 M sodium citrate and 1.2 M sodium chloride solution was added at the precipitation step and two final washes were performed. Approximately 2 µg of RNA was treated with DNase I (Invitrogen, Carlsbad, CA, USA) in 20 µL reactions to remove genomic DNA contamination and oligo-dT-primed reverse transcription was performed following manufacturer instructions (Tetro cDNA synthesis Kit, Bioline). Reactions were performed in a final volume of 20 µL and then diluted 1 : 10. The cDNA samples (cv. Gladius) were verified as genomic DNA free through lack of amplification in PCRs using primer sets specific to the noncoding microsatellite *Xcfb43* as well as a noncoding region of the *TaWIN1* gene (Table S3).

Subgenome-specific qRT-PCR primer pairs for each *TaNAS* gene were designed using the same methodology as described under 'Gene Sequencing' (Table 2). The specificity of each qRT-PCR primer pair was determined by melt curve analysis and sequencing. For each primer pair, primer efficiency was calculated using the formula $10^{(-1/m)} - 1$, where m corresponds to the slope of the standard curve generated using triplicate ten-fold serial dilutions (10^1–10^7) of purified template for each primer pair (Table 2). The qRT-PCR analysis of *TaNAS* gene expression was performed using the cDNA libraries of cvs. Chinese Spring and Gladius according to the protocols described in Vandesompele *et al.* (2002). The geometric mean of the three housekeeping genes *TaCyclophilin*, *TaGAPDH* and *TaEFA* was used as a normalization factor to normalize the expression of the *TaNAS* genes as described in Schreiber *et al.* (2009). A one-way ANOVA followed by a Tukey's test was performed within days (Days 0, 1, 5 and 7) and between treatments (+Fe and -Fe) to identify significant differences in the cv. Gladius qRT-PCR experiments. Normalization of cv. Gladius qRT-PCR gene expression was performed independently for the root and shoot tissues (Figure S2).

Acknowledgements

This research was supported by grants from the Australian Research Council (LP130100785) and the HarvestPlus Challenge Program. The authors would like to thank Margaret Pallotta for providing the wheat cv. Chinese Spring nulli-tetrasomic DNA used in this study.

References

Borg, S., Brinch-Pedersen, H., Tauris, B., Madsen, L.H., Darbani, B., Noeparvar, S. and Holm, P.B. (2012) Wheat ferritins: improving the iron content of the wheat grain. *J. Cereal Sci.* **56**, 204–213.

Budesinsky, M., Budzikiewicz, H., Prochazka, Z., Ripperger, H., Romer, A., Scholz, G. and Schreiber, K. (1980) Nicotianamine, a possible phytosiderophore of general occurrence. *Phytochemistry*, **19**, 2295–2297.

Chu, H., Chiecko, J., Punshon, T., Lanzirotti, A., Lahner, B., Salt, D. and Walker, E. (2010) Successful reproduction requires the function of Arabidopsis Yellow Stripe-Like1 and Yellow Stripe-Like3 metal-nicotianamine transporters in both vegetative and reproductive structures. *Plant Physiol.* **154**, 197–210.

Curie, C., Panaviene, Z., Loulergue, C., Dellaporta, S., Briat, J. and Walker, E. (2001) Maize yellow stripe1 encodes a membrane protein directly involved in Fe(III) uptake. *Nature*, **409**, 346–349.

Dash, S., Van Hemert, J., Hong, L., Wise, R.P. and Dickerson, J.A. (2012) PLEXdb: gene expression resources for plants and plant pathogens. *Nucleic Acids Res.* **40**, D1194–D1201.

Eagling, T., Neal, A.L., McGrath, S.P., Fairweather-Tait, S., Shewry, P.R. and Zhao, F.-J. (2014a) Distribution and speciation of iron and zinc in grain of two wheat genotypes. *J. Agric. Food Chem.* **62**, 708–716.

Eagling, T., Wawer, A.A., Shewry, P.R., Zhao, F.-J. and Fairweather-Tait, S.J. (2014b) Iron bioavailability in two commercial cultivars of wheat: comparison between wholegrain and white flour and the effects of nicotianamine and 2'-deoxymugineic acid on iron uptake into Caco-2 cells. *J. Agric. Food Chem.* **62**, 10320–10325.

Emanuelsson, O., Brunak, S., von Heijne, G. and Nielsen, H. (2007) Locating proteins in the cell using TargetP, SignalP and related tools. *Nat. Protoc.* **2**, 953–971.

Fujita, M., Horiuchi, Y., Ueda, Y., Mizuta, Y., Kubo, T., Yano, K., Yamaki, S. *et al.* (2010) Rice expression atlas in reproductive development. *Plant Cell Physiol.* **51**, 2060–2081.

Geldner, N. and Robatzek, S. (2008) Plant receptors go endosomal: a moving view on signal transduction. *Plant Physiol.* **147**, 1565–1574.

Genc, Y., McDonald, G.K. and Tester, M. (2007) Reassessment of tissue Na+ concentration as a criterion for salinity tolerance in bread wheat. *Plant, Cell Environ.* **30**, 1486–1498.

Guerinot, M.L. and Yi, Y. (1994) Iron: nutritious, noxious, and not readily available. *Plant Physiol.* **104**, 815–820.

Guindon, S., Dufayard, J.-F., Lefort, V., Anisimova, M., Hordijk, W. and Gascuel, O. (2010) New algorithms and methods to estimate maximum-likelihood phylogenies: assessing the performance of PhyML 3.0. *Syst. Biol.* **59**, 307–321.

Herbik, A., Koch, G., Mock, H., Dushkov, D., Czihal, A., Thielmann, J., Stephan, U. *et al.* (1999) Isolation, characterization and cDNA cloning of nicotianamine synthase from barley. A key enzyme for iron homeostasis in plants. *Eur. J. Biochem.* **265**, 231–239.

Hernandez, P., Martis, M., Dorado, G., Pfeifer, M., Gálvez, S., Schaaf, S., Jouve, N. *et al.* (2012) Next-generation sequencing and syntenic integration of flow-sorted arms of wheat chromosome 4A exposes the chromosome structure and gene content. *Plant J.* **69**, 377–386.

Higuchi, K., Kanazawa, K., Nishizawa, N.-K., Chino, M. and Mori, S. (1994) Purification and characterization of nicotianamine synthase from Fe-deficient barley roots. *Plant Soil*, **165**, 173–179.

Higuchi, K., Nakanishi, H., Suzuki, K., Nishizawa, N.K. and Mori, S. (1999a) Presence of nicotianamine synthase isozymes and their homologues in the root of graminaceous plants. *Soil Sci. Plant Nutr.* **45**, 681–691.

Higuchi, K., Suzuki, K., Nakanishi, H., Yamaguchi, H., Nishizawa, N. and Mori, S. (1999b) Cloning of nicotianamine synthase genes, novel genes involved in the biosynthesis of phytosiderophores. *Plant Physiol.* **119**, 471–480.

Higuchi, K., Watanabe, S., Takahashi, M., Kawasaki, S., Nakanishi, H., Nishizawa, N.K. and Mori, S. (2001) Nicotianamine synthase gene expression differs in barley and rice under Fe-deficient conditions. *Plant J.* **25**, 159–167.

Inoue, H., Higuchi, K., Takahashi, M., Nakanishi, H., Mori, S. and Nishizawa, N.K. (2003) Three rice nicotianamine synthase genes, *OsNAS1, OsNAS2*, and *OsNAS3* are expressed in cells involved in long-distance transport of iron and differentially regulated by iron. *Plant J.* **36**, 366–381.

IWGSC. (2014) A chromosome-based draft sequence of the hexaploid bread wheat (Triticum aestivum) genome. *Science*, **345**, 1–11.

Jain, M., Nijhawan, A., Arora, R., Agarwal, P., Ray, S., Sharma, P., Kapoor, S. *et al.* (2007) F-box proteins in rice. Genome-wide analysis, classification, temporal and spatial gene expression during panicle and seed development, and regulation by light and abiotic stress. *Plant Physiol.* **143**, 1467–1483.

Leach, L.J., Belfield, E.J., Jiang, C., Brown, C., Mithani, A. and Harberd, N.P. (2014) Patterns of homoeologous gene expression shown by RNA sequencing in hexaploid bread wheat. *BMC Genom.* **15**, 1–19.

Liu, C.J., Atkinson, M.D., Chinoy, C.N., Devos, K.M. and Gale, M.D. (1992) Nonhomoeologous translocations between group 4, 5 and 7 chromosomes within wheat and rye. *Theor. Appl. Genet.* **83**, 305–312.

Marschner, H. and Romheld, V. (1994) Strategies of plants for acquisition of iron. *Plant Soil*, **165**, 261–274.

Marschner, H., Romheld, V. and Kissel, M. (1986) Different strategies in higher-plants in mobilization and uptake of iron. *J. Plant Nutr.* **9**, 695–713.

Mizuno, D., Higuchi, K., Sakamoto, T., Nakanishi, H., Mori, S. and Nishizawa, N. (2003) Three nicotianamine synthase genes isolated from maize are differentially regulated by iron nutritional status. *Plant Physiol.* **132**, 1989–1997.

Noma, M. and Noguchi, M. (1976) Occurrence of nicotianamine in higher-plants. *Phytochemistry*, **15**, 1701–1702.

Noma, M., Noguchi, M. and Tamaki, E. (1971) A new amino acid, nicotianamine, from tobacco leaves. *Tetrahedron Lett.* **12**, 2017–2020.

Nozoye, T., Inoue, H., Takahashi, M., Ishimaru, Y., Nakanishi, H., Mori, S. and Nishizawa, N.K. (2007) The expression of iron homeostasis-related genes during rice germination. *Plant Mol. Biol.* **64**, 35–47.

Nozoye, T., Nagasaka, S., Bashir, K., Takahashi, M., Kobayashi, T., Nakanishi, H. and Nishizawa, N.K. (2014) Nicotianamine synthase 2 localizes to the vesicles of iron-deficient rice roots, and its mutation in the YXXφ or LL motif causes the disruption of vesicle formation or movement in rice. *Plant J.* **77**, 246–260.

Nussbaumer, T., Warth, B., Sharma, S., Ametz, C., Bueschl, C., Parich, A., Pfeifer, M. *et al.* (2015) Joint transcriptomic and metabolomic analyses reveal changes in the primary metabolism and imbalances in the subgenome orchestration in the bread wheat molecular response to Fusarium graminearum. *G3 (Bethesda)*, **5**, 2579–2592.

Pearce, S., Tabbita, F., Cantu, D., Buffalo, V., Avni, R., Vazquez-Gross, H., Zhao, R. *et al.* (2014) Regulation of Zn and Fe transporters by the GPC1 gene during early wheat monocarpic senescence. *BMC Plant Biol.* **14**, 368.

Perovic, D., Tiffin, P., Douchkov, D., Baumlein, H. and Graner, A. (2007) An integrated approach for the comparative analysis of a multigene family: the nicotianamine synthase genes of barley. *Funct. Integr. Genomics*, **7**, 169–179.

Robert, S., Bichet, A., Grandjean, O., Kierzkowski, D., Satiat-Jeunemaître, B., Pelletier, S., Hauser, M.-T. *et al.* (2005) An Arabidopsis endo-1,4-β-d-glucanase involved in cellulose synthesis undergoes regulated intracellular cycling. *Plant Cell*, **17**, 3378–3389.

Romheld, V., Marschner, H. and Kramer, D. (1982) Responses to Fe deficiency in roots of Fe-efficient plant-species. *J. Plant Nutr.* **5**, 489–498.

Roschzttardtz, H., Conejero, G., Divol, F., Alcon, C., Verdeil, J.L., Curie, C. and Mari, S. (2013) New insights into Fe localization in plant tissues. *Front. Plant Sci.* **4**, 1–11.

Scholz, G., Schlesier, G. and Seifert, K. (1985) Effect of nicotianamine on iron uptake by the tomato mutant chloronerva. *Physiol. Plant.* **63**, 99–104.

Scholz, G., Becker, R., Pich, A. and Stephan, U.W. (1992) Nicotianamine - a common constituent of strategies I and II of iron acquisition by plants: a review. *J. Plant Nutr.* **15**, 1647–1665.

Schreiber, A., Sutton, T., Caldo, R., Kalashyan, E., Lovell, B., Mayo, G., Muehlbauer, G. *et al.* (2009) Comparative transcriptomics in the Triticeae. *BMC Genom.* **10**, 1471–2164.

Sears, E.R. (1954) The aneuploids of common wheat. *Mo. Agr. Exp. Sta. Res. Bull.* **572**, 1–58.

Solovyev, V., Kosarev, P., Seledsov, I. and Vorobyev, D. (2006) Automatic annotation of eukaryotic genes, pseudogenes and promoters. *Genome Biol.* **7**(Suppl 1), S10.1–S10.12.

Stephan, U.W. and Scholz, G. (1993) Nicotianamine - mediator of transport of iron and heavy-metals in the phloem. *Physiol. Plant.* **88**, 522–529.

Takahashi, M., Terada, Y., Nakai, I., Nakanishi, H., Yoshimura, E., Mori, S. and Nishizawa, N.K. (2003) Role of nicotianamine in the intracellular delivery of metals and plant reproductive development. *Plant Cell*, **15**, 1263–1280.

Untergasser, A., Cutcutache, I., Koressaar, T., Ye, J., Faircloth, B.C., Remm, M. and Rozen, S.G. (2012) Primer3—new capabilities and interfaces. *Nucleic Acids Res.* **40**, e115.

Vandesompele, J., De Preter, K., Pattyn, F., Poppe, B., Van Roy, N., De Paepe, A. and Speleman, F. (2002) Accurate normalization of real-time quantitative RT-PCR data by geometric averaging of multiple internal control genes. *Genome Biol.* **3**, 1–11 ResearchCH0034.

Walker, E.L. and Waters, B.M. (2011) The role of transition metal homeostasis in plant seed development. *Curr. Opin. Plant Biol.* **14**, 318–324.

Wei, F., Coe, E., Nelson, W., Bharti, A., Engler, F., Butler, E., Kim, H. *et al.* (2007) Physical and genetic structure of the maize genome reflects its complex evolutionary history. *PLoS Genet.* **3**, e123.

von Wiren, N., Klair, S., Bansal, S., Briat, J.F., Khodr, H., Shioiri, T., Leigh, R.A. *et al.* (1999) Nicotianamine chelates both Fe-III and Fe-II. Implications for metal transport in plants. *Plant Physiol.* **119**, 1107–1114.

Zhou, M.-L., Qi, L.-P., Pang, J.-F., Zhang, Q., Lei, Z., Tang, Y.-X., Zhu, X.-M. *et al.* (2013a) Nicotianamine synthase gene family as central components in heavy metal and phytohormone response in maize. *Funct. Integr. Genomics*, **13**, 229–239.

Zhou, X., Li, S., Zhao, Q., Liu, X., Zhang, S., Sun, C., Fan, Y. *et al.* (2013b) Genome-wide identification, classification and expression profiling of nicotianamine synthase (NAS) gene family in maize. *BMC Genom.* **14**, 238.

Zuo, J.R., Niu, Q.W., Nishizawa, N., Wu, Y., Kost, B. and Chua, N.H. (2000) KORRIGAN, an arabidopsis endo-1,4-beta-glucanase, localizes to the cell plate by polarized targeting and is essential for cytokinesis. *Plant Cell*, **12**, 1137–1152.

7

Overexpression of GCN2-type protein kinase in wheat has profound effects on free amino acid concentration and gene expression

Edward H. Byrne[1,†], Ian Prosser[1], Nira Muttucumaru[1], Tanya Y. Curtis[1], Astrid Wingler[2], Stephen Powers[3] and Nigel G. Halford[1,*]

[1]Department of Plant Science, Rothamsted Research, Harpenden, Hertfordshire, UK
[2]Research Department of Genetics, Evolution and Environment, University College London, Gower Street, London, UK
[3]Department of Biomathematics and Bioinformatics, Rothamsted Research, Harpenden, Hertfordshire, UK

*Correspondence
email
nigel.halford@rothamsted.ac.uk
†Present address: Cereal Genomics, School of Biological Sciences, University of Bristol, Woodland Road, Bristol BS8 1UG, UK.
Accession number: The nucleotide sequence of *TaGCN2*, reported in this manuscript, has been submitted to the EMBL database and assigned the accession number FR839672.

Keywords: general control nonderepressible, sulphur signalling, asparagine, acrylamide, phosphorylation.

Summary

A key point of regulation of protein synthesis and amino acid homoeostasis in eukaryotes is the phosphorylation of the α subunit of eukaryotic translation initiation factor 2 (eIF2α) by protein kinase general control nonderepressible (GCN)-2. In this study, a GCN2-type PCR product (TaGCN2) was amplified from wheat (*Triticum aestivum*) RNA, while a wheat eIF2α homologue was identified in wheat genome data and found to contain a conserved target site for phosphorylation by GCN2. *TaGCN2* overexpression in transgenic wheat resulted in significant decreases in total free amino acid concentration in the grain, with free asparagine concentration in particular being much lower than in controls. There were significant increases in the expression of eIF2α and protein phosphatase PP2A, as well as a nitrate reductase gene and genes encoding phosphoserine phosphatase and dihydrodipicolinate synthase, while the expression of an asparagine synthetase (*AS1*) gene and genes encoding cystathionine gamma-synthase and sulphur-deficiency-induced-1 all decreased significantly. Sulphur deficiency–induced activation of these genes occurred in wild-type plants but not in *TaGCN2* overexpressing lines. Under sulphur deprivation, the expression of genes encoding aspartate kinase/homoserine dehydrogenase and 3-deoxy-D-arabino-heptulosonate-7-phosphate synthase was also lower than in controls. The study demonstrates that *TaGCN2* plays an important role in the regulation of genes encoding enzymes of amino acid biosynthesis in wheat and is the first to implicate GCN2-type protein kinases so clearly in sulphur signalling in any organism. It shows that manipulation of *TaGCN2* gene expression could be used to reduce free asparagine accumulation in wheat grain and the risk of acrylamide formation in wheat products.

Introduction

Translation initiation, the point at which a ribosome recruits an mRNA molecule, is a key control point for protein synthesis in all eukaryotic species. It is regulated by the phosphorylation of the α subunit of eukaryotic translation initiation factor 2 (eIF2α) (reviewed by Hershey and Merrick, 2000). eIF2 is a trimeric factor (subunits α, β and γ) that can bind either guanosine diphosphate (GDP) or triphosphate (GTP). Only when bound to GTP is it able to carry out its physiological function of binding Met-tRNA to the ribosome and transferring it to the 40S ribosomal subunit. Following attachment of the (eIF2.GTP.Met-tRNA) complex to the 40S subunit, the GTP is hydrolysed to GDP. Phosphorylation of eIF2α inhibits the conversion of eIF2-GDP to eIF2-GTP, preventing further cycles of translation initiation and suppressing protein synthesis (Wek *et al.*, 2006).

In budding yeast (*Saccharomyces cerevisiae*), phosphorylation of eIF2α not only causes a general reduction in protein synthesis, but also initiates a change in the expression of a large number of genes, most notably involved in amino acid biosynthesis. Thus, under conditions of amino acid starvation, yeast can switch on amino acid biosynthesis genes, helping the cell to maintain homoeostasis and survive. This 'general amino acid control' is orchestrated by the transcription factor general control nonderepressible 4 (GCN4) (Hinnebusch, 1997, 2005), the name arising from the fact that general amino acid control is in an irreversibly repressed state in *gcn4* and other *gcn* mutants. In budding yeast, GCN4 levels are regulated post-transcriptionally, the synthesis of GCN4 increasing when eIF2α is phosphorylated owing to translation proceeding from an initiation codon that is not used under normal conditions (Hinnebusch, 1992, 1994). GCN4 promotes the expression of genes encoding enzymes in every amino acid biosynthetic pathway except cysteine, as well as many other genes involved in a wide range of cellular processes (Natarajan *et al.*, 2001). In mammals, phosphorylation of eIF2α leads to an increase in the translation of ATF4, the functional orthologue of GCN4. Increased levels of ATF4 lead to the induction of additional bZIP transcription regulators, ATF3 and CHOP/GADD153 (Harding *et al.*, 2000).

The protein kinase that phosphorylates eIF2α was given the name GCN2 (Wek *et al.*, 1989). In yeast, GCN2 is a relatively large protein kinase (1659 amino acid residues; 190 kDa) that senses a reduction in cellular amino acid content through the interaction of its regulatory domain with uncharged tRNA, the cellular concentration of which increases under conditions of amino acid starvation (Wek et al., 1989, 2003; Zhu et al., 1996). The GCN2 regulatory domain has some amino acid sequence similarity with histidyl-tRNA synthetases and is sometimes called the histidyl-tRNA synthetase-like domain. Activation involves a conformational change in GCN2 and autophosphorylation at two threonine residues in the conserved activation loop of the kinase domain. GCN2 may also be activated and protein synthesis inhibited in response to purine deprivation, exposure to UV-B light, oxidative and osmotic stress or glucose deprivation (Yang et al., 2000; Hinnebusch, 2005; Mascarenhas et al., 2008).

Three other animal protein kinases are known to be able to phosphorylate eIF2α: double-stranded, RNA-dependent protein kinase (PKR), PKR-like endoplasmic reticulum kinase and haem-regulated inhibitor (Chen and London, 1995; Kaufman, 1999; Nanduri et al., 2000). The four eIF2α kinases share a highly conserved protein kinase domain but their regulatory domains differ, enabling each kinase to respond to a different stimulus.

The first plant GCN2 homologue to be identified was AtGCN2 from Arabidopsis (Arabidopsis thaliana) (Zhang et al., 2003). It is structurally similar to GCN2 from fungi and animals, with a characteristic eIF2α kinase domain adjacent to a histidyl-tRNA synthetase-like regulatory domain, and it complements the gcn2 mutation of yeast (Zhang et al., 2003). However, it is smaller than yeast GCN2 (1241 amino acid residues; 140 kDa). Arabidopsis mutants lacking AtGCN2 grow normally in compost but are more sensitive than wild type to herbicides such as glyphosate and chlorsulphuron that interfere with amino acid biosynthesis, an effect that can be reversed by feeding the plants with the appropriate amino acids (Zhang et al., 2008). These herbicides induce the phosphorylation of eIF2α in wild-type Arabidopsis but not in gcn2 mutants (Zhang et al., 2008). GCN2-like expressed sequence tags (ESTs) and genomic sequences have since been identified in a variety of plant species (Halford, 2006) but have not been characterized in any detail. In all plant species where full genome data are available, GCN2 is encoded by a single gene and is the only eIF2α kinase.

As in fungal systems, AtGCN2 may be activated in response to other stress stimuli, such as purine deprivation, UV light, cold shock and wounding (Lageix et al., 2008). AtGCN2 is also activated in response to treatment with methyl jasmonate or salicylic acid, which are involved in the activation of defence mechanisms in response to insect herbivores, and aminocyclopropane carboxylic acid (ACC), which is involved in ethylene biosynthesis and therefore ripening and senescence (Lageix et al., 2008).

The discovery of a plant GCN2 homologue was evidence that a general amino acid control system, similar to that of fungi and animals, might exist in plants, at least in part. Previous studies had suggested that this might be so. For example, blocking histidine biosynthesis in Arabidopsis with a specific inhibitor, IRL 1803, had been shown to increase the expression of eight genes involved not only in the synthesis of histidine but also in the synthesis of the aromatic amino acids (tyrosine, tryptophan and phenylalanine), lysine and purines (Guyer et al., 1995). Genes encoding tryptophan biosynthesis pathway enzymes had also been shown to be induced by amino acid starvation caused by glyphosate application and other treatments in Arabidopsis (Zhao et al., 1998). In another study, the contents of most minor amino acids had been shown to vary in concert in wheat, barley and potato leaves (Noctor et al., 2000). However, although Zhang et al. (2008) showed that the expression of key genes of amino acid biosynthesis was affected by treatment of Arabidopsis with herbicides that affected amino acid metabolism, this response was also seen in mutants lacking AtGCN2 (Zhang et al., 2008). The only exception was a nitrate reductase (NR) gene, NIA1, the expression of which was reduced in the mutant plants. Furthermore, no obvious candidate for a GCN4 homologue is identifiable in plants based on amino acid sequence similarity (Halford, 2006).

Wheat GCN2 has not been characterized previously but wheat eIF2α has been reported to contain a conserved GCN2 phosphorylation site, although its full amino acid sequence has not been described previously. Yeast GCN2 has been shown to phosphorylate wheat eIF2α in vitro at this site (Chang et al., 1999) and wheat eIF2α complements eIF2α deletion mutants of yeast, restoring a fully functional general amino acid control system (Chang et al., 2000).

Interest in the control of free amino acid accumulation in cereal grain and other important crop products has been stimulated in recent years because free amino acid concentrations have been shown to affect processing properties and product quality. Free amino acids react with reducing sugars in the Maillard reaction, a complex series of nonenzymatic reactions that occurs during frying, baking, roasting and high-temperature processing. The products of the Maillard reaction include melanoidin pigments and complex mixtures of compounds that impart flavour and aroma (Mottram, 2007; Halford et al., 2011). However, the Maillard reaction also produces undesirable compounds, and these include acrylamide, which was discovered in many popular foods in 2002 (Tareke et al., 2002). Acrylamide is formed if the amino acid that participates in the reaction's final stages is asparagine (Mottram et al., 2002; Stadler et al., 2002). Acrylamide is neurotoxic, carcinogenic and genotoxic in rodents and has been classified as a probable human carcinogen by the World Health Organisation (Friedman, 2003). The reduction of free amino acid and specifically free asparagine accumulation in wheat grain is therefore highly desirable. In wheat, sulphur deprivation has a dramatic effect particularly on free asparagine concentration in the grain, causing increases of up to 30-fold (Muttucumaru et al., 2006; Granvogl et al., 2007; Curtis et al., 2009).

In this study, a polymerase chain reaction (PCR) product derived from the transcript of a GCN2-related gene (TaGCN2) was amplified from wheat leaf RNA and transgenic wheat plants were produced in which TaGCN2 was overexpressed. Analysis of these plants showed dramatic effects on free amino acid levels and gene expression and placed TaGCN2 irrefutably in the sulphur signalling pathway. The study showed that manipulation of TaGCN2 gene expression could be used to reduce free asparagine accumulation in wheat grain and therefore the risk of acrylamide formation in wheat products.

Results

Molecular cloning of a wheat (Triticum aestivum) GCN2 homologue, TaGCN2

A GCN2-related polymerase chain reaction (PCR) product was amplified from wheat cv. Cadenza leaf RNA. A product of approximately 3.8 kb containing an open reading frame running from bases 1 to 3741 was cloned. This open reading frame encoded a protein having 52% amino acid sequence identity with Arabidopsis (Arabidopsis thaliana) AtGCN2 (Zhang et al., 2003) and 84% identity with a rice (Oryza sativa) GCN2-type protein kinase encoded by mRNA nucleotide sequence XM473001 from GenBank. The protein was given the name TaGCN2. Significantly higher degree of identity with the other cereal GCN2 homologue than with AtGCN2 should be noted.

Additional nucleotide sequence data from the 3′ end of the transcript were obtained by rapid amplification of the cDNA end (3′RACE). This showed the TaGCN2 transcript to have a 658-nucleotide untranslated region prior to a poly-adenosine tail of 22 nucleotides. The entire sequence of 4439 nucleotides was submitted to the EMBL database and has been assigned the accession number FR839672.

The TaGCN2 nucleotide sequence was used to mine the recently available wheat genomic sequence (http://www.cerealsdb.uk.net/), and three separate contigs that matched different parts of the TaGCN2 sequence were identified. The consensus sequence of one of these contigs aligned with the 5′ end of the TaGCN2 PCR product and extended a further 2 kb 'upstream' of the ATG translation start site. Another contig aligned with the 3′ end of the TaGCN2 PCR and 3′RACE products, with an intron in the 3′ untranslated region. The entire nucleotide and derived amino acid sequences of the TaGCN2 PCR product and the wheat genome sequence data that aligned with the 5′ and 3′ ends are given in Data S1.

The encoded protein consists of 1247 amino acid residues and has a molecular weight of 140 kDa. It contains a RING-finger, WD40, DEAD-box helicase domain (RWD-domain) at the N-terminus between residues 28 and 142, an eIF2α kinase domain between residues 422 and 738 and a histidyl-tRNA synthetase-like regulatory domain towards the C-terminal end of the protein between residues 799 and 1128 (Figure 1). An anticodon-binding subdomain of the regulatory domain was found at the extreme C-terminal end of the protein between residues 1129 and 1237, although it is truncated in TaGCN2 compared with yeast GCN2. The presence of these domains in the same protein is a defining characteristic of GCN2-type protein kinases (Wek et al., 1995).

In Arabidopsis, AtGCN2 has been shown to be expressed in all tissues (Zhang et al., 2003). Expression of TaGCN2 in flag leaves and grain through the period of grain development was analysed by real-time PCR. Transcripts were detectable in all of the samples, and no significant changes in transcript levels between tissues or at different developmental stages were evident.

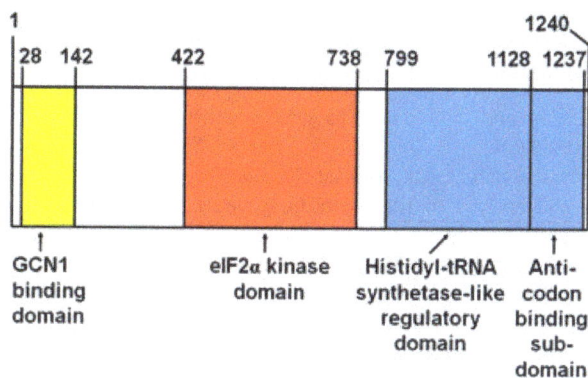

Figure 1 Schematic diagram representing the structure of wheat GCN2-related protein kinase, TaGCN2. The relative positions of the GCN1 binding domain (yellow), eIF2α kinase domain (red) and regulatory domain (including anticodon-binding subdomain) (blue) are shown.

In silico identification of a wheat (Triticum aestivum) eIF2α homologue and identification of a putative target site for phosphorylation by TaGCN2

A search of the wheat genome database (http://www.cerealsdb.uk.net/) was performed using a maize eIF2α nucleotide sequence, accession NP-001146159, and overlapping contigs were assembled. The derived amino acid sequence of the encoded protein is shown in Data S1; it comprised 340 amino acids with 95%–97% amino acid sequence identity with maize, sorghum and rice eIF2α proteins, accession numbers ACL53376, EER95021 and ABF95443. The putative target residue for phosphorylation by GCN2-type protein kinases is a serine residue in the N-terminal region, and it was readily identifiable at position 50 of the wheat eIF2α protein. This and the surrounding residues are absolutely conserved in organisms as diverse as yeast, humans and plants: NIEGMILF**S**ELSRRRIRSI (target serine in boldface and underlined).

Production of transgenic wheat plants overexpressing TaGCN2

TaGCN2 was overexpressed in transgenic wheat plants under the control of a rice actin gene promoter, which is constitutively active (McElroy et al., 1990). Three independent, homozygous lines, 395, 402 and 426, were produced. These lines came from separate transformation experiments, and TaGCN2 expression in 395 was measured using cyclophilin as a reference gene, while that in 402 and 426 was measured using glyceraldehyde 3-phosphate dehydrogenase (GAPDH) and succinate dehydrogenase (SDH) as reference genes. All three showed significantly higher ($P < 0.05$) levels of TaGCN2 expression than null segregant controls (Figure 2). The transgenic lines showed no visible alteration in phenotype compared with null segregant or wild-type plants.

Free amino acid concentrations in the seeds of homozygous T3 plants were measured by gas chromatography-mass spectrometry (GC-MS), and the results are given in Table 1. Analysis of variance (ANOVA) was applied to the data for each amino acid and the total free amino acids. Following an F-test result indicating significant ($P < 0.05$) overall differences between the lines, specific comparisons of transgenic lines with controls were made using the standard error of the difference (SED) (Table 1) in post-ANOVA t-tests. This showed total free amino acid and free asparagine concentrations to be significantly reduced ($P < 0.05$) in all three transgenic lines. In line 426, free asparagine concentration was 0.955 mmol/kg, compared with an average of 3.00 mmol/kg in null segregant controls, a reduction of more than two-thirds. The total free amino acid and free asparagine concentrations are shown graphically in Figure 3.

Effects of manipulating TaGCN2 gene expression on genes of amino acid biosynthesis under adequate nutrient supply and in response to sulphur deprivation

Transgenic wheat lines overexpressing TaGCN2 were used to investigate the role of TaGCN2 in regulating expression of key genes in amino acid metabolism under conditions of sulphur sufficiency and deficiency. Sulphur deprivation was used to perturb the system in this experiment because it has been shown to cause a massive increase in free amino acid accumulation in wheat, with free asparagine, which can increase 30-fold in concentration in wheat grain, and free glutamine accounting for

Analysis of variance was applied to assess the statistical significance of line and sulphur treatment as main effects and the interaction between these two factors using the F-test. The results are shown in Table 3A. Following an F-test result indicating significant ($P < 0.05$) or marginal ($0.05 < P < 0.10$) differences (genes shown in boldface in Table 3A), the least significant difference (LSD) at the 5% level of significance was used to separate pairs of means of interest in the appropriate table of means for each gene. The results are given in Table 3B for genes that showed no significant ($P > 0.05$) sulphur response and Table 3C for genes that showed a significant ($P < 0.05$) sulphur response. It should be noted that the ANOVA was applied to the log (to base 2)-transformed inverse of the normalized relative quantity (NRQ) data to ensure the homogeneity of variance across the line by sulphur treatment combinations and effectively to provide values back on the ct-scale. Therefore, as for ct values, a low $\log_2(1/NRQ)$ in Table 3B,C indicates high gene expression, whereas a high $\log_2(1/NRQ)$ indicates low gene expression. The expression levels of genes that showed significant differences ($P < 0.05$, LSD) between the control plants and both transgenic lines are shown graphically in Figure 4.

TaGCN2 was confirmed to be significantly ($P < 0.01$) overexpressed in both transgenic lines. Expression of translation initiation factor-2α (eIF2α) and protein phosphatase-2A (PP2A) was also significantly increased. eIF2α is the substrate for phosphorylation by GCN2-type protein kinases, while PP2A dephosphorylates eIF2α, thereby opposing the action of GCN2. The increase in the expression of these genes could therefore be interpreted as evidence of the plants compensating for the overexpression of *TaGCN2* by producing more substrate and more of the opposing phosphatase. There was also a significant ($P < 0.05$) increase in the expression of the NR gene. This is consistent with the finding of Zhang *et al.* (2008) that expression of a NR gene was reduced in an Arabidopsis mutant lacking AtGCN2 and is therefore further evidence of a role of GCN2 in regulating nitrogen assimilation in plants.

In the plants that were supplied with sulphur, there were significantly ($P < 0.05$) higher levels of expression of genes encoding phosphoserine phosphatase (PSP) and dihydrodipicolinate synthase (DHDPS) in the transgenic lines compared with controls, while the expression of an asparagine synthetase (*AS1*) gene and genes encoding cystathionine gamma-synthase (CGS) and sulphur-deficiency-induced-1 (SDI1) was significantly ($P < 0.05$) lower. SDI1 is involved in the utilization of stored sulphate pools under S-limiting conditions and is used as a marker for sulphur deficiency (Howarth *et al.*, 2009), while CGS is involved in the synthesis of the sulphur-containing amino acids, cysteine and methionine, as well as other aspects of sulphur metabolism.

This apparent link with sulphur was dramatically confirmed by the analysis of the sulphur-deprived plants. In the control lines, the expression of genes encoding SDI1 and AS1 increased significantly ($P < 0.05$) (Table 3A,C; Figure 4), whereas in the *TaGCN2* overexpressing lines, there was no increase in expression. The expression of two other genes, encoding aspartate kinase/homoserine dehydrogenase (AK/HSDH) and 3-deoxy-D-arabino-heptulosonate-7-phosphate synthase (DHS) (note the alternative abbreviation of DAHP), was significantly ($P < 0.05$) lower in the transgenic lines than in controls under sulphur deprivation. AK/HSDH is a bifunctional enzyme but the phosphorylation of aspartate by its AK activity is the first step in

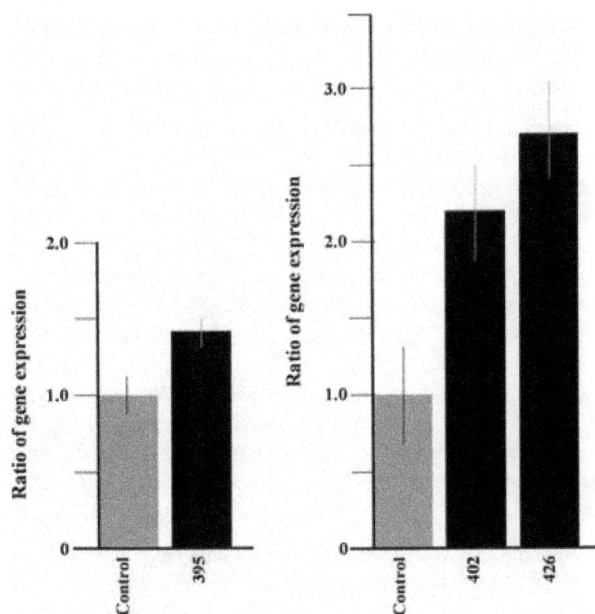

Figure 2 Relative expression of *TaGCN2* in the leaves of transgenic wheat lines in which *TaGCN2* was overexpressed under the control of a rice actin gene promoter, compared with controls. The analysis was carried out in two separate quantitative real-time polymerase chain reaction experiments using different reference genes. In each case, expression in the control lines is represented as 1. Error bars represent standard error of the mean from analyses of two biological replicates. Expression levels were significantly different between control and overexpressing lines ($P < 0.05$ for 395 and $P < 0.01$ for 402 and 426).

most of the increase (Muttucumaru *et al.*, 2006; Granvogl *et al.*, 2007; Curtis *et al.*, 2009). The plants were grown in vermiculite, which does not retain nutrients, and feeding was started 3 weeks after potting. There were two feeding regimes: one set of plants (S+) were watered with 'complete' medium containing 1.1 mM $MgSO_4$ (Muttucumaru *et al.*, 2006; Curtis *et al.*, 2009) and a second set (S−) were watered with the same medium containing one-tenth the concentration of $MgSO_4$.

The expression levels of *TaGCN2* and a suite of other genes (Table 2) in flag leaves of lines 402 and 426 were compared with those in wheat cv. Cadenza controls by real-time, quantitative PCR using genes encoding glyceraldehyde 3-phosphate dehydrogenase (GAPDH) and SDH as reference genes. The genes that were selected for study were those used previously in an analysis of an Arabidopsis mutant lacking GCN2 (Zhang *et al.*, 2008), with the addition of asparagine synthetase (*AS1*) and *sulphur deficiency-induced-1*, and encode enzymes in a range of amino acid biosynthetic pathways (Table 2). The target gene sequences were identified initially through searches of wheat ESTs using annotated Arabidopsis gene sequences and then checked against rice and *Brachypodium* genome data. In some cases, additional searches of the wheat genome database were carried out until the full-length gene sequence was obtained. With the exception of aspartate amino transferase (AAT) x, y and z, primers were designed to amplify a product from all three homeologues. In the case of AAT, primer pairs were designed for each different homeologue, but they were called x, y and z because it was not possible to assign the homeologues with certainty to the A, B and D genomes. Primer sequences are given in Table S1.

Table 1 Free amino acid concentrations (mmol kg^{-1}) in flour produced from grain of transgenic wheat lines in which gene *TaGCN2* was overexpressed, and null segregant controls for line 395 (control A) and for lines 402 and 426 (control B)

	Control A	Control B	Line 395	Line 402	Line 426	SED
Alanine	0.851 (−0.17)	0.682 (−0.38)	**0.488 (−0.72)**	**0.402 (−0.91)**	**0.369 (−1.00)**	0.141* 0.127†
Arginine	ND	ND	ND	ND	ND	−
Asparagine	3.83 (1.34)	2.17 (0.77)	**1.62 (0.49)**	**0.991 (−0.01)**	**0.955 (−0.05)**	0.137 0.122
Aspartic acid	3.07 (1.12)	2.12 (0.74)	**1.90 (0.64)**	2.03 (0.71)	1.943 (0.57)	0.175 0.157
Cysteine	0.061 (−3.11)	0.051 (−3.04)	0.066 (−2.73)	0.087 (−2.48)	0.077 (−2.57)	0.666 0.596
Glutamic acid	1.64 (0.49)	1.13 (0.11)	**0.855 (−0.16)**	0.898 (−0.12)	0.829 (−0.19)	0.178 0.159
Glutamine	0.891 (−0.13)	0.331 (−1.18)	**0.113 (−2.27)**	**0.114 (−2.21)**	**0.083 (−2.55)**	0.460 0.412
Glycine	0.281 (−1.28)	0.246 (−1.40)	0.204 (−1.59)	**0.169 (−1.78)**	**0.172 (−1.76)**	0.136 0.122
Histidine	0.203 (−1.80)	0.249 (−1.55)	0.134 (−2.09)	0.128 (−2.18)	0.097 (−2.35)	0.651 0.583
Isoleucine	0.121 (−2.12)	0.132 (−2.03)	**0.060 (−2.82)**	**0.067 (−2.71)**	**0.063 (−2.78)**	0.162 0.145
Leucine	0.219 (−1.52)	0.229 (−1.48)	**0.118 (−2.14)**	**0.093 (−2.39)**	**0.093 (−2.38)**	0.146 0.131
Lysine	0.278 (−1.32)	0.300 (−1.29)	0.196 (−1.67)	**0.137 (−2.05)**	**0.139 (−1.98)**	0.349 0.312
Methionine	0.020 (−4.00)	0.032 (−3.47)	0.016 (−4.19)	0.021 (−3.86)	0.018 (−4.02)	0.374 0.334
Phenylalanine	0.114 (−2.17)	0.105 (−2.26)	**0.073 (−2.62)**	0.095 (−2.37)	0.086 (−2.46)	0.165 0.148
Proline	0.385 (−0.958)	0.303 (−1.20)	0.129 (−2.06)	0.259 (−1.35)	**0.121 (−2.12)**	0.150 0.134
Serine	0.872 (−0.14)	0.999 (−0.08)	0.681 (−0.47)	0.652 (−0.43)	0.479 (−0.78)	0.817 0.578
Threonine	0.875 (−0.44)	0.921 (−0.17)	0.347 (−1.48)	**0.189 (−1.70)**	0.667 (−1.09)	0.727 0.650
Tryptophan	0.135 (−2.02)	1.34 (0.29)	**0.641 (−0.45)**	**0.851 (−0.17)**	1.37 (0.31)	0.171 0.153
Tyrosine	0.125 (−2.10)	0.127 (−2.09)	0.101 (−2.32)	0.108 (−2.27)	0.096 (−2.34)	0.268 0.240
Valine	0.327 (−1.12)	0.313 (−1.17)	**0.154 (−1.88)**	**0.132 (−2.03)**	**0.137 (−1.99)**	0.155 0.139
Total	13.4 (2.60)	10.8 (2.38)	**7.23 (1.98)**	**6.79 (1.91)**	**7.32 (1.99)**	0.169 0.152

Values are means of three replicates except for line 402 (two replicates), with (natural) log-transformed data values in parenthesis, which are used to compare lines using post-ANOVA *t*-tests based on the standard error of the difference (SED) value on 33 degrees of freedom (df). Readings in boldface indicate lines significantly different ($P < 0.05$) from the respective control.

*Comparisons with line 426.

†All other comparisons.

methionine, lysine and threonine synthesis. DHS is involved in the early stages of aromatic amino acid synthesis. Expression of PSP in the control plants increased significantly ($P < 0.05$) in response to sulphur deprivation to the levels seen in the over-expressing plants, which did not change in response to sulphur. In other words, overexpression of *TaGCN2* resulted in the expression of PSP being at the levels seen in sulphur-deprived control plants whether sulphur was supplied or not.

For the genes encoding AATx and y, there was a significant difference ($P < 0.05$) in the expression between the control and line 426, but no significant difference between the control and line 402. There was no significant difference ($P > 0.10$) in expression of AATz, alanine amino transferase (AlaAT), aceto-lactate synthase (ALS), histidinol dehydrogenase (HDH) or phos-phoribosylanthranilate transferase (PAT). The expression of a gene encoding a 14-3-3 protein that interacts with NR showed

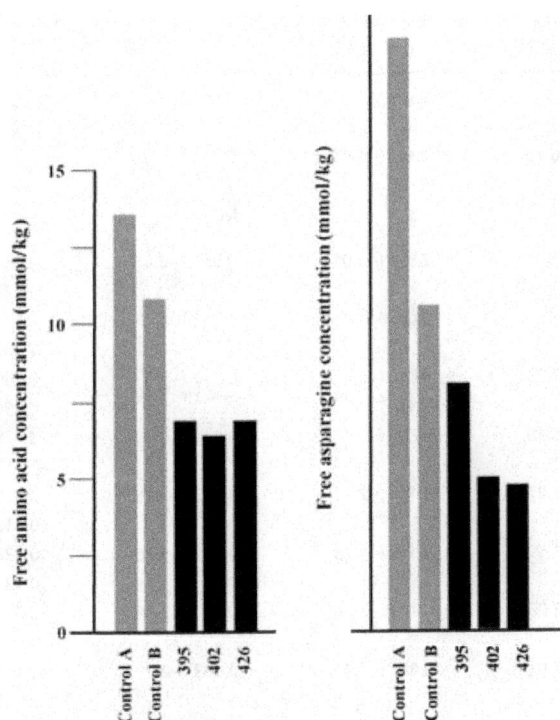

Figure 3 Concentrations (mmol kg^{-1}) of total free amino acids (left) and free asparagine (right) in the grain of transgenic wheat lines in which the expression of *TaGCN2* was increased by constitutive overexpression, compared with null segregant controls: 'Control A' for line 395 and 'Control B' for lines 402 and 426. The differences between the control and overexpressing lines were statistically significant ($P < 0.05$) for both total free amino acids and free asparagine. For statistical analyses, including the standard error of the differences for making comparisons of the lines, refer to Table 1.

a marginally significant ($P < 0.10$) response to sulphur but was not affected by *TaGCN2* overexpression.

Yield

The grain yield and 1000 grain weight of the *TaGCN2* overexpressing lines were measured and compared with wild-type (cv. Cadenza) controls (Table S2). There was a trend for the overexpressing lines to yield more grain weight per plant than the control plants, although 1000 grain weight was similar in control and transgenic lines, indicating that there was no difference in the size of individual grains. The nitrogen content of the grain from the transgenic lines was lower than that of controls (2.2 ± 0.3% dry weight compared with 2.7 ± 0.3% dry weight), while the carbon content was almost unchanged (45.6 ± 0.1% dry weight compared with 45.4 ± 0.1% dry weight), meaning that the transgenic lines had a higher ratio of carbon to nitrogen than the controls.

As genetically modified lines, European law required the plants to be kept in a containment glasshouse, and it was not possible to have sufficient replication of each line to ensure a robust assessment of the differences between them. Furthermore, trends in the yield of wheat under glasshouse conditions often disappear when the experiment is repeated under field conditions. However, it is important to note that there was no evidence of a negative effect of *TaGCN2* overexpression on grain yield.

Discussion

We have shown that overexpression of the wheat *GCN2* homologue, *TaGCN2*, has profound effects on free amino acid concentrations in wheat grain and on the expression of several genes encoding key enzymes in amino acid biosynthesis. Free amino acid concentrations in the grain of the transgenic lines were decreased, mainly as a result of substantial reductions in the concentrations of free asparagine. In one line, free asparagine concentration was reduced by more than two-thirds compared with controls. There was some evidence that *TaGCN2* overexpression could increase grain yield but statistical significance could not be established.

The data clearly showed *TaGCN2* to be involved in the regulation of gene expression in plants that were well nourished and also implicated *TaGCN2* in sulphur signalling. This was demonstrated dramatically in the analysis of asparagine synthetase (*AS1*) gene expression, which rose almost 10-fold in response to sulphur deprivation in wild-type plants but which was almost undetectable, with or without sulphur, in the transgenic lines. *AS1* gene expression has been shown to be induced by salinity and osmotic stress (Wang *et al.*, 2005) but has not previously been reported to increase in response to sulphur deprivation, although the fact that it does is not unexpected given the massive accumulation of asparagine seen in grain from sulphur-deprived wheat (Muttucumaru *et al.*, 2006; Granvogl *et al.*, 2007; Curtis *et al.*, 2009). *TaGCN2* overexpression also had profound effects on the expression of a gene used as a marker for sulphur deficiency, *sulphur deficiency-induced-1* (*SDI1*), and a gene encoding CGS. Expression of these genes was significantly reduced in the transgenic plants compared with controls when the plants were supplied with sulphur and *SDI1*, like *AS1*, also showed no induction in the transgenic lines in response to sulphur deficiency, whereas its expression increased significantly in controls.

The involvement of GCN2 or a related protein kinase in sulphur signalling has not been demonstrated so clearly in any organism before. However, phosphorylation of eIF2α, the substrate for GCN2, has been shown to be higher in liver cells of rats fed a diet deficient in sulphur-containing amino acids than in well-nourished rats (Sikalidis and Stipanuk, 2010). Fascinatingly, that study showed that asparagine synthetase gene expression was also increased.

The discovery of acrylamide in many popular foods (Tareke *et al.*, 2002) has stimulated great interest in the control of free amino acid and particularly free asparagine accumulation in grains, tubers and other crop products. Acrylamide forms as part of the Maillard reaction, a series of nonenzymatic reactions between reducing sugars and amino groups, principally those of amino acids. The Maillard reaction is an important one for the food industry because it produces the melanoidin compounds that give fried, roasted and baked products their colour and a host of volatiles that impart aroma and flavour. It is multistep, with amino groups participating in the first stage and the last, and is not one reaction but many. In the final stages, amino acids are deaminated and decarboxylated to give aldehydes (Strecker degradation) and the major route for acrylamide formation is a Strecker-type reaction involving asparagine (Mottram, 2007; Halford *et al.*, 2011). Asparagine concentration is the limiting factor for acrylamide formation in heated flour from wheat and rye grain (Muttucumaru *et al.*, 2006; Granvogl *et al.*, 2007; Curtis *et al.*, 2009, 2010). Aspara-

Table 2 List of genes that were analysed in transgenic wheat lines overexpressing *TaGCN2*

Abbreviation	Full name	Comment
GAPDH	Glyceraldehyde 3-phosphate dehydrogenase	Reference gene
SDH	Succinate dehydrogenase	Reference gene
AAT	Aspartate amino transferase	Responsible for the conversion of aspartate and α-ketoglutarate to oxaloacetate and glutamate
AK/HSDH	Aspartate kinase/homoserine dehydrogenase	Bifunctional: AK catalyses the phosphorylation of aspartate, first step in the synthesis of methionine, lysine and threonine. HSDH participates in glycine, serine and threonine metabolism and lysine synthesis
AlaAT	Alanine amino transferase	Responsible for the transfer of an amino group from alanine to α-ketoglutarate to give pyruvate and glutamate
ALS	Acetolactate synthase	Responsible for first step in synthesis of branched chain amino acids: valine, leucine and isoleucine
AS1	Asparagine synthetase	Responsible for the ATP-dependent transfer of amino group of glutamine to aspartate to generate glutamate and asparagine
CGS	Cystathionine gamma-synthase	Involved in methionine, cysteine, selenoamino acid and sulphur metabolism
DHDPS	Dihydrodipicolinate synthase	Responsible for first unique reaction of lysine synthesis
DHS2	3-Deoxy-D-arabino-heptulosonate-7-phosphate synthase	Involved in early stages of aromatic amino acid biosynthesis
eIF2α	Translation initiation factor 2α	Substrate of GCN2
GCN2	General control nonderepressible	Overexpressed in transgenic lines
HDH	Histidinol dehydrogenase	Involved in histidine metabolism
NR	Nitrate reductase	Responsible for key step in incorporation of inorganic nitrogen into amino acids
PAT1	Phosphoribosylanthranilate transferase	Involved in tryptophan biosynthesis
PP2A	Protein phosphatase 2A	Dephosphorylates eIF2α
PSP	Phosphoserine phosphatase	Involved in glycine, serine and threonine metabolism
SDI1	Sulphur-deficiency-induced-1	Marker for sulphur deprivation; unknown function
14-3-3NR	14-3-3 protein	Involved in interaction with NR, regulating activity

gine accumulates to high concentrations in plants in response to a variety of environmental and biotic stimuli (Lea *et al.*, 2007; Curtis *et al.*, 2009, 2010); in wheat, sulphur deprivation has a particularly dramatic effect, causing increases of up to 30-fold in free asparagine concentration in the grain (Muttucumaru *et al.*, 2006; Granvogl *et al.*, 2007; Curtis *et al.*, 2009). A two-thirds reduction in free asparagine concentration in wheat grain and a mitigation of the effects of sulphur deficiency would be of great benefit to the food industry.

The expression of a similar suite of genes in a *gcn2* mutant of Arabidopsis showed little change with wild type (Zhang *et al.*, 2008). However, the Arabidopsis study did not include an overexpression experiment or use sulphur deprivation to perturb the system. Nor did it include an analysis of *AS1* or *SDI1* genes and it was these genes that differed most between the *TaGCN2* overexpressing lines and the controls. Wheat appears to be extremely sensitive to sulphur deprivation and to respond with dramatic changes in free amino acid accumulation. The wheat system may therefore simply be a better one for demonstrating the role of *TaGCN2* in regulating gene expression.

The transgenic plants may have been compensating for *TaGCN2* overexpression by increasing expression of eIF2α, the substrate for GCN2-type protein kinases, and of a protein phosphatase 2A, which reverses the action of GCN2. This may explain why there was no evidence of a negative effect on yield in the overexpressing lines. The fact that there were such profound effects on the expression of other genes despite this leads us to speculate that GCN2 regulates gene expression in plants through a different mechanism from that described in budding yeast. In that organism, eIF2α phosphorylation by GCN2 controls the translation of transcription factor, GCN4. However, no

GCN4 homologue has been identified in plants, despite the extensive genome data that are now available (Halford, 2006). Animals, on the other hand, do have a GCN4 homologue, ATF4, but lack the ability to make many amino acids.

There are other differences between the regulatory system in yeast and the one that is being elucidated in plants. For example, overexpression of *TaGCN2* repressed the expression of genes encoding AS1, DHS and CGS, increased that of genes encoding AK/HSDH, PSP and DHDPS and had no consistent significant effect on genes encoding AAT, AlaAT, ALS, HDH and PAT. Obviously, this is not the same as the general amino acid control system of yeast, in which activation of GCN2 results in the translation of GCN4 and the promotion of expression of genes encoding enzymes in every amino acid biosynthetic pathway except cysteine (Natarajan *et al.*, 2001). The involvement of GCN2 in regulating genes in response to sulphur availability has also never been demonstrated in fungi. Clearly, while some of the components and mechanisms of the regulatory systems controlling protein synthesis and the expression of genes encoding enzymes of amino acid biosynthesis have been conserved as fungi and plants have diverged, others have changed substantially.

Experimental procedures

Isolation of wheat leaf RNA

Six-week-old wheat (*Triticum aestivum* cv. Cadenza) leaf material was snap-frozen in liquid nitrogen before being crushed to a fine powder using a chilled pestle and mortar. Total RNA was purified using the RNeasy Mini Kit (Qiagen Ltd, Crawley, UK) following the manufacturer's instructions. Alternatively, RNA

Table 3 Analysis of gene expression in *TaGCN2*-overexpressing lines 402 and 426 compared with wild-type wheat cv. Cadenza grown with sulphur supplied (S+) or withheld (S−). There were three biological replicates for each line by sulphur treatment combination. Genes encoding GAPDH and SDH were used as reference genes for all other genes. (A) *P*-values from the ANOVA for the result of the *F*-test on main effects and interactions between the line and sulphur treatment. The genes and *P*-values in boldface indicate that significant ($P < 0.05$) or marginal ($0.05 < P < 0.1$) differences were identified, warranting further investigation. Also shown are the value of residual variance (s^2) and the degrees of freedom (df). (B) Relevant means tables, on the $\log_2[1/$normalized relative quantity (NRQ)] scale, for comparison of overall line effects for genes showing no significant ($P > 0.05$) response to sulphur, using the least significant difference (LSD) (5%) values. Genes given in boldface show significant differences ($P < 0.05$) in expression between the control and both transgenic lines. Note that a low $\log_2(1/$NRQ) indicates high gene expression, whereas a high $\log_2(1/$NRQ) indicates low gene expression. (C) Relevant means tables, on the $\log_2(1/$NRQ) scale, for comparison of effect of line, sulphur, or line by sulphur interaction for genes showing a significant ($S < 0.05$) sulphur response, using the LSD (5%) values

Genes	Line	Sulphur	Line.Sulphur	s^2, residual df
A				
GAPDH	0.166	0.960	0.727	0.1261, 12
GAPDH2	0.131	0.772	0.952	0.1887, 12
SDH	0.166	0.960	0.727	0.1261, 12
AATx	0.006	0.926	0.948	0.1294, 12
AATy	0.080	0.862	0.489	0.2198, 12
AATz	0.237	0.967	0.518	0.08958, 12
AK/HSDH	0.131	0.292	0.084	0.1932, 12
AlaAT	0.397	0.167	0.215	0.2848, 12
ALS	0.030	0.381	0.790	0.7779, 12
AS1	<0.001	0.464	0.071	6.131, 12
CGS	0.026	0.645	0.207	0.1838, 12
DHDPS	0.012	0.607	0.837	0.3534, 12
DHS2	0.006	0.566	0.040	0.2601, 12
eIF2α	<0.001	0.503	0.640	0.09902, 12
GCN2	0.001	0.296	0.583	0.3432, 12
HDH	0.278	0.344	0.601	2.400, 12
NR	<0.001	0.517	0.155	0.3102, 12
PAT1	0.616	0.726	0.951	0.1285, 12
PP2A	0.001	0.532	0.502	0.1349, 12
PSP	0.076	0.138	0.071	0.1845, 12
SDI1	<0.001	0.434	0.055	2.256, 12
14-3-3	0.999	0.062	0.914	0.9885, 11

Genes	Line Control	402	426	SED	df	LSD (5%)	Comment
B							
GCN2	5.64	4.23	4.20	0.338	12	0.737	Control significantly different ($P < 0.05$) from 402 and 426
AATx	1.27	0.888	0.441	0.208	12	0.453	Control significantly different ($P < 0.05$) from 426 but not 402
AATy	1.54	1.18	0.86	0.271	12	0.590	Control significantly different ($P < 0.05$) from 426 but not 402
ALS	−0.02	−0.94	−1.58	0.509	12	1.11	Control significantly different ($P < 0.05$) from 426 but not 402
CGS	−1.41	−0.62	−1.04	0.248	12	0.539	Control significantly different ($P < 0.05$) from 402 and 426
DHDPS	1.43	0.25	0.12	0.534	12	1.16	Control significantly different ($P < 0.05$) from 402 and 426
eIF2α	5.19	3.96	3.52	0.182	12	0.396	Control significantly different ($P < 0.05$) from 402 and 426
NR	7.21	5.60	5.97	0.322	12	0.701	Control significantly different ($P < 0.05$) from 402 and 426
PP2A	0.557	−0.103	−0.475	0.212	12	0.462	Control significantly different ($P < 0.05$) from 402 and 426

Genes and S regime	Line Control	402	426	SED	df	LSD (5%)	Comment
C							
AK/HSDH							
S+	2.35	2.23	1.96	0.359	12	0.782	Control S+ significantly different ($P < 0.05$) from control S−. No S effect in 402 or 426

Table 3 Continued

Genes and S regime	Line			SED	df	LSD (5%)	Comment
	Control	402	426				
S−	1.44	2.55	1.88				
AS1							
S+	2.49	9.53	7.97	2.02	12	4.41	Control significantly different (P < 0.05) from 402 and 406 in S+ and S−.
S−	−2.55	9.89	10.0				Significant increase (P < 0.05) in expression in S− compared with
							S+ in control but not in 402 or 426
DHS2							
S+	0.08	−0.17	0.52	0.416	12	0.907	Control S+ significantly different (P < 0.05) from control S−. Control
S−	−0.77	0.51	1.13				significantly different (P < 0.05) from 402 and 426 in S−
PSP							
S+	2.93	2.04	1.75	0.351	12	0.764	Expression in control significantly higher (P < 0.05) in S− than in S+.
S−	1.96	1.74	2.06				Expression in control significantly different (P < 0.05) from 402 and
							426 in S+ but not S−
SDI1							
S+	9.71	12.5	12.9	1.23	12	2.67	Control significantly different (P < 0.05) from 402 and 406 in S+ and
S−	6.41	13.5	13.5				S−. Significant (P < 0.05) increase in expression in control in S− compared
							with S+ but not in 402 or 426

SED, standard error of the difference.

was extracted from leaf material with Trizol reagent (Invitrogen Ltd, Paisley, UK). RNA was treated with DNase (Promega, Southampton, UK) to prevent DNA contamination. RNA quality was checked using a spectrophotometer and in some cases by agarose gel electrophoresis.

Isolation of RNA from grain

Up to 250 mg of frozen grain material was allowed to thaw momentarily, squashed to rupture the structure, then re-frozen in liquid nitrogen and ground to a fine powder. RNA was extracted from powdered grain tissue using the CTAB method (Chang et al., 1993). RNA was further purified using the RNeasy MinElute clean-up column that included an on-column DNase treatment (Qiagen).

Molecular cloning of TaGCN2

The design of the antisense primer used to amplify a product from TaGCN2 mRNA, GCCAATCAGCTCCAGATTGTAGGA, was based on a wheat EST matching the 3′ end of the Arabidopsis AtGCN2 nucleotide sequence (Zhang et al., 2003), which was identified using an in silico search of the WhETS database (Mitchell et al., 2007). A similar search revealed a previously uncharacterized rice (Oryza sativa) GCN2-like nucleotide sequence (GenBank: XM473001) that was used to design a sense primer: ATGGGGCACAGCGCGAGGAAGAAGAA.

Complementary DNA was generated from the wheat RNA by reverse transcription using SuperScript III (Invitrogen). Amplification by PCR used Phusion High-Fidelity DNA Polymerase (Finnzymes, Vantaa, Finland). Cycling conditions were as follows: 98 °C for 30 s; 40 cycles of 98 °C for 10 s, 50 °C to 70 °C gradient for 20 s and 72 °C for 3 min.; final hold at 72 °C for 10 min.

Polymerase chain reaction products were cloned and nucleotide sequences were determined using a BigDye Terminator v3.1 Cycle Sequencing kit (Applied Biosystems, Life Technolo-

gies, Carlsbad, CA). The reaction conditions were as follows: 96 °C for 1 min., followed by 25 cycles of 96 °C for 10 s, 50 °C for 5 s and 60 °C for 4 min. Nucleotide sequence analysis was performed by Geneservice, Source Bioscience (Nottingham, UK) or MWG Biotech (Wolverhampton, UK).

Amplification of 3′ cDNA ends

The nucleotide sequence of the 3′ end of the TaGCN2 transcript was determined by rapid amplification of the cDNA end (RACE) using a GeneRacer kit (Invitrogen), which incorporates Phusion High-fidelity DNA polymerase. Three primers were used: AGTCTGTTCAAAGGGTGGCGGTGG, GGTGGACTCTTAAACGAGCGCATGGA and ACCAATAACACAGGCCGAAG. The PCR conditions were as follows: 98 °C for 30 s; 40 cycles of 98 °C for 10 s, 65 °C for 15 s and 72 °C for 20 s; final extension at 72 °C for 10 min. An aliquot of the reaction product was analysed by agarose gel electrophoresis; another aliquot was then used as the template for nested PCR.

Production of transgenic wheat plants

The full-length TaGCN2 open reading frame was spliced into a plasmid downstream of a rice actin gene promoter, which is constitutively active (McElroy et al., 1990). The termination signal was from the Agrobacterium tumefaciens nopaline synthase gene (nos) (Jefferson, 1987). The plasmid was introduced into wheat by particle bombardment of scutella tissue. Plasmid pAHC20 (Christensen and Quail, 1996), which conveys resistance to the herbicide phosphinothricin (PPT), was used for cotransformation. Following selection, PCR was used to establish the presence of the transgene. Transgenic plants were self-fertilized, and PPT-resistant progeny that tested positive for the presence of the construct were selected. This was repeated to the T3 generation, and three independent, homozygous lines, 395, 402 and 426, were produced. Embryo isolation and bombardment and plant regeneration and selection were performed

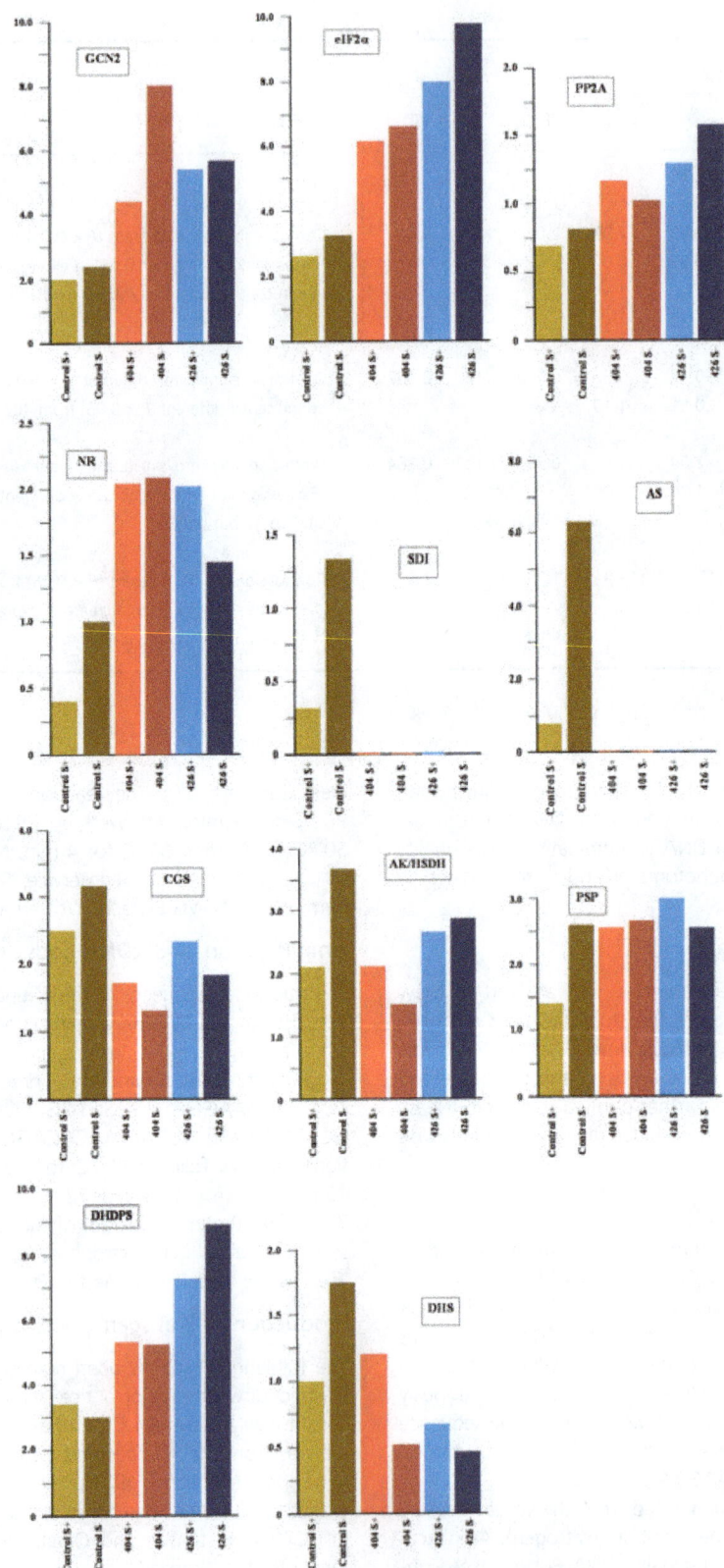

Figure 4 Expression [normalized relative quantities (NRQ)] values of a suite of genes (Table 3) in the leaves of transgenic wheat lines in which *TaGCN2* was overexpressed under the control of an actin gene promoter and in wild-type (cv. Cadenza) controls. The plants were grown with sulphur either supplied (S+) or withheld (S−). The analysis was carried out by quantitative real-time polymerase chain reaction using 3-phosphate dehydrogenase (GAPDH) and succinate dehydrogenase (SDH) as reference genes. Note that the graphs have different scales on the *y*-axis. In all cases shown, the levels of expression of the gene differed significantly (*P* < 0.05) between the transgenic and control lines in either S+ or S− conditions, or in both (Table 3).

within the Rothamsted Cereal Transformation Laboratory using the methods described by Sparks and Jones (2009). The presence of the transgene in individual plants was checked by PCR using genomic DNA as the template. The primers used were 5'-CAAGGACCACGCCGCGCAG, which anneals in exon 1, and 5'-GCTAAATCGGGTGTGAGGTGATTGTG, which anneals in exon 2. The product amplified from the endogenous gene therefore contained an approximately 0.8-kb intron that was not present in the transgene. Successive self-fertilization to the T_3 generation was carried out to achieve homozygosity.

Sulphur feeding

Transgenic and wild-type (cv Cadenza) wheat plants were grown in vermiculite in a glasshouse with a 16-h daylength (supplemental lighting was used as necessary) and a minimum temperature of 16 °C. Vermiculite does not retain nutrients, so once seed reserves were exhausted, the only nutrition available to the developing seedlings came from externally applied liquid feed solution. Feeding was started 3 weeks after potting and continued every 2 days until harvest. Distilled water was also supplied as required to prevent water stress. A completely randomized design was used for the pots in the glasshouse. Plants were supplied with either a medium containing a full nutrient complement of potassium, phosphate, calcium, magnesium, sodium, iron, nitrate (2 mM $Ca(NO_3)_2$ and 1.6 mM $Mg(NO_3)_2$) and sulphate ions (1.1 mM $MgSO_4$) (Muttucumaru et al., 2006; Curtis et al., 2009) or the same medium containing one-tenth the concentration of $MgSO_4$. RNA was prepared from flag leaves as described earlier.

Expression analyses by real-time quantitative PCR

First-strand cDNA synthesis was performed using SuperScript III (Invitrogen) to reverse-transcribe 1–2 µg DNase-treated RNA and was primed with an anchored dT_{20} primer in a final volume of 20 µL. The qPCR reaction mixture consisted of 10 µL SYBR Green JumpStart Taq ReadyMix (Sigma, Poole, UK), 5 µL diluted cDNA and 5 µL primers (125 nM final concentration). Samples were run in an ABI7500 real-time PCR system (Applied Biosystems), and the amplification conditions were 95 °C for 2 min, then 45 cycles of 95 °C for 15 s followed by 67 °C for 45 s. Primer nucleotide sequences are given in Table S1.

The efficiencies of the reactions were estimated using the LinReg PCR program (Ramakers et al., 2003), and the ct (at threshold fluorescence) and efficiency values were then used to calculate the NRQ with respect to the reference genes, cyclophilin, SDH and glyceraldehyde 3-phosphate dehydrogenase (GAPDH), for each sample/target gene combination.

$$NRQ = \frac{E_{target}^{-ct,target}}{\sqrt{E_{SDH}^{-ct,SDH} * E_{GAPDH}^{-ct,GAPDH}}}$$

where E_{target}, E_{SDH} and E_{GAPDH} are the estimated reaction efficiencies for a particular target gene and the two reference genes and where $ct,target$, ct,SDH and $ct,GAPDH$ are the corresponding ct values.

Statistical analysis of the gene expression data from the sulphur feeding experiment data was performed using the GenStat® (2010, Thirteenth Edition; VSN International Ltd, Hemel Hempstead, UK) statistical system. There were three biological replicates (leaf tissue samples) for each line by sulphur treat-

ment combination. Two genes, encoding GAPDH and SDH, were used as reference genes for all other genes. The stability of these genes across the line by sulphur treatment combinations was checked to confirm that they were suitable for this role. The NRQ for all genes were calculated. ANOVA was applied to the log (to base 2)-transformed inverse of the NRQ data. This transformation ensured homogeneity of variance across the line by sulphur treatment combinations and effectively provided values back on the ct-scale. Therefore, as for ct values, a low $log_2(1/NRQ)$ indicated a high gene expression, whereas a high $log_2(1/NRQ)$ indicated low gene expression. The analysis assessed the statistical significance of line and sulphur treatment main effects and the interaction between these two factors using the F-test. Following an F-test result indicating significant ($P < 0.05$) or marginal ($0.05 < P < 0.10$) differences, the least significant difference (LSD) at the 5% level of significance was used to separate pairs of means of interest in the appropriate table of means for each gene. Details of the genes analysed are given in Table 2. Primer sequences are given in Table S1.

Amino acid analyses

Free amino acid concentrations in mature grain of compost-grown plants were determined by gas chromatography-mass spectrometry (GC-MS) using methods described previously (Muttucumaru et al., 2006; Curtis et al., 2009). For each amino acid and the total, wheat lines were compared using ANOVA. Following an F-test result indicating significant ($P < 0.05$) overall differences between lines, specific comparisons of transgenic lines to controls were made using the SED in post-ANOVA t-tests based on the residual degrees of freedom (df). Analysis was performed using GenStat®.

Total nitrogen and carbon analysis

Measurements of total grain nitrogen and carbon were taken by the Analytical Unit of the Soil Science Department, Rothamsted Research, using the 'Dumas' digestion method and a LECO CNS 2000 Combustion Analyser (LECO Corporation, Saint Joseph, MI).

Acknowledgements

Rothamsted Research receives grant-aided support from the Biotechnology and Biological Sciences Research Council (BBSRC) of the United Kingdom. Edward Byrne was supported through a PhD studentship from the BBSRC, and the study was further supported through a grant from the BBSRC's 'Follow-on Fund'. The authors are grateful to staff of the cereal transformation unit at Rothamsted Research, including Huw Jones, Mandy Riley, Caroline Sparks, Angela Doherty and Melloney McCluskey.

References

Chang, S.J., Puryea, J. and Cairney, J. (1993) A simple and efficient method for isolating RNA from pine tree. Plant Mol. Biol. Rep. **11**, 113–116.

Chang, L.Y., Yang, W.Y., Browning, K. and Roth, D. (1999) Specific in-vitro phosphorylation of plant eIF2α by eukaryotic eIF2α kinases. Plant Mol. Biol. **41**, 363–370.

Chang, L.Y., Yang, W.Y. and Roth, D. (2000) Functional complementation by wheat eIF2α in the yeast GCN2-mediated pathway. Biochem. Biophys. Res. Commun. **279**, 468–474.

Chen, J.J. and London, I.M. (1995) Regulation of protein synthesis by heme-regulated eIF2α kinase. *Trends Biochem. Sci.* **20**, 105–108.

Christensen, A.H. and Quail, P.H. (1996) Ubiquitin promoter-based vectors for high-level expression of selectable and/or screenable marker genes in monocotyledonous plants. *Transgenic Res.* **5**, 213–218.

Curtis, T.Y., Muttucumaru, N., Shewry, P.R., Parry, M.A., Powers, S.J., Elmore, J.S., Mottram, D.S., Hook, S. and Halford, N.G. (2009) Evidence for genetic and environmental effects on free amino acid levels in wheat grain: implications for acrylamide formation during processing. *J. Agric. Food. Chem.* **57**, 1013–1021.

Curtis, T.Y., Powers, S.J., Balagiannis, D., Elmore, J.S., Mottram, D.S., Parry, M.A.J., Raksegi, M., Bedő, Z., Shewry, P.R. and Halford, N.G. (2010) Free amino acids and sugars in rye grain: implications for acrylamide formation. *J. Agric. Food. Chem.* **58**, 1959–1969.

Friedman, M. (2003) Chemistry, biochemistry and safety of acrylamide. A review. *J. Agric. Food. Chem.* **51**, 4504–4526.

Granvogl, M., Wieser, H., Koehler, P., von Tucher, S. and Schieberle, P. (2007) Influence of sulphur fertilization on the amounts of free amino acids in wheat. Correlation with baking properties as well as with 3-aminopropionamide and acrylamide generation during baking. *J. Agric. Food. Chem.* **55**, 4271–4277.

Guyer, D., Patton, D. and Ward, E. (1995) Evidence for cross-pathway regulation of metabolic gene expression in plants. *Proc. Natl Acad. Sci. USA*, **92**, 4997–5000.

Halford, N.G. (2006) Regulation of carbon and amino acid metabolism: roles of sucrose nonfermenting-1-related protein kinase-1 and general control nonderepressible-2-related protein kinase. *Adv. Bot. Res. Inc. Adv. Plant Path.* **43**, 93–142.

Halford, N.G., Curtis, T.Y., Muttucumaru, N., Postles, J. and Mottram, D.S. (2011) Sugars in crop plants. *Ann. Appl. Biol.* **158**, 1–25.

Harding, H.P., Novoa, I., Zhang, Y., Zeng, H., Wek, R., Schapira, M. and Ron, D. (2000) Regulated translation initiation controls stress-induced gene expression in mammalian cells. *Mol. Cell*, **6**, 1099–1108.

Hershey, J.W.B. and Merrick, W.C. (2000) The pathway and mechanism of initiation of protein synthesis. In *Translational Control of Gene Expression* (Sonenberg, N., Hershey, J.W.B. and Mathews, M.B., eds), pp. 33–88. New York, NY: Cold Spring Harbor Laboratory Press.

Hinnebusch, A.G. (1992) General and pathway-specific regulatory mechanisms controlling the synthesis of amino acid biosynthetic enzymes in *Saccharomyces cerevisiae*. In *Molecular and Cellular Biology of the Yeast Saccharomyces, Volume 2, Gene Expression* (Jones, E.W., Pringle, J.R. and Broach, J.B., eds), pp. 319–414. New York, NY: Cold Spring Harbor Laboratory Press.

Hinnebusch, A.G. (1994) Translational control of GCN4 – an *in vivo* barometer of initiation factor activity. *Trends Biochem. Sci.* **19**, 409–414.

Hinnebusch, A.G. (1997) Translational regulation of yeast GCN4 – A window on factors that control initiator-tRNA binding to the ribosome. *J. Biol. Chem.* **272**, 21661–21664.

Hinnebusch, A.G. (2005) Translational regulation of GCN4 and the general amino acid control of yeast. *Annu. Rev. Microbiol.* **59**, 407–450.

Howarth, J.R., Parmar, S., Barraclough, P.B. and Hawkesford, M.J. (2009) A sulphur-deficiency induced gene, sdi1, involved in the utilisation of stored sulphate pools under S-limiting conditions has potential as a diagnostic indicator of S-nutritional status. *Plant Biotechnol. J.* **7**, 200–209.

Jefferson, R.A. (1987) Assaying chimeric genes in plants: The GUS gene fusion system. *Plant Mol. Biol. Rep.* **5**, 387–405.

Kaufman, R.J. (1999) Stress signaling from the lumen of the endoplasmic reticulum: coordination of gene transcriptional and translational controls. *Genes Dev.* **13**, 1211–1233.

Lageix, S., Lanet, E., Pouch-Pelissier, M.N., Espagnol, M.C., Robaglia, C., Deragon, J.M. and Pelissier, T. (2008) Arabidopsis eIF2α kinase GCN2 is essential for growth in stress conditions and is activated by wounding. *BMC Plant Biol.* **8**, 134–142.

Lea, P.J., Sodek, L., Parry, M.A., Shewry, P.R. and Halford, N.G. (2007) Asparagine in plants. *Ann. Appl. Biol.* **150**, 1–26.

Mascarenhas, C., Edwards-Ingram, L.C., Zeef, L., Shenton, D., Ashe, M.P. and Grant, C.M. (2008) Gcn4 is required for the response to peroxide stress in the yeast *Saccharomyces cerevisiae*. *Mol. Biol. Cell*, **19**, 2995–3007.

McElroy, D., Zhang, W.G., Cao, J. and Wu, R. (1990) Isolation of an efficient actin promoter for use in rice transformation. *Plant Cell*, **2**, 163–171.

Mitchell, R.A.C., Castells-Brooke, N., Taubert, J., Verrier, P.J., Leader, D.J. and Rawlings, C.J. (2007) Wheat Estimated Transcript Server (WhETS): a tool to provide best estimate of hexaploid wheat transcript sequence. *Nucleic Acids Res.* **35**, W148–W151.

Mottram, D.S. (2007) The Maillard reaction: source of flavour in thermally processed foods. In *Flavours and Fragrances: Chemistry, Bioprocessing and Sustainability* (Berger, R.G., ed.), pp. 269–284. Berlin, Heidelburg: Springer-Verlag.

Mottram, D.S., Wedzicha, B.L. and Dodson, A.T. (2002) Acrylamide is formed in the Maillard reaction. *Nature*, **419**, 448–449.

Muttucumaru, N., Halford, N.G., Elmore, J.S., Dodson, A.T., Parry, M., Shewry, P.R. and Mottram, D.S. (2006) The formation of high levels of acrylamide during the processing of flour derived from sulfate-deprived wheat. *J. Agric. Food. Chem.* **54**, 8951–8955.

Nanduri, S., Rahman, F., Williams, B.R.G. and Qin, J. (2000) A dynamically tuned double-stranded RNA binding mechanism for the activation of antiviral kinase PKR. *EMBO J.* **19**, 5567–5574.

Natarajan, K., Meyer, M.R., Jackson, B.M., Slade, D., Roberts, C., Hinnebusch, A.G. and Marton, M.J. (2001) Transcriptional profiling shows that Gcn4p is a master regulator of gene expression during amino acid starvation in yeast. *Mol. Cell. Biol.* **21**, 4347–4368.

Noctor, G., Veljovic-Jovanovic, S. and Foyer, C.H. (2000) Peroxide processing in photosynthesis: antioxidant coupling and redox signalling. *Philos. Trans. R. Soc. London., B, Biol. Sci.* **355**, 1465–1475.

Ramakers, C., Ruijter, J.M., Deprez, R.H. and Moorman, A.F. (2003) Assumption-free analysis of quantitative real-time polymerase chain reaction (PCR) data. *Neurosci. Lett.* **339**, 62–66.

Sikalidis, A.K. and Stipanuk, M.H. (2010) Growing rats respond to a sulphur amino acid-deficient diet by phosphorylation of the α subunit of eukaryotic initiation factor 2 heterotrimeric complex and induction of adaptive components of the integrated stress response. *J. Nutr.* **140**, 1080–1085.

Sparks, C.A. and Jones, H.D. (2009) Biolistics transformation of wheat. *Methods Mol. Biol.* **478**, 71–92.

Stadler, R.H., Blank, I., Varga, N., Robert, F., Hau, J., Guy, P.A., Robert, M.C. and Riediker, S. (2002) Acrylamide from Maillard reaction products. *Nature*, **419**, 449–450.

Tareke, E., Rydberg, P., Karlsson, P., Eriksson, S. and Tornqvist, M. (2002) Analysis of acrylamide, a carcinogen formed in heated foodstuffs. *J. Agric. Food. Chem.* **50**, 4998–5006.

Wang, H., Liu, D., Sun, J. and Zhang, A. (2005) Asparagine synthetase gene TaASN1 from wheat is up-regulated by salt stress, osmotic stress and ABA. *J. Plant Physiol.* **162**, 81–89.

Wek, R.C., Jackson, B.M. and Hinnebusch, A.G. (1989) Juxtaposition of domains homologous to protein kinases and histidyl transfer RNA synthetases in GCN2 protein suggests a mechanism for coupling GCN4 expression to amino acid availability. *Proc. Natl Acad. Sci. USA*, **86**, 4579–4583.

Wek, S.A., Zhu, S.H. and Wek, R.C. (1995) The Histidyl-tRNA synthetase related sequence in the eIF2α protein kinase GCN2 interacts with tRNA and is required for activation in response to starvation for different amino acids. *Mol. Cell. Biol.* **15**, 4497–4506.

Wek, R.C., Ma, K., Vattem, K., Narasimhan, J., Staschke, K. and Jiang, H.Y. (2003) Regulation of eIF2 kinases in response to endoplasmic reticulum and nutritional stresses. *FASEB J.* **17**, A184.

Wek, R.C., Jiang, H.Y. and Anthony, T.G. (2006) Coping with stress: eIF2 kinases and translational control. *Biochem. Soc. Trans.* **34**, 7–11.

Yang, R.J., Wek, S.A. and Wek, R.C. (2000) Glucose limitation induces GCN4 translation by activation of GCN2 protein kinase. *Mol. Cell. Biol.* **20**, 2706–2717.

Zhang, Y., Dickinson, J.R., Paul, M.J. and Halford, N.G. (2003) Molecular cloning of an Arabidopsis homologue of GCN2, a protein kinase involved in co-ordinated response to amino acid starvation. *Planta*, **217**, 668–675.

Zhang, Y., Wang, Y., Kanyuka, K., Parry, M.A.J., Powers, S.J. and Halford, N.G. (2008) GCN2-dependent phosphorylation of eukaryotic translation initiation factor-2α in Arabidopsis. *J Exp Bot* **59**, 3131–3141.

8

Durable field resistance to wheat yellow mosaic virus in transgenic wheat containing the antisense virus polymerase gene

Ming Chen[1,†], Liying Sun[2,†], Hongya Wu[3], Jiong Chen[2], Youzhi Ma[1], Xiaoxiang Zhang[3], Lipu Du[1], Shunhe Cheng[3], Boqiao Zhang[3], Xingguo Ye[1], Junlan Pang[1], Xinmei Zhang[1], Liancheng Li[1], Ida B. Andika[2], Jianping Chen[2,*] and Huijun Xu[1,*]

[1]Institute of Crop Sciences, Chinese Academy of Agricultural Sciences/National Key Facility for Crop Gene Resources and Genetic Improvement, Key Laboratory of Biology and Genetic Improvement of Triticeae Crops, Ministry of Agriculture, Beijing, China
[2]State Key Laboratory Breeding Base for Zhejiang Sustainable Pest and Disease Control, MoA Key Laboratory for Plant Protection and Biotechnology, Zhejiang Provincial Key Laboratory of Plant Virology, Institute of Virology and Biotechnology, Zhejiang Academy of Agricultural Sciences, Hangzhou, China
[3]Institute of Agricultural Sciences of Lixiahe Districts, Jiangsu, China

*Correspondence
email jpchen2001@126.com
and

email xuhuijun@caas.cn
[†]These authors contributed equally to this work.

Keywords: transgenic wheat, wheat yellow mosaic virus, durable field resistance, RNA silencing, virus replicase gene.

Summary

Wheat yellow mosaic virus (WYMV) has spread rapidly and causes serious yield losses in the major wheat-growing areas in China. Because it is vectored by the fungus-like organism Polymyxa graminis that survives for long periods in soil, it is difficult to eliminate by conventional crop management or fungicides. There is also only limited resistance in commercial cultivars. In this research, fourteen independent transgenic events were obtained by co-transformation with the antisense NIb8 gene (the NIb replicase of WYMV) and a selectable gene bar. Four original transgenic lines (N12, N13, N14 and N15) and an offspring line (N12-1) showed high and durable resistance to WYMV in the field. Four resistant lines were shown to have segregated and only contain NIb8 (without bar) by PCR and herbicide resistance testing in the later generations. Line N12-1 showed broad-spectrum resistance to WYMV isolates from different sites in China. After growing in the infested soil, WYMV could not be detected by tissue printing and Western blot assays of transgenic wheat. The grain yield of transgenic wheat was about 10% greater than the wild-type susceptible control. Northern blot and small RNA deep sequencing analyses showed that there was no accumulation of small interfering RNAs targeting the NIb8 gene in transgenic wheat plants, suggesting that transgene RNA silencing, a common mechanism of virus-derived disease resistance, is not involved in the process of WYMV resistance. This durable and broad-spectrum resistance to WYMV in transgenic wheat will be useful for alleviating the damage caused by WYMV.

Introduction

Wheat yellow mosaic disease is characterized by mosaic or yellow-striped leaves and plant stunting and is one of the major threats to wheat production in Europe, North America and East Asia (Kühne, 2009). In China, this disease first occurred in the middle and lower reaches of the Changjiang River in 1970s and since the 1990s has spread to several provinces (Chen et al., 1989; Han et al., 2000). It results in reduction in grain yield by 20–70% in some individual fields (Liu et al., 2005b), and in China, it is predominantly caused by wheat yellow mosaic virus (WYMV) (Han et al., 2000). The virus is transmitted by an obligate soil-inhabiting fungus-like organism Polymyxa graminis (order Plasmodiophorales) and is protected inside the thick-walled resting spores and zoospores of the vector (Kühne, 2009). Because dormant spores of P. graminis can survive for long periods in the soil, the inoculum in contaminated fields is difficult to eliminate by conventional crop management or fungicides. The best effective method to control the disease is to grow resistant wheat varieties, but there are few known resistance genes in current wheat cultivars. It has been reported that WYMV resistance in wheat is controlled by one to three genes (Qin et al., 1986) with one major gene being associated with

homologous group 2 chromosome (Liu et al., 2005a; Nishio et al., 2010; Zhu et al., 2012). Furthermore, genetic analyses showed that the inheritance of this resistance is complex and influenced by many factors (Zhou et al., 2000). It therefore seems that genetic engineering may be more promising than conventional breeding for producing plants resistant to WYMV.

WYMV belongs to the genus Bymovirus (family Potyviridae). Its genome is encapsidated in filamentous particles and consists of two positive sense single-stranded RNA segments (RNA1, 7.5 kb and RNA2, 3.6 kb), each of which encodes a polyprotein that undergoes post-translational cleavage (Namba et al., 1998). The polyprotein encoded by RNA1 (269 kDa) produces eight proteins including the coat protein (CP) and a nuclear inclusion 'b' protein (NIb) that functions as an RNA dependent RNA polymerase (RdRp) and is important for virus replication (Figure 1a). RNA2 encodes a polyprotein (101 kDa) that give rise to two proteins of 28 kDa (P1) and 73 kDa (P2) (Namba et al., 1998).

Over the past two decades, several strategies have been developed based on the concept of pathogen-derived resistance (Sanford and Johnston, 1985). The integration of various viral sequences into their host plant genomes has proved to be effective in preventing or reducing the infection of many viruses (Gottula and Fuchs, 2009). The virus resistance in transgenic plants can be conferred by the expression of viral protein (protein-mediated) or viral RNA sequences alone (RNA-mediated) (Prins et al., 2008). RNA-mediated resistance can be achieved by transformation of plant with cDNA of the viral genome that is nontranslatable, in antisense orientation or arranged as an inverted repeat to produce hairpin RNA (Tenllado et al., 2004). RNA silencing, also called RNA interference (RNAi), has been demonstrated as a major mechanism of RNA-mediated resistance in transgenic plants carrying viral sequence (Simón-Mateo and García, 2011). Transgenic constructs capable of forming dsRNA (double-strand RNAs) transcripts were proved to be more effective in yielding high-level virus-resistant plants than the constructs producing either sense or antisense RNA alone (Pinto et al., 1999; Waterhouse et al., 1998). RNA silencing is an evolutionarily conserved process in a wide variety of eukaryotic organisms that is initiated when dsRNAs are processed by a ribonuclease III-like enzyme called Dicer into 21- to 24-nucleotide (nt) small interfering RNAs (siRNAs) (Baulcombe, 2004; Ding, 2010; Ding et al., 2004). The siRNAs are then incorporated into an RNA-induced silencing complex to guide the degradation or translational repression of homologous RNA targets in a sequence-specific manner (Ding and Voinnet, 2007).

Pathogen-derived resistance to viruses has been successfully engineered into a number of crop plants including wheat. Potyviridae is one of the largest plant virus families, and many of its members cause economically important crop diseases (Adams et al., 2011). The infection of many viruses belonging to the family Potyviridae has been successfully inhibited by viral-derived transgenes, and most of this engineered resistance is achieved through the RNA-mediated mechanism (Collinge et al., 2010). For examples, transgenic wheat plants carrying the complementary DNA of coat protein of wheat streak mosaic virus (WSMV, genus Tritimovirus) or carrying an RNAi construct that was designed from the nuclear inclusion protein 'a' (Nla) gene of WSMV to produce hairpin RNA were highly resistant to WSMV infection (Fahim et al., 2010; Sivamani et al., 2002). Likewise, RNA-mediated mechanism was implicated in the resistance of transgenic papaya against papaya ringspot virus (genus Potyvirus) (Gonsalves, 1998). There remain some problems hindering the use of pathogen-derived resistance in crop molecular breeding programmes. For example, most reported transgenic resistance has only been tested in the greenhouse, and there is little information from the field. The stability of transgenic resistance is not guaranteed, and only a few long-term tests have so far been reported.

In this article, we chose the Nlb8 region (1212 bp) as target sequence because it is one of the most conserved portions in WYMV genome and is therefore a good candidate for broad-spectrum resistance. The sequences of eight WYMV isolates sequenced in this part of their genome (one from Japan and seven from different parts of China) are >97% identity to one another (Figure S1). We report the production of wheat plants containing most of the antisense coding region of WYMV Nlb and the results of field testing for disease resistance. Field testing from 2000 to 2010 indicated that four original transgenic lines and a derived offspring line were highly resistant to WYMV infection and that the resistance was stably inherited, suggesting that this durable field resistance meets the requirements for practical application and could be therefore be used in breeding programmes for wheat disease resistance. Moreover, siRNAs analysis suggested that transgene RNA silencing, which is involved in virus resistance to other members of the family Potyviridae, is not part of the process of WYMV, suggesting that a novel transgenic virus resistance mechanism exists in these transgenic wheat plants.

Results

Selection and molecular identification of WYMV-resistant transgenic lines

The vector pubi-Nlb (Figure 1b) and assistant plasmid pAHC20 containing the selectable marker gene bar (Figure 1c) were co-transformed into wheat. After transformation, callus differentiation and regeneration of plants on the selective medium, 14 positive plants were identified by PCR from the T0 generation plants. The T1 generation plants derived from these 14 PCR-positive lines were tracked using PCR (Figure 2a and data not shown), and only PCR-positive plants were retained. The Nlb8 segment amplified from the transgenic wheat was sequenced, confirming that the Nlb8 inserted in the wheat genome was identical to that in the vector. RT-PCR showed that the Nlb8 gene was transcribed in some T2 generation transgenic lines including N12, N13, N14 and N15 (Figure 2b). At same time, WYMV resistance testing of the T2 transgenic wheat lines was performed in the disease nursery in 2000. Among the 14 T2 transgenic lines, four lines showed high resistance to WYMV (Figure 3a): N12, N13, N14 and N15. The numbers of virus-infected plants were, respectively, only 0, 0, 0 and 1, of 19, 48, 42 and 46, whereas about half of the susceptible control Yangmai 158 plants were infected (Table 1). To identify homozygousity of transgenic wheat

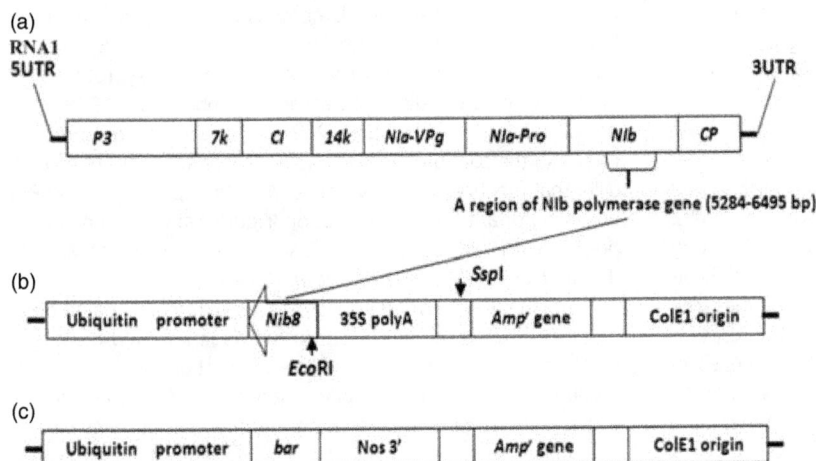

(a)
RNA1
5UTR   **3UTR**

P3 | 7k | CI | 14k | Nla-VPg | Nla-Pro | Nlb | CP

A region of Nlb polymerase gene (5284-6495 bp)

(b)

Ubiquitin promoter | Nlb8 | 35S polyA | Amp' gene | ColE1 origin

SspI
EcoRI

(c)

Ubiquitin promoter | bar | Nos 3' | Amp' gene | ColE1 origin

Figure 1 Structure of the wheat yellow mosaic virus genome (RNA1) and transformation plasmids. (a) The wheat yellow mosaic virus genome (RNA1). P3, P3 protein; 7k, 7 kDa protein; CI, cylindrical inclusion protein; 14k, 14 kDa protein; Nla-VPg, Nla-VPg protein; Nla-Pro, Nla-Pro protease; Nlb, Nlb polymerase; CP, coat protein. (b) The structure of vector pUbi-Nlb. Nlb8, Reverse complement sequence of Nlb polymerase gene (5284-6495 bp region); 35S polyA, Terminator. (c) Diagram of vector pAHC20 containing bar gene for plant selection in the co-transformation experiments. bar, herbicide-resistant gene; Nos 3', Terminator.

plants, 100 independent plants were randomly sampled from T3 population of the four lines N12, N13, N14 and N15, to detect the *Nlb8* gene by PCR, and 99% plants were tested to be positive, which indicated that these lines were homozygous in *Nlb8* insertion locus after T3 generation (data not shown).

To obtain marker-free transgenic wheat plants from the co-transformation experiments, the T3 plants derived from the positive N12 line were tested by PCR for the *Nlb8* and *bar* genes. Of the 268 plants tested, 28 had only *Nlb8* (without *bar*) and a marker-free line (N12-1) was developed from these plants (Figure 4). Later tests showed that the *bar* gene was also lost and the *Nlb8* retained in the T7 generations of N13, N14 and N15, whereas both genes were retained in N12 (Figure 4).

Southern blots of plant genomic DNA digested with *Eco*RI and *Ssp*I demonstrated single copy integration of the *Nlb8* gene in the five transgenic lines N12, N12-1, N13, N14 and N15 in the T6 generation (Figure 2c). To analyse the number of transgene locus and examine the heritability of *Nlb8* in N12-1 and N14, these two lines were crossed with WYMV-susceptible wheat variety Yangmai 15 (as female parent) and disease resistance of the F2 generation segregation population was tested in the field. In total, 693 resistant plants and 199 susceptible plants were found in the cross of Yangmai 15/N12-1, and 590 resistant plants and 183 susceptible plants in Yangmai 15/N14. The segregation ratios of 3.48 : 1 and 3.22 : 1, respectively, were consistent with Mendelian inheritance (Table 2), suggesting that *Nlb8* was integrated as a single locus in the N12-1 and N14 transgenic lines and that the resistance can be transferred to other wheat varieties by hybridization.

WYMV resistance of the transgenic wheat in field trials

The lines N12, N13, N14 and N15 from 2001 (T3) to 2005 (T7), and N12-1 from 2002 (T4) to 2005 (T7) were identified for disease resistance in a disease nursery in Jiangsu Province, and seed from the disease-free plants were harvested and bulked each year. WT controls were always heavily infected with Disease Index (DI, see methods) >90% in most seasons, whereas the transgenic lines had a much lower DI ($P < 0.01$), often at about or below 5% (Figure 3b,c and d, Table 3). PCR tests showed that the *Nlb8* gene was present in the plants of each generation. These results indicate that transformation using antisense-*Nlb8*

resulted in transgenic wheat plants with durable field resistance to WYMV.

To test the resistance to different virus strains, plants of N12-1 were grown on three different sites: Yantai (Shandong Province), Xiping (Henan Province) and Yangzhou (Jiangsu Province). The site at Yantai was mostly infested with WYMV, but Chinese wheat mosaic virus (CWMV), an unrelated virus (genus *Furovirus*, family *Virgaviridae*) transmitted by the same vector (Diao *et al.*, 1999), was occasionally detected, while only WYMV was detected in Xiping and Yangzhou. There are known to have differences between these WYMV populations (Sun *et al.*, 2013). N12-1 was highly resistant to WYMV in Yantai, Xiping and Yangzhou; a small number of transgenic plants developed disease symptom in Yantai, which is due to the CWMV infection (detected by ELISA). In contrast, the WT was severely diseased at all locations (Table 4).

Some major agronomic traits of the transgenic lines N12, N12-1, N13, N14 and N15 were compared to those of the WT Yangmai 158 in field disease nurseries in 2004 and 2005. In both years, the transgenic lines out yielded the WT, usually by at least 10% ($P < 0.05$) (Tables 5 and 6). Other apparent differences in agronomic traits were not statistically significant. In a field not infested with WYMV, there was no significant difference of grain yield or other agronomic traits between WT and transgenic wheat. It therefore appears that the increased yield of the transgenic wheat on the infested sites was a result of their excellent WYMV resistance.

Detection of WYMV accumulation in nontransgenic and transgenic plants

To investigate whether the absence of yellow mosaic symptoms in transgenic wheat plants is associated with the absence of virus accumulation, tissue printing and Western blot assays using an antibody specific for the WYMV CP were carried out. Leaves, stems and roots of WT Yangmai 158 (YM158) and transgenic line N12-1 were sampled 4 months after planting on WYMV-infested soil. In tissue printing assay, intense staining, indicating high levels of virus accumulation, was observed in upper leaves, lower leaves, stems and roots of WT plants, but not in those of N12-1 plants (Figure 5a). Similar results were also obtained from Western blot assays in which strong bands corresponding to

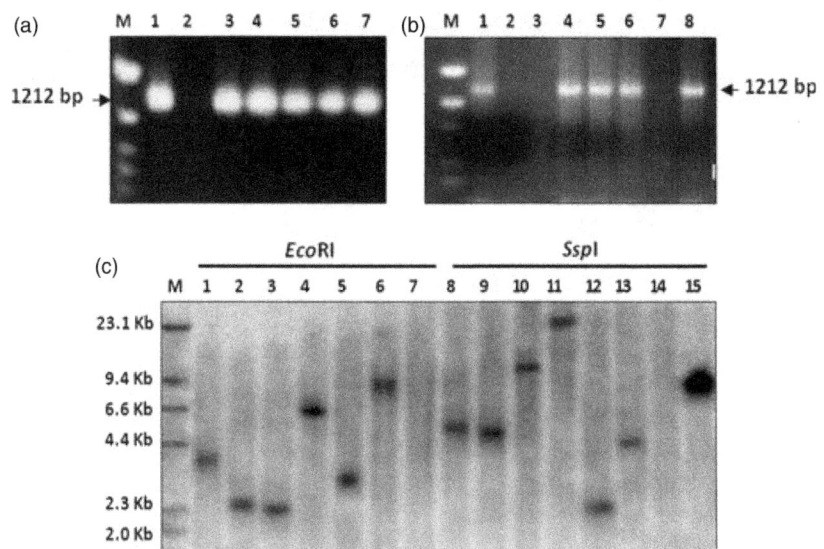

Figure 2 Molecular analysis of the transgenic wheat plants. (a) PCR testing for *Nlb8* gene in T1 generation plants. M, marker; 1, positive (plasmid) control; 2, negative control (Yangmai 158); 3, N12 line; 4, N13 line; 5, N14 line; 6, N15 line; 7, N17 line. (b) RT-PCR for *Nlb8* in T2 generation plants. M, marker; 1, positive (plasmid) control; 2 and 3, negative control (Yangmai 158); 4, N12 line; 5, N13 line; 6, N14 line; 7, N17 line; 8, N15 line. (c) Southern blotting for *Nlb8* gene in T6 generation lines using two restriction enzymes, *Eco*RI and *Ssp*I, to cut the genomic DNA. M, Marker; 2 and 8, N12 line; 3 and 9, N12-1 line; 1 and 10, N13 line; 4 and 11, N14 line; 5 and 12, N15 line; 6 and 13, N35 line; 7 and 14, Yangmai 158 (negative control); 15, positive control (plasmid).

Figure 3 Identification of transgenic wheat for disease resistance in field. Transgenic wheat in disease nursery at Yangzhou, Jiangsu Province, in March 2001 (a), Yangzhou in March 2002 (b), Yangzhou in March 2005 (c), and Yizheng, Jiangsu Province in March 2005 (d). Y158, wild-type Yangmai 158; N12, N12-1, N13, N14 and N15, transgenic wheat lines.

WYMV CP were detected in leaves, stems and roots of Yangmai 158 plants, but not in those of N12-1 plants (Figure 5b). Together, these results indicate that WYMV is unable to accumulate in N12-1 plants.

Table 1 Resistant testing of the T2 transgenic plants to the wheat yellow mosaic virus in field (2000)

Line	Total plants tested	Susceptible plants	DI (%)
N12	19	0	0.00
N13	48	0	0.00
N14	42	0	0.00
N15	46	1	2.17
Yangmai 158	46	22	47.83

A disease index (DI) was then calculated: DI (%) = $\sum$ (DS × Ni) × 100 / (3 × N), where DS is the disease scale (above), Ni the number of plants with this DS, and N the total number of plants observed.

The analysis of transgene- and WYMV-derived siRNAs in wheat plants

Next we analysed the accumulation of transgene-derived siRNAs in resistant transgenic wheat plant, because transgene RNA silencing is the most common mechanism in RNA-mediated virus resistance in transgenic plants (Simón-Mateo and García, 2011). First, we carried out Northern blot analysis using low molecular weight RNAs (30 µg) extracted from leaves of transgenic line N12-1. The blot was hybridized with a DIG-labelled DNA probe specific for WYMV RNA1 (position 5284-6495 nt) prepared by PCR. No transgene-derived siRNA was detected, whereas using the similar protocol with a DIG-labelled DNA probe, we were able to detect GFP transgene-derived siRNAs in an *Agrobacterium* co-infiltration assay where GFP was expressed in leaves of GFP-transgenic *N. benthamiana* line 16c (Voinnet *et al.*, 1998). In an attempt to increase the sensitivity of detection, we gel-purified small RNAs with size ranging from 18 to 30 nt from a larger quantity of the low molecular weight RNA fraction (300 µg) and then used these in Northern blot as described above. However, we were still unable to detect transgene-derived siRNA in N12-1 plants.

Recently, deep sequencing technology has provided a novel method to analyse siRNA (Thomas and Ansel, 2010). We therefore generated cDNA libraries from the small RNA fraction extracted from leaves of N12-1 and WYMV-infected Yangmai 158 plants and analysed these by Illumina sequencing. About 3.8 and 2.6 million reads were obtained from N12-1 and Yangmai 158 libraries, respectively. Computational analyses were then carried out to identify the WYMV-derived siRNAs in each library, allowing for two mismatches with the virus genome reference sequence. A large number of siRNAs (36 703 reads; 1.4%) in the Yangmai 158 library were found to match the WYMV genome compared with only a few (43 reads; 0.001%) from the N12-1 library. Importantly, no siRNA from the N12-1 library was found to match nucleotide position 5284-6495 of WYMV RNA1, which is the sequence corresponding to the transgene. Taken together with the results of Northern blot analysis, these observations strongly suggest that transgene RNA silencing, the common mechanism of virus-derived disease resistance, was not involved in the WYMV resistance of transgenic wheat plants. In the Yangmai 158 library, WYMV siRNAs were almost equally sense and antisense (Figure 6a), and 21-nt was the most abundant size (Figure 6b). These siRNAs were distributed throughout the positive- and negative-strands of WYMV RNA1 and 2 with some siRNA hotspots (Figure 6c,d). This result provides the first experimental evidence that WYMV infection in wild-type wheat plants induces an RNA-silencing-mediated antiviral defence.

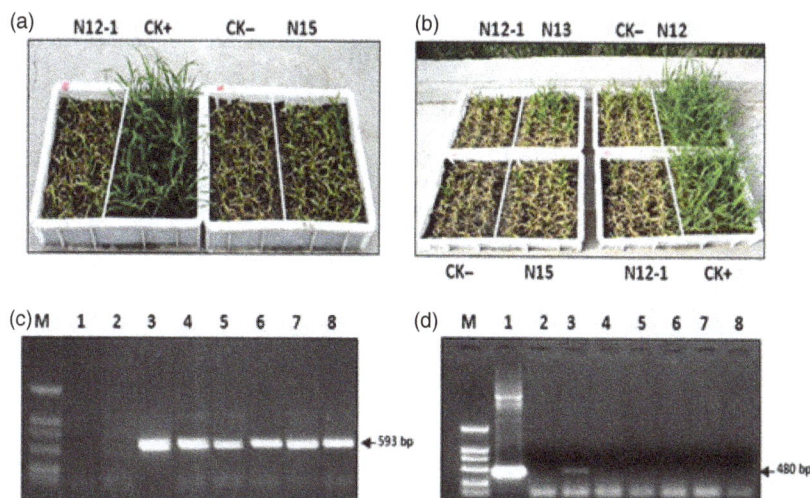

Figure 4 Selecting marker-free lines using PCR and herbicide-resistant tests. (a) Herbicide-resistant testing of T7 transgenic lines using 150 ppm Liberty (herbicide); CK+, positive control (transgenic wheat lines) with herbicide-resistant gene; CK−, negative control (Yangmai 158). (b) Herbicide-resistant testing of T7 transgenic lines using 100 ppm Basta (herbicide); CK−, negative control (Yangmai 158). (c) PCR detection of *Nlb8* gene of T7 generation transgenic lines; M, DNA marker; 1, negative control (H_2O); 2, negative control (Yangmai 158); 3, positive control (plasmid pUbi-Nlb); 4, N12-1; 5, N13; 6, N14; 7, N15; 8, N12. (d) PCR detection of *bar* gene; M, DNA marker; 1, positive control (plasmid pUbi-Nlb); 2, negative control (H_2O); 3, N12; 4, N12-1; 5, N13; 6, N14; 7, N15; 8, negative control (Yangmai 158).

Table 2 Segregation ratio of resistant and sensitive disease plants in F2 population of transgenic lines and Yangmai 15

Cross combination	Total plants	Resistant plants	Sensitive plants	Ratio between resistant and sensitive plants	Chi-square value*
Yangmai 15/N12-1	892	693	199	3.48:1	3.44
Yangmai 15/N14	773	590	183	3.22:1	0.72

*The critical value at 0.05 level is 3.84, and both of chi-square values are less than the critical value, which indicated that the segregation ratio between resistant and sensitive disease plants was 3 : 1 approximately.

Discussion

Novel transgenic wheat with durable field resistance to WYMV obtained by transformation of *Nlb8* gene from virus genome

There have been various reports that plant virus diseases can be controlled by transforming the plants with viral sequences and interfering with viral reproduction via the RNA interference pathway (Fahim *et al.*, 2010, 2012; Kung *et al.*, 2012; Pinto *et al.*, 1999; Simón-Mateo and García, 2011). Resistance has been obtained by inserting viral genes in positive orientation (Pinto *et al.*, 1999), in reverse (complementary) orientation (Sivamani *et al.*, 2002), as long hairpin ds-RNA sequences (Fahim *et al.*, 2010), as short hairpin ds-RNA sequences (Shimizu *et al.*, 2011) and by introducing multiple artificial microRNAs (Fahim *et al.*, 2012; Kung *et al.*, 2012). However, the transgenic resistance obtained by these techniques has mostly been tested only in the greenhouse, and where field testing has been carried out, results have not always been consistent with those in the greenhouse. The only example of transgenic wheat in which

resistance was maintained in both greenhouse and field is the expression of the viral origin antifungal protein KP4 to provide strong resistance to stinking smut (*Tilletia tritici*) (Schlaich *et al.*, 2006; Zhou *et al.*, 2000). However, transgenic wheat transformed with the replication enzyme gene or coat protein genes from wheat streak mosaic virus (WSMV; genus *Tritimovirus*, family *Potyviridae*) showed strong disease resistance in the greenhouse, but lost the resistance in field (Sharp *et al.*, 2002). Stability of transgenic disease resistance has rarely been reported, and it is therefore significant that transgenic resistance to WYMV was identified continuously for 5 years at three different sites in our experiments (Figure 3, Table 3). We believe that this is the first reported transgenic wheat with durable disease resistance in field, and it provides novel germplasm for the breeding of wheat resistant to WYMV.

The resistance mechanism in the transgenic wheat plants

Transformation of plants with a transgene designed to express antisense viral RNA has been one of the most effective methods to generate plants resistant to viral infection. In some cases, the resistance in the transgenic lines is associated with high accumulation of antisense RNA from transcription of the transgene (Bendahmane and Gronenborn, 1997; Hammond and Kamo, 1995). However, resistance has also been achieved by an RNA-silencing-based mechanism through the accumulation of transgene-derived siRNA in transgenic lines (Zhang *et al.*, 2005). Our analyses using Northern blot and deep sequencing of small RNAs showed that transgene-derived siRNAs did not accumulate in our resistant transgenic plants (Figure 6). These results suggest that transgene RNA silencing does not occur in this transgenic line and that resistance does not operate through the degradation of viral RNA by the homology-dependent mechanism that has been observed in many virus-resistant transgenic plants (Simón-Mateo and García, 2011). Southern blot analysis indicated that the four resistant transgenic lines each contain a single transgene copy and are

Table 3 Field investigation of transgenic plants for wheat yellow mosaic virus resistance (DI %) from 2001 to 2005

Line	2001 (T3)	2002 (T4)	2003 (T5)	2004 (T6)	2005 (T7)	Mean value		F value
N12	5.00 ± 4.58^B	5.50 ± 7.14^C	3.67 ± 5.19^B	4.60 ± 0.85^B	2.32 ± 0.44^B	4.22 ± 1.26^B	Intervarietal	288.41*
N13	8.00 ± 8.00^B	5.50 ± 6.56^C	6.09 ± 4.72^B	5.10 ± 0.57^B	3.35 ± 0.66^B	5.61 ± 1.68^B	Year	1.40
N14	6.33 ± 5.69^B	7.90 ± 3.43^{BC}	4.42 ± 6.45^B	4.30 ± 0.71^B	2.49 ± 0.74^B	5.09 ± 2.08^B		
N15	6.00 ± 5.20^B	16.00 ± 4.16^B	2.08 ± 4.17^B	4.50 ± 0.57^B	1.20 ± 1.06^B	5.96 ± 5.93^B		
N12-1	/	2.35 ± 2.73^C	0.00 ± 0.00^B	5.20 ± 0.99^B	1.58 ± 0.44^B	2.28 ± 2.18^B		
Yangmai 158	75.00 ± 18.52^A	95.00 ± 4.32^A	90.75 ± 5.38^A	99.60 ± 0.57^A	94.70 ± 3.82^A	91.01 ± 9.48^A		

The percentage of sensitive plants in total tested plants was used as DI (%) to evaluate the resistance of the transgenic lines. The capital letter of A, B, C or AB meant the significance difference at 0.01 level.

*Indicating significant difference at 0.01 level between wild-type and transgenic lines ($F_{0.01}$ = 4.10). The difference between years were not significant ($F_{0.05}$ = 2.87). The field testing was completed in disease nursery in Jiangsu Province by three times repeats.

Table 4 Disease-resistant evaluation of transgenic line N12-1 in three different regions of China

Region	Yantai* (WYMV + CWMV) DI (%)	Xiping* (WYMV) DI (%)	Yangzhou* (WYMV) DI (%)
Lines			
N12-1	$5.00^†$	$0.00^†$	$0.00^†$
Yangmai 158	100.00	100.00	52.30

*Yantai is in Shandong Province, Xiping in Henna Province and Yangzhou in Jiangsu Province.

†Indicating significant difference at 0.01 level between wild-type and the transgenic lines.

Table 5 Main agronomic characters and yields of T6 transgenic lines in field experiments in 2004 in Yangzhou

Lines	PH (cm)	SL (cm)	TN	NGPE	YT (t/h)	YI (%)
N12	77.32	10.75	14.63	34.79	7.49^a	12.03
N13	81.03	11.26	14.21	33.84	7.11^{ab}	6.31
N14	77.28	10.58	15.40	35.09	7.57^a	13.22
N15	80.07	10.90	17.48	36.41	7.54^a	12.72
N12-1	78.55	11.13	15.20	36.89	7.82^a	16.85
Yangmai 158	82.05	8.48	13.73	37.99	6.69^b	

The yield comparing experiments were completed in disease nursery in Yangzhou by three times repeats; PH, plant height; SL, spike length; TN, the tiller number; NGPE, number of grains per ears; YT (tons per hectare), yield-test results calculated as yield of grains in plot (6.67 m^2); YI, yield increased than that of Yangmai 158. The small letter of a, b or ab showed the significance difference at 0.05 level.

apparently derived from independent transformations (Figure 2c). It is therefore unlikely that the resistance occurred because the gene transformation caused the disruption of a host gene required for WYMV infection. The transgene transcript was detected by RT-PCR (Figure 2b), and it is therefore possible that the accumulation of antisense RNA in the cells inhibits WYMV infection, for example by interference in viral genome replication. Another possibility is that antisense transcripts anneal viral genome RNA to form dsRNAs, which are then digested by cellular RNase. Given that WYMV infects wheat plant at low temperature (Kühne, 2009) and that RNA silencing is less active at low temperature (Szittya et al., 2003; Zhang et al., 2012), it could be suggested that noninvolvement of RNA silencing pathway in the resistance mechanism may be one of the reasons for the stable and potent resistance of our transgenic plants.

Nlb8 confers resistance in various wheat cultivars

Transgenic disease resistance could be transferred from transgenic lines N12-1 and N14 into a susceptible variety Yangmai 15 (Table 2), indicating that Nlb8 can confers resistance in a different wheat cultivar. We have also been able to transfer the Nlb8 transgene from line N12-1 to other WYMV-susceptible commercial wheat varieties by multiple backcrossing and created some new lines with high resistance to WYMV and a single copy of the Nlb8 gene from N12-1 line, some of which also have good agronomic traits. WYM disease has spread gradually from the middle and lower reaches of the Yangtze River to the north since the 1990s, and losses are becoming more serious year by year.

Conventional resistance to WYMV is rare, but our results suggest that transgenic wheat has the potential to reduce the yield losses from this disease in China.

Experimental procedures

Construction of plasmids

The Nlb gene of WYMV isolated from Yangzhou, Jiangsu Province, China (Chen et al., 2000), was amplified by RT-PCR. The Nlb coding region of the RNA1 polyprotein was identified (nts 4912-6495), and sequencing analysis showed that it is highly homologous to the nucleotide sequences of the Nlb gene of Japanese and other Chinese isolates (Figure S1). A part of the Nlb gene (5284-6495) (Figure 1), designated Nlb8, was amplified from the virus genome using BamHI linker primers Nlb8F: 5'-CGGATCCATGATCAGAATGTTGGAAG-3' and Nlb8R: CGGATCCTTGGAG.

CTCAATGCTGCTATC-3'. The blunt-end fragment Nlb8 and linear vector pUbi-35S were then linked to produce constructs with the Nlb8 fragment inserted in either direction under the control of the ubiquitin promoter. The construct containing the reverse Nlb8 insert fragment was then identified by sequencing and named pubi-Nlb (Figure 1b). The assistant plasmid pAHC20 (Figure 1c) with selective marker gene bar (herbicide resistance gene) was prepared for co-transformation with plasmid pubi-Nlb.

Table 6 Main agronomic characters and yields of T7 transgenic lines in field experiments in 2005 in Yangzhou

Lines	M	PH (cm)	SL (cm)	NSS	NAPA	NGPE	TGW (g)	YT (t/h)	YI (%)
N12	5/28	88.0	8.9	18.8	6.4	52.3	40.6	7.7[a]	16.6
N13	5/28	89.0	9.4	19.1	6.6	54.1	40.3	7.3[ab]	10.8
N14	5/28	90.0	9.5	18.2	5.9	50.3	42.1	7.4[a]	12.0
N15	5/28	92.0	9.6	18.5	5.8	48.1	42.8	7.3[ab]	11.1
N12-1	5/28	89.0	9.3	18.5	6.2	51.7	40.5	7.5[a]	14.0
Yangmai 158	5/30	85.0	8.7	18.7	5.8	48.9	39.4	6.6[b]	

The yield comparing experiments were completed in disease nursery in Yangzhou by three times repeats; M, maturity (Month/day); PH, plant height; SL, spike length; NSS, number of seeded spikelet; NAPA, number of ears per plant; NGPE, number of grains per ears; TGW, 1000-grains weight; YT (tons per hectare), yield-test results calculated by the yield in plot (6.67 m^2); YI, yield increased than Yangmai 158. The small letter of a, b or ab showed the significance difference at 0.05 level.

Transformation of wheat using the immature embryos

Fourteen-day-old immature embryos (IEs) of wheat cv. Yangmai 158 were cultured on SD2 medium to induce primary callus at 25 °C under dark conditions. The IEs precultured for 7 days were pretreated on high-osmotic medium (SD2 medium supplemented with 0.2 M mannitol and 0.2 M sorbitol) (Zheng and He, 1994) for 4 h and then bombarded by PDS-1000/He (Bio-Rad, Hercules, CA), with a 1100-psi split membrane and gold particles coated with the plasmid DNAs. After bombardment, the tissues were post-treated on high-osmotic medium for 16–18 h and then transferred to SD2 medium and maintained in the dark for 2 weeks without selection. The transformed calli were transferred onto 1/2 MS medium containing 1 mg/L KT + 3 mg/L Bialaphos + 1 mg/L NAA for differentiation at 24 °C under light for 4 weeks. Green shoots were transferred to 1/2 MS + 5 mg/L Bialaphos medium and continuously cultured under the same conditions for another 4 weeks for elongation. When they had grown to about 2 cm, the plantlets were moved onto 1/2 MS medium containing 0.5 mg/L IAA. The regenerated plants were transferred into pots, and when large enough, the leaves were sampled and genomic DNA was isolated.

Molecular screening of the putative transgenic wheat plants

For PCR testing, genomic DNA was extracted as described previously (Murray and Tompson, 1980). The full-length primers without *Bam*HI sites were Nlb8F-WB 5'-ATGATCAGA-ATGTTGGAAGACGCC-3' and Nlb8R-WB 5'-TTGGAGCTCAA. TG CTGCTATCACC-3'. The amplification conditions were 94 °C 5 min for denaturing; 94 °C 30 s, 53 °C 1 min 45 s, 72 °C 1 min 30 s for 35 cycles for amplification; 72 °C 7 min for prolongation and subsequent storage at 4 °C. For RT-PCR assays, total RNA was extracted from the transgenic wheat and wild-type plants according to the manual of the TRIZOL Kit (GIBCO BRL) and used to synthesize cDNA with random primers (TaKaRa cDNA Synthesis Kit, Dalian, Liaoning province, China). The primers and reaction condition of RT-PCR were the same as for PCR testing.

Resistant assessment of the transgenic wheat to WYMV in field

The transgenic wheat lines were evaluated for WYMV resistance from 2001 to 2005 in three field disease nurseries of Yizheng, Jiangyan and Lixiahe in Jiangsu Province. Seeds were sown in autumn (23 October) and harvested in summer (early June) of the following year. The trials received natural rainfall and were fertilized three times during the season. In 2000, when there

were few seeds of the transgenic wheat lines, there was a single row (50 plants) of each test line. In 2001 and 2002 (T3 and T4 generation plants), the transgenic and wild-type wheat were sown in three replicate plots each of three rows of 1.4 m (40 plants). Disease incidence was evaluated on 100 plants per plot. From 2003 to 2005 (T5 to T7 generation plants), the plants were arranged in three replicate plots each of 6.7 m^2 (1000 plants), and disease incidence was evaluated on 20 plants from each of five sampling points in each plot. In 2004 and 2005, the main agronomic characters of the transgenic wheat lines were also investigated. The plants were scored for disease class on a scale of 0 to 3 to evaluate severity of the disease (Hou et al., 1985): 0 = no visible symptoms, 1 = lightly streak mosaic leaf but plant not stunted, 2 = distinct mosaic streak covering one-half of the diseased leaf, and 3 = mosaic area covering three quarters of the diseased leaf and the plant obviously stunted. A disease index (DI) was then calculated: DI (%) = $\sum$ (DS × Ni) × 100/(3 × N), where DS is the disease scale (above), Ni the number of plants with this DS, and N the total number of plants observed. The DI and phenotypic data for the WYM responses were analysed using Statistical Analysis System, version 8.1 (SAS v8.1) (SAS Institute Inc, Raleigh, NC).

In 2003–2005, similar trials were also conducted on infested sites at Yantai (Shandong Province), Xiping (Henan Province) and Yangzhou (Jiangsu Province).

Selection of marker-free transgenic wheat by PCR and herbicide spraying

As *Nlb8* gene and *bar* gene were co-transformed into wheat in our study, it was possible to obtain marker-free plants in the offspring of T0 transgenic wheat. PCR tests were used to test T3 plants for the *Nlb8* and *bar* genes. Primers Nlb8F-WB and Nlb8R-WB were used for detecting *Nlb8*, and primers Bar-1F 5'-CT GCACCATCGTCAACCACTACATC-3' and Bar-1R 5'-AGCTG CCAGAAAC.

CCACGTCAT-3' for the *bar* gene. The amplification conditions for *bar* gene were 94 °C 10 min for denaturing; 94 °C 30 s, 58 °C 1 min 30 s, 72 °C 1 min 15 s for 35 cycles for amplification; 72 °C 10 min for prolongation and storage at 4 °C. The transgenic offspring were also screened for marker-free plants by spraying herbicide (Liberty, 150 ppm or Basta, 100 ppm) at the seedling stage combined with PCR testing.

Copy number and segregation analysis of transgenic wheat plants

Genomic DNA was extracted from the leaves of the transgenic wheat plants. For Southern blotting, 35 µg genomic DNA was digested with *Eco*RI and *Ssp*I, separated by electrophoresis on

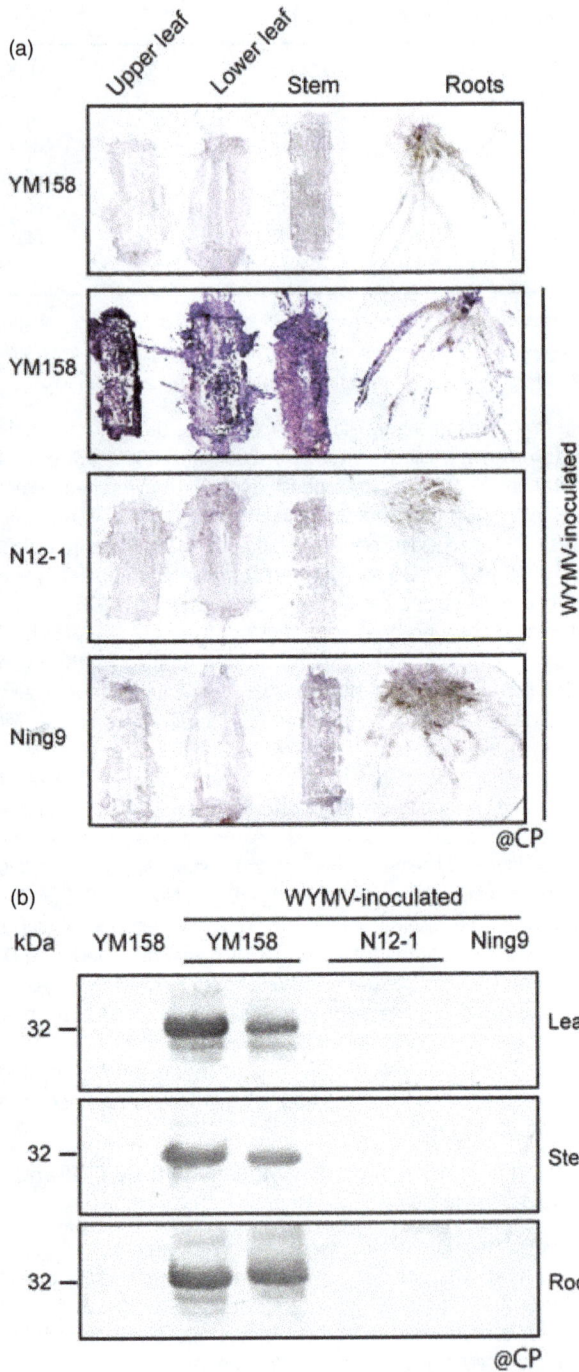

Figure 5 Wheat yellow mosaic virus (WYMV) accumulation in transgenic and nontransgenic wheat plants. (a) Tissue printing assay to detect WYMV accumulation in leaves stems and roots of susceptible nontransgenic control (Yangmai 158, YM158), resistant nontransgenic (Ning 9) control and transgenic (line N12-1) wheat plants grown on WYMV-infested soil. (b) Western blot analysis to detect WYMV accumulation in wheat plants described in panel A. Plants were sampled four months after planting. A polyclonal antibody specific for WYMV CP was used for the immunodetection.

0.8% agarose gels and transferred to N+ Hybond nylon membranes. The probe was amplified using primers NIb8F-WB and NIb8R-WB from the vector pubi-NIb and was labelled with α-^{32}P dCTP using TaKaRa Random Primer Labeling Kit. Southern blotting was carried out as described by Gao et al., (2005).

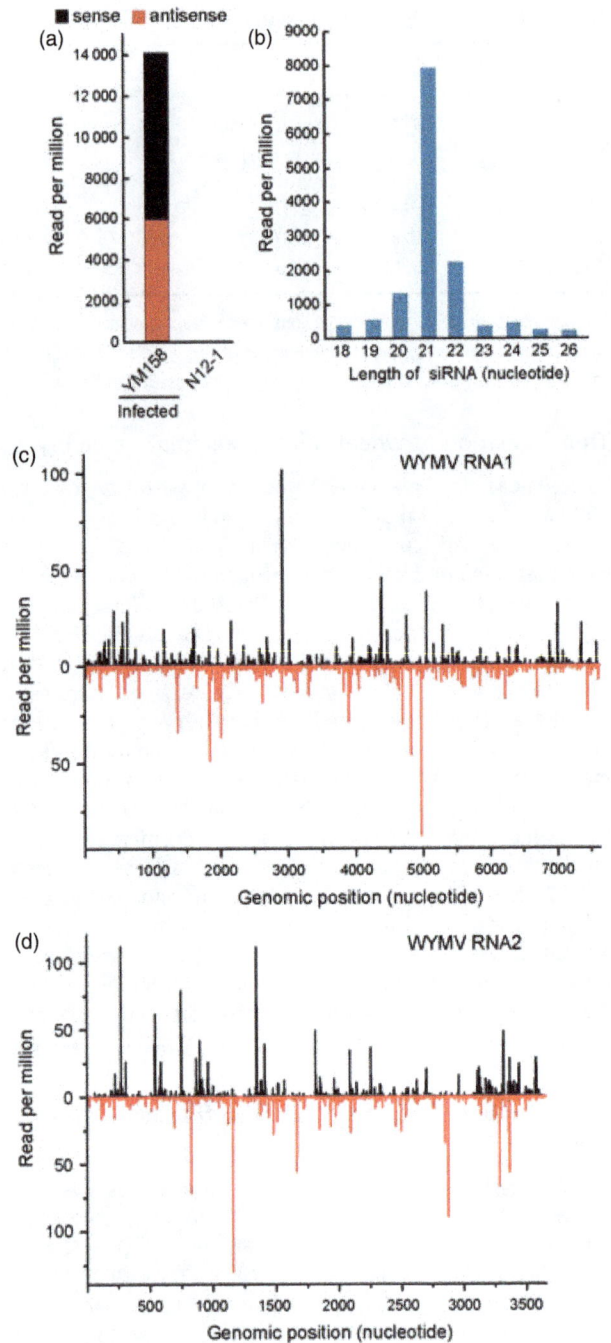

Figure 6 The presence of wheat yellow mosaic virus WYMV-derived siRNAs in inoculated nontransgenic and transgenic wheat plants. (a) Frequencies of WYMV siRNAs in nontransgenic (Yangmai 158, YM158) and transgenic (line N12-1) wheat plants grown on WYMV-infested soil. (b) Size distribution of WYMV siRNAs derived from the YM158 library. (c) and (d), Distribution of WYMV siRNAs along the RNA1 (c) and RNA2 (d) segments of the WYMV genome.

T4-generation plants of lines N12-1 and N14 were crossed with Yangmai 15 (a wheat variety susceptible to WYMV). 892 and 773 F2 plants from the respective crosses N12-1/Yangmai 15 and N14/Yangmai 15 were planted in WYMV nurseries in Yangzhou and evaluated for resistance.

Tissue printing and Western blot assays

Tissue printing assay was carried out as described previously (Andika *et al.*, 2005). Preparation of protein samples, SDS-PAGE, electroblotting and immunodetection for Western blot analysis were performed as described previously (Sun and Suzuki, 2008). WYMV CP was detected using primary anti-CP (1 : 5000) polyclonal serum and secondary polyclonal AP-conjugated goat anti-rabbit IgG (1 : 10 000) (Sigma, St. Louis, MO).

Northern blot analysis, deep sequencing and bioinformatic analysis of small RNAs

Total RNAs were extracted from the leaves pooled of the ten wheat plants with Trizol (Invitrogen, San Diego, CA) according to the manufacturer's instructions. Gel electrophoresis, blotting and detection of siRNAs using the DIG system were carried out as described previously (Goto *et al.*, 2003). A digoxigenin (DIG)-labelled DNA probe specific for WYMV RNA1 (nt 5284-6495) was prepared using the PCR DIG Probe Synthesis Kit (Roche, Mannheim, Germany).

The cDNAs of small RNA libraries were prepared using the Illumina TrueSeq Small RNA Sample Preparation Kit (Illumina, San Diego, CA) and then used for sequencing using the Illumina HiSeq2000 (Illumina). The adaptor sequences in the raw data sets derived from Illumina sequencing data were trimmed, and small RNA reads without an identifiable linker were removed. The reads of >30-nt or <18-nt were excluded for further analyses. Noncoding RNAs were then identified using the Rfam database and removed. Computational analyses were performed using Perl scripts. To search for viral siRNAs, reads from the two libraries were mapped to the WYMV RNA1 (AJ131981) and RNA2 (AJ131982) using BWA software (http://bio-bwa.sourceforge.net/) with two mismatches allowed.

Acknowledgements

This work was supported by the National Key Project for Research on Transgenic Biology (2013ZX08002-001) and the National 863 High-tech Project (2012AA10A309). We are grateful to Dr. Mike Adams for critically reading the manuscript.

References

Adams, M., Zerbini, F., French, R., Rabenstein, F., Stenger, D. and Valkonen, J. (2011) *Family potyviridae. Virus taxonomy, 9th report of the international committee for taxonomy of viruses*. Elsevier Academic Press, San Diego, USA, 1069-1089

Andika, I.B., Kondo, H. and Tamada, T. (2005) Evidence that RNA silencing-mediated resistance to Beet necrotic yellow vein virus is less effective in roots than in leaves. *Mol. Plant Microbe Interact.* **18**, 194–204.

Baulcombe, D. (2004) RNA silencing in plants. *Nature*, **431**, 356–363.

Bendahmane, M. and Gronenborn, B. (1997) Engineering resistance against tomato yellow leaf curl virus (TYLCV) using antisense RNA. *Plant Mol. Biol.* **33**, 351–357.

Chen J.P., Ruan, Y.L. and Dong, M.J. (1989) Study on the pathogen of a wheat soil-borne virus disease in China. *Virol. Sin.* **2**, 176–181.

Chen J., Chen, J.P., Yang, J.P., Cheng, Y., Diao, A., Adams, M.J. and Dua, J. (2000) Differences in cultivar response and complete sequence analysis of two isolates of wheat yellow mosaic bymovirus in China. *Plant. Pathol.* **49**, 370–374.

Collinge, D.B., Jørgensen, H.J., Lund, O.S. and Lyngkjær, M.F. (2010) Engineering field pathogen resistance in crop plants: current trends and future prospects. *Annu. Rev. Phytopathol.* **48**, 269–291.

Diao, A., Chen, J., Ye, R., Zheng, T., Yu, S., Antoniw, J. and Adams, M. (1999) Complete sequence and genome properties of Chinese wheat mosaic virus, a new furovirus from China. *J. Gen. Virol.* **80**, 1141–1145.

Ding, S.W. (2010) RNA-based antiviral immunity. *Nat. Rev. Immunol.* **10**, 632–644.

Ding, S.W. and Voinnet, O. (2007) Antiviral immunity directed by small RNAs. *Cell*, **130**, 413–426.

Ding, S.W., Li, H., Lu, R., Li, F. and Li, W.-X. (2004) RNA silencing: a conserved antiviral immunity of plants and animals. *Virus Res.* **102**, 109–115.

Fahim, M., Ayala-Navarrete, L., Millar, A.A. and Larkin, P.J. (2010) Hairpin RNA derived from viral NIa gene confers immunity to wheat streak mosaic virus infection in transgenic wheat plants. *Plant Biotechnol. J.* **8**, 821–834.

Fahim, M., Millar, A.A., Wood, C.C. and Larkin, P.J. (2012) Resistance to Wheat streak mosaic virus generated by expression of an artificial polycistronic microRNA in wheat. *Plant Biotechnol. J.* **10**, 150–163.

Gao, S.Q., Xu, H.J., Cheng, X.G., Chen, M., Xu, Z.S., Li, L.C., Ye, X.G., Du, L.P., Hao, X.Y. and Ma, Y.Z. (2005) Improvement of wheat drought and salt tolerance by expression of a stress-inducible transcription factor GmDREB of soybean (Glycine max). *Chinses Sci. Bull.* **50**, 2714–2723.

Gonsalves, D. (1998) Control of papaya ringspot virus in papaya: a case study. *Annu. Rev. Phytopathol.* **36**, 415–437.

Goto, K., Kanazawa, A., Kusaba, M. and Masuta, C. (2003) A simple and rapid method to detect plant siRNAs using nonradioactive probes. *Plant Mol. Biol. Rep.* **21**, 51–58.

Gottula, J. and Fuchs, M. (2009) Toward a quarter century of pathogen-derived resistance and practical approaches to plant virus disease control. *Adv. Virus Res.* **75**, 161–183.

Hammond, J. and Kamo, K. (1995) Effective resistance to potyvirus infection conferred by expression of antisense RNA in transgenic plants. *Mol. Plant Microbe Interact.* **8**, 674–682.

Han, C.G., Li, D.W., Xing, Y.M., Zhu, K., Tian, Z.F., Cai, Z.N., Yu, J.L. and Liu, Y. (2000) Wheat yellow mosaic virus widely occurring in wheat (*Triticum aestivum*) in China. *Plant Dis.* **84**, 627–630.

Hou, Q.S., Han, H., Zhou, Y.J. and Xiao, Q.P. (1985) Studies on a Soil-Borne Virus disease of wheat in Jangsu province I. Rules governing the incidence of the disease and identification of the pathogen. *Jiangsu J. Agric. Sci.* **1**, 26–28. (in Chinese).

Kühne, T. (2009) Soil-borne viruses affecting cereals—Known for long but still a threat. *Virus Res.* **141**, 174–183.

Kung, Y.J., Lin, S.S., Huang, Y.L., Chen, T.C., Harish, S.S., Chua, N.H. and Yeh, S.D. (2012) Multiple artificial microRNAs targeting conserved motifs of the replicase gene confer robust transgenic resistance to negative-sense single-stranded RNA plant virus. *Mol. Plant Pathol.* **13**, 303–317.

Liu, W.H., Nie, H., He, Z.T., Chen, X.L., Han, Y.P., Wang, J.R., Li, X., Han, C.G. and Yu, J.L. (2005a) Mapping of a Wheat Resistance Gene to Yellow Mosaic Disease by Amplified Fragment Length Polymorphism and Simple Sequence Repeat Markers. *J. Integr. Plant Biol.* **47**, 1133–1139.

Liu, W.H., Nie, H., Wang, S.B., Li, X., He, Z.T., Han, C.G., Wang, J.R., Chen, X.L., Li, L.H. and Yu, J.L. (2005b) Mapping a resistance gene in wheat cultivar Yangfu 9311 to yellow mosaic virus, using microsatellite markers. *Theor. Appl. Genet.* **111**, 651–657.

Murray, M.G. and Tompson, W.F. (1980) Rapid isolation of high molecular weight plant DNA. *Nucleic Acids Res.* **8**, 4321–4325.

Namba, S., Kashiwazaki, S., Lu, X., Tamura, M. and Tsuchizaki, T. (1998) Complete nucleotide sequence of wheat yellow mosaic bymovirus genomic RNAs. *Arch. Virol.* **143**, 631–643.

Nishio, Z., Kojima, H., Hayata, A., Iriki, N., Tabiki, T., Ito, M., Yamauchi, H. and Murray, T.D. (2010) Mapping a gene conferring resistance to Wheat yellow mosaic virus in European winter wheat cultivar 'Ibis'(*Triticum aestivum* L.). *Euphytica*, **176**, 223–229.

Pinto, Y.M., Kok, R.A. and Baulcombe, D.C. (1999) Resistance to rice yellow mottle virus (RYMV) in cultivated African rice varieties containing RYMV transgenes. *Nature biotechnol.* **17**, 702–707.

Prins, M., Laimer, M., Noris, E., Schubert, J., Wassenegger, M. and Tepfer, M. (2008) Strategies for antiviral resistance in transgenic plants. *Mol. Plant Pathol.* **9**, 73–83.

Qin, J., Li, Z., Tao, J. and Qin, Y. (1986) Primary study on resistance inheritance to yellow mosaic disease of wheat. *J. Sichuan Agric. Univ.* **4**, 17–28. (in Chinese).

Sanford, J.C. and Johnston, S.A. (1985) The concept of parasite-derived resistance—Deriving resistance genes from the parasite's own genome. *J. Theor. Biol.* **113**, 395–405.

Schlaich, T., Urbaniak, B.M., Malgras, N., Ehler, E., Birrer, C., Meier, L. and Sautter, C. (2006) Increased field resistance to Tilletia caries provided by a specific antifungal virus gene in genetically engineered wheat. *Plant Biotechnol. J.* **4**, 63–75.

Sharp, G.L., Martin, J., Lanning, S., Blake, N., Brey, C., Sivamani, E., Qu, R. and Talbert, L. (2002) Field Evaluation of Transgenic and Classical Sources of Resistance. *Crop Sci.* **42**, 105–110.

Shimizu, T., Nakazono-Nagaoka, E., Uehara-Ichiki, T., Sasaya, T. and Omura, T. (2011) Targeting specific genes for RNA interference is crucial to the development of strong resistance to Rice stripe virus. *Plant Biotechnol. J.* **9**, 503–512.

Simón-Mateo, C. and García, J.A. (2011) Antiviral strategies in plants based on RNA silencing. *Biochim. Biophys. Acta*, **1809**, 722–731.

Sivamani, E., Brey, C.W., Talbert, L.E., Young, M.A., Dyer, W.E., Kaniewski, W.K. and Qu, R. (2002) Resistance to wheat streak mosaic virus in transgenic wheat engineered with the viral coat protein gene. *Transgenic Res.* **11**, 31–41.

Sun, L. and Suzuki, N. (2008) Intragenic rearrangements of a mycoreovirus induced by the multifunctional protein p29 encoded by the prototypic hypovirus CHV1-EP713. *RNA*, **14**, 2557–2571.

Sun, B., Sun, L., Tugume, A., Adams, M., Yang, J., Xie, L. and Chen, J. (2013) Selection pressure and founder effects constrain genetic variation in differentiated populations of a soil-borne bymovirus Wheat yellow mosaic virus (Potyviridae) in China. *Phytopathology*, **103**, 949–959.

Szittya, G., Silhavy, D., Molnar, A., Havelda, Z., Lovas, A., Lakatos, L., Banfalvi, Z. and Burgyan, J. (2003) Low temperature inhibits RNA silencing-mediated defence by the control of siRNA generation. *EMBO J.* **22**, 633–640.

Tenllado, F., Llave, C. and Díaz-Ruíz, J.R.. (2004) RNA interference as a new biotechnological tool for the control of virus diseases in plants. *Virus Res.* **102**, 85–96.

Thomas, M.F. and Ansel, K.M. (2010) Construction of small RNA cDNA libraries for deep sequencing. In: *MicroRNAs and the Immune System* (Silvia, M., ed), pp. 93–111. New York: Springer.

Voinnet, O., Vain, P., Angell, S. and Baulcombe, D.C. (1998) Systemic spread of sequence-specific transgene RNA degradation in plants is initiated by localized introduction of ectopic promoterless DNA. *Cell*, **95**, 177–187.

Waterhouse, P.M., Graham, M.W. and Wang, M.B. (1998) Virus resistance and gene silencing in plants can be induced by simultaneous expression of sense and antisense RNA. *Proc. Natl Acad. Sci. USA*, **95**, 13959–13964.

Zhang, P., Vanderschuren, H., Fütterer, J. and Gruissem, W. (2005) Resistance to cassava mosaic disease in transgenic cassava expressing antisense RNAs targeting virus replication genes. *Plant Biotechnol. J.* **3**, 385–397.

Zhang, X., Singh, J., Li, D. and Qu, F. (2012) Temperature-dependent survival of Turnip crinkle virus-infected Arabidopsis plants relies on an RNA silencing-based defense that requires dcl2, AGO2, and HEN1. *J. Virol.* **86**, 6847–6854.

Zheng, H.J. and He, S.J. (1994) Some improvements of the biolistic transformation system for *Oryza sativa* L. *China Biotechnol.* **12**, 111–115. (in Chinese).

Zhou, Y., Cheng, Z., Hou, Q., Fan, Y. and Wu, S. (2000) Resistance of wheat varieties to wheat spindle streak mosaic disease. *Acta Phytophylacica Sin.* **27**, 102–106. (In Chinese).

Zhu, X.B., Wang, H.Y., Guo, J., Wu, Z.Z., Cao, A.Z., Bie, T.D., Nie, M.J., You, F.M., Cheng, Z.B., Xiao, J., Liu, Y.Y., Cheng, S.H., Chen, P.D. and Wang, X.E. (2012) Mapping and validation of quantitative trait loci associated with wheat yellow mosaic bymovirus resistance in bread wheat. *Theor. Appl. Genet.* **124**, 177–188.

9

^{1}H-NMR screening for the high-throughput determination of genotype and environmental effects on the content of asparagine in wheat grain

Delia I. Corol[1], Catherine Ravel[2], Marianna Rakszegi[3], Gilles Charmet[2], Zoltan Bedo[3], Michael H. Beale[1], Peter R. Shewry[1] and Jane L. Ward[1],*

[1]*Department of Plant Biology and Crop Science, Rothamsted Research, Harpenden, Hertfordshire, UK*

[2]*INRA-UBP, UMR1095 GDEC, Clermont-Ferrand Cedex, France*

[3]*Agricultural Institute, Centre for Agricultural Research of the Hungarian Academy of Sciences, Martonvásár, Hungary*

*Correspondence

email jane.ward@rothamsted.ac.uk

Keywords: Asparagine, ^{1}H-NMR, wheat, metabolite profiling, G × E, heritability.

Summary

Free asparagine in cereals is known to be the precursor of acrylamide, a neurotoxic and carcinogenic product formed during cooking processes. Thus, the development of crops with lower asparagine is of considerable interest to growers and the food industry. In this study, we describe the development and application of a rapid ^{1}H-NMR-based analysis of cereal flour, that is, suitable for quantifying asparagine levels, and hence acrylamide-forming potential, across large numbers of samples. The screen was applied to flour samples from 150 bread wheats grown at a single site in 2005, providing the largest sample set to date. Additionally, screening of 26 selected cultivars grown for two further years in the same location and in three additional European locations in the third year (2007) provided six widely different environments to allow estimation of the environmental (E) and G x E effects on asparagine levels. Asparagine concentrations in the 150 genotypes ranged from 0.32 to 1.56 mg/g dry matter in wholemeal wheat flours. Asparagine levels were correlated with plant height and therefore, due to recent breeding activities to produce semi-dwarf varieties, a negative relationship with the year of registration of the cultivar was also observed. The multisite study indicated that only 13% of the observed variation in asparagine levels was heritable, whilst the environmental contribution was 36% and the GxE component was 43%. Thus, compared to some other phenotypic traits, breeding for low asparagine wheats presents a difficult challenge.

Introduction

Free amino acids generally account for about 5% or less of the total nitrogen content of wheat grain, with asparagine accounting for 10% or less of the total (reviewed by Lea *et al.*, 2007). Consequently this amino acid was of little scientific interest or practical importance until the demonstration that it is a precursor of acrylamide which is formed in processed foods by a Maillard reaction with reducing sugars (Mottram *et al.*, 2002; Stadler *et al.*, 2002). Thus, acrylamide, which has neurotoxic and carcinogenic properties, may be present in cooked foods at concentrations up to 1 mg/kg (Friedman, 2003; Tareke *et al.*, 2002). Furthermore, the formation of acrylamide is correlated with the free asparagine content of wheat flour, rather than the content of reducing sugars (Muttucumaru *et al.*, 2006, 2008). Consequently the control of asparagine synthesis and accumulation is of considerable current interest to food processors and in contemporary crop science. The development of high-throughput analytical screens to monitor asparagine levels in crops and crop products is essential to support breeding and selection programmes aimed at reducing the risk of acrylamide formation.

Most recent studies have utilized GC, GC-MS or LC-MS of partially purified, and derivitised, amino acid fractions, as analytical techniques for asparagine quantitation. An early study indicated that asparagine accumulated in cereal grain under conditions of sulphur deficiency (Shewry *et al.*, 1983) and this has been confirmed by more recent studies (Muttucumaru *et al.*, 2006). The asparagine content of wheat grain also increases with protein content (Claus *et al.*, 2008) raising particular concerns for the production of bread-making wheats. However, substantial variation in the asparagine content of wheat grain has been reported which does not appear to be related to nutritional status alone (Baker *et al.*, 2006; Claus *et al.*, 2006), indicating the impact of other environmental factors. Furthermore, although variation has been reported in the contents of asparagine in different wheat cultivars grown at six locations in the UK (Curtis *et al.*, 2009), the relative effects of genotype and environment, and possible interactions between these, have not been quantified in large scale studies across diverse locations and environments. However, analysis of five rye cultivars grown in six European environments showed that only 23% of the total variance in free asparagine concentration was attributed to the effect of variety (Curtis *et al.*, 2010). More recently, a detailed comparison was reported of 92 wheat varieties grown in two glasshouse experiments. This showed that the broad sense heritability for asparagine was low (32%) but nevertheless identified possible SNP markers for breeding (Emebiri, 2014).

The rye samples analysed by Curtis *et al.* (2010) were provided by the EU FP6 HEALTHGRAIN programme which was focussed on providing health benefits to consumers by increasing the consumption of protective compounds in whole grains or their fractions (Poutanen *et al.*, 2008). This study included a detailed analysis of genetic diversity in the composition of wheat grain, with 150 bread wheat lines and 50 lines of other cereals (including the 'ancient' wheats einkorn, emmer and spelt) being initially grown on a single site in Hungary in 2005 (Ward *et al.*, 2008). A smaller set of 26 lines, including 23 from the initial screen, were then grown on the same site for two further years (2006, 2007) and on three further sites (in France, UK and Poland) in 2007 only (Shewry *et al.*, 2010). This material has been analysed for a wide range of phytochemicals, vitamins, minerals and dietary fibre components (Shewry *et al.*, 2010, 2011; Ward *et al.*, 2008; Zhao *et al.*, 2009), providing data on genetic diversity and heritability. We have also reported on the analysis of methyl donors such as choline and glycine betaine (Corol *et al.*, 2012) by ^{1}H-NMR. This method gives quantitative data on a number of compounds, and in this study, we report on the further application of the high-throughput ^{1}H-NMR screen to the HEALTHGRAIN wheat material to determine the concentration of asparagine which is the main factor determining the formation of acrylamide in baked cereal products. The data allowed a comprehensive study of the genotype *versus* environmental (GxE) effects on the content of asparagine, which is considered to be the main target for wheat crop improvement towards lower acrylamide potential (Halford *et al.*, 2012).

Results and discussion

Development of the screen

^{1}H-NMR profiling of unpurified extracts made directly into deuterated aqueous methanol is a well-established technique in plant metabolomics (Baker *et al.*, 2006; Ward and Beale, 2006; Ward *et al.*, 2003). This method is capable of generating metabolite fingerprinting data from large numbers (1000s) of samples. Furthermore, the data are free from alignment and reproducibility problems that are commonly encountered with chromatography-based analytical techniques, and thus, data from multiple batches of samples collected over several weeks or months can be batch processed with high confidence. Extraction of metabolites involves relatively simple procedures, and there is no need for further derivitisation to 'view' metabolites of interest. [^{1}H]-NMR is also absolutely quantitative, irrespective of compound properties, and does not rely on the use of calibration curves. Individual metabolite concentrations can be deduced despite the complex spectra obtained from typical plant or grain samples. Accurate quantitation of individual metabolites in such complex metabolite fingerprints is, however, reliant on the presence of molecule-specific resonances that are nonoverlapping with other signals in the spectra. A typical NMR spectrum obtained from wheat flour extract is shown in Figure 1, and it can be seen that the characteristic, geminally coupled, signals of the asparagine C-3 hydrogen atoms fall in a relatively clear area of the spectrum and thus are suitable for integration. Although both of the diastereotopic hydrogens at position C-3 of asparagine are clearly separated and visible in the NMR spectra

of cereal flour extracts, the cleanest signal of the pair at $\delta 2.95$ (dd, J = 17 and 4 Hz) arising from 3-H_b was used for integration and quantitation. The signal for 3-H_a at $\delta 2.83$ was not utilized due to small interfering signals from aspartate and other metabolites.

After automated alignment and normalization to the d_4-TSP internal standard, spectra were reduced to equally sized bins and integration of the selected peaks was accomplished by comparison to the known concentration of the d_4-TSP standard. The region taken for asparagine quantitation is $\delta 2.9755$–2.9255 representing 3-H_b (1 hydrogen). The high reproducibility of NMR is well documented (Viant *et al.*, 2009; Ward *et al.*, 2010), and therefore, errors due to instrument drift are minimal. Across three separate extraction replicates, typical relative standard deviations were below 10%.

Comparison of asparagine contents in wholemeal flour of different wheat species

Integration of binned NMR spectra (circa 10 000 bins of 0.001 ppm each) allowed batch processing of the large number of spectra and accurate quantitation of asparagine. For the lines grown at a single site in 2005, the average contents of asparagine in wholemeal flour samples varied from 0.56 ± 0.14 mg/g d.m. in emmer (*T. turgidum* var. *dicoccum*) (*n* = 5) to 1.12 ± 0.25 mg/g d.m. in einkorn (*T. monococcum* var. *monococcum*) (*n* = 5) (Table 1 and Figure S1). The average contents in winter (*n* = 130) and spring (*n* = 20) bread wheat (*T. aestivum*) genotypes were remarkably similar and were 0.73 ± 0.25 mg/g d.m. and 0.75 ± 0.21 mg/g d.m., respectively, with the contents in the 150 bread wheat lines ranging from 0.32 to 1.56 mg/g d.m. Similarly, the average asparagine content in spelt (*T. aestivum* var. spelta) wholemeal (*n* = 5) was similar to the winter and spring wheat genotypes at 0.72 mg/g d.m, although the range observed in concentration was narrower with variation, in the 5 genotypes studied, of 0.6–0.79 mg/g d.m. Durum wheat

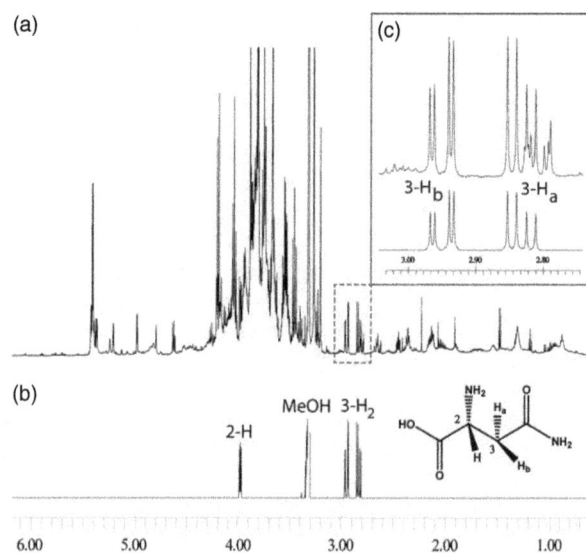

Figure 1 NMR quantitation of asparagine in cereal flours. (a) NMR spectrum of typical wheat flour extract made in $CD_3OD:D_2O$ (1:4). (b) spectrum of pure asparagine made in $CD_3OD:D_2O$ (1:4). (c) expansion of the 3-H_2 signal (top) and illustration of 3-H_b of asparagine standard (bottom) which was utilized for quantitation.

(*T. turgidum* var. durum) samples had intermediate levels with an average asparagine concentration of 0.88 mg/g d.m.

Survey of 150 Bread Wheat Genotypes grown together at a single site

Bread wheats make up the greatest proportion of the samples under study. As all lines had been grown at a single location, in the same year, it could be assumed that most of the variation in composition of these samples could be ascribed to the genotype allowing a comparison of asparagine content to be made. The concentration of asparagine in bread wheats typically followed a unimodal distribution which is slightly skewed right (Figure 2). Mean asparagine content of the 150 bread wheats was 0.73 mg/g d.m., whilst the median value was 0.67 mg/g d.m. Fifty (of 150) genotypes contained asparagine contents of between 0.62 and 0.78 mg/g d.m and a further 43 genotypes contained lower concentrations of asparagine, between 0.46 and 0.62 mg/g d.m. A total of 16 genotypes had asparagine levels in excess of 1.1 mg/g d.m., whilst 10 genotypes had very low asparagine contents of between 0.3 and 0.46 mg/g d.m. A full listing of the genotypes grouped into the eight concentration ranges is given in Table S1. The majority of genotypes had asparagine levels between 0.46 and 0.94 mg/g d.m. Those genotypes which did not fall within this range included 3 spring wheat and 31 winter wheat cultivars (Table 2). Of the 10 genotypes with the lowest asparagine content (0.3–0.46 mg/g d.m.), only 1 (Chinese Spring) was spring type, whilst the others (Alba, Bilancia, Blasco, Granbel, Mv-Emese, Nomade, Palesio, Soissons and Valoris) were winter type. However, it should be noted that only 20 spring wheats were analysed compared with 130 winter wheats. The variation in asparagine content observed here (0.3–1.1 mg/g d.m.) was similar in extent (by over 3.5-fold) but the values generally higher than those reported for 92 wheat varieties grown in the glasshouse (0.137–0.471 mg/g d.m) (Emebiri, 2014).

Correlation with grain quality

A number of metabolite classes, such as phenolics, sterols and alkylresorcinols, have previously been shown to correlate with grain quality parameters in the HEALTHGRAIN sample set (Rakszegi et al., 2008; Shewry et al., 2010 and Ward et al., 2008). The asparagine concentrations determined in this study were therefore examined against the parameters described in Rakszegi et al., 2008; to determine relationships with yield and quality parameters (Table 3). No correlations of asparagine levels were observed between yield, test weight, thousand kernel weight, kernel diameter or mean kernel weight, indicating that final asparagine concentration was not determined by grain size or the total amount of grain produced by the plant. Similarly, the total yields of flour and bran, which will be affected by grain weight, showed no significant relationship with asparagine

content. However, a positive correlation ($r = 0.413$, $P < 0.0001$) was observed between asparagine concentration and plant height, with grain from taller plants containing higher final levels of grain asparagine. Breeders have significantly reduced the height of wheats since the second half of the last century, with most modern cultivars being semi-dwarf. Hence, an inverse correlation was also observed between asparagine concentration and year of cultivar registration with a clear reduction in mean asparagine content being evident from the 1940s onwards (Figure S3).

Bread-making quality is determined by a number of parameters, with grain texture and protein content and composition being the most important. In particular, the gluten proteins confer unique rheological properties (visco-elasticity) to doughs, with strong (highly elastic) doughs being preferred for bread making. We found that the asparagine concentration in whole grain was correlated with total protein per grain ($r = 0.45$, $P \leq 0.0001$) and also with the % protein in both white flour ($r = 0.38$, $P \leq 0.0001$) and wholemeal ($r = 0.51$, $P \leq 0.0001$). Additionally, positive correlations were seen with total gluten (which is itself correlated with grain protein) ($r = 0.44$, $P \leq 0.0001$) and Zeleny sedimentation (a measure of gluten content and quality) ($r = 0.37$, $P \leq 0.0001$).

The gluten proteins are poor in asparagine (up to 1 mol% in glutenin subunits, 0.8–2.6 mol% in gliadins) (Shewry et al., 2009a), and it is likely that the pool size of free asparagine is regulated at low levels under normal conditions of plant growth. However, asparagine is known to accumulate under conditions of restricted protein synthesis or stress (Lea et al., 2007). It is therefore crucial that the developing grain should not be subjected to conditions that may lead to asparagine accumulation, such as restriction of protein synthesis by sulphur limitation, drought or heat stress.

Hard grain texture is required for bread-making quality as it results in greater starch damage and higher water absorption. Although we observed no correlation between asparagine concentration and grain hardness (measured as hardness index), a positive correlation ($r = 0.35$, $P \leq 0.0001$) was observed with % water absorption. A weak negative correlation was also observed between asparagine concentration and starch content ($r = 0.32$, $P \leq 0.0001$). Although some α-amylase activity is required for bread making, high α-amylase activity, resulting from prematurity amylase production (PMA) or preharvest sprouting (PHS), adversely affects baking quality. No correlation was observed between falling number, a measure of α-amylase activity, and asparagine concentration. It is clear, therefore, that asparagine accumulation is positively correlated with factors affecting bread making: protein content and quality and water absorption. It is therefore important for breeders to identify lines which are outside this correlation.

Table 1 Asparagine Concentrations (mg/g d.m.) in Wholemeal samples from different cereals.

Cereal	Number of samples analysed	Average (mg/g d.m.)	Max	Min	CV(%)
Dicoccum	5	0.56 ± 0.14	0.79	0.43	24.00
Durum wheat	10	0.88 ± 0.2	1.32	0.69	22.85
Monococcum	5	1.12 ± 0.25	1.50	0.94	21.88
Spelt	5	0.72 ± 0.08	0.79	0.60	10.85
Spring wheat	20	0.75 ± 0.21	1.40	0.43	27.72
Winter wheat	130	0.73 ± 0.25	1.56	0.32	34.32

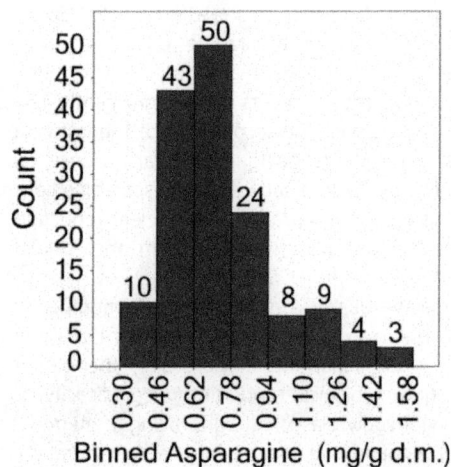

Figure 2 Frequency distribution of 151 bread wheat genotypes based on their asparagine concentration.

Correlation with other amino acids and carbohydrates

^{1}H-NMR has the advantage over targeted methods for amino acid analysis, such as GC-MS, of being able to analyse compounds from multiple compound classes in the same analysis. In wheat, the major classes of metabolites are soluble carbohydrates and amino acids. Levels of the most abundant of these compounds were therefore examined to determine whether asparagine concentration was correlated with any of the other major co-extracted components from wheat flour. Figure 3 represents a heatmap of Pearson correlations constructed from the full NMR metabolomics dataset, showing that asparagine concentrations were highly correlated with levels of glutamine, aspartate and alanine. This is perhaps to be expected as asparagine is synthesized from oxaloacetate via aspartate catalysed by a transaminase enzyme. Asparagine synthetase then converts aspartate to asparagine using glutamine, AMP and pyrophosphate. Further but weaker correlations were observed with a number of other amino acids such as GABA, threonine, leucine,

Table 2 Groupings of bread wheat genotypes showing the highest and lowest concentrations of free asparagine. Data selected from a comparison of 150 bread wheats grown on a single site in the same year (2005)

	Cultivar	Winter (W) or Spring (S) wheat	Asparagine Concentration (mean ± SD)
Low asparagine			
0.32–0.43 mg/g d.m.	Chinese Spring	S	0.43 ± 0.13
	Palesio	W	0.43 ± 0.04
	Blasco	W	0.43 ± 0.03
	Mv-Emese	W	0.41 ± 0.02
	Bilancia	W	0.40 ± 0.04
	Granbel	W	0.35 ± 0.08
	Soissons	W	0.35 ± 0.04
	Nomade	W	0.33 ± 0.04
	Valoris	W	0.32 ± 0.02
	Alba	W	0.32 ± 0.02
High asparagine			
1.50–1.56 mg/g d.m.	Fleischmann 401	W	1.56 ± 0.06
	Spark	W	1.52 ± 0.07
	Kirkpinar 79	W	1.50 ± 0.11
1.28–1.40 mg/g d.m.	Mexique 50	S	1.40 ± 0.05
	Renan	W	1.37 ± 0.06
	Bankuti 1201	W	1.35 ± 0.07
	Alabasskaja	W	1.28 ± 0.09
1.10–1.25 mg/g d.m	Kirac 66	W	1.25 ± 0.02
	Qualital	W	1.22 ± 0.06
	Blue/A	W	1.16 ± 0.05
	Mv-Magdalena	W	1.15 ± 0.06
	Tamaro	W	1.15 ± 0.08
	Gerek 79	W	1.15 ± 0.06
	Atlas-66	W	1.15 ± 0.12
	NS Rana 1	W	1.11 ± 0.11
	Hana	W	1.10 ± 0.04
0.95 1.06 mg/g d.m.	GK-Tiszataj	W	1.06 ± 0.07
	Karl 92	W	1.06 ± 0.02
	Seu Seun 27	W	1.03 ± 0.08
	Key	W	1.03 ± 0.05
	Probstdorfer Perlo	W	1.02 ± 0.02
	Mv-Suba	W	0.99 ± 0.03
	Lona	S	0.95 ± 0.04
	Sava	W	0.95 ± 0.06

valine, isoleucine and proline. There was little correlation with levels of glutamate, and a negative correlation was observed with tryptophan concentration. The relationship of asparagine concentration with soluble sugars was also explored. No correlation was observed between asparagine concentration and the levels of sucrose or maltose, the most abundant sugars in the wheat ^{1}H-NMR spectrum. A weak positive correlation, however, was seen between glucose and asparagine concentrations, whilst the trisaccharide raffinose showed a weak inverse relationship.

Effect of environmental conditions and growing location on asparagine

Twenty-six genotypes were selected and grown for two further years in the same location and in three additional locations in the third year. This allowed the effects of environment on asparagine concentration to be explored across diverse growing locations (Table 4).

When grown at Martonvásár, (Hungary) for 3 years (2005, 2006 and 2007), the concentration of asparagine ranged from 0.32 ± 0.02 mg/g d.m. (cv. Valoris, 2005) to 1.15 ± 0.12 mg/g d.m. (cv Atlas 66, 2005). Values of the means across the 26 genotypes ranged from 0.615 to 0.764 mg/g d.m. and showed a high coefficient of variation (CV = 19.6–26.7%). The variance of individual genotypes due to year of growth ranged from 3.00 to 37.6% (Table 5) with Atlas 66 and Lynx having the highest contents of asparagine (0.834 ± 0.277 and 0.836 ± 0.232 mg/g d.m., respectively) when the data were averaged over the 3 years. It is notable that Atlas 66 is a high protein line which has been used as a source of this trait in breeding programmes (Johnson et al., 1985). Similarly, Valoris had the lowest mean concentration (0.414 ± 0.09 mg/g d.m.) over the 3-year period. The effects of environment were studied further by analysis of the 26 selected lines grown in 2007 in the UK, France and Poland as well as Hungary. Comparison of these samples (grown at 4 locations in a single year) showed a similar pattern in variation in asparagine content (Table 4), from 0.31 mg/g d.m. (Valoris, UK) to 1.12 mg/g d.m. (Lynx, Hungary). The mean values for the 26 lines across the growing locations ranged from 0.47 (UK) to 0.80 (Hungary) mg/g d.m. and showed a high coefficient of variation (CV = 20.7–26.0%). The variance of individual genotypes (Table 5) due to location was typically higher than that observed for the single site comparison over three growing years and ranged from 6.4% (Gloria) to 52.6% (Estica). The contents of asparagine were generally highest in the samples grown in Hungary and France and significantly ($P < 0.001$) lower in those grown in the UK and Poland.

Taking all 6 environments (i.e. multiple sites and years) into account, the genotype with the highest mean asparagine concentration was Lynx (0.79 ± 0.25 mg/g d.m.) (Table 5). Conversely, Valoris contained the lowest asparagine concentration over the 6 environments (0.408 ± 0.08 mg/g d.m.) Despite the large variation due to environment (CV ranged from 9.2% to 45.1%), some genotypes appeared to be more 'stable' than others. Figure 4 illustrates the mean asparagine contents across the 6 environmental conditions years for each genotype. Plots are ordered by the observed concentration ranges across the 3 year, multi-environment study. The most stable lines (exhibiting the lowest concentration range across six conditions) were Gloria, Spartanka and Obrii, whilst those showing the highest variation included Estica, Atlas 66 and Crousty. Thus, despite Valoris having the lowest mean concentration of asparagine across the 6 environments, this genotype is more susceptible to effects of the environment than some of the other genotypes. However, Valoris did show the lowest 'minimum concentration' in four of the six environments and thus remains a genotype of interest for lower asparagine content. Likewise, Gloria, Mv-Emese and Isengrain are potentially interesting to breeders. These, together with Valoris, have significantly lower asparagine levels than other genotypes and under the environments examined in this study, also show

Table 3 Correlations of grain quality parameters with mean asparagine concentration across 150 genotypes grown in Hungary in 2005

Quality parameter	r	P-value	Slope
Yield (kg/plot)	0.140	0.0929	Negative
Thousand kernel weight (g/1000 kernel)	0.027	0.7479	Positive
Flour yield (%)	0.138	0.0923	Negative
Chopin bran yield (%)	0.041	0.6225	Positive
Year of registration	**0.255**	**0.0019**	Negative
Plant height (cm)	**0.413**	**<0.0001**	Positive
Mean kernel diameter (mm)	0.129	0.1142	Positive
Mean kernel weight (mg)	0.061	0.4536	Positive
Hardness Index	0.034	0.6753	Positive
Total protein (%)	**0.449**	**<0.0001**	Positive
Wholemeal protein content (%)	**0.507**	**<0.0001**	Positive
Flour protein content flour (%)	**0.376**	**<0.0001**	Positive
Gluten content (%)	**0.437**	**<0.0001**	Positive
Starch content (%)	**0.321**	**<0.0001**	Negative
Moisture content (%)	0.119	0.1443	Positive
Water absorption (%)	**0.354**	**<0.0001**	Positive
Gluten index	0.123	0.133	Positive
Falling number (s)	0.070	0.3916	Positive
Zeleny sedimentation (mL)	**0.366**	**<0.0001**	Positive

Bold values indicate a significant correlation where $p < 0.05$.

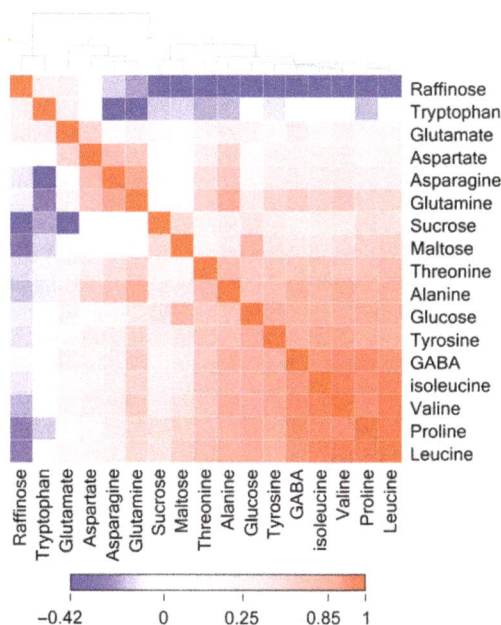

Figure 3 Heatmap of Pearson correlation coefficients of amino acid and soluble carbohydrate levels derived from ^{1}H-NMR metabolite profiling of wholemeal samples of 150 bread wheat genotypes grown in Martonvasar, Hungary in 2005.

Table 4 Mean asparagine concentrations (mg/g d.m.) of 26 wheat lines grown in six different environments. Values represent the mean ($n = 3$) ± standard deviation

Genotype	Asparagine Concentration (mg/g d.m.)					
	Hungary 2005	Hungary 2006	Hungary 2007	France 2007	UK 2007	Poland 2007
Claire	0.626 ± 0.079	0.542 ± 0.238	1.062 ± 0.075	0.596 ± 0.021	0.493 ± 0.106	0.509 ± 0.009
Lynx	0.728 ± 0.132	0.656 ± 0.141	1.123 ± 0.065	1.061 ± 0.099	0.667 ± 0.071	0.509 ± 0.035
Tommi	0.858 ± 0.014	0.648 ± 0.049	0.740 ± 0.022	0.848 ± 0.065	0.346 ± 0.051	0.523 ± 0.133
CF99105	0.775 ± 0.027	0.601 ± 0.015	0.727 ± 0.067	0.789 ± 0.020	0.477 ± 0.043	0.635 ± 0.042
Obrii	0.575 ± 0.034	0.689 ± 0.079	0.709 ± 0.054	0.669 ± 0.077	0.534 ± 0.039	0.625 ± 0.049
Malacca	0.641 ± 0.073	0.655 ± 0.128	0.958 ± 0.066	0.718 ± 0.072	0.425 ± 0.023	0.477 ± 0.017
Riband	0.664 ± 0.064	0.653 ± 0.033	1.094 ± 0.139	1.021 ± 0.054	0.410 ± 0.036	0.686 ± 0.058
Estica	0.774 ± 0.026	0.570 ± 0.037	0.665 ± 0.038	1.353 ± 0.092	0.336 ± 0.044	0.893 ± 0.029
Rialto	0.694 ± 0.009	0.586 ± 0.049	0.974 ± 0.093	0.627 ± 0.036	0.389 ± 0.069	0.516 ± 0.055
Campari	0.677 ± 0.038	0.777 ± 0.073	0.973 ± 0.059	0.812 ± 0.051	0.439 ± 0.029	0.498 ± 0.015
Avalon	0.649 ± 0.052	0.590 ± 0.032	0.696 ± 0.087	0.765 ± 0.069	0.398 ± 0.087	0.680 ± 0.062
Maris-Huntsman	0.540 ± 0.089	0.570 ± 0.007	0.586 ± 0.034	0.831 ± 0.062	0.573 ± 0.051	0.468 ± 0.094
Spartanka	0.555 ± 0.023	0.656 ± 0.030	0.707 ± 0.033	0.735 ± 0.064	0.619 ± 0.026	0.627 ± 0.038
Disponent	0.712 ± 0.056	0.747 ± 0.062	0.898 ± 0.052	1.109 ± 0.046	0.436 ± 0.008	0.678 ± 0.062
Atlas-66	1.146 ± 0.117	0.617 ± 0.075	0.740 ± 0.041	0.519 ± 0.057	0.357 ± 0.019	0.615 ± 0.039
Valoris	0.321 ± 0.024	0.420 ± 0.086	0.501 ± 0.058	0.475 ± 0.025	0.313 ± 0.070	0.419 ± 0.074
Tremie	0.614 ± 0.052	0.727 ± 0.112	0.881 ± 0.071	0.728 ± 0.058	0.531 ± 0.074	0.511 ± 0.026
Gloria	0.472 ± 0.059	0.556 ± 0.011	0.525 ± 0.024	0.599 ± 0.049	0.584 ± 0.112	0.607 ± 0.061
San-Pastore	0.605 ± 0.052	0.688 ± 0.070	0.718 ± 0.060	1.000 ± 0.022	0.712 ± 0.080	0.800 ± 0.034
Herzog	0.618 ± 0.050	0.661 ± 0.065	0.814 ± 0.041	0.768 ± 0.053	0.463 ± 0.077	0.415 ± 0.037
Mv-Emese	0.406 ± 0.018	0.415 ± 0.022	0.595 ± 0.073	0.612 ± 0.051	0.621 ± 0.046	0.535 ± 0.059
Isengrain	0.464 ± 0.005	0.569 ± 0.065	0.512 ± 0.078	0.708 ± 0.050	0.386 ± 0.039	0.410 ± 0.039
Crousty	–	0.597 ± 0.014	0.623 ± 0.058	1.132 ± 0.112	0.430 ± 0.034	0.578 ± 0.052
Tiger	–	0.339 ± 0.073	0.598 ± 0.079	0.680 ± 0.060	0.404 ± 0.053	0.522 ± 0.029
Cadenza	0.816 ± 0.034	0.520 ± 0.031	0.695 ± 0.053	0.933 ± 0.065	0.562 ± 0.081	–
Chinese Spring	0.431 ± 0.129	0.939 ± 0.035	0.756 ± 0.052	0.743 ± 0.120	0.428 ± 0.023	–
Mean	0.640 ± 0.171	0.615 ± 0.120	0.764 ± 0.180	0.801 ± 0.208	0.474 ± 0.107	0.572 ± 0.119
CV (%)	26.7	19.6	23.6	26.0	22.6	20.7
Range	0.321–1.146	0.339–0.939	0.501–1.123	0.475–0.801	0.313–0.712	0.410–0.893

the least variation with respect to their growing environment. Conversely, Spartanka and Obrii, which showed low influence of the environment, had higher mean asparagine levels across 6 environments. Thus, when selecting genotypes for low concentrations of asparagine, it is necessary to consider not only the mean concentrations observed but also the range of concentrations observed when grown under different environmental conditions, including trials at multiple locations in different years. This finding is consistent with data reported in Curtis et al. (2009) where reports of sevenfold differences were observed in certain genotypes grown across different UK locations in successive years, a range that was larger than that obtained when the same genotypes were grown in the glasshouse.

Comparison of the ^{1}H-NMR derived data with existing data from GC-MS

To compare the suitability of ^{1}H-NMR as a method for asparagine quantitation, we compared relevant data in this study to that published on similar lines but analysed using alternate methods such as GC-MS. Muttucumaru et al. (2006) reported asparagine concentrations of 3.07 and 4.43 mmol/kg f.w. for grain derived the cultivar Hereward when grown in field conditions. In our study, using ^{1}H-NMR, we measured an asparagine concentration of 0.59 mg/g dry matter in wholemeal from Hereward plants grown in Hungary in 2005 (Table S1). This equates to 3.90 mmol/

kg f.w. and is thus comparable to the Muttucumaru et al. study. Grain from the cultivar Malacca has been reported to contain 5.20 mmol/kg f.w. asparagine when grown in a glasshouse (Muttucumaru et al., 2006) but significantly less when grown in the field across 6 UK sites in 2006 (0.68–2.79 mmol/kg f.w.) and 2007 (1.65–3.87 mmol/kg f.w.) (Curtis et al., 2009). Again, both studies used GC-MS to determine amino acid concentration. Data from our ^{1}H-NMR study, after conversion to comparable units, returned an asparagine concentration of 2.81 ± 0.15 mmol/kg f.w for grain from this genotype when grown in the UK in 2007. This is therefore within the range of the Curtis et al. study. The Curtis et al. (2009) study also reported asparagine concentrations for the cultivar Claire of 0.82–2.5 mmol/kg f.w. (across 6 UK sites in 2006) and 2.04–2.68 mmol/kg f.w. (across 6 UK sites in 2007). In our own study, we measured a slightly higher level of asparagine (equivalent to 3.25 mmol/kg f.w.) which although elevated with respect to the Curtis et al. (2009) study is still less than the Muttucumaru et al. (2006) report of 4.12 mmol/kg asparagine content for this line when grown under glasshouse conditions.

Effect of environment on the asparagine content of white flours

White flour samples were also analysed for the 26 wheat genotypes grown in 2007 in the 4 locations, and ^{1}H-NMR data of

Table 5 Statistical comparison of asparagine concentration (mg/g d.m.) of 26 wheat lines grown in 6 environments

	3 Years at a single location (Hungary 2005–2007)			4 Locations in 2007 (Hungary, France, Poland, UK)			6 Environments (Hungary 2005–2007 & France, Poland and UK in 2007)		
	Mean (mg/g d.m.)	Range (mg/g d.m.)	CV (%)	Mean (mg/g d.m.)	Range (mg/g d.m.)	CV (%)	Mean (mg/g d.m.)	Range (mg/g d.m.)	CV (%)
Atlas 66	0.834 ± 0.277	0.617–1.146	33.1	0.558 ± 0.161	0.357–0.740	28.9	0.666 ± 0.267	0.357–1.146	40.1
Avalon	0.645 ± 0.053	0.590–0.696	8.2	0.634 ± 0.162	0.398–0.765	25.6	0.629 ± 0.127	0.398–0.765	20.2
Cadenza	0.677 ± 0.149	0.520–0.816	22.0	0.730 ± 0.188	0.562–0.933	25.7	0.705 ± 0.172	0.520–0.933	24.4
Campari	0.809 ± 0.150	0.677–0.973	18.6	0.680 ± 0.255	0.439–0.973	37.4	0.696 ± 0.201	0.439–0.973	28.9
CF99105	0.701 ± 0.090	0.601–0.775	12.8	0.657 ± 0.135	0.477–0.789	20.6	0.667 ± 0.120	0.477–0.789	17.9
Chinese Spring	0.709 ± 0.257	0.431–0.939	36.3	0.642 ± 0.186	0.428–0.756	28.9	0.659 ± 0.224	0.428–0.939	33.9
Claire	0.743 ± 0.279	0.542–1.062	37.6	0.665 ± 0.269	0.493–1.062	40.4	0.638 ± 0.214	0.493–1.062	33.5
Crousty	0.610 ± 0.018	0.597–0.623	3.0	0.691 ± 0.306	0.430–1.132	44.2	0.672 ± 0.268	0.430–1.132	39.9
Disponent	0.786 ± 0.099	0.712–0.898	12.6	0.780 ± 0.289	0.436–1.109	37.1	0.763 ± 0.226	0.436–1.109	29.6
Estica	0.670 ± 0.102	0.570–0.774	15.3	0.811 ± 0.427	0.336–1.353	52.6	0.765 ± 0.345	0.336–1.353	45.1
Gloria	0.518 ± 0.042	0.472–0.556	8.2	0.579 ± 0.037	0.525–0.607	6.4	0.557 ± 0.051	0.472–0.607	9.2
Herzog	0.698 ± 0.103	0.618–0.814	14.7	0.615 ± 0.205	0.415–0.814	33.3	0.623 ± 0.160	0.415–0.814	25.6
Isengrain	0.515 ± 0.053	0.464–0.570	10.2	0.504 ± 0.146	0.386–0.708	29.0	0.508 ± 0.118	0.386–0.708	23.3
Lynx	0.836 ± 0.252	0.656–1.123	30.1	0.840 ± 0.299	0.509–1.123	35.6	0.791 ± 0.245	0.509–1.123	31.0
Malacca	0.751 ± 0.179	0.641–0.958	23.8	0.644 ± 0.245	0.425–0.958	38.0	0.735 ± 0.231	0.509–1.123	31.4
Maris–Huntsman	0.565 ± 0.023	0.540–0.586	4.1	0.615 ± 0.153	0.468–0.831	24.9	0.595 ± 0.123	0.468–0.831	20.7
MV-Emese	0.472 ± 0.107	0.406–0.595	22.6	0.591 ± 0.039	0.535–0.621	6.5	0.531 ± 0.098	0.406–0.621	18.4
Obrii	0.658 ± 0.073	0.575–0.709	11.1	0.634 ± 0.075	0.534–0.709	11.8	0.633 ± 0.069	0.534–0.709	10.8
Rialto	0.751 ± 0.201	0.586–0.974	26.7	0.627 ± 0.251	0.389–0.974	40.1	0.631 ± 0.198	0.389–0.974	31.3
Riband	0.803 ± 0.252	0.653–1.094	31.3	0.803 ± 0.316	0.410–1.094	39.4	0.754 ± 0.256	0.410–1.094	34.0
San-Pastore	0.670 ± 0.059	0.605–0.718	8.8	0.808 ± 0.135	0.712–1.000	16.7	0.754 ± 0.136	0.605–1.000	18.0
Spartanka	0.639 ± 0.078	0.555–0.707	12.1	0.672 ± 0.058	0.619–0.735	8.6	0.650 ± 0.065	0.555–0.735	10.0
Tiger	0.469 ± 0.183	0.339–0.598	39.1	0.551 ± 0.117	0.404–0.680	21.3	0.509 ± 0.139	0.339–0.680	27.3
Tommi	0.749 ± 0.106	0.648–0.858	14.1	0.614 ± 0.224	0.346–0.848	36.5	0.660 ± 0.199	0.346–0.858	30.2
Tremie	0.740 ± 0.134	0.614–0.881	18.1	0.663 ± 0.175	0.511–0.881	26.4	0.665 ± 0.140	0.511–0.881	21.1
Valoris	0.414 ± 0.090	0.321–0.501	21.8	0.427 ± 0.083	0.313–0.501	19.5	0.408 ± 0.077	0.313–0.501	19.0

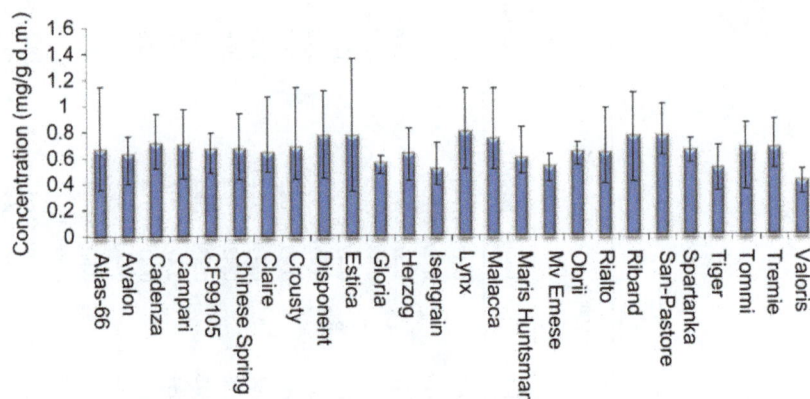

Figure 4 Average asparagine concentrations (mg/g d.m.) over 6 environments (Hungary (2005–2007), France (2007), UK (2007) and Poland (2007). Error bars represent observed asparagine concentration range across the 6 years/locations.

these samples were also collected to determine whether asparagine content located within the starchy endosperm was also subject to similar levels of variation due to environment as that observed in bran-containing samples of wholemeal. Observed asparagine concentrations (Table S2) in white flour varied from 0.105 ± 0.05 mg/g d.m. (Valoris, 2007, UK) to 0.730 ± 0.06 mg/g d.m. (Estica, 2007, France). The concentration in white flour as a proportion of that in wholemeal varied from 15% (Lynx, 2007, UK) to 57% (Tiger, 2007, France), which agrees with a previous study which showed that asparagine is concentrated in the bran fractions (Shewry et al., 2009b). In general, however, the amounts of asparagine present in white flour correlated well ($r^2 = 0.81$) with those observed in the corresponding wholemeal samples (Figure 5a). The effect of environment also followed the pattern observed for wholemeal samples (Figure 5b). Across the 26 lines grown in 4 European locations, the mean asparagine levels were highest in the material grown in France and Hungary (0.33 ± 0.14 and 0.34 ± 0.13 mg/g d.m., respectively) and lowest when the same genotypes were grown in the UK (0.18 ± 0.04 mg/g d.m). The mean levels in material grown in Poland were 0.20 ± 0.07 mg/g d.m. Inspection of the asparagine concentration in white flour for each of the 26 lines, averaged across the 4 locations, compared with the corresponding plots for wholemeal samples, showed that lines which had previously been identified as 'low asparagine' in the wholemeal samples also contained the lowest mean level of asparagine when white flour was analysed (Figure 6). Examples included Gloria, Isengrain and Valoris genotypes. The genotypes which showed low environmental variation in the asparagine content of wholemeal (Gloria, MV-Emese, Obrii, Spartanka, Valoris) also showed the lowest variation in asparagine content in the corresponding white flours. Hence, it should be possible for breeders to select for low asparagine content of white flour by analysis of wholemeal.

Heritability of asparagine concentration

The 26 lines described here were grown under a wide range of conditions including different soil types, rainfall and soil water availability, temperature and agricultural practices. Analysis of the grain therefore allowed us to partition the variation between the effects of genotype (G), environment (E), G × E interactions and that which cannot be explained by these factors (termed error). For the asparagine content of wholemeal samples, the ratio of genetic variance to total variance was 0.128, which indicates that only 13% of the observed variation in asparagine content is heritable (Figure 7). By contrast, the variance due to the

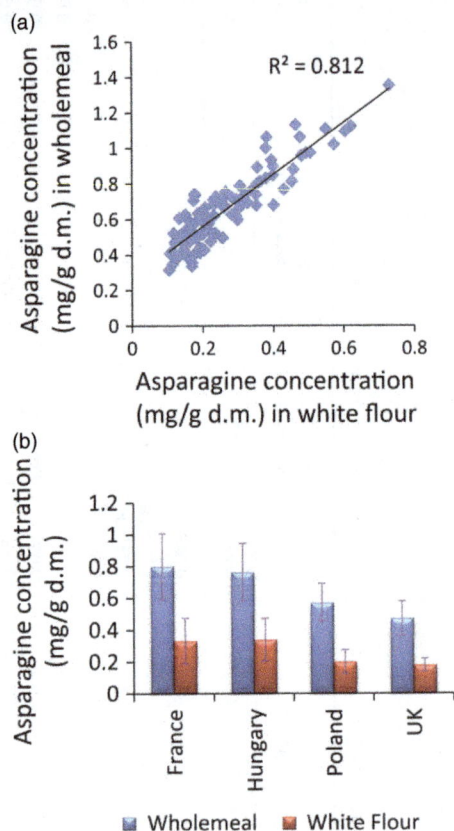

Figure 5 Comparison of asparagine concentration (mg/g d.m.) in white flour and wholemeal samples grown at 4 locations (France, Hungary, UK and Poland) in 2007. (a) correlation of asparagine concentrations (mg/g d.m.) in white flours against wholemeals; (b) mean levels of measured asparagine concentration in white flour and wholemeal samples from each of 4 growing locations. Error bars represent standard deviation of 3 replicate samples.

environment was 36%, whilst that apportioned to genotype x environment was higher at 43%. In fact, the variance due to genotype (13%) was the lowest apart from that assigned to error (9%).

This low 'broad sense heritability' is an important observation as it shows that breeders will find it difficult to select for low asparagine content without the availability of specific molecular markers which are not sensitive to environmental factors. Emeberi (2014) reported higher heritability (32%) for asparagine content

(a)

(b)

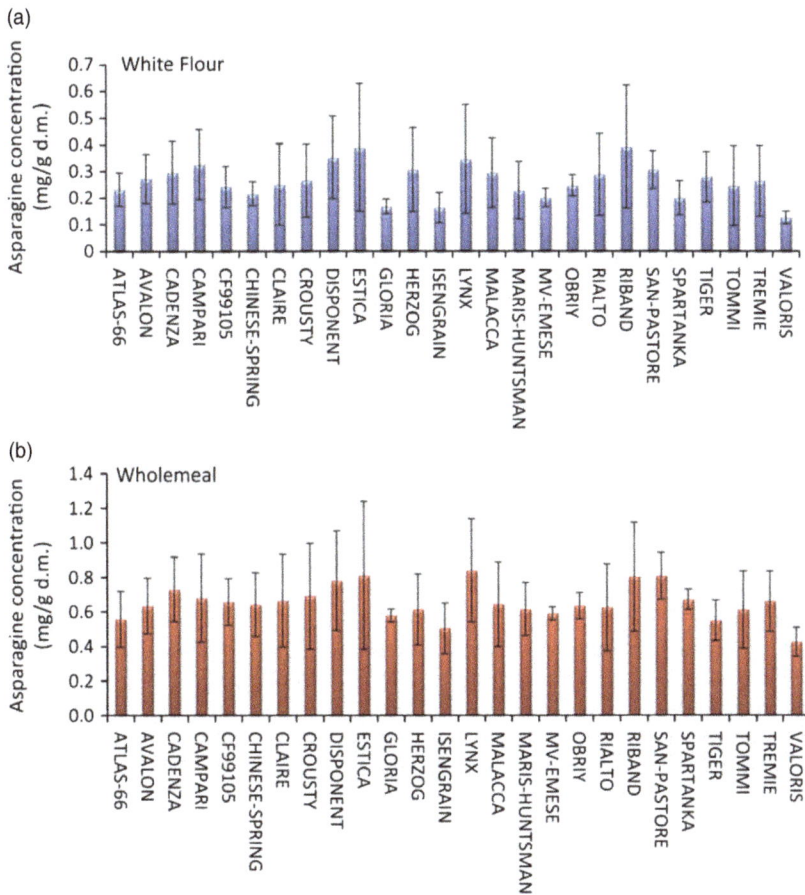

Figure 6 Average asparagine concentrations (mg/g d.m.) from wheat lines grown in 2007 at 4 environments (Hungary, France, UK and Poland). Error bars represent observed asparagine concentration range across the 4 locations. (a) data from white flour samples; (b) data from wholemeal samples.

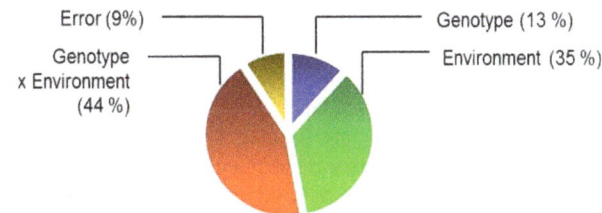

Figure 7 Pie chart showing computed variance components for asparagine concentration in wholemeal samples from 26 bread wheats grown in 6 environments.

in wholegrain of 92 wheat varieties. However, these varieties were grown in replicated glasshouse trials in which environmental effects were likely to be much less than in the six highly contrasting environments studied here. The same author also used a genomewide scan to identify 9 single nucleotide polymorphisms (SNPs) which explained between 14% and 24% of the observed variation in asparagine content. This again is indicative of low heritability and suggests that selection for low asparagine will be a challenge for wheat breeders.

Correlations with Environmental Parameters

Temperature and precipitation data were collected during each growth season (2005–2007) for each of the locations (Figure S2). The mean concentrations of asparagine in wholemeal of the 26 wheat genotypes were used to explore correlations with weather conditions over the growth period at each location (Table 6), with statistically significant correlations being shown in bold. The mean concentration of asparagine showed strong negative correlations ($r = -0.85$, $P = 0.03$) with the total precipitation between heading and harvest dates and also with the total precipitation during the period 3 months before heading date to the harvest date. However, there was no significant correlation with total precipitation during the 3 months before the heading date, indicating that the key time period for effects on asparagine accumulation is during grain development (between the heading and harvest dates). The concentration of asparagine also correlated positively with the mean temperature between heading and harvest dates although the P-value indicated a lower level of significance ($r = 0.74$, $P = 0.095$). As temperatures are generally lower during higher periods of rainfall, it is to be expected that correlations with both temperature and rainfall will be observed. The correlation of asparagine with temperature and inverse correlation with precipitation agree with the data from the 2007 multisite trial where asparagine concentrations were higher in flours from grain grown in Hungary and France (both having higher mean temperatures and lower levels of precipitation) compared to those grown in Poland and the UK, the latter having the lowest mean temperature and the highest amount of rainfall.

Conclusions

We have used ^{1}H-NMR profiling to generate quantitative data on asparagine concentrations in large sample sets, using a simple, quick and cheap sample preparation protocol. Analysis by ^{1}H-NMR provided reproducible data with low errors between

Table 6 Correlations of temperature, precipitation and soil parameters with mean asparagine concentration across 6 different growing environments

Environmental parameter	r	Slope	P-value
Total precipitation from heading date to harvest date (mm)	**0.849**	Negative	**0.033**
Total precipitation from 3 months before heading date to harvest date (mm)	**0.849**	Negative	**0.033**
Mean temperature from heading date to harvest date (°C)	**0.737**	Positive	0.095
Average maximum temp for any 10-day period between heading and harvest (°C)	**0.541**	Positive	0.268
Average minimum temp for any 10-day period between heading and harvest (°C)	0.240	Negative	0.646
Total precipitation during 3 months before heading date (mm)	0.074	Positive	0.890

Bold values indicate a significant correlation where $p < 0.05$.

technical replicates. Comeasurement of a number of different metabolites within the ^{1}H-NMR metabolomics dataset of wholemeal samples of grain allowed levels of asparagine to be compared with other abundant metabolites. We observed a wide variation in asparagine concentration that was highly correlated with certain bread-making quality parameters such as protein content and water absorption. The asparagine content of white flour was correlated with that of wholemeal with some genotypes having low contents in both fractions. Comparison of 150 bread wheat genotypes, the largest reported study to date, showed that asparagine was typically lower in shorter wheat plants and this was further endorsed by an observed negative relationship between asparagine concentration and the year of registration. This shows that breeding for reduced stature, to maximize yield and harvest index, has also resulted in reduced total grain asparagine.

Comparison of genotypes grown in multiple locations showed a large impact of the environment on asparagine concentration in mature grain with the level of precipitation and temperature during grain development (between heading and harvest) having the greatest effect. Whilst some genotypes could be identified which showed less impact due to the environmental conditions, the overall heritability of asparagine content in whole grain was low (13%), indicating that breeding for low asparagine lines will be more difficult than for many other phenotypic traits. Hence, selection for low asparagine will be a challenge for wheat breeders.

Materials and methods

Material

The first field experiment was carried out at Martonvásár (near Budapest, Hungary) in 2004–2005. A total of 150 bread wheat lines were selected to represent the range of diversity in the gene pool available for plant breeders, including wide geographical diversity in origin (from Europe to East Asia, the Americas, and Australia) and including landraces, breeding lines and modern and older cultivars. One hundred and thirty were winter type and 20 were spring type. Five modern cultivars of spelt (a hulled form of hexaploid wheat, *T. aestivum* var. *spelta*), 10 lines of tetraploid durum wheat (*T. turgidum* var. *durum*), five lines each of two early cultivated forms of wheat, diploid einkorn (*T. monococcum* var. *monococcum*) and tetraploid emmer (*T. turgidum* var. *dicoccum*), 10 lines of rye (*Secale cereale*), five lines of oats (*Avena sativa*) and 10 lines of barley (*Hordeum vulgare*) were also included. Full details are given by Ward *et al.* (2008). Twenty-three of the wheat lines and five rye lines were selected for further studies, together with three additional wheat lines and one additional rye line. These were grown at Martonvásár again

in 2005–2006 and 2006–2007 and at Nickerson Seeds UK (Saxham, near Bury St Edmunds, UK), DankoPlant Breeders Ltd. (Choryn, near Poznan, Poland) and the INRA experimental station at Clermont Ferrand (France). Agronomic treatments were standard for the individual sites, with 110 kg of N/Ha being applied in Poland, 204 kg of N/Ha in the United Kingdom, 200 kg of N/Ha in France and 140 kg of N/Ha in Hungary and appropriate use of agrochemicals. Winter, spring and durum wheats were conditioned to 15.5% moisture content before milling, whereas other species were conditioned to 14% moisture content. Milling was carried out using a Perten Laboratory Mill 3100 (with 0.5 mm sieve) and Retsch ZM100 (for *T. monococcum* and oats) to produce wholemeal. Samples were immediately cooled to −20 °C and stored at the same temperature in sealed bags.

^{1}H-NMR profiling

NMR sample preparation was carried out according to the procedures described in Ward *et al.* (2003) and Baker *et al.* (2006). NMR extractions into 80:20 $D_2O:CD_3OD$ containing 0.05% d_4-trimethylsilylpropionate (TSP) (1 mL) were performed for three technical replicates, of 30 mg, for each biological sample.

^{1}H-NMR spectra were acquired under automation at 300°K using an Avance Spectrometer (Bruker BioSpin, Coventry, UK) operating at 600.0528 MHz and equipped with a 5 mm selective inverse probe. Spectra were collected using a water suppression pulse sequence with a 90° pulse and a relaxation delay of 5 s. Each spectrum was acquired using 128 scans of 64 000 data points with a spectral width of 7309.99 Hz. Spectra were automatically Fourier transformed using an exponential window with a line broadening value of 0.5 Hz. Phasing and baseline correction were carried out within the instrument software. [1]H chemical shifts were referenced to d_4-TSP at δ0.00.

Data processing and statistical analysis

^{1}H-NMR spectra were automatically reduced, using Amix (Analysis of MIXtures software; Bruker BioSpin, Rheinstetten, Germany), to ASCII files containing integrated regions or 'buckets' of equal width (0.001 ppm for quantitation of asparagine and 0.01 ppm for correlation analysis). Spectral intensities were scaled to the d_4-TSP region (δ0.05 to −0.05). The ASCII file was imported into Microsoft Excel for the addition of sampling/treatment details. Regions for individual metabolites were identified via comparison to a library of known standards run under identical conditions. The region used for the quantitation of asparagine was δ2.9755–2.9255. Peaks in this region were integrated against the known concentration of TSP in the sample (0.05% w/v).

Calculations of mean, standard deviations and coefficients of variation were carried out using Microsoft Excel. To relate asparagine concentration values to the physical parameters of the wheat genotypes and to environmental conditions, Pearson correlation coefficients were calculated (from data on a dry weight basis) using Spotfire Decision Site (v. 9.1.2., TIBCO, Somerville, MA).

G × E analyses

Variance due to genotype and environment was calculated according to methods used in Corol et al., 2012. Datasets from the 26 wheat varieties grown in the six different environments were used in statistical models with all effects considered as random to estimate variance components with SAS software (proc VARCOMP). Three technical replicates were used as error terms in the following model: $X = \mu + E + G + G \times E + \varepsilon$.

As replicates were technical and not true field replicates, the error term is likely to be an underestimate of the true error. Therefore, we used the ratio $\sigma_g^2/(\sigma_g^2 + \sigma_E^2 + \sigma_{G \times E}^2)$ as a surrogate to heritability h^2. Indeed, this parameter, although likely to be an underestimate of h^2, is a suitable parameter for plant breeders, as a high value indicates that the trait is mostly affected by the genotype.

Acknowledgements

Generation of wheat tissues was carried out during the HEALTH-GRAIN project (FOOD-CT-2005-514008) and was a project funded by the European Commission in the Communities 6th Framework Programme. Rothamsted Research receives grant-aided support from the Biotechnology and Biological Sciences Research Council of the UK.

References

Baker, J.M., Hawkins, N.D., Ward, J.L., Lovegrove, A., Napier, J.A., Shewry, P.R. and Beale, M.H. (2006) A metabolomic study of substantial equivalence of field-grown genetically modified wheat. Plant Biotechnol. J. **4**, 381–392.

Claus, A., Schreiter, P., Weber, A., Graeff, S., Herrmann, W., Claupein, W., Schieber, A. and Carle, R. (2006) Influence of agronomic factors and extraction rate on the acrylamide contents in yeast-leavened breads. J. Agric. Food Chem. **54**, 8968–8976.

Claus, A., Carle, R. and Schieber, A. (2008) Acrylamide in cereal products: a review. J. Cereal Sci. **47**, 118–133.

Corol, D.I., Ravel, C., Raksegi, M., Bedo, Z., Charmet, G., Beale, M.H., Shewry, P.R. and Ward, J.L. (2012) Effects of genotype and environment on the contents of betaine, choline and trigonelline in cereal grains. J. Agric. Food Chem. **60**, 5471–5481.

Curtis, T., Muttucumaru, N., Shewry, P.R., Parry, M.A.J., Powers, S.J., Elmore, J.S., Mottram, D.S., Hook, S. and Halford, N.G. (2009) Effects of genotype and environment on free amino acid levels in wheat grain: implications for acrylamide formation during processing. J. Agric. Food Chem. **57**, 1013–1021.

Curtis, T.Y., Powers, S.J., Balagiannis, D., Elmore, J.S., Mottram, D.S., Parry, M.A.J., Rakszegi, M., Bedö, Z., Shewry, P.R. and Halford, N.G. (2010) Free amino acids and sugars in rye grain: implications for acrylamide formation. J. Agric. Food Chem. **58**, 1959–1969.

Emebiri, L. (2014) Genetic variation and possible SNP markers for breeding wheat with low-grain asparagine, the major precursor for acrylamide formation in heat-processed products. J. Sci. Food Agric. **94**, 1422–1429.

Friedman, M. (2003) Chemistry, biochemistry and safety of acrylamide, a review. J. Agric. Food Chem. **51**, 4504–4526.

Halford, H.G., Curtis, T.Y., Muttucumaru, N., Postles, J., Elmore, J.S. and Mottram, D.S. (2012) The acrylamide problem: a plant and agronomic science issue. J. Exp. Bot. **63**, 2841–2851.

Johnson, V.A., Mattern, P.J., Peterson, C.J. and Kuhr, S.L. (1985) Improvement of wheat protein by traditional breeding and genetic techniques. Cereal Chem. **62**, 350–355.

Lea, P.J., Sodek, L., Parry, M.A.J., Shewry, P.R. and Halford, N.G. (2007) Asparagine in plants. Ann. Appl. Biol. **150**, 1–26.

Mottram, D.S., Wedzicha, B.L. and Dodson, A.T. (2002) Acrylamide is formed in the Maillard reaction. Nature, **419**, 448–449.

Muttucumaru, N., Halford, N.G., Elmore, J.S., Dodson, A.T., Parry, M.A.J., Shewry, P.R. and Mottram, D.S. (2006) The formation of high levels of acrylamide during the processing of flour derived from sulfate-deprived wheat. J. Agric. Food Chem. **54**, 8951–8955.

Muttucumaru, N., Elmore, J.S., Curtis, T., Mottram, D.S., Parry, M.A.J. and Halford, N.G. (2008) Reducing acrylamide precursors in raw materials derived from wheat and potato. J. Agric. Food Chem. **56**, 6167–6172.

Poutanen, K., Shepherd, R., Shewry, P.R., Delcour, J.A., Björck, I. and van der Kamp, J.-W. (2008) Beyond whole grain: the European HEALTHGRAIN project aims at healthier cereal foods. Cereal Foods World, **53**, 32–35.

Rakszegi, M., Boros, D., Kuti, C., Lang, L., Bedo, Z. and Shewry, P.R. (2008) Composition and end-use quality of 150 wheat lines selected for the HEALTHGRAIN diversity screen. J. Agric. Food Chem. **56**, 9750–9757.

Shewry, P.R., Franklin, J., Parmar, S., Smith, S.J. and Miflin, B.J. (1983) The effects of sulphur starvation on the amino acid and protein compositions of barley grain. J. Cereal Sci. **1**, 21–31.

Shewry, P.R., D'Ovidio, R., Lafiandra, D., Jenkins, J.A., Mills, E.N.C. and Bekes, F. (2009a) Wheat grain proteins. Wheat: Chemistry and Technology, 4th edn (Khan, K. and Shewry, P.R., eds), pp. 223–298. St Paul, MN, USA: AACC.

Shewry, P.R., Zhao, F.J., Gowa, G.B., Hawkins, N.D., Ward, J.L., Beale, M.H., Halford, N.G., Parry, M.A. and Abecassis, J. (2009b) Sulphur nutrition differentially affects the distribution of asparagine in wheat grain. J. Cereal Sci. **50**, 407–409.

Shewry, P.R., Piironen, V., Lampi, A.-M., Edelmann, M., Kariluoto, S., Nurmi, T., Fernandez-Orozco, R., Ravel, C., Charmet, G., Andersson, A.A.M., Åman, P., Boros, D., Gebruers, K., Dornez, E., Courtin, C.M., Delcour, J.A., Rakszegi, M., Bedő, Z. and Ward, J.L. (2010) The HEALTHGRAIN wheat diversity screen: effects of genotype and environment on phytochemicals and dietary fiber components. J. Agric. Food Chem. **58**, 9291–9298.

Shewry, P.R., Van Schaik, F., Ravel, C., Charmet, G., Rakszegi, M., Bedo, Z. and Ward, J.L. (2011) Genotype and environment effects on the contents of vitamins B1, B2, B3, and B6 in wheat grain. J. Agric. Food Chem. **59**, 10564–10571.

Stadler, R.H., Blank, I., Varga, N., Robert, F., Hau, J., Guy, P.A., Robert, M.C. and Riediker, S. (2002) Acrylamide from Maillard reaction products. Nature, **419**, 449–450.

Tareke, E., Rydberg, P., Karlsson, P., Eriksson, S. and Törnqvist, M. (2002) analysis of acrylamide, a carcinogen formed in heated foodstuffs. J. Agric. Food Chem. **5**, 4998–5006.

Viant, M.R., Bearden, D.W., Bundy, J.G., Burton, I.W., Collette, T.W., Ekman, D.R., Ezernieks, V., Karakach, T.K., Lin, C.Y., Rochfort, S., De Ropp, J.S., Teng, Q., Tieerdema, R.S., Walter, J.A. and Wu, H. (2009) International NMR-based environmental metabolomics intercomparison exercise. Environ. Sci. Tech. **43**, 219–225.

Ward, J.L. and Beale, M.H. (2006) NMR spectroscopy in plant metabolomics. In Biotechnology in Agriculture and Forestry, Vol 57 Plant Metabolomics (Saito, K., Dixon, R.A. and Willmitzer, L., eds), pp. 81–91. Berlin Heidelberg: Springer-Verlag.

Ward, J.L., Harris, C., Lewis, J. and Beale, M.H. (2003) Assessment of ^{1}H NMR spectroscopy and multivariate analysis as a technique for metabolite fingerprinting of Arabidopsis thaliana. Phytochemistry, **62**, 949–957.

Ward, J.L., Poutanen, K., Gebruers, K., Piironen, V., Lampi, A.-M., Nyström, L., Andersson, A.A.M., Åman, P., Boros, D., Rakszegi, M., Bedő, Z. and Shewry, P.R. (2008) The HEALTHGRAIN cereal diversity screen: concept, results and prospects. J. Agric. Food Chem. **56**, 9699–9709.

10

Bread matters: a national initiative to profile the genetic diversity of Australian wheat

David Edwards[1], Stephen Wilcox[2], Roberto A. Barrero[3], Delphine Fleury[4], Colin R. Cavanagh[5,6], Kerrie L. Forrest[7], Matthew J. Hayden[7], Paula Moolhuijzen[3], Gabriel Keeble-Gagnère[3], Matthew I. Bellgard[3], Michał T. Lorenc[1], Catherine A. Shang[8], Ute Baumann[4], Jennifer M. Taylor[5], Matthew K. Morell[5], Peter Langridge[4], Rudi Appels[3] and Anna Fitzgerald[8,*]

[1]Australian Centre for Plant Functional Genomics and University of Queensland, St. Lucia, Qld, Australia
[2]Australian Genome Research Facility, The Walter and Eliza Hall Institute of Medical Research, Parkville, Vic., Australia
[3]Centre for Comparative Genomics, Murdoch University, Perth, WA, Australia
[4]Australian Centre for Plant Functional Genomics, University of Adelaide, Urrbrae, SA, Australia
[5]CSIRO Plant Industry, Black Mountain Laboratories, Canberra, ACT, Australia
[6]CSIRO Food Future Flagship, Black Mountain Laboratories, Canberra, ACT, Australia
[7]Department of Primary Industries, Victorian AgriBiosciences Centre, Bundoora, Vic., Australia
[8]Bioplatforms Australia, Macquarie University, North Ryde, NSW, Australia

*Correspondence

email afitzgerald@bioplatforms.com

Keywords: bread wheat, whole genome sequencing, single nucleotide polymorphisms.

Summary

The large and complex genome of wheat makes genetic and genomic analysis in this important species both expensive and resource intensive. The application of next-generation sequencing technologies is particularly resource intensive, with at least 17 Gbp of sequence data required to obtain minimal (1×) coverage of the genome. A similar volume of data would represent almost 40× coverage of the rice genome. Progress can be made through the establishment of consortia to produce shared genomic resources. Australian wheat genome researchers, working with Bioplatforms Australia, have collaborated in a national initiative to establish a genetic diversity dataset representing Australian wheat germplasm based on whole genome next-generation sequencing data. Here, we describe the establishment and validation of this resource which can provide a model for broader international initiatives for the analysis of large and complex genomes.

Introduction

On a 4-year average (2006–2010), Australia was the ninth largest producer of wheat, 16th largest consumer of wheat, and fifth largest exporter of wheat in the world (United States Department of Agriculture Foreign Agricultural Service, 2010). Intensive breeding has led to significant increases in yield and productivity, although various biotic and abiotic stresses can cause major yield loss in Australia. While many crops such as rice and maize have benefited from advanced genomic tools and complete genome sequences, the size and complexity of the wheat genome have limited genomic applications for wheat improvement. The lack of supportive knowledge on crop genomics is a serious impediment towards tapping potential biotechnological tools for crop improvement. Hence, concerted efforts are required to characterize genetic diversity in Australian bread wheat varieties to enable the generation of superior genotypes underpinning crop improvement, productivity and resilience.

Wheat has a very large genome, estimated to be 17 Gbp in size (Paux et al., 2008). The large size of the wheat genome is in part attributable to being an allohexaploid, meaning that it contains three distinct genomes. The diploid donor species, AA, BB and DD are thought to have diverged between 2.5 and 4.5 MYA and combined to produce Triticum aestivum in two distinct hybridization events. First, Triticum urartu (AA) and an unknown relative of Aegilops speltoides (BB) are believed to have produced the tetraploid Triticum turgidum ssp. dicoccoides around 0.2–0.5 MYA (Huang et al., 2002). This was followed by hybridization with Aegilops tauschii (DD) around 8500 years ago to produce the hexaploid T. aestivum (Kihara, 1944; McFadden and Sears, 1946). In addition to polyploidy, the wheat genome has experienced significant proliferation of repetitive elements, resulting in a composition of between 75% and 90% repetitive DNA sequences (Flavell et al., 1977; Wanjugi et al., 2009). This level of complexity hinders the development and application of genomic tools for wheat crop improvement.

The application of molecular markers to advance cereal breeding is now well established (Edwards, 2007; Edwards and Batley, 2008; Gupta et al., 2001). Modern cereal breeding is dependent on molecular markers for the rapid and precise analysis of germplasm, trait mapping and marker assisted selection (Lai et al., 2012b). Molecular markers can be used to select parental genotypes in breeding programmes, eliminate linkage drag in backcrossing and select for traits that are difficult to measure using phenotypic assays (Duran et al., 2010). Molecular markers have many other uses in genetics, such as the discovery of alleles associated with agronomic traits, verification of variety distinctness, uniformity and stability assessment, and inferences of population history (Duran et al., 2009b). Furthermore, molecular markers are invaluable as a tool for genome mapping in all systems, offering the potential for generating very high-density genetic maps that can be used to develop haplotypes for genes or regions of interest (Duran et al., 2009a;

Rafalski, 2002). SNPs represent the most frequent type of genetic polymorphism and may therefore provide a high density of markers and therefore increased mapping resolution near a locus of interest (Duran et al., 2010).

In November 2003, a USDA-NSF funded international workshop of wheat geneticists and sequencing specialists identified the first objectives towards sequencing the hexaploid wheat genome, that is, physical mapping and assessment of sequencing strategies. To capitalize on the momentum of this workshop, the International Wheat Genome Sequencing Consortium (IWGSC, http://www.wheatgenome.org) was established in January 2005 with the goal of coordinating the international effort to build the foundation for and lead the sequencing of the bread wheat genome. The IWGSC has achieved success in engaging countries worldwide in tackling the wheat genome through an approach to sequence flow-sorted chromosomes and thus reduce the complexity of the genome, followed by the construction of DNA BAC libraries from the purified chromosome arm for detailed analysis (Doležel et al., 2007; Molnár et al., 2011; Safar et al., 2004; Šafář et al., 2010). The first BAC library has been used successfully in a project to establish a sequence-ready physical map of chromosome 3B, the largest wheat chromosome (2× the rice genome), and was published by Paux et al. (2008).

In parallel to the BAC-based analysis of the wheat genome, the larger data volumes from the Illumina sequencing platform, combined with advanced bioinformatics provide the potential to gain insight into complex plant genomes (Berkman et al., 2012a; Lee et al., 2012; Marshall et al., 2010). This data has been applied for rapid genome sequencing (Batley and Edwards, 2009; Edwards and Batley, 2010; Imelfort and Edwards, 2009) as well as to discover very large numbers of genome-wide SNPs (Imelfort et al., 2009). More than one million SNPs have been identified between six inbred maize lines (Lai et al., 2010). This study also identified a large number of presence/absence variations, which may be associated with heterosis in this species. More recently, Allen et al. (2011) identified 14 078 putative SNPs in 6255 distinct reference sequences with Illumina GAIIx data from wheat lines Avalon, Cadenza, Rialto, Savannah and Recital. The validation rate from a subset of 1659 was 67%. A pipeline package called AGSNP has been applied to identify SNPs between two accessions of one of the diploid progenitors of bread wheat, A. tauschii (Luo et al., 2009). Roche 454 sequencing of A. tauschii accession AL8/78 has since been combined with Applied Biosystems SOLiD sequencing of genomic DNA and cDNA from A. tauschii accession AS75 using AGSNP to identify a total of 497 118 candidate A. tauschii SNPs (You et al., 2011).

Given the progress in the application of next-generation sequencing in other complex crop species, there is a significant opportunity to apply these approaches to understand genomic diversity in hexaploid bread wheat. However, the size of the genome presents challenges in terms of meeting the cost and sequencing throughput requirements. A large national initiative was established in Australia in 2010, to coordinate diverse wheat genetic and genomic activities and establish a resource for Australian crop improvement. With investment from Bioplatforms Australia and support from the Australian Genome Research Facility, the consortium has succeeded in generating between 5× and 10× coverage of 16 varieties chosen to represent the diversity of Australian wheat germplasm. This resource promises to be a foundation for SNP discovery, supporting Australian wheat crop improvement in the coming decades and

provides a model for other national and international crop genomics initiatives. Here, we describe the coordinated development of this resource together with preliminary analysis and quality assessment of the data.

Dataset design and generation

Method for selection

The wheat cultivars were chosen according to three criteria: they represent genetic diversity and have an economic impact in Australia; they are used in building genetic resources such as genetic populations or biotechnologies (parental lines and transformation); and they are key varieties that are both internationally studied and relevant to research in Australia (Rocca-Serra et al., 2010; Sansone et al., 2012; Taylor et al., 2008). After categorization and ranking of 46 suggested lines based on input from breeders, researchers and other stakeholders, a total of 16 lines were selected (Table 1).

Five to ten plants of each line were grown in a growth chamber or glasshouse. DNA was extracted from leaf samples of each plant using a standard phenol/chloroform method as described in Pallotta et al. (2000). Each plant was fingerprinted using a set of 10–20 molecular markers to verify the consistency of the germplasm with known genetic resources: parental lines were compared with derived segregating populations; highly variable microsatellites markers were also used to discriminate known versions of some cultivars. Several biotypes have been previously identified within cv. Wyalkatchem. We chose a biotype that has been used as a recurrent parent in backcrossing projects at the University of Adelaide and has been characterized for a number of known loci (seed and information kindly provided by Howard Eagles). For other varieties, such as Chara and Baxter, known to have biotypes, selection was based on an individual plant used in generating the mapping populations currently being utilized.

The sequencing was performed on one DNA sample from a single plant with the corresponding consistent fingerprint. The same plant was seed multiplied by bagging each spike to ensure pure self-crossing. The seed stocks are available through the Australian Pre-breeding Alliance database at the Australian Winter Cereals Collection (http://www2.dpi.qld.gov.au/extra/asp/AusPGRIS/). Each strain will be designated with the cultivar name followed by the BPA suffix (Table 1).

Data generation

Wheat genomic DNA was assessed for quality using the Nano-Drop ND-1000 spectrophotometer (ThermoScientific, Willmington, DE) and standard agarose gel electrophoresis. DNA libraries for sequencing were prepared using Illumina TruSeq DNA Library Preparation Kits (Cat. No. FC-390-1021, Illumina Inc., San Diego, CA) and associated recommended protocol. The protocol required that 1 µg of input genomic DNA be sheared using the Covaris S2 (Covaris Inc., Woburn, MA) which resulted in a peak fragment size of 200 bp. Fragmented DNA samples then underwent end-repair to generate blunt ends followed by an A-Tailing reaction to create a uniform 3′ overhang. This overhang was used to ligate the Illumina adapters and index sequences required for the sequencing reaction and subsequent variety identification.

The ligated DNA fragments were purified using the Qiagen MinElute Gel Extraction Kit (Cat. No. 28604, Qiagen,

Table 1 Wheat varieties selected for whole genome shotgun sequencing

	Wheat variety	A	B	C	Gbp	Pedigree
1	AC Barrie		×	×	181	NEEPAWA/COLUMBUS//BW-90
2	Alsen		×	×	129	ND-674/ND-2710/ND-688
3	Baxter	×	×		135	INIA-66/GAMUT//COOK/4/JUPATECO/3/LERMA-ROJO-64/SONORA-64-A// (SIB)TIMGALEN
4	Chara	×	×		273	BD-225/CD-87
5	Drysdale		×		160	HARTOG*3/QUARRION
6	Excalibur		×		171	RAC-I77(Sr26)/UNICULM-492//RAC-311-S
7	Gladius	×	×		201	RAC-875/KRICHAUFF//EXCALIBUR/KUKRI/3/RAC-875/KRICHAUFF/4/RAC-875// EXCALIBUR/KUKRI
8	H45	×	×		189	KALYANSONA/BLUEBIRD//ANZA*3/WW-80/3/OLYMPIC*2/CIANO-67
9	Kukri		×		173	CO-1213/RAC-549
10	Pastor		×	×	214	PFAU/SERI-82//BOBWHITE
11	RAC875		×		159	RAC-655/3/Sr21/4*LANCE//4*BAYONET
12	VolcaniDDI (V761-28-J4-B2-NZ8[†])		×	×	168	BTL/3/NURSIT-163/G-25/M-708
13	Westonia	×	×		123	SPICA/TIMGALEN//TOSCA/3/CRANBROOK//BOBWHITE*2/JACUP
14	Wyalkatchem	×	×		332	MACHETE/3/(84-W-129-504)GUTHA//JACIP*2/11th-ISEPTON-135
15	Xiaoyan 54[‡]		×	×	243	ST-2422-464/XIAOYAN-86
16	Yitpi	×	×		222	C-8-MMC-8-HMM/FRAME

Varieties were selected on the basis of A: genetic diversity, B: availability of derived genetic resources and C: potential international interest. Pedigrees: 1–14, 16 as documented by the Genetic Resources Information System for Wheat and Triticale (http://wheatpedigree.net) and 15 selection history as per Grama et al. (1984) and Grama et al. (1987).

[†]Selection history of line sequenced.

[‡]Selection from Xiaoyan 6.

Germantown, MD) and size selected by agarose gel to isolate fragments in the range of 300–400 bp. These fragments were amplified with ten cycles of PCR according to the TruSeq protocol, and the size and concentration of the final library were measured using a Bioanalyser DNA 1000 chip and fluorimetry (PicoGreen QuantIT assay, Molecular Probes Inc., Eugene, OR).

Completed libraries were denatured and diluted to 7 pM for clonal bridge amplification on the Illumina cBot. To obtain the required 10× raw coverage (calculated at 160 Gbp of data) for each variety, on average, three varietal DNA libraries were sequenced across two Illumina HiSeq flowcells according to manufacturer's protocols using multiplex indexes. Base-calling was processed with Illumina RTA software v1.10.36 (currently 1.12, Illumina Inc.). De-multiplexing and conversion to FastQ format were performed with CASAVA v1.7 (Illuminia Inc.) and the later runs with the upgraded v1.8.

Data quality control

To ensure high-quality reads are available for downstream analyses, sequenced datasets were subjected to a series of processing steps. First, poor-quality reads were removed using the following criteria: (i) contain ≥5% of bases as ambiguous calls (Ns), (ii) consist entirely of adenosines (poly A), (iii) the base quality for ≥50% of bases is lower than 7, (iv) both reads of the mate-pair are identical (PCR duplications), (v) after adaptor trimming reads are <50% of initial length. Next, possible mate-pair overlaps were evaluated, where the last ten bases of the first mate were aligned onto the second mate. If a perfect alignment was found, then the alignment was extended allowing up to 10% sequence divergence. Overlapping mate-pairs were joined as extended single end reads, and

both fasta and fastq files for these sequences were generated. Finally, poor-quality bases ($Q \leq 10$) at the 3' end of the reads were trimmed and reads with ≥50% of initial length were removed.

Australian wheat varieties database, data sharing and accessibility

The Australian wheat varieties sequencing data have been curated using an experimental metadata relational database based on Investigation Study Assay (ISA) infrastructure (Rocca-Serra et al., 2010; http://www.isa-tools.org/). The database captures relevant metadata and can be searched and browsed using a web-based interface that also provides links directly to the externally stored data files for download.

We have used ISA-tab format and the minimum information framework defined by (MIBBI) Minimum Information for Biological and Biomedical Investigations (Taylor et al., 2008) to report necessary metadata to facilitate reproducibility and reuse of this Australian wheat varieties reference dataset. Adoption of the ISA infrastructure to curate this data connects this initiative to a growing ISA data commons that promote public data sharing between diverse research domains and maximizes the potential benefit of this dataset to the greater scientific community (Sansone et al., 2012; http://isacommons.org/).

All raw and quality controlled data are publicly available through the Australian wheat varieties database (http://www.bioplatforms.com.au/datasets/wheat/wheat-sequencing/variety-sequencing). We request that analysed data be contributed back to the database to maximize the usefulness of the resource. Users of this data are required to act responsibly and ethically and to adhere to the Bioplatforms Australia Data Release Policy (http://www.bioplatforms.com.au/datasets/wheat).

Data validation

SNP concordance with the 9k SNP assay

To assess the utility of the Bioplatforms Australia wheat genomic resource for SNP discovery, the accuracy for genotype-calling at 1600 characterized SNP loci was investigated. The 1600 SNPs were chosen as they had known genotypes in a subset of the sequenced BPA wheat varieties, determined from genotyping of the DNA samples supplied for genomic sequencing using an Illumina 9000-feature Infinium SNP assay developed by the International Wheat SNP Working Group (http://wheat.pw.usda.gov/ggpages/9K_assay_available.html), and genotyping-by-sequencing using transcriptome sequence generated for the varieties.

Genomic sequence for varieties AC Barrie, Baxter and Chara was used to assess genotype-calling accuracy at the 1600 validated SNP loci. Following quality trimming, the processed raw sequence reads for each variety (corresponding to about 6×, 7× and 11× genome coverage for AC Barrie, Baxter and Chara, respectively) were mapped against reference sequence for the 1600 SNPs using BWA software (Li and Durbin, 2009). SNP genotypes were called using custom scripts when the minimum coverage at the sequence variant position was seven or ten reads. Assuming a homoeolog-specific reference sequence, a minimum coverage of seven and ten reads at the SNP position corresponds to 87.5% and 97.9% statistical confidence for the genotype call (Galan et al., 2010).

With a minimum coverage of ten reads at the SNP position, genotype calls from the genomic sequence had 91%, 85% and 96% concordance with the validated SNP genotypes for AC Barrie, Baxter and Chara, respectively. However, only 488, 217 and 761 SNP genotypes were called from the sequence data in total, respectively. When the minimum coverage at the SNP position was reduced to seven reads, 637, 392 and 844 genotypes were called with similar concordance, respectively. These results indicate that the Bioplatforms Australia genomic resource can be used to reliably call SNP genotypes with high statistical confidence at homoeolog-specific loci and imply its utility for de novo SNP discovery. Examination of the genomic distribution of SNPs (which have been genetically mapped, n = 901) across chromosome groups and sub-genomes showed no bias for SNP discovery across the wheat genome (Table 2). The results further suggest that at 6–10× genome coverage, reliable SNP discovery and genotype-calling can be typically achieved for about 24%–52% of SNPs and indicate that increased genome coverage would further improve the accuracy of SNP discovery. This observation is consistent with the requirement of the SNP discovery pipeline for a minimum coverage of seven reads at a sequence variant position for genotype-calling. A detailed description for de novo SNP discovery using the Bioplatforms Australia genomic resource will be published elsewhere.

SNP discovery by genome mapping

Where reference genomic sequence assemblies are available, it is possible to predict genomic SNPs by mapping paired sequence reads from whole genome shotgun sequencing to the reference. Consistent variety-specific sequence variation within the aligned reads is indicative of variety-specific SNPs. As wheat varieties are predominantly homozygous across their genomes, relatively low coverage is required compared to similar approaches using heterozygous populations. However, SNP discovery can be confounded by the presence of multiple genomes, and reference sequences representing each of the sub-genomes are required to differentiate between homologous (inter genomic) and homoeologous (inter varietal) SNPs. We have developed SGSautoSNP (Second Generation Sequencing autoSNP) software specifically to predict SNPs from whole genome Illumina shotgun sequence data from homozygous polyploid species, and this has been successfully applied to identify more than 1.5 million SNPs across the canola genome with an accuracy of >96% (D. Edwards, personal communication). Reference genomic templates are currently only available for wheat group 7 chromosomes (Berkman et al., 2011, 2012b; Lai et al., 2012a). To assess whether these could be used for genome mapping-based SNP discovery, whole genome shotgun data for an initial four Australian varieties were mapped using SOAP (Li et al., 2008). No pre-filtering of the data was performed with the exception of duplicate read removal. This initial test identified more than 900 000 SNPs between four Australian varieties along this chromosome group and suggests that this approach could be applied to the complete wheat variety dataset. SNP density varied between the genomes, with fewer SNPs identified on 7D (0.40 SNPs per Kbp) compared to 7A (1.69 SNPs per Kbp) and 7B (1.39 SNPs per Kbp). These preliminary results are presented within a GBrowse database and are available publicly at http://www.wheatgenome.info.

Conclusions

Through the establishment of a national initiative, we have produced whole genome shotgun data for 16 wheat varieties representing the diversity within Australian cultivated bread wheat. Preliminary validation suggests that this data is suitable for the identification of genome-wide sequence polymorphisms. This data is publicly accessible and presents a valuable resource for wheat crop improvement in both Australia and internationally.

Acknowledgements

This project was funded by Bioplatforms Australia through the Australian Government's Education Investment Fund (EIF) Super

Table 2 Genomic distribution of SNPs used for validation

Chromosome group	Sub-genome		
	A (%)	B (%)	D (%)
1	75 (88)	54 (87)	10 (89)
2	57 (92)	98 (91)	12 (73)
3	52 (95)	60 (89)	3 (100)
4	46 (87)	26 (90)	7 (100)
5	64 (88)	101 (94)	14 (100)
6	56 (89)	73 (95)	9 (85)
7	47 (92)	33 (91)	4 (100)

Per cent in bracket indicates average genotype concordance between known and predicted genotypes for AC Barrie, Baxter and Chara.

Science Initiative. iVEC for providing access to computational resources and storage.

References

Allen, A.M., Barker, G.L.A., Berry, S.T., Coghill, J.A., Gwilliam, R., Kirby, S., Robinson, P., Brenchley, R.C., D'Amore, R., McKenzie, N., Waite, D., Hall, A., Bevan, M., Hall, N. and Edwards, K.J. (2011) Transcript-specific, single-nucleotide polymorphism discovery and linkage analysis in hexaploid bread wheat (*Triticum aestivum* L.). *Plant Biotechnol. J.* **9**, 1086–1099.

Batley, J. and Edwards, D. (2009) Genome sequence data: management, storage, and visualization. *Biotechniques*, **46**, 333–336.

Berkman, P.J., Skarshewski, A., Lorenc, M.T., Lai, K., Duran, C., Ling, E.Y.S., Stiller, J., Smits, L., Imelfort, M., Manoli, S., McKenzie, M., Kubalakova, M., Simkova, H., Batley, J., Fleury, D., Dolezel, J. and Edwards, D. (2011) Sequencing and assembly of low copy and genic regions of isolated *Triticum aestivum* chromosome arm 7DS. *Plant Biotechnol. J.* **9**, 768–775.

Berkman, P.J., Lai, K., Lorenc, M.T. and Edwards, D. (2012a) Next generation sequencing applications for wheat crop improvement. *Am. J. Bot.* **99**, 365–371.

Berkman, P.J., Skarshewski, A., Manoli, S., Lorenc, M.T., Stiller, J., Smits, L., Lai, K., Campbell, E., Kubalakova, M., Simkova, H., Batley, J., Dolezel, J., Hernandez, P. and Edwards, D. (2012b) Sequencing wheat chromosome arm 7BS delimits the 7BS/4AL translocation and reveals homoeologous gene conservation. *Theor. Appl. Genet.* **124**, 423–432.

Doležel, J., Kubaláková, M., Paux, E., Bartoš, J. and Feuillet, C. (2007) Chromosome-based genomics in the cereals. *Chromosome Res.* **15**, 51–66.

Duran, C., Appleby, N., Edwards, D. and Batley, J. (2009a) Molecular genetic markers: discovery, applications, data storage and visualisation. *Curr. Bioinform.* **4**, 16–27.

Duran, C., Edwards, D. and Batley, J. (2009b) Molecular marker discovery and genetic map visualisation. In *Bioinformatics, tools and applications* (Edwards, D., Hanson, D. and Stajich, J., eds), pp. 165–190, New York: Springer.

Duran, C., Eales, D., Marshall, D., Imelfort, M., Stiller, J., Berkman, P.J., Clark, T., McKenzie, M., Appleby, N., Batley, J., Basford, K. and Edwards, D. (2010) Future tools for association mapping in crop plants. *Genome*, **53**, 1017–1023.

Edwards, D. (2007) Bioinformatics and plant genomics for staple crops improvement. In *Breeding Major Food Staples* (Kang, M.S. and Priyadarshan, P.M., eds), pp. 93–106, Ames: Blackwell.

Edwards, D. and Batley, J. (2008) Bioinformatics: fundamentals and applications in plant genetics, mapping and breeding. In *Principles and Practices of Plant Genomics* (Kole, C. and Abbott, A.G., eds), pp. 269–302, Enfield: Science Publishers, Inc.

Edwards, D. and Batley, J. (2010) Plant genome sequencing: applications for crop improvement. *Plant Biotechnol. J.* **7**, 1–8.

Flavell, R.B., Rimpau, J. and Smith, D.B. (1977) Repeated sequence DNA relationships in four cereal genomes. *Chromosoma*, **63**, 205–222.

Galan, M., Guivier, E., Caraux, G., Charbonnel, N. and Cosson, J.-F. (2010) A 454 multiplex sequencing method for rapid and reliable genotyping of highly polymorphic genes in large-scale studies. *BMC Genomics*, **11**, 296.

Grama, A., Gerecther-Amitai, Z.K., Blum, A. and Rubenthaler, G. L. (1984) Breeding bread wheat cultivars for high protein content by transfer of protein genes from Triticum dicoccoides. In *Cereal Grain Protein Improvement*, pp. 145–153. Vienna: International Atomic Energy Agency, Series 681-E.

Grama, A., Porter, N.G. and Wright, D.S.C. (1987) Hexaploid wild emmer wheat derivatives grown under New Zealand conditions 2. Effect of foliar urea sprays on plant and grain nitrogen and baking quality. New Zealand *J. Agri. Res.* **30**, 45–51.

Gupta, P.K., Roy, J.K. and Prasad, M. (2001) Single nucleotide polymorphisms: a new paradigm for molecular marker technology and DNA polymorphism detection with emphasis on their use in plants. *Curr. Sci.* **80**, 524–535.

Huang, S., Sirikhachornkit, A., Su, X.J., Faris, J., Gill, B., Haselkorn, R. and Gornicki, P. (2002) Genes encoding plastid acetyl-CoA carboxylase and 3-phosphoglycerate kinase of the *Triticum/Aegilops* complex and the evolutionary history of polyploid wheat. *Proc. Natl Acad. Sci. USA*, **99**, 8133–8138.

Imelfort, M. and Edwards, D. (2009) De novo sequencing of plant genomes using second-generation technologies. *Brief. Bioinform.* **10**, 609–618.

Imelfort, M., Duran, C., Batley, J. and Edwards, D. (2009) Discovering genetic polymorphisms in next-generation sequencing data. *Plant Biotechnol. J.* **7**, 312–317.

Kihara, H. (1944) Discovery of the DD-analyzer, one of the ancestors of vulgare wheats. *Agric. Hortic.* **19**, 2.

Lai, J.S., Li, R.Q., Xu, X., Jin, W.W., Xu, M.L., Zhao, H.N., Xiang, Z.K., Song, W.B., Ying, K., Zhang, M., Jiao, Y.P., Ni, P.X., Zhang, J.G., Li, D., Guo, X.S., Ye, K.X., Jian, M., Wang, B., Zheng, H.S., Liang, H.Q., Zhang, X.Q., Wang, S.C., Chen, S.J., Li, J.S., Fu, Y., Springer, N.M., Yang, H.M., Wang, J.A., Dai, J.R., Schnable, P.S. and Wang, J. (2010) Genome-wide patterns of genetic variation among elite maize inbred lines. *Nat. Genet.* **42**, 1027–1030.

Lai, K., Berkman, P.J., Lorenc, M.T., Duran, C., Smits, L., Manoli, S., Stiller, J. and Edwards, D. (2012a) WheatGenome.info: An integrated database and portal for wheat genome information. *Plant Cell Physiol.* **53**, 1–7.

Lai, K., Lorenc, M.T. and Edwards, D. (2012b) Genomic databases for crop improvement. *Agronomy*, **2**, 62–73.

Lee, H., Lai, K., Lorenc, M.T., Imelfort, M., Duran, C. and Edwards, D. (2012) Bioinformatics tools and databases for analysis of next generation sequence data. *Brief. Funct. Genomics*, **2**, 12–24.

Li, H. and Durbin, R. (2009) Fast and accurate short read alignment with Burrows-Wheeler transform. *Bioinformatics*, **25**, 1754–1760.

Li, R.Q., Li, Y.R., Kristiansen, K. and Wang, J. (2008) SOAP: short oligonucleotide alignment program. *Bioinformatics*, **24**, 713–714.

Luo, M.C., Deal, K.R., Akhunov, E.D., Akhunova, A.R., Anderson, O.D., Anderson, J.A., Blake, N., Clegg, M.T., Coleman-Derr, D., Conley, E.J., Crossman, C.C., Dubcovsky, J., Gill, B.S., Gu, Y.Q., Hadam, J., Heo, H.Y., Huo, N., Lazo, G., Ma, Y., Matthews, D.E., McGuire, P.E., Morrell, P.L., Qualset, C.O., Renfro, J., Tabanao, D., Talbert, L.E., Tian, C., Toleno, D.M., Warburton, M.L., You, F.M., Zhang, W. and Dvorak, J. (2009) Genome comparisons reveal a dominant mechanism of chromosome number reduction in grasses and accelerated genome evolution in Triticeae. *Proc. Natl Acad. Sci. USA*, **106**, 15780–15785.

Marshall, D., Hayward, A., Eales, D., Imelfort, M., Stiller, J., Berkman, P., Clark, T., McKenzie, M., Lai, K., Duran, C., Batley, J. and Edwards, D. (2010) Targeted identification of genomic regions using TAGdb. *Plant Methods*, **6**, 19.

McFadden, E. and Sears, E. (1946) The origin of *Triticum spelta* and its free-threshing hexaploid relatives. *J. Hered.* **37**, 81–89.

Molnár, I., Kubaláková, M., Šimková, H., Cseh, A., Molnár-Láng, M. and Doležel, J. (2011) Chromosome Isolation by Flow Sorting in Aegilops umbellulata and Ae. comosa and Their Allotetraploid Hybrids Ae. biuncialis and Ae. geniculata. *PLoS ONE*, **6**, e27708.

Pallotta, M.A., Graham, R.D., Langridge, P., Sparrow, D.H.B. and Barker, S.J. (2000) RFLP mapping of manganese efficiency in barley. *Theor. Appl. Genet.* **101**, 1100–1108.

Paux, E., Sourdille, P., Salse, J., Saintenac, C., Choulet, F., Leroy, P., Korol, A., Michalak, M., Kianian, S., Spielmeyer, W., Lagudah, E., Somers, D., Kilian, A., Alaux, M., Vautrin, S., Berges, H., Eversole, K., Appels, R., Safar, J., Simkova, H., Dolezel, J., Bernard, M. and Feuillet, C. (2008) A physical map of the 1-gigabase bread wheat chromosome 3B. *Science*, **322**, 101–104.

Rafalski, A. (2002) Applications of single nucleotide polymorphisms in crop genetics. *Curr. Opin. Plant Biol.* **5**, 94–100.

Rocca-Serra, P., Brandizi, M., Maguire, E., Sklyar, N., Taylor, C., Begley, K., Field, D., Harris, S., Hide, W., Hofmann, O., Neumann, S., Sterk, P., Tong, W. and Sansone, S.-A. (2010) ISA software suite: supporting standards-compliant experimental annotation and enabling curation at the community level. *Bioinformatics*, **26**, 2354–2356.

Safar, J., Bartos, J., Janda, J., Bellec, A., Kubalakova, M., Valarik, M., Pateyron, S., Weiserova, J., Tuskova, R., Cihalikova, J., Vrana, J., Simkova, H., Faivre-Rampant, P., Sourdille, P., Caboche, M., Bernard, M., Dolezel, J. and Chalhoub, B. (2004) Dissecting large and complex genomes: flow sorting and BAC cloning of individual chromosomes from bread wheat. *Plant J.* **39**, 960–968.

Šafář, J., Šimková, H., Kubaláková, M., Číhalíková, J., Suchánková, P., Bartoš, J. and Doležel, J. (2010) Development of chromosome-specific BAC resources for genomics of bread wheat. *Cytogenet. Genome Res.* **129**, 211–223.

Sansone, S.-A., Rocca-Serra, P., Field, D., Maguire, E., Taylor, C., Hofmann, O., Fang, H., Neumann, S., Tong, W., Amaral-Zettler, L., Begley, K., Booth, T., Bougueleret, L., Burns, G., Chapman, B., Clark, T., Coleman, L.-A., Copeland, J., Das, S., de Daruvar, A., de Matos, P., Dix, I., Edmunds, S., Evelo, C.T., Forster, M.J., Gaudet, P., Gilbert, J., Goble, C., Griffin, J.L., Jacob, D., Kleinjans, J., Harland, L., Haug, K., Hermjakob, H., Sui, S.J.H., Laederach, A., Liang, S., Marshall, S., McGrath, A., Merrill, E., Reilly, D., Roux, M., Shamu, C.E., Shang, C.A., Steinbeck, C., Trefethen, A., Williams-Jones, B., Wolstencroft, K., Xenarios, I. and Hide, W. (2012) Toward interoperable bioscience data. *Nat. Genet.* **44**, 121–126.

Taylor, C.F., Field, D., Sansone, S.-A., Aerts, J., Apweiler, R., Ashburner, M., Ball, C.A., Binz, P.-A., Bogue, M., Booth, T., Brazma, A., Brinkman, R.R., Michael Clark, A., Deutsch, E.W., Fiehn, O., Fostel, J., Ghazal, P., Gibson, F., Gray, T., Grimes, G., Hancock, J.M., Hardy, N.W., Hermjakob, H., Julian, R.K., Kane, M., Kettner, C., Kinsinger, C., Kolker, E., Kuiper, M., Novere, N.L., Leebens-Mack, J., Lewis, S.E., Lord, P., Mallon, A.-M., Marthandan, N., Masuya, H., McNally, R., Mehrle, A., Morrison, N., Orchard, S., Quackenbush, J., Reecy, J.M., Robertson, D.G., Rocca-Serra, P., Rodriguez, H., Rosenfelder, H., Santoyo-Lopez, J., Scheuermann, R.H., Schober, D., Smith, B., Snape, J., Stoeckert, C.J., Tipton, K., Sterk, P., Untergasser, A., Vandesompele, J. and Wiemann, S. (2008) Promoting coherent minimum reporting guidelines for biological and biomedical investigations: the MIBBI project. *Nat. Biotechnol.* **26**, 889–896.

Wanjugi, H., Coleman-Derr, D., Huo, N., Kianian, S., Luo, M., Wu, J., Anderson, O. and Gu, Y. (2009) Rapid development of PCR-based genome-specific repetitive DNA junction markers in wheat. *Genome*, **52**, 576–587.

You, F., Huo, N., Deal, K., Gu, Y., Luo, M.-C., McGuire, P., Dvorak, J. and Anderson, O. (2011) Annotation-based genome-wide SNP discovery in the large and complex Aegilops tauschii genome using next-generation sequencing without a reference genome sequence. *BMC Genomics*, **12**, 59.

11

Resistance to *Wheat streak mosaic virus* generated by expression of an artificial polycistronic microRNA in wheat

Muhammad Fahim[1,2,†], Anthony A. Millar[2], Craig C. Wood[1] and Philip J. Larkin[1,*]

[1]CSIRO Plant Industry, Canberra, ACT, Australia
[2]Division of Plant Sciences, Research School of Biology, Australian National University, Canberra, ACT, Australia

*Correspondence

email Philip.larkin@csiro.au
[†]Present address: Lecturer, Department of Microbiology, Hazara University Mansehra, KPK, Pakistan.
email mfahim@hu.edu.pk

Keywords: *wheat streak mosaic virus*, artificial microRNA, multiplex amiRNA, transgenic resistance, *Triticum aestivum*, *Tritimovirus*, Potyviridae.

Summary

Wheat streak mosaic virus (WSMV) is a persistent threat to wheat production, necessitating novel approaches for protection. We developed an artificial miRNA strategy against WSMV, incorporating five amiRNAs within one polycistronic amiRNA precursor. Using miRNA sequence and folding rules, we chose five amiRNAs targeting conserved regions of WSMV but avoiding off-targets in wheat. These replaced the natural miRNA in each of five arms of the polycistronic rice miR395, producing amiRNA precursor, *FanGuard* (FGmiR395), which was transformed into wheat behind a constitutive promoter. Splinted ligation detected all five amiRNAs being processed in transgenic leaves. Resistance was assessed over two generations. Three types of response were observed in T_1 plants of different transgenic families: completely immune; initially resistant with resistance breaking down over time; and initially susceptible followed by plant recovery. Deep sequencing of small RNAs from inoculated leaves allowed the virus sequence to be assembled from an immune transgenic, susceptible transgenic, and susceptible non-transgenic plant; the amiRNA targets were fully conserved in all three isolates, indicating virus replication on some transgenics was not a result of mutational escape by the virus. For resistant families, the resistance segregated with the transgene. Analysis in the T_2 generation confirmed the inheritance of immunity and gave further insights into the other phenotypes. Stable resistant lines developed no symptoms and no virus by ELISA; this resistance was classified as immunity when extracts failed to transmit from inoculated leaves to test plants. This study demonstrates the utility of a polycistronic amiRNA strategy in wheat against WSMV.

Introduction

Wheat streak mosaic virus (WSMV; Genus *Tritimovirus*; Family Potyviridae) has remained a threat to wheat production wherever it occurs, and its distribution is expanded as evidenced by its confirmed presence in Australia in 2003. Although no data are available for seed transmission of other isolates of WSMV, the Australian isolate is transmitted both through its natural vector wheat curl mite *Aceria tosichella* (Slykhuis, 1955; Harvey and Seifers, 1991; Seifers *et al.*, 1998) and through seed (Jones *et al.*, 2005). The virus spread rapidly across the Australian continent between 2003 and 2007 (Dwyer *et al.*, 2007). The virus poses a new threat to the wheat production and required development of new bio-security practices. The widespread occurrence of the virus and its vector, the potential major impacts on yield, and the impracticality of managing the mites add to the priority of breeding virus-resistant varieties and developing alternative methods of virus control through development of virus-resistant transgenic wheat.

Pathogen-derived resistance was pioneered with the expression of viral coat protein in transgenic tobacco plants (Abel *et al.*, 1986) and developed into more efficient and effective transgenic protection against viruses in plants utilizing double-stranded RNA (dsRNA)-induced RNA interference (RNAi) (Abbott

et al., 2002; Smith *et al.*, 2000; Waterhouse *et al.*, 1998). It is now established that RNAi is a natural surveillance mechanism conserved across eukaryotic organisms, where small RNAs either repress or cleave the complementary mRNAs in sequence-specific manner (Baulcombe, 2004). Since then, the strategy has been successfully employed to confer resistance in various plants against invading pathogens.

Previously, we have shown that transgene constructs capable of forming dsRNA transcripts are more likely to result in immunity against WSMV (Fahim *et al.*, 2010) than either of the previous two strategies that involved sense expression of the nuclear inclusion b or coat protein genes (Sivamani *et al.*, 2000, 2002). However, the use of long hairpin RNA (hpRNA) from conventional RNAi vectors as in Fahim *et al.*'s (2010) study theoretically entails an increased risk of 'off-target' effects, i.e. silencing of unintended genes (Jackson *et al.*, 2003). Furthermore, some express concern that agricultural-scale deployment of antivirus hpRNA-expressing transgenic plants might lead to evolution of new virus biotypes via heterologous recombination or complementation between the relatively long viral sequences expressed from the transgene and RNA from a non-target virus infecting the same plant. Although the likelihood of such events seems remote and could be further reduced by judicious selection of smaller sequences for the hpRNA constructs, nevertheless the utility of other approaches is worth exploring. One such approach, amiRNA-mediated gene silencing, has recently been developed specifically to address the risk of off-target effects and transgene–virus recombination to form new biotypes (Schwab et al., 2006).

The amiRNA approach utilizes a naturally occurring miRNA precursor as a backbone, with the mature miRNA sequence being replaced to gain new targeting ability (Ossowski et al., 2008; Vaucheret et al., 2004). In plants, amiRNAs have been successfully used to down-regulate endogenous genes (Alvarez et al., 2006; Khraiwesh et al., 2008; Molnar et al., 2009; Schwab et al., 2006; Warthmann et al., 2008) and also for developing transgenic virus resistance against Turnip yellow mosaic virus (TYMV), Turnip mosaic virus (TuMV) (Niu et al., 2006), Cucumber mosaic virus (CMV) (Duan et al., 2008), Potato virus X (PVX), and Potato virus Y (PVY) (Ai et al., 2011) in Arabidopsis; against CMV (Qu et al., 2007) in tobacco; against CMV (Zhang et al., 2011) in tomato; and against Cassava brown streak virus (CBSV) and Cassava brown streak Uganda virus (CBSUV) in cassava (Wagaba et al., 2010).

It has been argued that the use of short viral sequences in this amiRNA approach is less likely to enable the emergence of novel viral entities through recombination and trans-encapsidation (Schnippenkoetter et al., 2001). However, when only a small viral sequence is used, the virus is more likely to evolve in the amiRNA target sequence via transition mutation and enable avoidance of amiRNA complementarity and defence (Simon-Mateo and Garcia, 2006; Lin et al., 2009). Other examples include HIV escape mutants to avoid RNAi (Boden et al., 2003; Das et al., 2004; Westerhout et al., 2005). To substantially reduce the risk of viruses evolving to avoid degradation, a strategy would be very useful where an amiRNA precursor gene expressed multiple amiRNAs targeting different conserved structural and functional portions of the viral genome. Resistance to this protection would require simultaneous mutations to avoid all the amiRNA sequences. A similar rationale was invoked in the work of Israsena et al. (2009) where they designed precursor genes encoding three amiRNA against rabies virus and tested them in cell culture. Likewise, multiple siRNA were developed from a multiplex miRNA directed against HIV in cultured cells (Liu et al., 2008; ter Brake et al., 2006).

miRNA precursors that have been used for the delivery of amiRNAs in plants include miR159a (Niu et al., 2006); miR171a (Qu et al., 2007), miR172a (Schwab et al., 2006), miR30 (Zeng et al., 2002), miR528 (Warthmann et al., 2008), and miR167b (Ai et al., 2011). In plants, Niu et al. (2006) demonstrated that a dimeric amiRNA precursor in Arabidopsis could be effective against two different viruses. Others have successfully targeted two endogenous transcripts with dimeric amiRNA precursors in Arabidopsis (Park et al., 2009) and Chlamydomonas (Zhao et al., 2009) using different miRNA precursors. Here, in our studies, we used the multiplex precursor of rice miR395 family of miRNAs that was identified in both Arabidopsis thaliana and Oryza sativa computationally and was later experimentally verified (Guddeti et al., 2005; Kawashima et al., 2009). OsmiR395 targets ATP sulphurylases that are involved in sulphate assimilation (Rotte and Leustek, 2000) and is induced in sulphur starvation to regulate a low-affinity sulphate transporter and two ATP sulphurylases (Allen et al., 2005; Jones-Rhoades and Bartel, 2004). The rice miR395 is a single ~1-kb transcript that generates a convoluted RNA structure that generates seven fully processed miRNA (Jones-Rhoades and Bartel, 2004; S. Belide, J. R.

Petrie, P. Shrestha, M. Fahim, Q. Liu, C. C. Wood and S. P. Singh, unpublished).

Here, we expressed five pre-amiRNA, potentially generating ten amiRNA species, to different conserved regions of the WSMV genome from a modified version of rice miRNA precursor miR395. The polycistronic amiRNA strategy is able to produce marker-free transgenic wheat plants immune to WSMV, demonstrating that this is a viable strategy for a major crop species. Moreover, it alleviates the concerns of recombinant novel viral entities forming and also produces plants predicted to avoid the loss of resistance caused by virus mutation. The resolution of technological problems and concerns implies this strategy has a strong biotechnological potential for agriculture.

Results

Design of polycistronic amiRNA construct

We chose pre-miR395 as the backbone for simultaneous expression of multiple amiRNAs targeting various conserved regions in the WSMV genome. We combined published amiRNA selection criterion (http://wmd3.weigelworld.org/cgi-bin/webapp.cgi) into a software application we call 'miR Mate'. Applying this to the available full WSMV genome sequences (five at the time) identified approximately 120 potential target sites for amiRNA. As a final selection filter, we searched Wheat TIGR mRNA databases for potential off-targets of these potential WSMV amiRNAs using the online miRU program (http://plantgrn.noble.org/psRNATarget/). No expressed sequences were found as potential targets when three mismatches were allowed. Through this process, five amiRNA were chosen and designated amiRNA-1, amiRNA-2, amiRNA-3, amiRNA-4, and amiRNA-5; their WSMV genome targets and their target coordinates are given in Table 1 and Figure 1. The target of amiRNA-1 lies in 5' UTR region, amiRNA-2 targets the newly described open reading frame (ORF) pipo region of P3 cistron (Chung et al., 2008), amiRNA-3 targets P1 gene, amiRNA-4 targets P3 cistron (upstream of pipo), and amiRNA-5 targets the HCpro gene on WSMV genome (Figure 1). We deliberately chose a mix of targets on the genomic and replicative strands of the virus in case one strand was more available for the amiRNA surveillance than the other. While further good targets could be identified in the 3' genes, the bias to the 5' genes is simply a result of beginning the screening for potential off-targets from that end. Surprisingly, no 21-nt sequence in the extreme 3' region of the virus could be identified, which was both conserved and met the design rules. The endogenous miRNAs and miRNA* that are derived from the miR395 precursor were replaced with these amiRNA and amiRNA* sequences to conserve the secondary structure of the transcript. The predicted secondary structure of the polycistronic miR395 and the modified artificial miR395 were almost identical (Figure 2), presumably enhancing the prospect of the predicted biogenesis of mature amiRNAs. This artificial polycistronic precursor was named FanGuard395 (FGmiR395).

Generation of wheat carrying the FanGuard395 transgene

FGmiR395 was synthesized by Geneart and cloned into pWubi vector, to generate FGWS-pWubi, where FGmiR395 was behind a constitutive maize polyubiquitin promoter separated from the transgene by a spliceable ubiquitin intron. FGWS-pWubi was cobombarded into wheat immature embryos along with

Table 1 Conservation of amiRNA targets in WSMV Genome. The alignment used to design amiRNA against conserved targets in WSMV genome. AlignX was used with default settings using Vector NTI 10. (a) Alignment of the five chosen target regions in the five published WSMV genomes at the time of the design of the amiRNA. (b) Alignment in the five target regions in the ten new WSMV genomes that became available subsequent to amiRNA design. (c) Alignment of the amiRNA target regions in the WSMV-ACT isolate (unpublished and obtained subsequent to the design of FGmiR395 and production of the transgenic plants). The mismatched nucleotides with other isolates are highlighted. amiRNA-1 and amiRNA-2 target the replicating strand, signified by numbering 1–21 from left to right; amiRNA-3, amiRNA-4 and amiRNA-5 target the genomic strand of the WSMV, signified by numbering 1–21 from right to left. The chosen targets remain absolutely conserved in all five target regions of the WSMV-ACT isolate, when virus population RNA was resequenced from inoculated plants in the study including –S, +R, and the transgenic breakdown plants +S

Accession	Description	Origin	amiRNA-1 Target sequence	amiRNA-2 Target sequence	amiRNA-3 Target sequence	amiRNA-4 Target sequence	amiRNA-5 Target sequence
			AGCTCTCGCATAGAGATAAGC (1→21)	TCGAGCAAGATCTTTCACACG (1→21)	GAAGATTCCATTATGTGCCGA (21→1)	CCAGGAAGCATTTTCTGGTCA (21→1)	CCGCGAACGTCTTGCAAGTTA (21→1)
(a)							
AF057533	Sydney 81	Nebraska	AGCTCTCGCATAGAGATAAGC	TCGAGCAAGATCTTTCACACG	GAAGATTCCATTATGTGCCGA	CCAGGAAGCATTTTCTGGTCA	CCGCGAACGTCTTGCAAGTTA
1AF285169	Type Strain	Kansas	AGCTCTCGCATAGAGATAAGC	TCGAGCAAGATCTTTCACACG	GAAGATTCCATTATGTGCCGA	CCAGGAAGCATTTTCTGGTCA	CCGCGAACGTCTTGCAAGTTA
AF285170	El Batan	Mexico	AGCTCTCGCATAGAGATAAGC	TCGAGCAAGATCTTTCACACG	GAAGATA**C**CAT**A**TGTGCCGA	CCAGGAAGCATTTTCTGGTCA	CCGCGAACGTCTTGCAAGTTA
AF454454	Czech	Czech	AGCTCTCGCATAGAGATAAGC	TCGAGCAAGATCTTTCACACG	GAAGAT**T**CCATTATGTGCCGA	CCAGGAAGCATTTTCTGGTCA	CCGCGAACGTCTTGCAAGTTA
AF454455	Turkish	Turkey	AGCTCTCGCATAGAGATAAGC	TCGAGCAAGATCTTTCACACG	GAAGATTCCATTATGTGCCGA	CCAGGAAGCATTTTCTGGTCA	CCGCGAACGTCTTGCAAGTTA
(b)							
AF511614	H95S	Kansas	AGCTCTCGCATAGAGATAAGC	TCGAGCAAGATCTTTCACACG	GAAG**G**TTCCATTATGTGCCGA	CCAGGAAGCATTTTCTGGTCA	CCGCGAACGTCTTGCAAGTTA
AF511615	H98	Kansas	AGCTCTCGCATAGAGATAAGC	TCGAGCAAGATCTTTCACACG	GAAGATTCCATTATGTGCCGA	CCAGGAAGCATTTTCTGGTCA	C**T**GCGAACGTCTTGCAAGTTA
AF511618	ID96	Idaho	AGCTCTCGCATAGAGATAAGC	TCGAGCAAGATCTTTCACACG	GAAGATTCCATTATGTGCCGA	CCAGGAAGCATTTTCTGGTCA	CCGCGAACGTCTTGCAAGTTA
AF511619	ID99	Idaho	AGCTCTCGCATAGAGATAAGC	TCGAGCAAGATCTTTCACACG	GAAGATTCCATTATGTGCCGA	CCAGGAAGCATTTTCTGGTCA	CCGCGAACGTCTTGCAAGTTA
AF511630	Mon96	Montana	AGCTCTCGCATAGAGATAAGC	TCGAGCAAGATCTTTCACACG	GAAGATTCCATTATGTGCCGA	CCAGGAAGCATTTTCTGGTCA	CCGCGAACGTCTTGCAAGTTA
EU914917	Naghadeh	Iran	AGCTCTCGCATAGAGATAAGC	TCGAGCAAGATCTTTCACACG	GAAGAT**T**CATTATGT**ACCA**A	CCAGGAAGCATTTTCTGGTCA	CCGCGAACGTCTTGCAAGTTA
EU914918	Sadat-Saher	Iran	AGCTCTCGCATAGAGATAAGC	TCGAGCAAGATCTTTCACACG	GAAGAT**T**CCATTATGTGCCGA	CCAGGAAGCATTTTCTGGTCA	CCGCGAACGTCTTGCAAGTTA
F511643	WA99	WA, USA	AGCTCTCGCATAGAGATAAGC	TCGAGCAAGATCTTTCACACG	GAAGAT**T**CATTATGTGCCGA	CCAGGAAGCATTTTCTGGTCA	CCGCGAACGTCTTGCAAGTTA
FJ348358	WA94	WA, USA	AGCTCTCGCATAGAGATAAGC	TCGAGCAAGATCTTTCACACG	GAAGAT**T**CATTATGTGCCGA	CCAGGAAGCATTTTCTGGTCA	CCGC**A**AACGTCTTGCAAGTTA
FJ348359	ARG2	Argentina	AGCTCTCGCATAGAGATAAGC	TCGAGCAAGATCTTTCACACG	GAAGATTCCATTATGTGCCGA	CCAGGAAGCATTTTCTGGTCA	CCGCGAACGTCTTGCAAGTTA
(c)							
Unpublished	ACT	Australia	AGCTCTCGCATAGAGATAAGC	TCGAGCAAGATCTTTCACACG	GAAGATTCCATTATGTGCCGA	CCAGGAAGCATTTTCTGGTCA	CCGCGAACGTCTTGCAAGTTA

WSMV, *Wheat streak mosaic virus*.

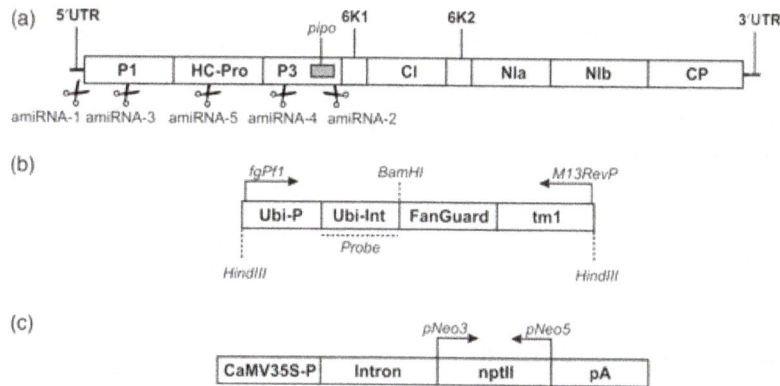

Figure 1 Structure of the *Wheat streak mosaic virus* (WSMV) genome (approximately 9400 nt), the target sites for amiRNAs and the *FGmiR395* transgene. (a) Genome map of WSMV showing the five conserved regions (indicated by scissors) targeted by amiRNAs, amiRNA-1 to amiRNA-5. (b) Design of *FGmiR395* construct (1400 nt) used to transform wheat using biolistics; shown are the probe region for Southern blot and primer sequences FgPf1 and M13RevP used in PCR. (c) Diagram of pCMneoSTLS2 containing the *nptII* gene for geneticin resistance, used in the cotransformation of immature wheat embryos and showing the position of the PCR primers pNeo3 and pNeo5.

Figure 2 amiRNA Secondary Structure. A comparison of truncated *Osa-miR395* and *FGmiR395* secondary structures. (a) Predicted secondary structure of *miR395* truncated to include only the first five native miRNAs. (b) Predicted secondary structure of *FGmiR395* replacing the first five natural miRNA sequences with amiRNAs design against *Wheat streak mosaic virus*, numbered 1–5. These are the predicted fold structures of transcripts using *RNAfold*. Bars showing regions corresponding to amiRNA guide strand. Secondary structure probabilities are indicated by heat map (Blue, weak; Red, strong).

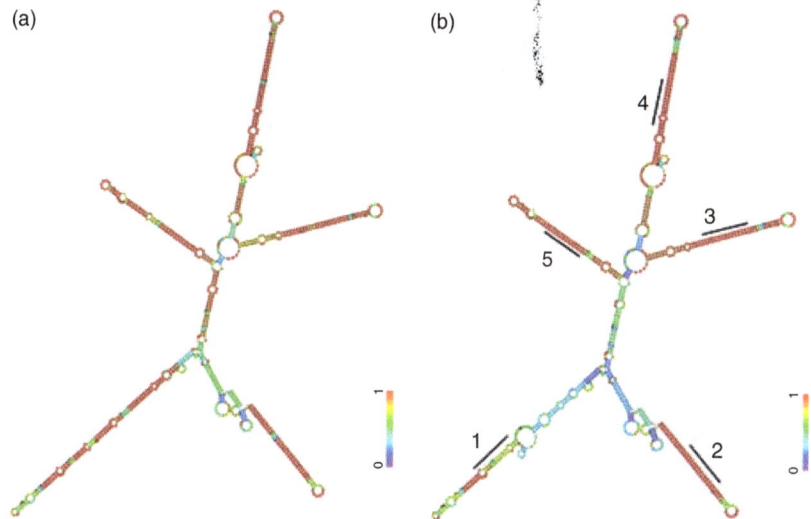

plasmid pCMneoSTLS2 (Maas *et al.*, 1997) that contained neomycin phosphotransferase gene (*nptII*), conferring resistance to geneticin (G418). From a wheat transformation of 379 embryos, a total of 23 T_0 transgenic wheat plants were generated from 16 different embryos; therefore, there were at least 16 independent events.

The transgenic lines were designated FanGuard plasmid (FGP) plus a number corresponding to the bombarded embryo. Where multiple T_0 plants were obtained from a single embryo, they were distinguished with lower-case letters, e.g. FGP1a and FGP1b. All T_0, T_1, and T_2 transgenic plants were morphologically indistinguishable from the wild-type parental cultivar, Bob White selection 26 (BW26), implying that the FGmiR395 does not influence growth or development (Figure 3b).

Plants recovered from the transformation selection cultures were screened through PCR for the *nptII* selectable marker from the pCMneoSTLS2 cotransformation vector (Figure 1c). This confirmed that all 23 plants coming through the antibiotic selection were true transgenics carrying the selectable marker. Genomic PCR screening for the presence of *FGmiR395* was carried out using primers, fgPf1 and M13RevP, that span from within the promoter region, full ubiquitin intron and all of the *FGmiR395* transgene including the *nos* terminator (Figure 1b). PCR analysis of T_1 families confirmed that 14 of the 23 families had *FGmiR395* as well as *nptII* (Figure 3a, PCR data not shown) and 10 of these were from different embryos and were therefore independent transgenic events. Southern blot analysis of T_2 plants from a subset of these lines confirmed the presence of transgene (Figure 4). Nine families were negative for *FGmiR395* and positive for *nptII*. The latter lines were discarded after the preliminary assessment for resistance (next section).

Preliminary assessment of *FGmiR395* transgenic wheat in T_1

The T_1 generation was subsequently challenged with WSMV through mechanical inoculation at the three-leaf stage (4–17 plants per family) using the spray gun. Wheat streak mosaic disease is characterized by light-green-to-faint-yellow streaks in wheat leaves parallel to the veins. The virus arrests growth, and plants show moderate-to-severe stunting with prostrated tillers often with empty spikes or spikes with shrivelled kernels. Serological characterization of the transgenic families involved

(a)

(b)

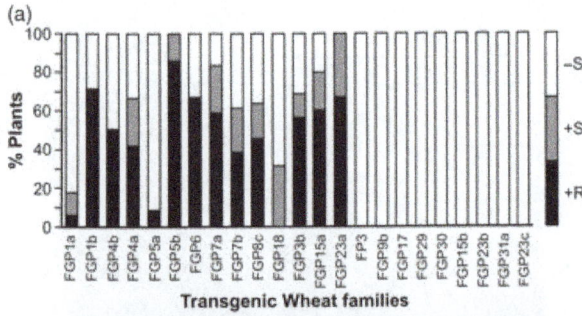

Figure 3 Preliminary assessment of resistance in T_1 families of FGmiR395 expressing transgenic wheat. (a) *Wheat streak mosaic virus* ELISA-based bioassay analysis of resistance in segregating populations. Virus levels were detected by double-antibody sandwich enzyme-linked immunosorbent assay at 14 d p.i. (days postinoculation). −S indicated *FGmiR395* transgene–negative susceptible segregants, +S indicated transgene-carrying susceptible segregants, while +R indicated the transgene-carrying resistant segregants. (b) Resistance phenotypes: Hea., healthy control; Inf., infected control; R., FGP15a2.10, an inoculated transgenic immune; M., FGP15a2.7, an inoculated moderate resistant; S., FGP15a2.2, an inoculated susceptible negative segregant (−S).

inoculating each individual plant with WSMV and assaying with a double-antibody sandwich enzyme-linked immunosorbent assay (DAS-ELISA) at 14 days postinoculation (d p.i.) and 28 d

p.i. Virus accumulation in leaves was determined using ELISA and expressed as a ratio of inoculated plants to non-inoculated controls.

The progeny of nine geneticin-resistant *nptII* positive lines were negative by PCR for *FGmiR395* and were completely susceptible to the challenged virus. The T_1 families from all the other transgenic lines included plants inheriting *FGmiR395* and plants resistant to the virus. The transgenic (signified by +), resistant (signified by R) segregants were designated as *FGmiR395+R*. These plants were completely free of virus symptoms at all four data points 7, 14, 21, and 28 d p.i. They were indistinguishable from uninfected wild-type BW26 plants. Such phenotypes were observed in segregating families of all *FGmiR395*-carrying events except FGP18.

In several transgenic families, some *FGmiR395*-carrying individuals were incompletely resistant, characterized as either: (i) intermediate phenotype between susceptible and resistant arising from either resistance breakdown or plant recovery from virus infection, or (ii) fully susceptible phenotype identical to infected wild-type BW26. The intermediate phenotype was characterized by intermediate plant height and a lower virus titre compared to susceptible control BW26; these types of transgenic segregants were designated as moderately resistant or *FGmiR395+MR* (Figure 3b). The fully susceptible *FGmiR395*-carrying plants were designated as *FGmiR395+S*; these were indistinguishable from infected BW26 or the null segregant *FGmiR395−S* phenotype.

In this preliminary analysis, most *FGmiR395* families segregated for all three phenotypes (Figure 3a); however, in four events, FGP1b, FGP4b, FPG5a, and FGP6, all *FGmiR395*-inheriting segregants were fully resistant. The numbers of plants were variable and low (4–17), so little attention was paid to the segregation ratios at this stage.

All segregants in FGP18 displayed either *FGmiR395+S* or *FGmiR395−S* phenotype in this preliminary assessment and showed no resistance. These susceptible transgenic plants along with other *FGmiR395-S* from other families exhibited characteristic virus symptoms and were comparable with virus-infected wild-type control BW26 plants.

The initial analysis revealed a variety of phenotypes in response to WSMV inoculation. Transgenic families FGP4a, 6, 8c, 15a, and 18 were selected as representative of the range of

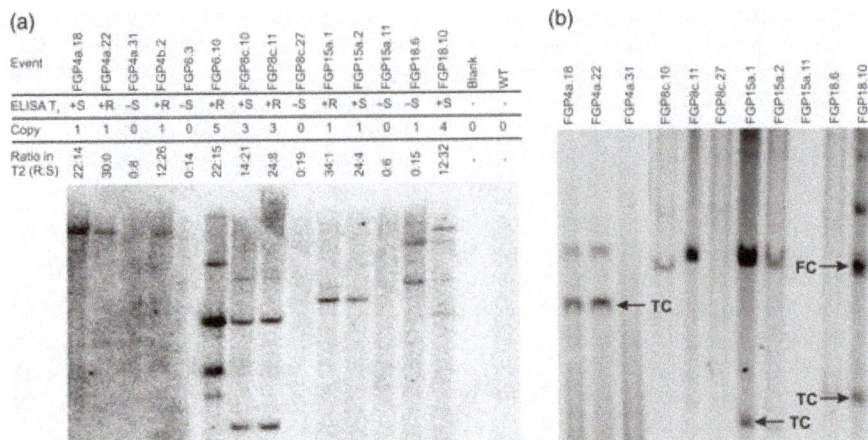

Figure 4 Southern blot analysis of families in T_2 generation. (a) *Bam*HI digested wheat DNA for transgene copy number. (b) *Hind*III-digested wheat DNA for transgene copy size. FC is full copy insert size 3 KB; TC is truncated copy inserts. Plant FGP18.6 is labelled −S by PCR and bioassay; it has FG sequence present, but that sequence is truncated so that it is not detected by the PCR.

Analysis of transgenic family FGP6 for resistance to WSMV

The preliminary analysis indicated that all *FGmiR395*-positive segregants in FGP1b, FGP4b, FGP5a, and FGP6 were resistant (Figure 3a). Subsequently, a bioassay on 26 T_1 individuals of FGP6 showed that all 13 *FGmiR395*-carrying segregants were symptom free at all four time points (7, 14, 21, and 28 d p.i.) and had background ELISA ratio at 14 and 28 d p.i. One of the major effects of virus infection on plant physiology is the severe stunting and extreme reduction in plant height. When we plotted the virus concentration (ELISA ratio of virus inoculated plant) in segregating FGP6 progeny, we found that they clearly grouped into two clusters (Figure 5b), showing that the plant height is inversely related to the virus concentration.

PCR analysis of selectable marker *nptII* in this family revealed that 17 of 26 plants carried the selectable marker (Table 2). One immune *FGmiR395* transgenic plant FGP6.22 was negative for selectable marker *nptII*, while five *FGmiR395*-negative segregants were carrying *nptII*, and eight segregants were negative for both transgenes. It is worth noting that the approach and technique utilized in this work can yield marker-free immune transgenics.

Southern blot analysis carried out on T_2 segregants confirmed stable integration of *FGmiR395* transgene(s) into the wheat genome (Table 2) and apparently multiple copies of the transgene (*Bam*HI digest: Figure 4a).

Efficacy of viral suppression in transgenic families

The analyses reported thus far for the resistant transgenic families showed the complete absence of symptoms in inoculated transgenic individuals and ELISA readings very similar to the ELISA readings of uninoculated controls, suggesting the complete absence of virus from the inoculated transgenic plants. Experiments were conducted to see whether infectious virus or viral RNA could be recovered from the resistant inoculated transgenic plants. Leaf sap from inoculated plants in three transgenic families (FGP8c, FGP13b, and 15a) was extracted and inoculated onto test plants of control BW26 at 1/10 (w/v) dilution to investigate the presence of any infectious WSMV particles. Results from these test inoculation experiments revealed that all *FGmiR395*+R phenotypes were immune to WSMV, as no infectious virus could be recovered and carried over to control wheat through mechanical inoculation with the most concentrated leaf extract inoculum. Sap from inoculated segregants with *FGmiR395* transgenes failed to transmit infection to susceptible BW26 as judged by symptoms and ELISA, whereas sap from segregants with no transgene and non-transformed controls (BW26) did transmit infection in every case. Examples of these tests are shown in Figure 6.

The formation of amiRNAs from the *FGmiR395* transcript

The expression of the amiRNAs was analysed in virus-free transgenic wheat leaves and detected by splinted ligation using miRtect IT (Maroney *et al.*, 2007). Potentially, from the five duplex arms of the precursor *FGmiR395*, five guide strands (amiRNA) could be produced; moreover, if one also considers the loading of passenger strand (amiRNA*) into RNA Induced Silencing Complex (RISC), then a total of ten amiRNA could potentially be produced against the virus. To detect both amiRNA and amiRNA* sequences using splinted ligation, we

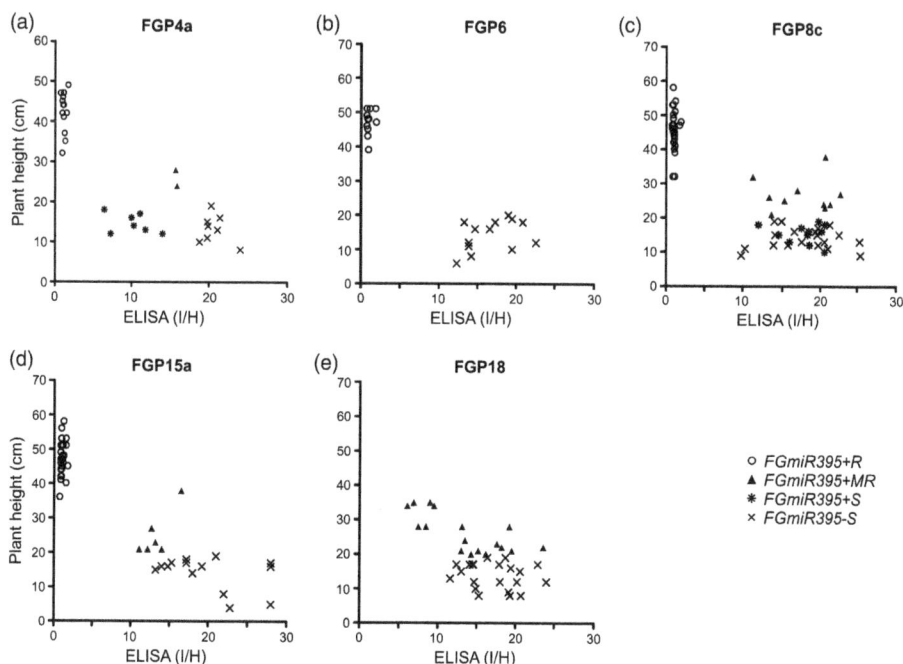

Figure 5 Segregation of resistance in FGP families expressing FGmiR395 transgene. Virus levels were detected by double-antibody sandwich enzyme-linked immunosorbent assay at 28 d p.i. postinoculation and expressed as ratio of inoculated and healthy control. The plant height (in cm) was measured at the heading stage and plotted against the corresponding ELISA ratio. −S indicated the *FGmiR395* transgene−negative susceptible segregant, +S indicated transgene-carrying susceptible segregant, +MR indicated segregants that carried the transgene but accumulated virus titre owing to resistance breakdown, while +R indicated the transgene-carrying resistant segregants. Transgenic families are shown as (a) FGP4a, (b) FGP6, (c) FGP8c, (d) FGP15a, and (e) FGP18. MR, moderate resistance.

Table 2 Segregation of transgene and resistance in T_1 and T_2. Analysis of six selected FGmiR395 families in T_1 and T_2. Recorded for T_1 are the results of Southern hybridisation, segregation for genomic PCR, and resistance. Segregation for resistance in T_2 is also shown for a series of selected derivative families. Selected T_2 families were derived as shown from T_1 individuals that were transgene carrying and resistant (+R); transgene-carrying susceptible (+S); and transgene-negative susceptible (−S). Southern blots are shown in Figure 4

TO (NT)	T_1						T_2		
TO Parent	*FGmiR395* Copy no. Southern blot (examples Figure 4a)	*FGmiR395* no. of loci (segregation)	PCR *ptII* + : −	PCR *GmiR395* + : −	ELISA WSMV R : S		T_1 parent	ELISA-based phenotype in T_1	ELISA phenotype in T_2 segregants R : S
FGP4a	2	1	21 : 11	24 : 8	15 : 9		4a.18	+S	20 : 15
							4a.22	+R	29 : 0
							4a.31	−S	0 : 8
FGP6	4	−	17 : 9	13 :173	13 : 13		6.3	−S	0 : 14
							6.10	+R	22 : 15
FGP8c	3	1	49 : 21	51 : 19	30 : 21		8c.10	+S	14 : 21
							8c.11	+R	24 : 8
							8c.27	−S	0 : 19
FGP15a	1	1	39 : 13	39 : 13	32 : 7		15a.1	+R	34 : 1
							15a.2	+S	24 : 4
							15a.11	−S	0 : 6
FGP18	3	−	26 : 12	17 : 21	0 : 38		18.6	−S	0 : 15
							18.10	+S	12 : 32

WSMV, *Wheat streak mosaic virus.*

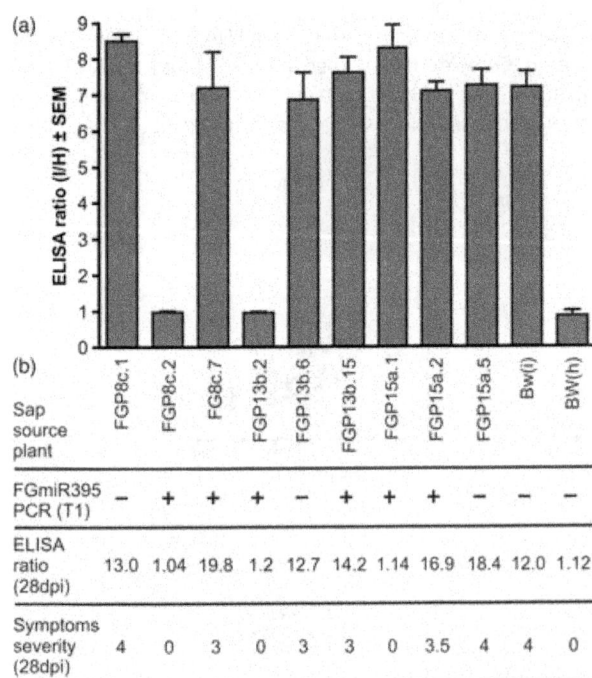

Figure 6 Virus transmission from the inoculated transgenic plants onto Bobwhite26. Sap was extracted from inoculated transgenic plants at 1/10 (w/v). Each sap extract was inoculated onto three wild-type BW26 plants. At 28 d p.i., samples were collected and processed for *Wheat streak mosaic virus* ELISA. (a) Plotted is the average ELISA ratio (inoculated divided by healthy) for the three test-inoculated plants. (b) Also tabulated are the molecular analyses of the inoculated T_1 individuals whose sap extracts were being tested. Shown for each sap donor is the *FGmiR395* PCR result, the bioassay result for that donor plant (as 28 d p.i. ELISA ratio), and the symptom score at 28 d p.i. (0–4 scale). Control BW(i) is inoculated BW26. BW(h) is the uninoculated BW26.

designed the ten bridging oligos accordingly (Table S1). All five designed anti-WSMV miRNAs, amiRNA-1, amiRNA-2, amiRNA-3, amiRNA-4, and amiRNA5, were expressed in the plant. Using an immune plant in T_2 generation, seven of the potential ten amiRNA were generated from *FGmiR395* and accumulated sufficiently to be readily detected (Figure 7). At least these seven amiRNAs would be loaded into RISC and would be expected to target the invading virus at both the genomic RNA and replicative RNA levels in four viral regions, the 5′ UTR, P1, P3, and HCpro regions.

Investigating the resistance breakdown phenotype

In addition to stable immunity *FGmiR395*+R, other phenotypes were observed in some transgenic events described previously as *FGmiR395*+MR and *FGmiR395*+S. These also were studied in more detail in the five families chosen for closer examination. Negative segregants *FGmiR395*−S along with wild-type BW26 were used as negative (susceptible) controls in all assays.

Transgenic event FGP4a

In the T_1 generation of FGP4a, 24 of 32 plants carried the transgene; however, only 15 of these showed strong resistance throughout the experiment. The resistance in nine individuals broke down with time, and symptoms began to appear as faint streaks at 14 d p.i. Two of these nine individuals showed a moderate resistance (MR) as judged by no symptoms at 14 d p.i., but the gradual development of leaf symptoms, growth to only intermediate height, and accumulated virus titre at 28 d p.i.; these segregants were characterized as *FGmiR395*+MR (Figure 5a). PCR, ELISA ratio, and plant height plot suggested four groups in this family. The third cluster (*FGmiR395*+S) accumulated less virus than the null segregants but was equally affected in height.

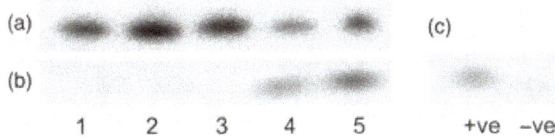

Figure 7 Expression analysis of *Wheat streak mosaic virus*-specific amiR-NAs: Levels of the amiRNAs in immune transgenic segregant FGP4a.22 were examined using the MiRtect IT splinted ligation assay. Assays with bridge oligonucleotides specific for (a) amiRNA, (b) amiRNA*, and (c) internal controls. The numbers in the bottom panel correspond to amiRNA-1 to -5, respectively.

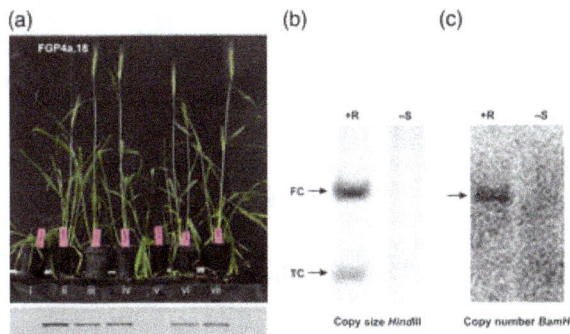

Figure 8 Segregation of resistance in T_2 progeny of T_1 individual FGP4a.18: (a) Inoculated T_2 plants and FGmiR395 PCR results. (i) and (v) transgene-negative susceptibles; (ii) transgene carrying moderately resistant; (iii) (iv) (vi) and (vii) transgene-carrying immunes. (b) Southern blot for copy size of FGP4a (+R) and FGP4a (−S), where FC shows the expected full size, and TC shows a truncated copy. (c) Southern blot for copy number FGP4a (+R) and FGP4a (−S).

Seeds collected from three FGP4a T_1 individuals, representing immune (*FGmiR395*+R), resistance breakdown (*FGmiR395*+MR), and susceptible phenotypes (*FGmiR395*−S) were analysed in T_2 generation (Table 2 and Table S2).

A total of 29 plants were grown and bioassayed from the seed collected from T_1 segregant FGP4a.22 (+R), and when challenged with WSMV, all 29 T_2 individuals proved to be immune to WSMV and no virus was retrieved on BW26 through back-assay. The progeny of intermediate phenotype FGP4a.18 (+MR) with resistance breakdown phenotype in T_1 produced 15 resistant progeny in T_2 plants (of 35 total). This T_2 segregation for resistance in FGP4a.18 is also illustrated in Figure 8a. We hypothesize this may indicate the segregation of an antagonistic factor such as an interfering truncated transgene copy away from a functional copy of *FGmiR395* resulting in more stable resistance. Southern blot analysis for copy size did reveal the presence of a truncated copy (Figure 8b). Further-more, it is also possible that the zygosity of *FGmiR395* might be involved in the amiRNA production where the homozygous level of expression may result in more stable resistance. The T_2 prog-eny propagated from a negative segregant parent FGP4a.31 (*FGmiR395*−S) were fully susceptible and indistinguishable from the wild-type susceptible BW26 controls.

Transgenic event FGP15a

In FGP15a, a total of 52 individuals were analysed in T_1 genera-tion. There appears to be one copy of *FGmiR395* and one copy

of *nptII*, and both segregate in simple Mendelian proportions (Table 2). Most of 32 tested *FGmiR395* + segregants were strongly resistant (+R), but six had moderate resistance (+MR) (Figure 5d). These six segregants had lower virus accumulation and were less stunted compared to the *FGmiR395* null segre-gants (Figure 5d).

Three different FGP15a T_1 segregants were selected for anal-ysis in T_2, and once again, resistant phenotypes were recovered from a susceptible transgene-positive T_1 parent (FGP15a.2). The family had one site of insertion of the transgene; however, a strong signal might suggest concatamerization at this site; there was also evidence of a truncated copy (Figure 4). From the +R T_1 individual FGP15a.1, only one of 35 T_2 was not immune.

Transgenic event FGP8c

In FGP8c, four classes of phenotype were observed. A total of 70 individuals were analysed for the transgenes, and 51 were assayed in T_1 for WSMV resistance (Table 2). Of T_1 plants tested (Figure 5c), 30 were +R with average plant height of 45.56 cm; seven were +MR where resistance held up at 14 d p.i. but grad-ually broke down with the virus titre ratio above ten at 28 d p.i. and average plant height of 28.5 cm; 14 segregants were +S despite carrying a full copy of *FGmiR395*, with average plant height of 16.92 cm and indistinguishable from the 19 −S segre-gants with average plant height 14.26 cm.

Seeds were collected from three T_1 individuals representing +R, +S, and −S classes (Table 2). From the immune +R individ-ual FGP8c.11, resistance segregated normally (24 : 8) in T_2. Again, the T_2 progeny of a +S plant, FGP8c.10, segregated some strongly resistant plants (14 : 21). The Southern blot anal-ysis revealed the presence of three copies (Table 2, Figure 4) that appeared to be of equal size but segregated together at one insertion locus.

Leaf saps from the strongly resistant segregants of event FGP 8c were back-inoculated to susceptible controls and found to be immune; this is illustrated in Figure 6 by T_1 individual FGP8c.2. However, individuals such as FGP8c.7 showed some breakdown of resistance after 14 d p.i., and sap from this plant was infective to the same extent as the null segregant FGP 8c.1 (Figure 6).

Transgenic event FGP18

Event FGP18 was examined further because the apparent full-length insertions of the *FGmiR395* gene failed to confer any plants with a high level of resistance in T_1 (Figure 3a, Figure 5e). In 38 T_1 segregants, the *nptII* gene segregated simply, but segregation was significantly distorted away from the *FGmiR395* transgene (17 : 21, Table 2). While none of the T_1 individuals were strongly resistant, Figure 5e shows that the *FGmiR395*-carrying plants clustered distinctly from the null segregants, with average plant height of +S being 26.1 cm, compared to 14.6 cm in −S segregants (Figure 5e, Table S2). Southern blot analysis revealed the presence of four copies (Fig-ure 4a) including truncated copies (Figure 4b). The T_2 progeny from a +S and −S T_1 parent were analysed for virus resistance. Resistance was evident in 12 T_2 individuals from FGP18.10 of 44 tested (Table 2). It is noteworthy that even in this multi-insertion transgenic event, which appeared in T_1 generation to be ineffective, by T_2, it was possible to identify fully resistant transgenic segregants. Plant FGP18.6 is noteworthy. Using PCR and bioassay, it was classified as −S; it did have transgenic sequence present (Figure 4a), but the sequence is truncated

(Figure 4b). Presumably, the truncation has eliminated one or both of the primer sites so that the PCR gave no product. When only the truncated copy is present, the plants are susceptible in both T_2 and T_3 generations (Table 2).

The possibility that the virus was evolving to avoid all five amiRNA species can be dismissed based on two lines of evidence. T_2 transgenics were challenged with the infected sap derived from both susceptible and transgene-carrying susceptible segregants in T_1. Leaf sap collected from *FGmiR395+S* T_1 plants was used to inoculate T_2 progeny. These potentially evolved virus populations continued to be controlled by resistant transgenics. Furthermore, in a number of families (Table 2), +R and +MR phenotypes emerged from +S T_1 parents. Therefore, the original +S susceptibility was not the result of virus mutation; otherwise, resistance would not have emerged in T_2 when challenged with the putatively mutant virus preparations. In addition, deep sequencing was carried out of small RNA fractions extracted from virus-inoculated plants of the three phenotypes, +R, +S, and −S. This allowed the full virus genomic sequence to be reconstructed from each of these virus populations. As reported in Table 1, the virus sequence of WSMV-ACT recovered from the non-transgenic plant (−S), in each of the five target regions, was exactly the sequence as used in the construction of the amiRNA. Furthermore, the sequence in these regions was entirely unchanged in the virus population from both the transgenic resistant plant (+R) and the transgenic susceptible plant (+S).

The recovery phenotype

Individuals with a *recovery* phenotype were observed in transgenic family FGP13b. These plants had virus symptoms at 14 d p.i. and a high virus titre; however, when assayed at 28 d p.i., virus was not detected through ELISA nor were there any symptoms observed on the newly emerged leaves. An example of this phenomenon is displayed in transgenic segregant FGP13b.2 (Figure 9), where the new leaf displayed characteristic virus symptoms at 14 d p.i. (still evident at 28 d p.i., Figure 9b) and had an ELISA ratio of 2.5. However, as the plant developed further, no virus symptoms were observed on the newly emerged leaves at 28 d p.i. (Figure 9c). Moreover, no WSMV-specific PCR product could be amplified from reverse-transcribed RNA from the newly emerged leaves. This suggests that the amiRNA was ineffective in the young plant but expressed better and became effective in overcoming virus multiplication as the plant developed further.

Sap was extracted from the newly emerged leaves and inoculated onto susceptible BW26 plants to test whether the virus has been completely eliminated from the newly emerged leaves. No infectious virus could be recovered in this back-assay (Figure 6, FGP13b.2). This lead to the conclusion that in some resistant transgenic events, the virus was able to get away to initial establishment but that the amiRNA expression subsequently was able to completely eliminate the virus from the recovering plant.

Discussion

We engineered a complex rice-derived miR395 with five artificial miRNA precursors designed to target WSMV genome, to achieve amiRNA-mediated resistance in wheat. We retained the predicted secondary structure of miR395 in the synthetic *FGmiR395* transgene and hypothesized that the FanGuard

Figure 9 Recovery phenotype. (a) Inoculated leaf (b) leaf showing symptoms (c) newly emerged healthy leaf. In this transgenic FGP13b.2 individual, we inoculated the leaf with spray gun (necrotic spots shows mechanical injury resulting from spray inoculation); at 14 d p.i., the plants showed symptoms and accumulated virus by ELISA ratio. However, at 28 d p.i., newly emerged leaves were completely symptom-free with no virus detected by ELISA. When sap from new leaves of this plant was test-inoculated onto BW26, no infectious virus could be recovered (Figure 5).

transcript would be processed upon expression to produce five 21-bp amiRNA duplexes and then up to ten species of amiRNA, because secondary structure of the plant precursor miRNAs appears to be more important for processing by DCL1 than the sequence of the mature miRNA itself (Schwab *et al.*, 2006; Ossowski *et al.*, 2008). WSMV has a +ssRNA single-stranded (monopartite) genome that replicates through a dsRNA intermediate to generate a negative (−) strand during this replication process. This monopartite genome provides the opportunity to target the virus at any accessible position that would result in homology-dependent degradation of the viral genome or the replicative strand. amiRNA-1 and amiRNA-2 were designed to be complementary to and target the replicating strand, while amiRNA-3, amiRNA-4, and amiRNA-5 were designed to be complementary to and target the genomic strand of the virus. It has been shown that both positive and negative strands of hepatitis C virus may be targeted by siRNAs (Wilson and Richardson, 2005). The amiRNA* (passenger strand) might also get loaded into RISC and thus mediate the degradation of the opposite strand of the virus (either genomic or replicative strand) than the predicted one. In fact, our results showed that seven of the potential ten types of amiRNA could be readily detected.

Resistance cosegregated with the transgene in most of the transgenic events of this study. However, analysis of T_1 families revealed a range of phenotypes. (i) Immune individuals were obtained, which remained symptomless and with no viral coat protein accumulation all the way through to maturity. Sap prepared from these inoculated plants failed to transmit infection to susceptible controls. Such immunity was evident across two generations. (ii) In other cases, the plants were resistant without symptoms or ELISA detected coat protein at 14 d p.i., but subsequently, the resistance broke down, allowing virus to accumulate by 28 d p.i., (iii) A third transgenic phenotype may be called *plant recovery*, in which early susceptibility is followed by full recovery and resistance, (iv) The forth phenotype is where the presence of multiple copies of the transgene confers only moderate resistance. We challenged the transgenic families with high-titre virus inoculum at two time points 0 and 10 d p.i. and therefore suggest that the breakdown of resistance in some events might result from an excessive virus pressure that overwhelms the amiRNA-mediated resistance. Such a phenomenon has been observed in transgenic barley against *Barley yellow dwarf virus* (BYDV) where an increase in viruliferous aphid infestation resulted in breakdown of RNAi resistance to BYDV (M.-B. Wang, pers. commun.). Ai *et al.* (2011) showed that although amiRNAs were detectable and resistance to PVX or PVY evident in their transgenic plants, the resistance was overcome by reinoculation at 35 d p.i., resulting in increased viral pressure. The pressure of inoculation in our experiments was very high and may be responsible for some of the more complex resistance phenotypes. It would be interesting to investigate the efficiency of the *FGmiR395*-expressing wheat plants in the field under the milder pressures expected from natural wheat curl mite infestations.

It could also be informative to quantify the level of amiRNA expression in the various transgenic phenotypes to observe whether there is a correlation with the degree or stability of resistance. Such an analysis would need to follow all ten of the potential amiRNA species and would ideally include the various phases of resistance breakdown and various phases of recovery. Previous studies have shown some degree of correlation between amiRNA expression and virus resistance (PVX and PVY in *Arabidopsis*, Ai *et al.*, 2011; CMV in tobacco, Qu *et al.*, 2007). When there are at least five and potentially ten species of amiRNA attacking the virus, as with *FGmi395*, the analysis of correlation with resistance will be complex.

One potential risk with amiRNA-mediated resistance is the generation of virus mutants that escape the amiRNA surveillance (Simon-Mateo and Garcia, 2006; Lin *et al.*, 2009). We addressed this issue in our studies by selecting amiRNA targets based upon conserved regions in five full WSMV genome sequences available. We observed no evidence that the virus was evolving (mutating) during the course of the experiment. This was evident because virus populations collected from *FGmiR395+S* T_1 plants was used to inoculate T_2 progeny. Not only did the resistant transgenic continue to be resistant to these inoculations, but progeny from some +S T_1 parents included +R and +MR phenotypes. We take this as evidence that the original +S susceptibility was not the result of virus mutation; otherwise, resistance would not have emerged in T_2 when challenged with the putatively mutant virus preparation. Furthermore, deep sequencing of virus populations from infected transgenic plants confirmed no mutations in the target sequences. By the end of this study, the number of full-length WSMV genomes available

online (NCBI) had grown from five to 13. An alignment of all 13 revealed that three of five chosen amiRNA targets are still completely conserved in the WSMV genome (Table 1). This highlights the importance of having multiple targets in polycistronic amiRNA and the importance of aligning as many virus genomes as possible to select highly conserved regions.

One of the targets in *FGmiR395* was the functional region called *pipo* (pretty interesting Potyviridae ORF) within the gene encoding the P3 protein. *Pipo* was initially identified as a small ORF embedded in the P3 cistron of *Turnip mosaic virus* (TuMV; genus *Potyvirus*; family Potyviridae) (Chung *et al.*, 2008), and its presence was confirmed in 48 viruses representing all genera in the family Potyviridae, including WSMV. Mutation in *pipo* hinders various important functions; in the case of WSMV that includes effects on replication and movement in the plant. When mutations are introduced into the *pipo* region of P3, without affecting the amino acid sequence of the translated protein, the virus loses the ability to replicate in protoplasts (Chung *et al.*, 2008) or it is restricted to only a few cells in inoculated plants (Wen and Hajimorad, 2010).

As one would expect, the presence of one full copy of the transgene can be enough to confer resistance (immunity) against WSMV. However, we saw evidence in some transgenic events of additional truncated *FGmiR395* insertions and behaviour in T_1 and T_2 generations suggestive that the truncated copy may be interfering with the expression of the full-length copy; subsequent loss of the truncated copy restores effective resistance. This interpretation will require further experimentation to confirm. We sometimes observed resistance segregation ratios inconsistent with Mendelian expectation based on the inferred insert locus number. Usually, this involved lower than expected numbers of resistant individuals and is likely associated with some inserts of the transgene being ineffective or conditionally effective (Matzke *et al.*, 2009).

Previously, we have reported the use of long hairpin dsRNA-mediated WSMV immunity in wheat (Fahim *et al.*, 2010). The comparison to the present study is a rare opportunity to contrast the two approaches in the same genetic background against the same virus. Compared to amiRNA, long hairpin RNA had the advantage of a very high frequency of insertion events with stable heritable immunity. In this respect, long hairpin RNAi is very attractive from a biotechnology application perspective. However, long hpRNA would have a higher probability of unintended silencing of off-target genes in the host (Xu *et al.*, 2006; Duan *et al.*, 2008; Khraiwesh *et al.*, 2008). Furthermore, long hpRNA approaches are seen by some as posing a risk in the field of heterologous recombination with other virus genomes and resulting in new virus biotypes. Low temperatures can also compromise the efficacy of RNAi silencing strategies (Szittya *et al.*, 2003). On the other hand, miRNAs appear to be completely temperature independent, and the transgenic lines expressing virus-derived amiRNA retain their resistance at low temperatures (Niu *et al.*, 2006; Szittya *et al.*, 2003). We were able to achieve immunity in wheat to WSMV from both the long hairpin dsRNA and the amiRNA strategies. We conclude that amiRNA-based viral resistance, especially polycistronic amiRNA as advocated here, deserves and needs further in-depth studies to improve the amiRNA efficiency. An even better comparison would be achieved if the long hpRNA was designed to cover the same regions as the amiRNA used in this study.

The work described here exemplifies the utility of miR395 and similar miRNA clusters as a carrier of multiple amiRNAs.

They can be used to target multiple regions of the one virus (as here), multiple viruses, or multiple endogenous mRNA species (S. Belide, J. R. Petrie, P. Shrestha, M. Fahim, Q. Liu, C. C. Wood and S. P. Singh, unpublished). Mixed viral infections are common in the field especially in fruits and vegetables. Using the polycistronic amiRNA, it will be possible to target highly conserved regions of multiple viruses. Similarly, polycistronic pre-amiRNA genes will be effective in targeting multiple plant endogenous genes for functional genomics and in applications such as redirecting plant metabolism into novel products.

We conclude that polycistronic amiRNAs can be utilized to induce virus resistance in commercially valuable plants, where there are limited options of natural resistance. We anticipate ongoing improvements in the understanding of miRNA biogenesis and design of amiRNA to further enhance the utility for virus resistance and engineering other agronomically important traits. Furthermore, the expression of multiple amiRNAs from a single precursor transgene will minimize the difficulties of repeated transformations, need for multiple selectable markers, and the constraint of breeding with multiple independent loci.

Experimental procedures

Designing WSMV-specific amiRNA

To select conserved regions in WSMV genome as targets for artificial miRNAs, full-genome sequences of WSMV were retrieved from NCBI (Table 1). The sequences were aligned with Clustal W/AlignX (a component of Vector NTI Advance® 10.3.0) to screen for highly conserved regions in the viral genome. The possible amiRNA sequences were generated from the highly conserved regions (20 nt or more in length) using the basic criteria defined at WMD3 (http://wmd3.weigelworld.org/, a web microRNA design tool) and incorporated into a software algorism called *miR Mate* developed specifically for this study. The algorithm was developed using Microsoft. NET Framework and also incorporates the Vienna RNA Package 1.7 algorithm RNAfold.exe (Hofacker et al., 1994; McCaskill, 1990; Zuker and Stiegler, 1981). The miRNA design criteria used include A/U at position 1 (Mi et al., 2008; Eamens et al., 2009; Takeda et al., 2008), A at position 10 (Reynolds et al., 2004; Mallory et al., 2004), and G/C at position 21 (P. Waterhouse, pers. commun.). *miR Mate* utilizes the *RNAfold* algorithm to calculate minimal free energy (mfe) values for the formation of the candidate miRNA's folded structure; values of ≤-30 kcal/mol represent optimal stability. The negative values reflect the fact that stored energy is released during the formation of the structure; the more negative the value, the more energy is released and the more favourable is the formation of the structure. Candidate amiRNAs with the lowest mfe value (the highest stability) were then assessed for potential off-targets in wheat and barley.

The set of potential virus target sequences were used to search for genes that may be potential off-targets in wheat or barley, using miRU: Plant microRNA Potential Target Finder http://bioinfo3.noble.org/miRNA/miRU.htm (a recent version *psRNATarget: A Plant Small RNA Regulator Target Analysis Server* is available at http://bioinfo3.noble.org/psRNATarget/) (Brennecke et al., 2005; Jones-Rhoades and Bartel, 2004; Lim et al., 2005; Mallory et al., 2004). WSMV-derived amiRNAs were selected having the least probability of targeting any sequence in the gene or EST databases of wheat or barley.

The stemloop backbone

For the delivery of the final five amiRNAs as a polycistronic transgene, we selected the precursor of rice miR395 that is expressed under sulphur stress conditions (Guddeti et al., 2005; Jones-Rhoades and Bartel, 2004; Kawashima et al., 2009). A synthetic gene called *FanGuard* (FG) was designed by replacing the five native miRNA sequences in the first five duplex arms of native miR395 with five amiRNA designed to target WSMV. In a parallel study, a similar construct was used to simultaneously silence five endogenous genes in Arabidopsis (Belide et al. submitted). The designed *FGmiR395* was synthesized through GENEART® GmbH (http://www.geneart.com) flanked by restriction sites for *BamHI* and *KpnI* in the carrier plasmid. The FG gene was excised from the carrier plasmid using appropriate restriction enzymes and ligated between the Ubiquitin promoter and tm1 terminator of vector pWubi-tm1 vector (Wang and Waterhouse, 2000) generating cereal transformation plasmid FG-pWubi.

Wheat transformation

Transgenic wheat plants were generated following microparticle bombardment of 186 immature cv. Bob White 26 (BW26) wheat embryos. The embryos were cobombarded with two plasmids, FG-pWubi and a selectable marker plasmid pCMneoSTLS2, as described previously (Fahim et al., 2010; Pellegrineschi et al., 2002).

Analysis of T_0 transgenic plants—PCR

DNA extraction was carried out using DNAeasy Plant Mini Kit following manufacturer's instructions (Qiagen Inc, Valencia, CA). For PCR-based genotyping of the *FGmiR395* transgene, DNA was amplified (Figure 1b) using FgPf1 5'-TGCAGCATC-TATTCATATGC-3' and M13RevP 5'-CATGGTCATAGCTGTT-3', that generated approximately 1.4 kb of FG-pWubi amplicon covering promoter, transgene, and terminator regions, under the following thermocycler conditions 94 °C for 30 s, 60 °C for 45 s, 72 °C for 60 s for 35 cycles with a final extension at 72 °C for 10 min. For the selectable marker *nptII* (Figure 1c), a 700-bp *nptII* fragment was amplified using the forward primer Neo3 5'-TACGGTATCGCCGCTCCCGAT-3' and reverse primer Neo5 5'-GGCTATTCGGCTATGACTG-3', both sequences being in the *nptII* coding region, using the following thermocycler conditions: 94 °C for 30 s, 55 °C for 30 s, 72 °C for 60 s for 40 cycles with a final extension at 72 °C for 10 min.

Analysis of T_1 and T_2 transgenic plants—Virus bioassay

Virus inoculum was prepared by grinding WSMV-infected tissue in a mortar and pestle at a 1 : 10 w/v ratio in 0.02 M potassium phosphate buffer (pH 7). After filtering through four layers of Miracloth® (Calbiochem, La Jolla, CA), abrasive celite (Johns-Manville, Denver, CO) was added at 2% w/v to serve as an abrasive. For the analysis in T_1 plants, the inoculum was prepared from virus-infected non-transgenic BW26; for the analysis of T_2 plants, sap from a mixture of *FGmiR395* carrying susceptible and *FGmiR395*-negative segregant susceptible plants was used. The sap–celite mixture was first applied with an air-powered spray gun, and then, leaves were gently rubbed with fingers to ensure the infection of plant by the virus. At 10 d p.i., plants were reinoculated with the virus-infected sap to ensure high inoculum pressure.

The plants were scored for symptoms at 7, 14, 21, and 28 d p.i. on a scale of 0–4 with 0 as healthy, 1 as mild with very few streaks, 2 as moderate with streaks that coalesce, 3 as severe with approximately 50% leaf area with streaks, and 4 as the most severe or lethal symptoms where the streaks develop into chlorosis of more than 70% of leaf area based on visual observation. WSMV-specific ELISA was performed on leaf samples collected at 14 d p.i. and 28 d p.i., using Agdia reagents (Agdia, Elkhart, IN) following the manufacturer's instructions. Plates were read at A_{405nm} in ELISA Reader Spectra Max 340 PC (Molecular Devices, Sunnyvale, CA) 60 min after the addition of substrates. Healthy controls were included on every plate, every sample was duplicated, and duplicate value means were used in calculating the ELISA value ratio between inoculated and healthy controls.

Segregation analysis of FanGuard transgene and resistance in T_1 and T_2 generations

For detailed segregation analysis of selected events in T_1 generation, 25–35 seeds were germinated in pots. Leaf samples were collected, and DNA was extracted as described previously, and genomic PCR was conducted to detect both *FGmiR395* and *npt*II amplicons. The cosegregation of resistance with the transgene was assessed by challenging with WSMV as described earlier. ELISA was performed 14 and 28 d p.i. in T_1 and only 28 d p.i. for T_2 generations on inoculated plants. Plant heights and symptoms were recorded at 7, 14, 21, and 28 d p.i.

Test inoculation to detect infectious virus in leaf sap

Sap was extracted at 28 d p.i. from each inoculated transgenic plant to be tested, using 0.02 M potassium phosphate buffer at 1 : 10 (1 g leaf per 10 mL buffer) concentration and mixed with celite abrasive and then inoculated onto three control BW26 plants. This test inoculation (or back-inoculation) method was used to evaluate the effectiveness of the FanGuard (*FGmiR395*) transgene in eliminating viral replication and preventing the formation of infectious particles. Symptoms were scored on the test-inoculated plants and leaf samples collected 28 d p.i. for ELISA as described previously.

Analysis of transgenic plants—Southern hybridization

Southern hybridization was carried out as described previously (Fahim *et al.*, 2010; Lagudah *et al.*, 1991). Instead of T_0 plants, a pool of 8–44 T_1 individuals per family were used for the analysis of transgene copy number and copy size. This method was used so as not to compromise the initial transgenic T_0 plant yet to capture all the insertion events likely to have been present in the T_0 plant. DNA was digested with *Bam*HI to determine the number of independent insertions; there is only one site for *Bam*H1 in *FGmi395* (Figures 1b and 4). DNA was digested with *Hind*III to assess whether the inserted copies were full length or truncated.

amiRNA analysis in immune transgenic plants

The small RNA fraction was enriched from a fraction of total RNA extracted from 100 mg of transgene-carrying immune FGP4a.22 T_2 plants, using miRvana Kit (Ambion, Austin, TX) following manufacturer instructions. The extracted total RNA was run on denaturing 17% PAGE and stained with EtBr. Using 100-bp RNA ladder as reference, the region corresponding to

15–50 bp was dissected and small RNA was extracted from excised gel overnight in 4 M NaCl. The RNA concentration was measured in 1 μL of solution using Nanodrop (Thermo Scientific, Wilmington). The splinted ligation was performed on the purified fraction with miRtect-IT™ miRNA Labeling and Detection Kit (USB, Cleveland, OH) (Maroney *et al.*, 2007). Specific bridge oligonucleotides (Table S1) were designed according to the manufacturer's directions. Using 50 ng of enriched smRNA per reaction, amiRNAs were captured by a specific bridge oligonucleotide and ligated to the P^{32}-labelled detection oligonucleotide with T4 DNA ligase. Ligated products were separated on 17% urea–polyacrylamide gel and visualized using Fujifilm FLA-5000 phosphor imager.

Small RNA library preparation and deep sequencing

Small RNAs were enriched using the mirVana miRNA Isolation Kit (Invitrogen) following manufacturer instructions. Small RNA-Seq libraries were prepared based on Illumina's alternative v1.5 protocol and a published method (Lu *et al.*, 2007) and run on the Illumina's GAIIx platform at the Genome Discovery Unit of Australian National University. WSMV sequences were assembled with assistance from Dr Stephen Ohms, JCSMR, The Australian National University.

Acknowledgements

We are thankful to Dr Peter Waterhouse and Dr Ming-Bo Wang for their technical advice at various stages of these studies, Mr. Zarman Mazhar Rizvi for his assistance with putting together various algorithms under one miR Mate, Drs. James Petrie and Qing Liu for technical supervision in amiRNA detection, and Dr Jun Fan, The John Curtin School of Medical Research, The Australian National University, for providing excellent training and technical supervision in small RNA library preparation. Dr Stephen Ohms is gratefully acknowledged for his assistance with bioinformatic analysis of small RNAs. The first author thankfully acknowledges AusAID for financial assistance as PhD studentship.

References

Abbott, D., Wang, M.-B. and Waterhouse, P.M. (2002) A single copy of virus-derived transgene-encoding hairpin RNA confers BYDV immunity. In *Barley Yellow Dwarf Disease: Recent Advances and Future Strategies* (Henry, M. and McNab, A., eds), pp. 22–26. Mexico, DF: CIMMYT.

Abel, P.P., Nelson, R.S., De, B., Hoffmann, N., Rogers, S.G., Fraley, R.T. and Beachy, R.N. (1986) Delay of disease development in transgenic plants that express the tobacco mosaic virus coat protein gene. *Science*, **232**, 738–743.

Ai, T., Zhang, L., Gao, Z., Zhu, C.X. and Guo, X. (2011) Highly efficient virus resistance mediated by artificial microRNAs that target the suppressor of PVX and PVY in plants. *Plant Biol. (Stuttg)*, **13**, 304–316.

Allen, E., Xie, Z.X., Gustafson, A.M. and Carrington, J.C. (2005) microRNA-directed phasing during trans-acting siRNA biogenesis in plants. *Cell*, **121**, 207–221.

Alvarez, J.P., Pekker, I., Goldshmidt, A., Blum, E., Amsellem, Z. and Eshed, Y. (2006) Endogenous and synthetic microRNAs stimulate simultaneous, efficient, and localized regulation of multiple targets in diverse species. *Plant Cell*, **18**, 1134–1151.

Baulcombe, D. (2004) RNA silencing in plants. *Nature*, **431**, 356–363.

Boden, D., Pusch, O., Lee, F., Tucker, L. and Ramratnam, B. (2003) Human immunodeficiency virus type 1 escape from RNA interference. *J. Virol.* **77**, 11531–11535.

ter Brake, O., Konstantinova, P., Ceylan, M. and Berkhout, B. (2006) Silencing of HIV-1 with RNA Interference: A Multiple shRNA Approach. *Molecular Therapy*, **14**, 883–892.

Brennecke, J., Stark, A., Russell, R.B. and Cohen, S.M. (2005) Principles of microRNA-target recognition. *PLoS Biol.* **3**, 404–418.

Chung, B.Y.W., Miller, W.A., Atkins, J.F. and Firth, A.E. (2008) An overlapping essential gene in the Potyviridae. *Proc. Natl Acad. Sci. USA*, **105**, 5897–5902.

Das, A.T., Brummelkamp, T.R., Westerhout, E.M., Vink, M., Madiredjo, M., Bernards, R. and Berkhout, B. (2004) Human immunodeficiency virus type 1 escapes from RNA interference-mediated inhibition. *J. Virol.* **78**, 2601–2605.

Duan, C.G., Wang, C.H., Fang, R.X. and Guo, H.S. (2008) Artificial microRNAs highly accessible to targets confer efficient virus resistance in plants. *J. Virol.* **82**, 11084–11095.

Dwyer, G.I., Gibbs, M.J., Gibbs, A.J. and Jones, R.A.C. (2007) Wheat streak mosaic virus in Australia: relationship to isolates from the Pacific Northwest of the USA and its dispersion via seed transmission. *Plant Dis.* **91**, 164–170.

Eamens, A.L., Smith, N.A., Curtin, S.J., Wang, M.B. and Waterhouse, P.M. (2009) The *Arabidopsis thaliana* double-stranded RNA binding protein DRB1 directs guide strand selection from microRNA duplexes. *RNA*, **15**, 2219–2235.

Fahim, M., Ayala-Navarrete, L., Millar, A.A. and Larkin, P.J. (2010) Hairpin RNA derived from viral NIa gene confers immunity to *Wheat streak mosaic virus* infection in transgenic wheat plants. *Plant Biotechnol. J.* **8**, 821–834.

Guddeti, S., Zhang, D.C., Li, A.L., Leseberg, C.H., Kang, H., Li, X.G., Zhai, W.X., Johns, M.A. and Mao, L. (2005) Molecular evolution of the rice miR395 gene family. *Cell Res.* **15**, 631–638.

Harvey, T.L. and Seifers, D.L. (1991) Transmission of wheat streak mosaic virus to sorghum by the wheat curl mite (Acari, Eriophyidae). *J. Kans. Entomol. Soc.* **64**, 18–22.

Hofacker, I.L., Fontana, W., Stadler, P.F., Bonhoeffer, L.S., Tacker, M. and Schuster, P. (1994) Fast folding and comparison of RNA secondary structures. *Monatshefte Für Chemie*, **125**, 167–188.

Israsena, N., Supavonwong, P., Ratanasetyuth, N., Khawplod, P. and Hemachudha, T. (2009) Inhibition of rabies virus replication by multiple artificial microRNAs. *Antiviral Res.* **84**, 76–83.

Jackson, A.L., Bartz, S.R., Schelter, J., Kobayashi, S.V., Burchard, J., Mao, M., Li, B., Cavet, G. and Linsley, P.S. (2003) Expression profiling reveals off-target gene regulation by RNAi. *Nat. Biotechnol.* **21**, 635–637.

Jones, R.A.C., Coutts, B.A., Mackie, A.E. and Dwyer, G.I. (2005) Seed transmission of *Wheat streak mosaic virus* shown unequivocally in wheat. *Plant Dis.* **89**, 1048–1050.

Jones-Rhoades, M.W. and Bartel, D.P. (2004) Computational identification of plant MicroRNAs and their targets, including a stress-induced miRNA. *Mol. Cell*, **14**, 787–799.

Kawashima, C.G., Yoshimoto, N., Maruyama-Nakashita, A., Tsuchiya, Y.N., Saito, K., Takahashi, H. and Dalmay, T. (2009) Sulphur starvation induces the expression of microRNA-395 and one of its target genes but in different cell types. *Plant J.* **57**, 313–321.

Khraiwesh, B., Ossowski, S., Weigel, D., Reski, R. and Frank, W. (2008) Specific gene silencing by artificial MicroRNAs in *Physcomitrella patens*: an alternative to targeted gene knockouts. *Plant Physiol.* **148**, 684–693.

Lagudah, E.S., Appels, R. and Mcneil, D. (1991) The nor-D3 locus of *Triticum tauschii* – natural variation and genetic-linkage to markers in chromosome-5. *Genome*, **34**, 387–395.

Lim, L.P., Lau, N.C., Garrett-Engele, P., Grimson, A., Schelter, J.M., Castle, J., Bartel, D.P., Linsley, P.S. and Johnson, J.M. (2005) Microarray analysis shows that some microRNAs downregulate large numbers of target mRNAs. *Nature*, **433**, 769–773.

Lin, S.S., Wu, H.W., Elena, S.F., Chen, K.C., Niu, Q.W., Yeh, S.D., Chen, C.C. and Chua, N.H. (2009) Molecular evolution of a viral non-coding sequence under the selective pressure of amiRNA-mediated silencing. *Plos Pathog.* **5**, e1000312.

Liu, Y.P., Haasnoot, J., ter Brake, O., Berkhout, B. and Konstantinova, P. (2008) Inhibition of HIV-1 by multiple siRNAs expressed from a single microRNA polycistron. *Nucleic Acids Res.* **36**, 2811–2824.

Lu, C., Meyers, B.C. and Green, P.J. (2007) Construction of small RNA cDNA libraries for deep sequencing. *Methods*, **43**, 110–117.

Maas, C., Simpson, C.G., Eckes, P., Schickler, H., Brown, J.W.S., Reiss, B., Salchert, K., Chet, I., Schell, J. and Reichel, C. (1997) Expression of intro modified nptII genes in monocotylenonous and dicotyledonous plants cells. *Mol. Breed.* **3**, 15–28.

Mallory, A.C., Reinhart, B.J., Jones-Rhoades, M.W., Tang, G.L., Zamore, P.D., Barton, M.K. and Bartel, D.P. (2004) MicroRNA control of PHABULOSA in leaf development: importance of pairing to the microRNA 5′region. *EMBO J.* **23**, 3356–3364.

Maroney, P.A., Chamnongpol, S., Souret, F. and Nilsen, T.W. (2007) A rapid, quantitative assay for direct detection of microRNAs and other small RNAs using splinted ligation. *RNA*, **13**, 930–936.

Matzke, M., Kanno, T., Daxinger, L., Huettel, B. and Matzke, A.J. (2009) RNA-mediated chromatin-based silencing in plants. *Curr. Opin. Cell Biol.* **21**, 367–376.

McCaskill, J.S. (1990) The equilibrium partition-function and base pair binding probabilities for RNA secondary structure. *Biopolymers*, **29**, 1105–1119.

Mi, S., Cai, T., Hu, Y., Chen, Y., Hodges, E., Ni, F., Wu, L., Li, S., Zhou, H., Long, C., Chen, S., Hannon, G.J. and Qi, Y. (2008) Sorting of small RNAs into Arabidopsis argonaute complexes is directed by the 5′ terminal nucleotide. *Cell*, **133**, 116–127.

Molnar, A., Bassett, A., Thuenemann, E., Schwach, F., Karkare, S., Ossowski, S., Weigel, D. and Baulcombe, D. (2009) Highly specific gene silencing by artificial microRNAs in the unicellular alga *Chlamydomonas reinhardtii*. *Plant J.* **58**, 165–174.

Niu, Q.W., Lin, S.S., Reyes, J.L., Chen, K.C., Wu, H.W., Yeh, S.D. and Chua, N.H. (2006) Expression of artificial microRNAs in transgenic Arabidopsis thaliana confers virus resistance. *Nat. Biotechnol.* **24**, 1420–1428.

Ossowski, S., Schwab, R. and Weigel, D. (2008) Gene silencing in plants using artificial microRNAs and other small RNAs. *Plant J.* **53**, 674–690.

Park, W., Zhai, J.X. and Lee, J.Y. (2009) Highly efficient gene silencing using perfect complementary artificial miRNA targeting AP1 or heteromeric artificial miRNA targeting AP1 and CAL genes. *Plant Cell Rep.* **28**, 469–480.

Pellegrineschi, A., Noguera, L.M., Skovmand, B., Brito, R.M., Velazquez, L., Salgado, M.M., Hernandez, R., Warburton, M. and Hoisington, D. (2002) Identification of highly transformable wheat genotypes for mass production of fertile transgenic plants. *Genome*, **45**, 421–430.

Qu, J., Ye, J. and Fang, R.X. (2007) Artificial microRNA-mediated virus resistance in plants. *J. Virol.* **81**, 6690–6699.

Reynolds, A., Leake, D., Boese, Q., Scaringe, S., Marshall, W.S. and Khvorova, A. (2004) Rational siRNA design for RNA interference. *Nature Biotechnology*, **22**, 326–330.

Rotte, C. and Leustek, T. (2000) Differential subcellular localization and expression of ATP sulfurylase and 5′-adenylylsulfate reductase during ontogenesis of arabidopsis leaves indicates that cytosolic and plastid forms of ATP sulfurylase may have specialized functions. *Plant Physiol.* **124**, 715–724.

Schnippenkoetter, W.H., Martin, D.P., Willment, J.A. and Rybicki, E.P. (2001) Forced recombination between distinct strains of *Maize streak virus*. *J. Gen. Virol.* **82**, 3081–3090.

Schwab, R., Ossowski, S., Riester, M., Warthmann, N. and Weigel, D. (2006) Highly specific gene silencing by artificial microRNAs in Arabidopsis. *Plant Cell*, **18**, 1121–1133.

Seifers, D.L., Harvey, T.L., Martin, T.J. and Jensen, S.G. (1998) Partial host range of the high plains virus of corn and wheat. *Plant Dis.* **82**, 875–879.

Simon-Mateo, C. and Garcia, J.A. (2006) MicroRNA-Guided processing impairs Plum pox virus replication, but the virus readily evolves to escape this silencing mechanism. *J. Virol.* **80**, 2429–2436.

Sivamani, E., Brey, C.W., Dyer, W.E., Talbert, L.E. and Qu, R.D. (2000) Resistance to *Wheat streak mosaic virus* in transgenic wheat expressing the viral replicase (NIb) gene. *Mol. Breed.* **6**, 469–477.

Sivamani, E., Brey, C.W., Talbert, L.E., Young, M.A., Dyer, W.E., Kaniewski, W.K. and Qu, R.D. (2002) Resistance to *Wheat streak mosaic virus* in transgenic wheat engineered with the viral coat protein gene. *Transgenic Res.* **11**, 31–41.

Slykhuis, J.T. (1955) *Aceria tulipae* Keifer (Acarina, Eriophyidae) in relation to the spread of wheat streak mosaic. *Phytopathology*, **45**, 116–128.

Smith, N.A., Singh, S.P., Wang, M.B., Stoutjesdijk, P.A., Green, A.G. and Waterhouse, P.M. (2000) Gene expression – total silencing by intron-spliced hairpin RNAs. *Nature*, **407**, 319–320.

Szittya, G., Silhavy, D., Molnar, A., Havelda, Z., Lovas, A., Lakatos, L., Banfalvi, Z. and Burgyan, J. (2003) Low temperature inhibits RNA silencing-mediated defence by the control of siRNA generation. *EMBO J.* **22**, 633–640.

Takeda, A., Iwasaki, S., Watanabe, T., Utsumi, M. and Watanabe, Y. (2008) The mechanism selecting the guide strand from small RNA duplexes is different among argonaute proteins. *Plant Cell Physiol.* **49**, 493–500.

Vaucheret, H., Vazquez, F., Crete, P. and Bartel, D.P. (2004) The action of ARGONAUTE1 in the miRNA pathway and its regulation by the miRNA pathway are crucial for plant development. *Genes Dev.* **18**, 1187–1197.

Wagaba, H., Basavaprabhu, P.L., Jitender, Y.S., Nigel, T.J., Alicai, T., Baguma, Y., Settumba Mukasa, B. and Fauquet, C.M. (2010). Testing the efficacy of artificial microRNAs to control cassava brown streak disease. Second RUFORUM Biennial Meeting 20–24 September 2010, Entebbe, Uganda, 287–291.

Wang, M.B. and Waterhouse, P.M. (2000) High-efficiency silencing of a beta-glucuronidase gene in rice is correlated with repetitive transgene structure but is independent of DNA methylation. *Plant Mol. Biol.* **43**, 67–82.

Warthmann, N., Chen, H., Ossowski, S., Weigel, D. and Herve, P. (2008) Highly specific gene silencing by artificial miRNAs in rice. *PLoS ONE*, **3**, e1829.

Waterhouse, P.M., Graham, H.W. and Wang, M.B. (1998) Virus resistance and gene silencing in plants can be induced by simultaneous expression of sense and antisense RNA. *Proc. Natl Acad. Sci. USA*, **95**, 13959–13964.

Wen, R.H. and Hajimorad, M.R. (2010) Mutational analysis of the putative pipo of soybean mosaic virus suggests disruption of PIPO protein impedes movement. *Virology*, **400**, 1–7.

Westerhout, E.M., ter Brake, O., Ooms, M., Vink, M., Das, A.T. and Berkhout, B. (2005) HIV-1 can evade RNAi-mediated inhibition by altering the secondary structure of its RNA genome. *J. Biotechnol.* **118**, S69–S69.

Wilson, J.A. and Richardson, C.D. (2005) Hepatitis C Virus replicons escape RNA interference induced by a short interfering RNA directed against the NS5b coding region. *J. Virol.* **79**, 7050–7058.

Xu, P., Zhang, Y., Kang, L., Roossinck, M.J. and Mysore, K.S. (2006) Computational estimation and experimental verification of off-target silencing during posttranscriptional gene silencing in plants. *Plant Physiol.* **142**, 429–440.

Zeng, Y., Wagner, E.J. and Cullen, B.R. (2002) Both natural and designed micro RNAs technique can inhibit the expression of cognate mRNAs when expressed in human cells. *Mol. Cell*, **9**, 1327–1333.

Zhang, X., Li, H., Zhang, J., Zhang, C., Gong, P., Ziaf, K., Xiao, F. and Ye, Z. (2011). Expression of artificial microRNAs in tomato confers efficient and stable virus resistance in a cell-autonomous manner. *Transgenic Res.* **20**, 569–581.

Zhao, T., Wang, W., Bai, X. and Qi, Y. (2009) Gene silencing by artificial microRNAs in Chlamydomonas. *Plant J.* **58**, 157–164.

Zuker, M. and Stiegler, P. (1981) Optimal computer folding of large RNA sequences using thermodynamics and auxiliary information. *Nucleic Acids Res.* **9**, 133–148.

12

Metabolomic profiling and genomic analysis of wheat aneuploid lines to identify genes controlling biochemical pathways in mature grain

Michael G. Francki[1,2,*], Sarah Hayton[3], Joel P. A. Gummer[3,4,5], Catherine Rawlinson[3,5] and Robert D. Trengove[3,4,5]

[1]Department of Agriculture and Food Western Australia, Grains Industry, South Perth, WA, Australia

[2]State Agricultural Biotechnology Centre, Murdoch University, Murdoch, WA, Australia

[3]Separation Science and Metabolomics Laboratory, Research and Development, Murdoch University, Murdoch, WA, Australia

[4]School of Veterinary and Life Sciences, Murdoch University, Murdoch, WA, Australia

[5]Metabolomics Australia, Murdoch University Node, Murdoch, WA, Australia

*Correspondence

email michael.francki@agric.wa.gov.au

Summary

Metabolomics is becoming an increasingly important tool in plant genomics to decipher the function of genes controlling biochemical pathways responsible for trait variation. Although theoretical models can integrate genes and metabolites for trait variation, biological networks require validation using appropriate experimental genetic systems. In this study, we applied an untargeted metabolite analysis to mature grain of wheat homoeologous group 3 ditelosomic lines, selected compounds that showed significant variation between wheat lines Chinese Spring and at least one ditelosomic line, tracked the genes encoding enzymes of their biochemical pathway using the wheat genome survey sequence and determined the genetic components underlying metabolite variation. A total of 412 analytes were resolved in the wheat grain metabolome, and principal component analysis indicated significant differences in metabolite profiles between Chinese Spring and each ditelosomic lines. The grain metabolome identified 55 compounds positively matched against a mass spectral library where the majority showed significant differences between Chinese Spring and at least one ditelosomic line. Trehalose and branched-chain amino acids were selected for detailed investigation, and it was expected that if genes encoding enzymes directly related to their biochemical pathways were located on homoeologous group 3 chromosomes, then corresponding ditelosomic lines would have a significant reduction in metabolites compared with Chinese Spring. Although a proportion showed a reduction, some lines showed significant increases in metabolites, indicating that genes directly and indirectly involved in biosynthetic pathways likely regulate the metabolome. Therefore, this study demonstrated that wheat aneuploid lines are suitable experimental genetic system to validate metabolomics–genomics networks.

Keywords: wheat, metabolomics, genomics, aneuploidy, seed, quality.

Introduction

Detailed knowledge of biological processes can significantly enhance our ability to manipulate desirable phenotypes for crop improvement. Small molecules resulting from metabolism (i.e. metabolites) are an important link between genes and phenotypes as they represent a nearer biological end point to the desired trait than either genes or their encoded protein (Fiehn, 2002; Hall, 2006). The complement of metabolites (metabolome) in any particular plant tissue has the potential to provide a diagnostic and predictive phenotypic trait value. However, changes of the metabolome during plant growth and in response to environmental signals (Hall et al., 2006; Schauer and Fernie, 2006) may resultantly render subsequent comparison of profiles between genotypes unrelated to the genetic differences, but rather reveal metabolites more closely correlated with physiological differences and environmental responses. Therefore, metabolite profiles in mature grain would be most suitable in this regard as there are no further developmental changes within the plant, and association of compounds to phenotypic traits can be made specifically to genotype and environment responses.

Metabolomics is an impartial technology and, when integrated with complementary disciplines, contributes towards the interpretation of interconnecting biological processes associated with phenotypes. Metabolic profiling therefore is becoming an increasingly popular tool in functional genomics (Bino et al., 2004). The co-occurrence of gene transcripts and small-molecule metabolites (associated with the metabolomics discipline) provides a basis for generating data-driven theoretical models of biological networks (Saito and Matsuda, 2010; Saito et al., 2008; Yuan et al., 2008). Although 'guilt by association' of transcripts and metabolites is a widely accepted principle for assuming gene function, the active state of proteins through post-translational modification or the influence of substrates or cofactors by undisclosed interconnecting genes and biological pathways may have a significant influence in metabolite variation (Fridman and Pichersky, 2005; Saito et al., 2008). To this end, a holistic analysis of genes involved in discrete primary and interconnecting secondary pathways and their interplay would significantly contribute towards understanding the biological networks regulating the plant metabolome and phenotypic variation.

The most comprehensive reconstruction and modelling of a metabolic network based on multi-omics approach was achieved in the filamentous fungus, *Aspergillus niger*, providing a detailed understanding of genes regulating metabolism and new information on physiological traits (Andersen *et al.*, 2008). However, integration of 'omics technologies in crop species is not as well advanced, posing the next major experimental challenge to model metabolic networks that give rise to phenotypes. The genomes of major crop species have been sequenced or are in the process of completion (Feuillet *et al.*, 2011), paving the way to develop the resources needed towards understanding the link between genes of interconnecting biological pathways with phenotypes. Allohexaploid bread wheat ($2n = 6\times = 42$, genomes AABBDD) is one of the more complex crop genomes whereby similar genes on homoeologous chromosomes could pose a significant challenge in reconstructing biological networks. An ordered draft wheat genome sequence has recently been completed with >124 000 gene loci distributed across all chromosomes of the A, B and D subgenomes (International Wheat Genome Sequencing Consortium, 2014) that will assist in identifying genes controlling biological processes and metabolite abundances responsible for phenotypes and trait variation. A preliminary analysis of gene content neither showed a bias in gene composition nor transcription wide global dominance by any particular subgenome but, rather, each had a higher degree of regulatory and transcriptional autonomy (The International Wheat Genome Consortium (IWGSC), 2014). The availability of the draft sequence of the wheat genome enabled the analysis of gene interaction in wheat grain. Expression analysis of a subset of genes confirmed a lack of global dominance from any of the subgenomes during wheat grain development, but rather cell type- and stage-dependent genome dominance, with inter- and intragenomic regulation of gene expression (Pfeifer *et al.*, 2014). Therefore, transcript accumulation is a result of the interplay between subgenomes and amongst individual cell types giving rise to a particular function, confirming the complex regulation of gene expression adding to the multifaceted processes of interacting biological pathways leading to phenotypes in wheat grain.

Data-driven theories of metabolic networks require validation by forward or reverse genetics (Saito *et al.*, 2008), but often the chosen experimental system neither ratifies nor refutes existing hypotheses on key genetic determinants controlling biological processes. Modifying transcript expression using transgenics or mutations, for instance, may not affect metabolite or trait variation if a targeted gene identified from theoretical models is not a rate-limiting step in the primary biochemical pathway or, indeed, other interconnecting biological pathways affect transcriptional regulation or post-translational modification. In this regard, experimental genetic systems capable of simultaneously associating large sets of genes with a metabolite profile are preferred to validate known and discover new interconnecting biological pathways that determine the metabolome. The polyploid nature of bread wheat genome is a particularly unique genetic system as it can tolerate substantial chromosomal aberrations without compromising plant survival, allowing analysis of phenotypic changes caused by multiple gene loss. In particular, aneuploid lines with missing chromosome arms (ditelosomics) or smaller deleted segments have been well characterized (Endo and Gill, 1996) and shown to be useful in identifying

genes controlling phenotypes (Erayman *et al.*, 2004). Therefore, metabolic profiling of ditelosomic and deletion lines could provide a powerful genetic system and an appropriate supporting tool to identify genes responsible for primary and interconnecting biological pathways that corroborate networks controlling metabolite accumulation.

Metabolite profiling of mature wheat grain has identified both polar and nonpolar compounds (Bellegia *et al.*, 2013; Lee *et al.*, 2013; Matthews *et al.*, 2012) with significant differences between durum and bread wheat (Matthews *et al.*, 2012). Therefore, it appears that polyploidy has a significant effect on accumulation and composition of metabolites in wheat grain. However, the contribution of the gene content of each chromosome is yet to be realized, although the use of aneuploid lines as a genetic system could greatly enhance our ability to understand and validate interconnecting networks that control the accumulation of compounds of the wheat metabolome. The general aim of this study therefore was to determine the feasibility of using wheat aneuploid lines to identify genes of biological pathways that control the accumulation of metabolites in mature wheat grain. Specifically, this study aimed to (i) develop an untargeted metabolite profile of mature grain of wheat with a full complement of chromosomes and compare metabolite content and composition from selected wheat ditelosomic lines; (ii) identify known biochemical pathways in wheat and ditelosomic lines and interrogate the draft wheat genome sequence to reveal underlying genes that control metabolite accumulation; and (iii) assess the suitability of aneuploid lines as a genetic system to validate genes controlling the abundance and composition of metabolites in mature seed. The study focused on mature grain of Chinese Spring (the genotype from which aneuploid lines were derived) and ditelosomic lines where genes on the short and long arms of A, B and D genomes of homoeologous group 3 chromosomes were deleted, and the resultant effects on seed metabolites. The outcome of this study will determine whether wheat aneuploid lines are appropriate to identify and validate genes controlling metabolite accumulation in mature grain that could be used to manipulate grain quality traits.

Results

Metabolite profiling of mature seed of Chinese Spring and ditelosomic lines

The analyses resolved a total of 412 analytes between the metabolite profiles of mature wheat seed and ditelosomic lines from homoeologous group 3 chromosomes (see Table S1 for metabolite specification). PCA described 53% of the data variance between principal components one and two (Figure 1; 33% and 20% for PC1 and PC2, respectively). The clustering of individual replicates ($n = 4$–5) observed within genotypes, relative to that observed between genotypes, identified distinct groupings correlated to the different genetic backgrounds. Of most interest were those differences in metabolite profiles between the Chinese Spring and ditelosomic lines; however, it was between Chinese Spring, DT3AL, DT3BS and DT3BL that the PCA model described the least variation amongst metabolites (Figure 1). PC1 described the ditelosomic line DT3DS as most distinct from the remaining ditelosomics or Chinese Spring. PC2 described the ditelosomic lines DT3AS and DT3DL to be the next most distinct, describing the difference between these two lines and all others, including Chinese Spring. Overall, the model identified the ditelosomic lines DT3AS, DT3DS and DT3DL as having the largest

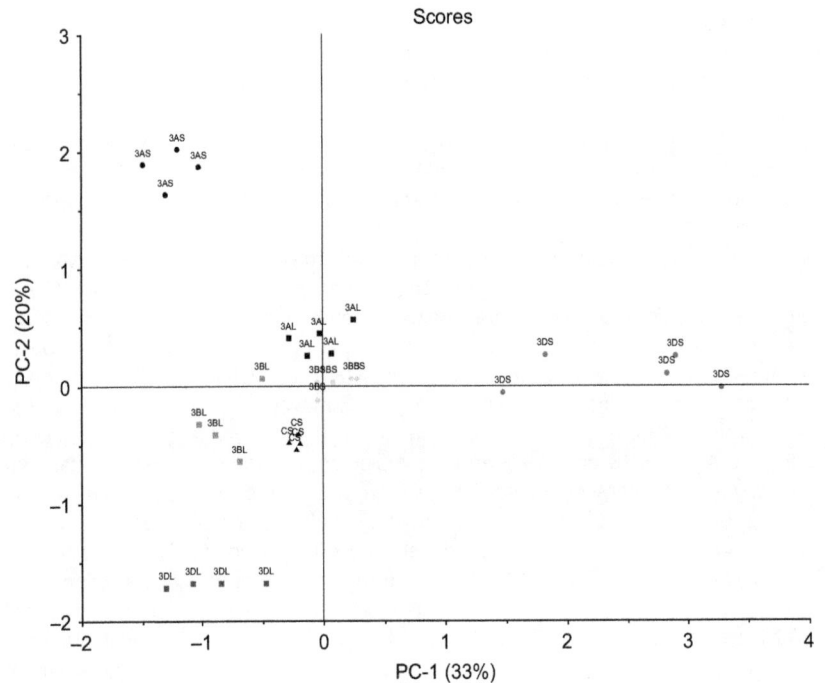

Figure 1 Principal component analysis plot of metabolites for wheat cv Chinese Spring (CS) and genotypes ditelosomic for 3AS, 3AL, 3BS, 3BL, 3DS, 3DL (n = 4–5). The total metabolite variation (53%) represented in the first and second component is 33% and 20%, respectively.

differences between metabolite profiles (Figure 1), indicative of the genes on chromosomes 3AL, 3DL and 3DS, respectively, having a major influence on the variation in polar metabolite composition of mature seed.

The deconvoluted mass spectra from the profiles of Chinese Spring and each ditelosomic lines were compared against the mass spectra and chromatographic features of authentic metabolite standards in combination with the NIST 2011 mass spectral library, to identify individual compounds in mature seed. All identities were determined by a match score of 80% [as scored for uracil (2TMS_18.47_1342)] or above. Succinic acid (2TMS_17.92_1321) and tyrosine (3TMS_30.86_1934) scored the highest match of 96%. The score reflected the relative similarity of ion intensities of the metabolite to that of the library entry. The MS library matches revealed a total of 55 identified metabolite features (Table 1) from the 412 measured analytes (Table S1). Those identified were categorized into one of seven major metabolite classes including amino acids, amino alcohols, fatty acids, nucleosides, organic acids, sugars and sugar alcohols, with the largest number of identified compounds classed as amino acids (Table 1). The comparative metabolite abundances between lines were expressed as a significant ($P < 0.01$) fold change difference to the same compound identified in Chinese Spring. Eighty per cent of the identified metabolites show a significant difference between at least one ditelosomic line and Chinese Spring (Table 1), indicating that genes on the short and long arm of homoeologous group 3 have a major effect on the metabolite profile of mature wheat grain. The significant differences in biochemical profiles between Chinese Spring and ditelosomic lines provided a means to identify putative genes controlling variation in metabolites. Specific metabolites having significant fold -change difference between Chinese Spring and ditelosomic lines or that share common biochemical pathways were selected for further investigation for underlying genes controlling metabolite accumulation.

Metabolite profile and putative genes controlling trehalose accumulation

The MST trehalose (8TMS_42.71_2728) was selected for further analysis as it had the highest fold change difference of all metabolites, with a significant two-fold decrease for DT3BS ($P < 0.01$) and a highly significant 11-fold increase detected for DT3DS ($P < 0.001$) compared to Chinese Spring (Table 1). Consistently, the (non-scaled) chromatograms showed an accumulation of trehalose in DT3DS relative to Chinese Spring and the remaining ditelosomic lines (Figure 2). Trehalose is controlled by a relatively simple biochemical pathway involving three enzymatic steps where UDP-glucose and glucose-6-phosphate are substrates for conversion to trehalose-6-phosphate by trehalose-6-phosphate synthase (TPS) which, in turn, is used to synthesize trehalose through trehalose-6-phosphate phosphatase (TPP) activity (Figure 3). Furthermore, trehalose is converted to form two glucose molecules by trehalase (Figure 3), and putative wheat genes encoding the three enzymes were searched in the wheat genome survey sequence. For comparative purposes, the identification of wheat cDNAs encoding TPS, TPP and trehalase was revealed by TBLASTX analysis using annotated FL-cDNAs from Arabidopsis and rice as query sequences. Full-length wheat cDNA sequences with e-values <4e^{-105} were identified, including two with homology to TPS, one cDNA encoding trehalase and six with significant homology to TPP (Table 2). BLASTN analysis of the draft wheat genome sequence using wheat FL-cDNA as query sequences identified four copies of TPS and three copies of trehalase-related genes on the long arm of homoeologous chromosomes 1 and 5 (Table 2). Genes related to TPP, however, represented a larger multigene family, consisting of at least 13 copies with genes located on homoeologous chromosomes 1, 5 and 6 in addition to copies located on 3AL, 3BL and 3DL (Table 2). Of the ditelosomic lines that lacked the TPP-related sequences, DT3BS showed the expected decrease in metabolite levels (Table 1), indicating that TPP on 3BL may be a

Table 1 Summary of identified metabolites detected in mature wheat grain

Metabolite	DT3AS	DT3AL	DT3BS	DT3BL	DT3DS	DT3DL
Unidentified pentose sugars						
Unknown aldopentose, 5 TMS, 26.79, 1724	1.15	0.98	0.57**	0.96	3.30**	0.85
Unknown aldopentose, 5 TMS, 26.83, 1724	1.06	0.96	1.05	0.86	4.10**	1.60**
Amino acid						
Aspartic acid, 3 TMS, 22.67, 1516	0.30*	0.66	0.55	0.70	2.40	0.28
GABA, 3 TMS, 22.87, 1527	4.88**	0.62	0.61	0.92	1.41	1.55
Glutamic acid, 3 TMS, 24.79, 1623	0.74	1.25	1.04	1.10	2.26	1.07
Glycine, 2 TMS, 12.72, 1110	0.91	0.77	0.60**	0.66*	0.95	0.84
Glycine, 3 TMS, 17.63, 1308	1.31*	0.75	0.60**	0.79	0.81	0.76*
Isoleucine 1 TMS, 14.36, 1175	0.89	0.76	0.48*	0.54*	0.81	0.46*
Alanine, 2 TMS, 12.11, 1085	1.64**	0.67*	0.65	0.92	1.03	0.83
Glutamic acid, 2 TMS, 22.7, 1519	1.36**	0.81*	0.62**	0.72**	0.92	1.01
Glutamine, 3 TMS, 27.89, 1777	1.75	2.07	1.70	0.85	4.90	5.69
Isoleucine, 2 TMS, 17.32, 1295	1.11	0.61*	0.46**	0.77	1.07	0.41**
Leucine, 1 TMS, 13.94, 1159 (putative)	1.01	0.70	0.53*	0.58*	0.84	0.48*
Methionine, 1 TMS, 20.27, 1416	4.00**	3.02**	1.06	2.41**	0.58	1.57*
Phenylalanine, 1 TMS, 23.37, 1550	1	2.23	0.63	0.71	0.80	0.65
Phenylalanine, 2 TMS, 24.97, 1630	0.92	1.53	0.85	0.86	2.31	0.54
Proline, 2 TMS, 17.43, 1300	2.03**	0.61	0.36*	0.69	0.79	0.95
Proline, × TMS, 24.06, 1585	1.49	0.77	0.42*	0.52	0.85	0.86
Threonine, 3 TMS, 19.59, 1387	1.4	0.86	0.81	1.22	1.96	0.95
Tryptophan, 1 TMS, 34.84, 2172	0.47	1.05	0.84	0.24	3.89	0.01*
Tyrosine, 3 TMS, 30.86, 1934	0.71	0.77	0.77	0.56*	1.61	0.63
Valine, 1 TMS, 12.13, 1085	0.98	0.71	0.47*	0.47*	0.7	0.53*
Serine, 2 TMS, 16.43, 1260	2.32**	1.12	0.92	1.23	0.66	0.91
Serine, 3 TMS, 18.98, 1363	3.11**	0.97	0.86	1.87*	1.21	0.79
Amino alcohol						
Ethanolamine, 3 TMS, 16.6, 1266	1.57**	0.99	1.15	1.16	1.03	0.91
Fatty acid						
Arachidic acid, 1 TMS, 38.86, 2444	1.33	1.09	0.97	0.98	1.06	1.1
Stearic acid, 1 TMS, 35.95, 2244	1.52	1.08	1.06	1.2	0.88	1.07
Nucleoside						
Adenosine, × TMS, 40.95, 2603	0.99	0.62	0.31*	0.68	1.42	0.6
Uracil 2 TMS, 18.47, 1342	1.16	0.46**	0.47**	0.48**	0.62**	0.57**
Uridine, 3 TMS, 38.98, 2462	1.32	0.96	0.64**	0.81	1.52*	0.58**
Organic acids						
Azelaic acid, 2 TMS, 28.33, 1801	0.99	0.87	0.59*	0.59*	0.79	0.54*
Benzoic acid, 1 TMS, 16.26, 1254	0.52	0.85	0.58	0.48	1.24	0.76
Citric acid, 4 TMS, 28.69, 1817	0.59*	0.46**	0.35**	0.30**	0.65*	0.33**
Gluconic acid, 6 TMS, 31.84, 1989	2.50*	1.48*	0.81*	1.05	0.94	1.36
Malonic acid, 2 TMS, 15.13, 1208	1.09	1.06	0.94	1.21	1.31*	1.1
Oxalic acid, 2 TMS, 13.13, 1125	2.87	6.45	0.96	0.8	1.17	0.77
Quinic acid, × TMS, 29.38, 1851	6.76**	1.35*	0.93	2.29**	0.46**	0.89*
Shikimic acid, 28.47, 1809	2.48**	0.98	0.94	1.1	0.55**	0.65
Succinic acid, 2 TMS, 17.92, 1321	0.70**	0.82*	0.71**	0.70**	0.95	0.64**
Fumaric acid, 2 TMS, 18.29, 1357	0.91	1.13	0.66*	0.87	1.07	0.70*
Others						
Tocopherol, 1 TMS, 47.69, 3141	0.96	0.99	0.93	0.82	0.74**	0.9
Squalene, 43.91, 2818	1.2	1.33	1.26	1.17	1.02	1.06
Sugars						
Cellobiose, × TMS, 42.6, 2721	0.64	1.03	0.73	0.8	1.37	1.07
Fructose, 5 TMS, MEOX, 29.60, 1862	2.56**	1.28*	1.38**	1.42**	0.78**	0.82*
Fructose, 5 TMS, MEOX, 29.78, 1871	3.14**	1.47**	1.58**	1.71**	0.74**	0.80*
Ribose, 4 TMS, MEOX, 25.89, 1678	1.68**	1.14	1.15	0.83	1.29	0.74
Glucose, 5 TMS, MEOX, 30.05, 1885	5.19**	1.58**	1.64**	2.24**	0.88	0.94
Mannose, 5 TMS, MEOX, 30.12, 1889	6.86**	1.63*	1.63*	2.39**	0.81	0.99
Trehalose, 8 TMS, 42.71, 2728	1.48	0.93	0.56*	0.67	11.60**	0.91

Table 1 Continued

Metabolite	DT3AS	DT3AL	DT3BS	DT3BL	DT3DS	DT3DL
Xylose, 4 TMS, 25.42, 1656	1.67**	1.14	1.09	1.14	1.59*	0.71*
Stachyose, × TMS, 61.39, 4464	0.82	1.19	1.09	1.05	1.35	0.76
Sucrose, 8 TMS, 41.32, 2630	1.05	1.25	0.93	0.91	1.09	0.98
Sugar alcohols						
Mannitol, 6 TMS, 30.6, 1915	1.46*	1.59**	1.04	1.34	4.96**	1.90**
Myo-inositol, 6 TMS, 33.38, 2081	1.03	0.59**	0.52**	0.60**	1.03	0.46**
Scyllo-inositol, 6 TMS, 32.28, 2020	0.60*	0.8	0.58**	0.75*	1.29*	1.02

The fold change difference in metabolite accumulation compared to Chinese Spring is shown for each ditelosomic (DT) line with significant ($P < 0.01$) and highly significant ($P < 0.001$) differences indicated by * and **, respectively.

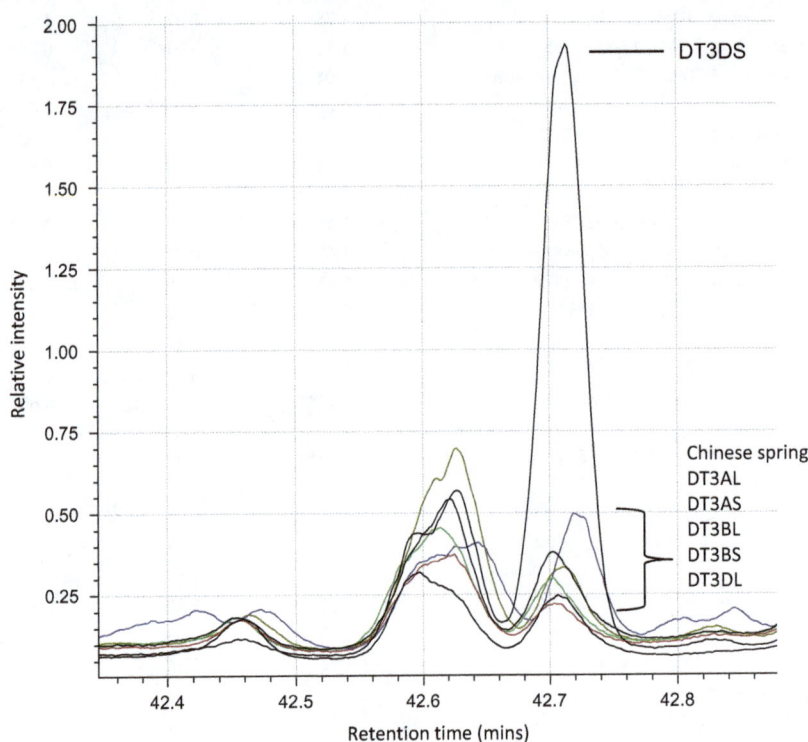

Figure 2 Total ion chromatogram overlays for trehalose 8TMS_42.71_2728 between a representative replicate for Chinese Spring and each ditelosomic line

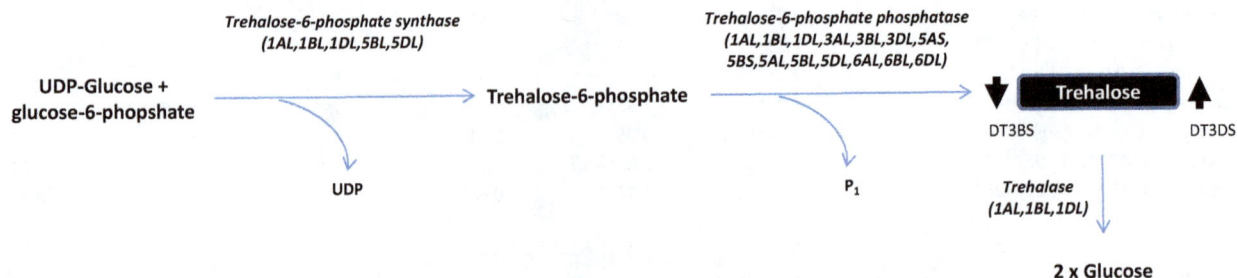

Figure 3 Schematic diagram of the biochemical pathway for trehalose accumulation. Blue box highlights trehalose detected in the untargeted analysis with black arrows indicating increase and decrease in trehalose for DT3BS and DT3DS, respectively. Enzyme names are shown in blue with the chromosomal location of corresponding wheat genes shown in parentheses.

rate-limiting step in trehalose accumulation. However, the absence of either TPP or trehalase did not show a similar effect in DT3AS and DT3DS, indicating that TPP genes on 3AL and 3DL may have an alternative role other than trehalose accumulation in mature grain (Figure 3). On the contrary, the 11-fold increase in trehalose for DT3DS was unlikely to be attributed to any gene in the primary biochemical pathway for trehalose, indicating that other unknown genes of intercon-

Table 2 Summary of rice and wheat FL-cDNA sequences with annotation and amino acid identity to enzymes of the trehalose and branched-chain amino acid biosynthetic pathways. The chromosomal locations of wheat cDNA are based on identity with DNA sequences in the survey sequence from International Wheat Genome Sequencing Consortium (IWGSC)

Metabolite	Enzyme	Annotated FL-cDNA*	Wheat FL-cDNA	TBLASTX e-value	Wheat chromosome locations
Trehalose	Trehalose-6-phosphate synthase	Y08568 (A.t.)*	FJ167677	e = 0.0	1AL, 1BL, 1DL, 5DL
		AF370287 (A.t.)	AK331389	e = 0.0	1AL, 1BL, 1DL, 5DL
		AY063055 (A.t.)	FJ167677	e = 0.0	1AL, 1BL, 1DL, 5DL
		AK103775	AK331389	e = 0.0	1AL, 1BL, 1DL, 5DL
			FJ167677	e = 0.0	1AL, 1BL, 1DL, 5DL
			AK331389	e = 0.0	1AL, 1BL, 1DL, 5DL
			FJ167677	e = $2e^{-106}$	1AL, 1BL, 1DL, 5BL
			AK331389	e = $4e^{-105}$	1AL, 1BL, 1DL, 5BL
	Trehalose-6-phosphate phosphatase	AK072132	AK333853	e = 0.0	1AL, 1BL, 1DL, 3AL, 3BL, 3DL, 5AS, 5BS, 5BL, 5DL
			AK334843	e = 0.0	1AL, 1BL, 1DL, 3AL, 3BL, 3DL, 5AS, 5BS, 5BL, 5DL
			FN564426	e = 0.0	1AL, 1BL, 1DL, 3AL, 3BL, 3DL, 5AS, 5BS, 5BL, 5DL
			AK332212	e = 0.0	1AL, 1DL, 3AL, 3BL, 5AL, 5BS, 5BL, 5DL
			AK331757	e = 0.0	1AL, 1BL, 1DL
			BT009244	e = 0.0	6AL, 6BL, 6DL
	Trehalase	BT010732 (A.t)	AK331310	e = $2e^{-177}$	1AL, 1BL, 1DL
		AK108163	AK331310	e = 0.0	1AL, 1BL, 1DL
Aspartate	Asparagine synthetase	D83378	AK333183	e = 0.0	1AL,1BL,1DL
			AY621539	e = 0.0	5AL,5BL,5DL
			AK334107	e = 0.0	5AL,5BL,5DL
			BT009245	e = 0.0	5AL,5BL,5DL
			BT009049	e = 0.0	3AS,3DS
Glutamate	Aspartate transaminase	AK069075	AK331565	e = $3e^{-150}$	1AS,1BS,1DS
		AK067732	AK331959	e = $3e^{-150}$	1AS,1BS,1DS
			AK333562	e = $3e^{-127}$	5AS,5BS,5DS
			AK333743	e = $2e^{-124}$	5AL,5L,5DL
		AK068200	BT009428	e = $5e^{-180}$	1AS,1BS,1DS
		AK103586	AK332497	e = 0.0	3AL,3BL,3DL
			BT009009	e = 0.0	3AL,3BL,3DL
			EU346759	e = 0.0	3AL,3BL,3DL
			AK332709	e = $1e^{-144}$	6AL,6BL,6DL
			EU885207	e = $1e^{-136}$	6BL,6DL
			AK333705	e = $7e^{-136}$	6AS,6BS,6DS
Methionine	Aspartate kinase	AK121930	AK333665	e = 0.0	4AL,5BL,5DL
		AK073189	BT009484	e = 0.0	4AL,5BL,5DL
			AK334445	e = 0.0	3AL,3BL,3DL
	Aspartate semialdehyde dehydrogenase	AK060701	BT008970	e = 0.0	5AL,5BL,5DL
			BT009463	e = 0.0	5AL,5BL,5DL
	Homoserine dehydrogenase	AK068391	AK335256	e = 0.0	2AL,2BL,2DL
	Homoserine kinase	AK060519	AK333708	E = $7e^{-136}$	4AS,5AL,5BL,5DL
	Cystathionine-γ-synthase	NM_001071075	AK335253	e = $3e^{-137}$	7AS,7BS,7DS
	Cystathionine-β-lyase	NM_001063486	AK335253	e = 0.0	7AS,7BS,7DS
			BT009509	e = 0.0	7AS,7BS,7DS
	Methionine synthase	AF439723 (Z.m.)	AK335562	e = 0.0	4AL,4DS,4BS,6BS,5DS,5BS,5AS
			BT009353	e = 0.0	4AL,4DS,4BS,6BS,5DS,5BS
			AK335485	e = 0.0	5DS,5BS,5AS,4BS,4DS,4AL,6B
Threonine	Threonine synthase	AK101669	AK330620	e = 0.0	1AL,3AL,3BL,3DL
Isoleucine/Valine	Threonine dehydratase	XM_006650431	tplb0062e09	e = 0.0	4AL,4DS,4BS,5AS,5DS
	Acetolactate synthase	AB049823	AY210406	e = 0.0	6BS,6AL,6BL,6DL
	Ketoacid reductoisomerase	AK072075	BT009123	e = 0.0	1AL
		AK065295	BT009123	e = 0.0	1AL
		AK061892	BT009123	e = 0.0	1AL
	Dihydroxy acid dehydratase	AK102083	AK335234	e = 0.0	7AS,7BS,7DS
	Amino acid aminotransferase	AK120579	AK335425	e = $3e^{-123}$	1AL,1BL,1DL
			AK330986	e = $8e^{-113}$	2AL, 2BL,2DL

Table 2 Continued

Metabolite	Enzyme	Annotated FL-cDNA*	Wheat FL-cDNA	TBLASTX e-value	Wheat chromosome locations
			BT009368	$e = 8e^{-112}$	2AL, 2BL,2DL
		AK108687	AK335425	$e = 0.0$	1AL,1BL,1DL
			AK330986	$e = 2e^{-176}$	2AL, 2BL,2DL
			BT009368	$e = 2e^{-174}$	2AL, 2BL,2DL
		AK106376	AK330986	$e = 0.0$	2AL, 2BL,2DL
			BT009368	$e = 0.0$	2AL, 2BL,2DL
			AK335425	$e = 4e^{-165}$	1AL,1BL,1DL
Alanine	Alanine aminotransferase	AK107237	AK333743	$e = 0.0$	5AL,5BL,5DL
			AK331959	$e = 0.0$	1AS,1BS,1DS
			AK331565	$e = 0.0$	1AS,1BS,1DS
			AK333562	$e = 9e^{-170}$	5AS,5BS,5DS
		AK119373	AK333743	$e = 0.0$	5AL,5BL,5DL
			AK331959	$e = 0.0$	1AS,1BS,1DS
			AK331565	$e = 0.0$	1AS,1BS,1DS
			AK333562	$e = 9e^{-170}$	5AS,5BS,5DS
Leucine	Isopropylmalate synthase	AK066890	AK332549	$e = 0.0$	5AL,5BL,5DL
		AK243491	AK332549	$e = 0.0$	5AL,5BL,5DL
	Isopropylmalate isomerase	NM_129871 (A.t.)	BT009140	$e = 6e^{-100}$	6AL,6BL,6DL
	Isopropylmalate dehydrogenase	AK059596	BT009215	$e = 0.0$	2AL,2BL,2DL,6AL,6BL,6DL
			AK331720	$e = 0.0$	2AL,2BL,2DL,6AL,6BL,6DL
			BT009114	$e = 0.0$	2AL,2BL,2DL,6AL,6BL,6DL
			AK331640	$e = 0.0$	1BL,2AL,2BL,2DL,6DL
		AK120254	AK334888	$e = 0.0$	2AL,2BL,2DL
			BT009017	$e = 1e^{-126}$	2AL,2BL,2DL
	Amino acid aminotransferase	AK120579	AK335425	$e = 3e^{-123}$	1AL,1BL,1DL
			AK330986	$e = 8e^{-113}$	2AL, 2BL,2DL
			BT009368	$e = 8e^{-112}$	2AL, 2BL,2DL
		AK108687	AK335425	$e = 0.0$	1AL,1BL,1DL
			AK330986	$e = 2e^{-176}$	2AL, 2BL,2DL
			BT009368	$e = 2e^{-174}$	2AL, 2BL,2DL
		AK106376	AK330986	$e = 0.0$	2AL, 2BL,2DL
			BT009368	$e = 0.0$	2AL, 2BL,2DL
			AK335425	$e = 4e^{-165}$	1AL,1BL,1DL

*GenBank Accession Numbers from *Oryza sativa*, *Arabidopsis thaliana* (A.t.) or *Zea mays* (Z.m.)

necting pathways play a significant role in controlling trehalose in mature grain.

Variation for branched-chain amino acids and associated genes

Branched-chain amino acids in ditelosomic lines were selected for further analysis because of their variable metabolite profiles (Table 1) and interconnecting biochemical pathways that link genes regulating amino acid accumulation. Aspartate is the precursor for methionine and threonine and isoleucine, whereas valine, alanine and leucine are derived from a common precursor, pyruvate (Figure 4). Although the aspartate-derived amino acids are known to be amenable to the analytical methods, only a putative identity could be given to leucine. The putative leucine (1TMS_13.94_1159) matched RI criteria, and the identifying ions consistent with this metabolite were observed, but co-elution prevented clean deconvolution or background ion subtraction and therefore confident identification. The corresponding genes for the biosynthesis of leucine were no longer investigated in this study. Nevertheless, the abundance of other amino acids

detected in ditelosomic lines relative to Chinese Spring showed unambiguous chromatographic resolution and with a significant decrease in aspartate (aspartic acid 2 and 3 TMS; Table 1) for DT3AS only, whereas the remaining showed either an increase or decrease for at least two ditelosomic lines (Figure 4). Threonine (3 TMS, 19.59, 1387) was the exception where no significant difference was detected between any of the ditelosomic lines and Chinese Spring (Figure 4). Genes on homoeologous group 3 chromosomes, therefore, appear to have a significant effect on the accumulation of many aspartate-derived amino acids and were traced to identify those associated with the primary biochemical pathway having potential regulatory roles.

The interconnected biochemical pathway to convert aspartate to methionine, threonine, isoleucine, valine, alanine and leucine involves 18 enzymes (Figure 4). Although not classified as aspartate-derived amino acids, the biosynthesis of glutamine and asparagine involves two additional enzymes, aspartate transaminase and asparagine synthetase, respectively (Figure 4). Therefore, genes encoding 20 enzymes were analysed for their

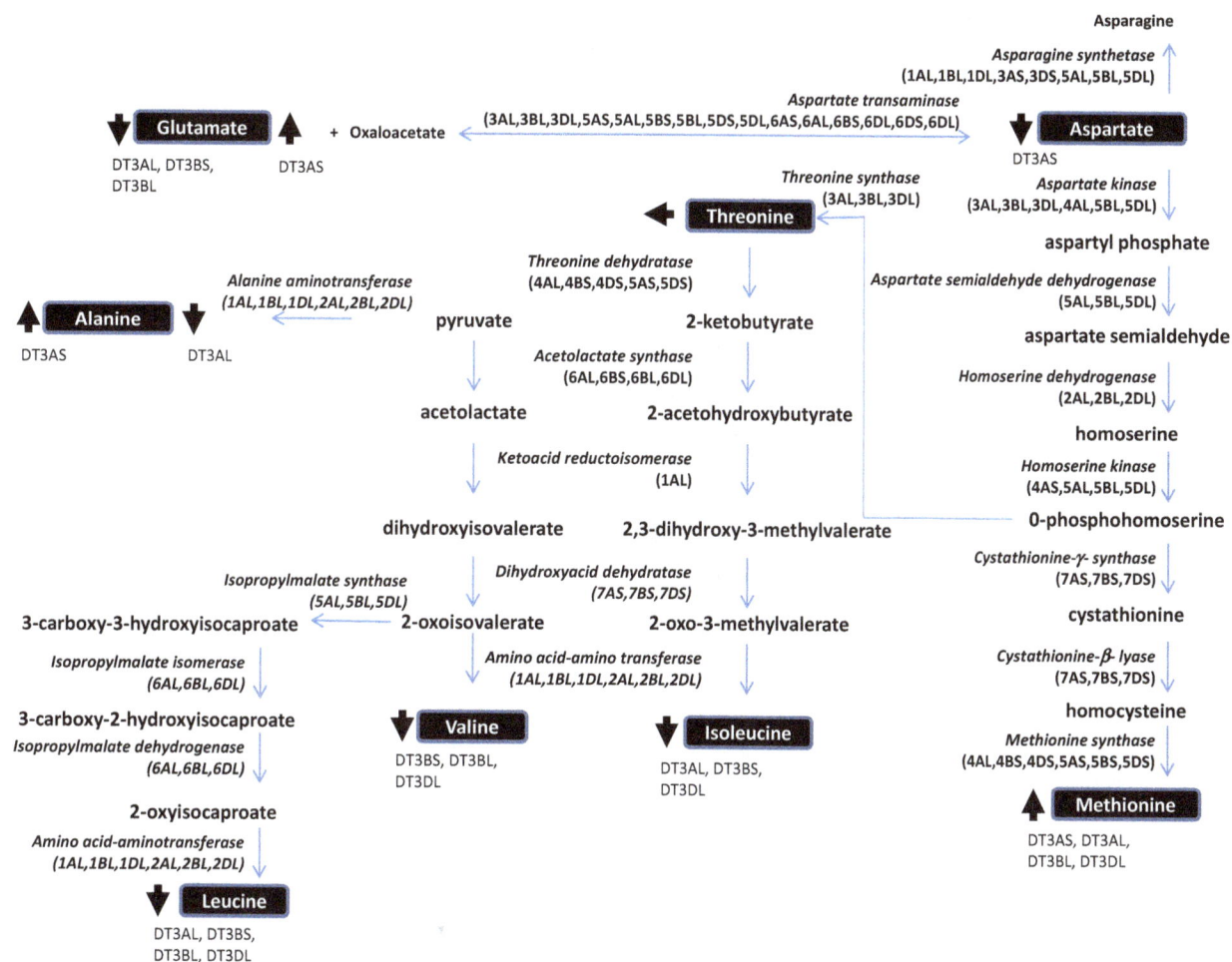

Figure 4 Schematic diagram of the biochemical pathway for the branched-chain amino acids. Blue boxes represent amino acids detected in the untargeted metabolite analysis with black arrows indicating increase or decrease in amino acid for corresponding ditelosomic lines compared with Chinese Spring. Enzymes names are shown in blue with the chromosomal location of their corresponding genes in parentheses.

potential role in controlling amino acid accumulation. The search for annotations from predominantly rice and other plant species identified FL-cDNA for each of the 20 enzymes and was used as query sequences in subsequent TBLASTX analysis to ascertain corresponding wheat FL-cDNA. At least one wheat FL-cDNA was identified for each respective enzyme, but most were represented by several FL-cDNA, indicating that most enzymes were encoded by multigene families and their chromosomal location ascertained by BLASTN search of the draft wheat genome sequence (Table 2 and Figure 4). Interestingly, genes encoding aspartate kinase, an enzyme involved in the first committed step to produce methionine from aspartate and threonine synthase that converts threonine from O-phosphohomoserine, were located on homoeologous group 3 chromosomes. As there is a significant decrease in aspartate for DT3AS (Table 1, Figure 4), the genes controlling its accumulation may either be aspartate transaminase or aspartate kinase located on chromosome 3AL, whereas the deletion of threonine synthase from 3AL, 3BL or 3DL in DT3AS, DT3BS and DT3DS lines, respectively, has no effect on threonine accumulation. The remaining 16 enzymes involved in the biosynthesis of aspartate-derived amino acids were located on chromosomes other than homoeologous group 3 despite increases and decreases in abundance of amino acids in ditelosomic lines relative to Chinese Spring (Table 2 and Figure 4). Therefore,

genes from other interacting pathways and located on homoeologous group 3 chromosomes may be responsible for controlling the abundance of methionine, isoleucine, valine, leucine and alanine.

Discussion

This study demonstrated that aneuploid lines were a suitable genetic system to identify changes in metabolite profiles in mature grain when compared with the standard wheat genotype, Chinese Spring. Grain metabolite composition is impacted by different genotypes and genotype-by-environment interaction (Bellegia et al., 2013), whereas differences in extraction processes and analytical instrumentation, even in nontargeted metabolomics, will favour some metabolite classes over others (Khakimov et al., 2013). Nevertheless, grain of the genotypes analysed in this study was comparable; having been grown and stored in the same conditions and metabolites extracted and analysed using the same procedures and instrumentation to minimize any effects caused by the environment or detection methodologies. Additionally, having focussed on the grain rather than plant tissues, metabolism (in terms of normal physiological processes) is assumed to have ceased. Therefore, significant differences in compounds in ditelosomic lines with near identical background to

Chinese Spring were confidently attributed to genes missing from respective chromosome arms.

Methods consistent with untargeted metabolomics, including metabolite isolation and MS acquisition, were used to obtain an unbiased measurement of the metabolites in mature grain. Although classes of compounds identified in this study were similar to those previously reported (Bellegia et al., 2013; and Lee et al., 2013), the specific metabolites identified differed between these studies. For instance, metabolites related to unsaturated fatty acids, fatty alcohols, flavonols, phenolics, phytosterols and vitamins reported in Bellegia et al. (2013) and Lee et al. (2013) were not detected in this study and can be attributed to differences either in genotypes, environmental effects or disparate metabolite extraction and detection methods. It is interesting to note, however, that no untargeted metabolite profile for wheat grain has reported the detection of compounds associated with carotenoids, which give rise to flour yellowness, important for wheat end products (for review, see Ficco et al., 2014), reflecting variation in compound resolution across studies and the requirement for complementary analyses for greater metabolome coverage (Gummer et al., 2009; Wishart et al., 2009). No single analytical method is capable of resolving the complete metabolome of any given tissue or system, due to the dynamic differences in metabolite chemistries, more specifically chemical structure, polarity, solubility and chromatographic behaviour (Ward et al., 2003). This is a characteristic of small-molecule analysis reflected in metabolomics, more so than other 'omics' disciplines. Therefore, the inclusion of a targeted metabolomics approach, or complementary nontargeted analytical methods, together with different metabolite isolation procedures would provide a more complete measure of the metabolome (Harrigan et al., 2007; Khakimov et al., 2013).

Despite different biological and technical parameters that may affect metabolic profiles, this study used Chinese Spring as a reference genotype for qualitative and quantitative comparison of compounds in ditelosomic lines, to identify chromosome regions affecting metabolite accumulation in wheat seed. Significant differences between Chinese Spring and the ditelosomic lines indicated aneuploid lines are suitable to investigate the underlying genes controlling metabolite variation. It was expected that if genes encoding enzymes directly related to the corresponding biosynthetic pathways were located on homoeologous group 3 chromosomes, then ditelosomic lines would have a significant reduction in corresponding metabolites compared with Chinese Spring. Although a proportion showed a reduction, some ditelosomic lines showed significant increases in specific metabolites, indicating that genes directly and indirectly involved in biosynthetic pathways likely regulate metabolite accumulation.

An example of extreme metabolite variation in this study was in trehalose accumulation. The decrease of trehalose in DT3BS was presumably through the absence of the gene encoding TPP on chromosome 3BL; however, a reduction in trehalose was not detected for DT3AS and DT3DS, indicating that members of the TPP gene family on 3AL and 3DL may not serve a similar function to the gene on 3BL. Members of gene families on homoeologous chromosomes have previously been reported to differ in function on the basis of aneuploidy analysis, such as those encoding proteins regulating Na^+/K^+ accumulation (Ariyarathna et al., 2014), and therefore, it is reasonable to assume that TPP genes on 3AL and 3DL have alternative functions. Trehalose has been reported to accumulate in roots and shoots of wheat (El-Bashiti et al., 2005), and therefore, TPP genes on 3AL and 3DL may

function to control trehalose accumulation in specific tissue. However, the contrasting 11-fold accumulation in DT3DS is extraordinary, and it was presumed that genes encoding trehalase that normally reduce trehalose when the enzyme converts it to glucose molecules (Müller et al., 2001) may, indeed, increase trehalose in their absence. Genes related to trehalase were not located on wheat homoeologous group 3 chromosomes and, therefore, assumed that trehalose is regulated by genes of other pathways. Trehalose metabolism in plants is highly regulated through an intricate network of interconnecting pathways involved in post-translational modification, such as AMP-activated protein kinases and Snf-related protein kinases known to affect the altered state of TPS (Halford et al., 2003; Harthill et al., 2006; Martínez-Barajas et al., 2011; Paul et al., 2010; Zhang et al., 2009). Therefore, genes involved in these or similar intricate pathways may be located on 3DL that have an effect on enzyme activity that would normally down-regulate trehalose in mature wheat grain. As trehalose accumulation has been implicated in providing protective mechanisms against stress tolerance in plants (Fernandez et al., 2010; Garg et al., 2002; Iordachescu and Imai, 2008; Penna, 2003), further investigations of interconnecting but yet undefined pathways are certainly warranted to identify key genes on chromosome 3DL that normally inhibit trehalose accumulation. Metabolic profiling of deletion lines (Endo and Gill, 1996) in subsequent studies would assist in ascertaining the key genetic determinants from a smaller pool of candidates on 3DL and develop strategies to manipulate elevated levels of trehalose that may lead to improved stress tolerance during grain filling.

The regulation of aspartate-derived amino acids is of particular interest in this study, not only for their interconnecting biological pathways but their importance in the human diet. Methionine, isoleucine and threonine are essential amino acids, not synthesized in animals, and, therefore, are important for improving the nutritional value of cereal grain (Ufaz and Galili, 2008). Aspartate is the primary amino acid by which these essential amino acids are synthesized, and its accumulation is affected by aspartate kinase for the production of methionine as a substrate for aspartate transaminase to produce glutamate. In this study, the reduction of asparate in DT3AS corresponds to either the loss of aspartate kinase or the aspartate transaminase genes on 3AL, where the latter is complemented by an increase in glutamate in the same ditelosomic line. However, decreases in aspartate or increases in glutamate were not observed for DT3BS and DT3DS, so it is likely that aspartate kinase or aspartate transaminase on chromosome 3BL and 3DL may have a different role other than regulating amino acid accumulation in wheat gain, potentially regulating metabolite accumulation in other tissue. Metabolite profiling of aneuploid lines from other tissue will provide further information on alternative roles of aspartate kinase and aspartate transaminase in regulating amino acid accumulation during plant growth and development.

A notable feature of the branched-chain amino acid pathway is the nonsignificant difference in levels of threonine in ditelosomic lines relative to Chinese Spring despite its biosynthesis from O-phosphohomoserine through threonine synthase; an enzyme encoded by three genes located on the long arm of homoeologous group 3 chromosomes. Therefore, it appears that threonine synthase genes on chromosomes 3AL, 3BL and 3DL encode an active enzyme capable of maintaining threonine homoeostasis during grain development despite the absence of any particular gene in a corresponding ditelosomic line. O-phosphohomoserine is a common precursor and significant increases in methionine were detected in some ditelosomic lines; however, no genes encoding enzymes controlling methionine or, indeed, accumulation of other amino acids were identified on group 3 chromosomes. Therefore, genes from unidentified interconnecting pathways are likely to control methionine accumulation and the remaining branched-chain amino acids including isoleucine, valine, leucine and alanine. Amino acid accumulation can be affected by other biological processes, including proteins involved in subcellular localization and transport mechanisms, feedback inhibition and activation, post-translation regulation through allosteric regulation of enzymes and transcriptional regulation (Jander and Joshi, 2010; Joshi et al., 2010; Ortiz-Lopez et al., 2000). Indeed, these biological processes integrate in a complex manner to control branched-chain amino acid synthesis whereby some of the underlying but, as yet, unidentified genes may be located on homoeologous group 3 chromosomes. The wheat aneuploid lines would provide an appropriate experimental system in future studies to support the functional analysis of alternative genes involved in the network of numerous biological processes controlling branched-chain amino acid accumulation.

This study has strategically used ditelosomic lines to provide information on genes encoding enzymes of biosynthetic pathways that control metabolite accumulation in a tissue-specific manner. The role of TPP on 3BL reducing trehalose and aspartate kinase on 3AL decreasing aspartate accumulation are good examples on the use of aneuploid lines to discriminate functional roles of genes on homoeologous chromosomes in controlling metabolite accumulation in mature grain. Moreover, the analysis of ditelosomic lines has uncovered a plethora of unidentified biological networks other than genes encoding enzymes of the primary biosynthetic pathway that controls metabolites. The future challenge will be to discover the intricate components of these pathways and their precise role in controlling metabolite accumulation. The completion of the wheat genome sequence including the annotation of pseudomolecules, similar to that for rice and maize (Ouyang et al., 2007; Zhou et al., 2009), coupled with untargeted and targeted metabolite analysis of seeds of wheat aneuploid lines with small deleted regions for all wheat chromosomes (Endo and Gill, 1996) will be a powerful metabolomics–genomics strategy and supporting genetic system to ascertain interconnecting biological networks and underlying genes regulating metabolite and trait variation in wheat grain.

Experimental procedures

Plant material

Seeds of wheat line Chinese Spring and ditelosomic lines, DT3AS, DT3AL, DT3BS, DT3BL, DT3DS, DT3DL, were kindly provided by Dr Jon Raupp, Wheat Genetic and Genomics Resource Center, Kansas State University, Kansas, USA. Seeds were sown in pots and plants grown to maturity in the glasshouse in 2013. Harvested grain from four individual plants of Chinese Spring and each ditelosomic line was pooled and stored at 4 °C in an airtight container with silica gel for 3 months until used for metabolite extraction.

Metabolite extraction

Harvested grain was retrieved from 4 °C storage, and 10–15 seeds per technical replicate (five replicates total) were lyophilized for 16 h in a LABCONCO Freezone 2.5 Plus (Labconco Corp Kansas City, MO). Seeds were ground to a fine powder in a mortar and pestle, chilled with dry ice and 25 mg per replicate transferred to a 2.0 mL tube. Methanol was added to each tube, together with 650 ng $^{13}C_6$-sorbitol (ISTD; in methanol) to a combined volume of 500 μL, and vigorously agitated within a Precellys 24 lysis cryo-mill tissue lyser (Bertin Technologies, Aix-en-Provence, France) at ~5000 g for two subsequent rounds of 20 s. The suspension was agitated in a thermomixer (Eppendorf, South Pacific Pty. Ltd., North Ryde, Australia) at ~1000 g for 15 min at 10 °C and sample particulate collected by centrifuge at 10 000 g. The supernatant was transferred to a fresh tube and the extraction repeated with another 500 μL methanol, without any further addition of ISTD. The supernatants were combined and dried in preparation for derivatization by vacuum removal of the organic solvent, followed by drying by lyophilization, as described by Gummer et al. (2013). This required concentration of the extract to <100 μL volume in an Eppendorf Concentrator Plus vacuum concentrator (Eppendorf, South Pacific Pty. Ltd., North Ryde, Australia) and the subsequent addition of 300 μL of LC-MS grade water. The sample was then frozen in liquid nitrogen and dried by lyophilization in a LABCONCO Freezone 2.5 Plus (Labconco Corp Kansas City, MO). The dried extracts were stored at −80 °C until metabolite derivatization.

GC-MS analysis of metabolites

The metabolites were derivatized by a combination of methoximation and silylation reactions. To the dried metabolites was added 20 μL of methoxyamine HCl (Sigma-Aldrich, Castle Hill, NSW, Australia) [20 mg/mL in pyridine (UNIVAR)], followed by brief mixing by vortex and incubation at 30 °C for 90 min with agitation at ~800 g in an Eppendorf thermomixer. Fourty microlitres of MSTFA (Sigma-Aldrich, Castle Hill, NSW, Australia) was then added and mixed briefly by vortex before incubation at 37 °C for 30 min with agitation at 300 rpm. The entire volume was then transferred to a 200-μL glass insert within a 2-mL analytical vial and, five microlitres of n-alkanes [(C_{10}, C_{12}, C_{15}, C_{19}, C_{22}, C_{28}, C_{32} and C_{36}); Sigma-Aldrich] in hexane [for retention index (RI) calculation] was added and mixed. Samples were loaded on to the GC-MS in a randomized sequence for analysis.

Metabolites were analysed using a Shimadzu QP2010 Ultra GC-MS with AOC-20i Autosampler and injector unit (Shimadzu, Kyoto, Japan) equipped with an Agilent Factor Four fused silica capillary column (VF-5 ms 30 × 0.25 mm × 0.25 μm + 10 m EZ-Guard; Agilent Technologies, Santa Clara, CA). Helium was used as the carrier gas at constant flow. One microlitre of sample was injected into a split/splitless GC inlet, held at 230 °C using a splitless mode of injection, with an initial GC column temperature held at 70 °C. The oven temperature was initially ramped one °C/min for five minutes before a final ramp of 5.6 °C/min to a final temperature of 320 °C. The transfer line and ionization source were held at 280 °C and 230 °C, respectively. The mass spectrometer was set to scan a mass range of 40–600 m/z at 10 scans/s using a 70 eV electron beam.

Metabolomics data analyses and metabolite identification

GC-MS data were analysed using Shimadzu GC-MS solution 2.61 (Shimadzu, Kyoto, Japan). A target list of detected analytes was assembled from the collected GC-MS data for development of a processing method. Each detected metabolite was assigned three (unique, where possible) ions: one quantifier and two qualifier ions. The ions were recognized within a five-second retention time (RT) window. Each metabolite entry was checked for the presence of conflicting ions within the assigned RT deviation. Relative quantitation was determined by calculation and comparison of quantifier ion peak areas. Raw peak areas were normalized to the ISTD ($^{13}C_6$ sorbitol).

Metabolites were identified by mass spectral match to an in-house library, generated from the analysis of authentic metabolite standards. Identification required a minimum forward match percentage of 80% or higher, to be within 5 retention indices (RIs) of the analysed standard compound. Putative identification of metabolites was carried out using the National Institute of Standards and Technology (NIST) 2011 mass spectral library. Metabolites were assigned a mass spectral tag (MST) describing the respective identification and analytical features of the analyte of ontology 'metabolite ID_RT_RI'.

Multivariate and statistical analyses were performed using the Unscrambler X software, version 10.1 (CAMO Software, Oslo, Norway). The ISTD-corrected data matrix was scaled by $log_{10}(x + 1)$ transformation prior to principal component analysis (PCA), using noniterative partial least squares algorithm, cross-validation with no rotation. Significant differences in metabolite abundance between metabolite profiles were determined using an independent, two-tailed Student's t-test and were deemed to be significant or highly significant when $P \leq 0.01$ or $P \leq 0.001$, respectively.

BLAST similarity searching, gene identification and location in the wheat genome

Full-length (FL-) cDNA from rice (The Rice Full-Length cDNA Consortium, 2013) annotated to encode enzymes of biochemical pathways was retrieved from National Center for Biotechnology Information (NCBI) database (http://www.ncbi.nlm.nih.gov/) using key word searching. In the event that rice cDNA sequences were not identified, key word searches were extended to identify annotated FL-cDNA from *Zea mays* or *Arabidopsis thaliana*. Annotated cDNA sequences were used as query sequences in TBLASTX search to identify corresponding wheat FL-cDNA from the Chinese Spring collection (Kawaura et al., 2009). Wheat sequences were identified as orthologs of the annotated rice cDNA when e-values of TBLASTX hits were $<6e^{-100}$. The wheat FL-cDNA sequences were used as a query in BLASTN searching against the wheat genome survey sequence (http://wheat-urgi.versailles.inra.fr/Seq-Repository/) and assigned chromosomal location based on 90% sequence identity threshold value.

References

Andersen, M.R., Nielsen, M.L. and Nielsen, J. (2008) Metabolic model integration of the bibliome, genome, metabolome, and reactome of *Aspergillus niger*. *Mol. Syst. Biol.* **4**, 178.

Ariyarathna, H.A.C.K., Ul-Haq, T., Colmer, T.D. and Francki, M.G. (2014) Characterization of the multigene family *TaHKT2;1* in bread wheat and the role of gene members in plant Na+ and K+ status. *BMC Plant Biol.* **14**, 159.

Bellegia, R., Platani, C., Nigro, F., De Vita, P., Cattivelli, L. and Papa, R. (2013) Effect of genotype, environment, and genotype-by-environment interaction on metabolite profiling in durum wheat (*Triticum durum* Desf.) grain. *J. Cereal Sci.* **57**, 183–192.

Bino, R.J., Hall, R.D., Fiehn, O., Kopka, J., Saito, K., Draper, J., Nikolau, B.J., Mendes, P., Roessner-Tunali, U., Beale, M.H., Trethewey, R.N., Lange, B.M., Wurtele, E. and Sumner, L.W. (2004) Potential of metabolomics as a functional genomics tool. *Trends Plant Sci.* **9**, 418–425.

El-Bashiti, T., Hamamci, H., Öktem, H. and Yücel, M. (2005) Biochemical analysis of trehalose and its metabolizing enzymes in wheat under abiotic stress conditions. *Plant Sci.* **169**, 47–54.

Endo, T.R. and Gill, B.S. (1996) The deletion stocks of common wheat. *J. Hered.* **87**, 295–307.

Erayman, M., Sandhu, D., Sidhu, D., Dilbirligi, M., Baenziger, P.S. and Gill, K.S. (2004) Demarcating the gene rich regions of the wheat genome. *Nucleic Acids Res.* **32**, 3546–3565.

Fernandez, O., Béthencourt, L., Quero, A., Sangwan, R.S. and Clément, C. (2010) Trehalose and plant stress responses: friend or foe? *Trends Plant Sci.* **15**, 409–417.

Feuillet, C., Leach, J.E., Rogers, J., Schnable, P.S. and Eversole, K. (2011) Crop genome sequencing: lessons and rationales. *Trends Plant Sci.* **16**, 77–88.

Ficco, D.B.M., Mastrangelo, A.M., Trono, D., Borrelli, G.M., De Vita, P., Fares, C., Beleggia, R., Platani, C. and Papa, R. (2014) The colors of durum wheat: a review. *Crop Past. Sci.* **65**, 1–5.

Fiehn, O. (2002) Metabolomics- the link between genotypes and phenotypes. *Plant Mol. Biol.* **48**, 155–171.

Fridman, E. and Pichersky, E. (2005) Metabolomics, genomics, proteomics, and the identification of enzymes and their substrates and products. *Curr. Opin. Plant Biol.* **8**, 242–248.

Garg, A.K., Kim, J.-K., Owens, T.G., Ranwala, A.P., Choi, Y.D., Kochian, L.D. and Wu, R.J. (2002) Trehalose accumulation in rice plants confers high tolerance levels to abiotic stresses. *Proc. Natl Acad. Sci. USA*, **99**, 15898–15903.

Gummer, J., Banazis, M., Maker, G., Solomon, P., Oliver, R. and Trengove, R. (2009) Use of mass spectrometry for metabolite profiling and metabolomics. *Aust. Biochem.* **40**, 5–16.

Gummer, J.P.A., Trengove, R.D., Oliver, R.P. and Solomon, P.S. (2013) Dissecting the role of G-protein signalling in primary metabolism in the wheat pathogen *Stagonospora nodorum*. *Microbiology*, **159**, 1972–1985.

Halford, N.G., Hey, S., Jhurreea, D., Laurie, S., McKibbin, R.S., Paul, M. and Zhang, Y. (2003) Metabolic signalling and carbon partitioning: role of Snf1-related (SnRK1) protein kinase. *J. Exp. Bot.* **54**, 467–475.

Hall, R.D. (2006) Plant metabolomics: from holistic hope, to hype, to hot topic. *New Phytol.* **169**, 453–468.

Harrigan, G.C., Martino-Catt, S. and Glenn, K.C. (2007) Metabolomics, metabolic diversity and genetic variation in crops. *Metabolomics*, **3**, 259–272.

Harthill, J.E., Meek, S.E.M., Morrice, N., Peggie, M.W., Borch, J., Wong, B.H.C. and MacKintosh, C. (2006) Phosphorylation and 14-3-3 binding of Arabidopsis trehalose-phosphate synthase 5 in response to 2-deoxyglucose. *Plant J.* **47**, 211–223.

Iordachescu, M. and Imai, R. (2008) Trehalose biosynthesis in response to abiotic stresses. *J. Integr. Plant Biol.* **50**, 1223–1229.

Jander, G. and Joshi, V. (2010) Recent progress in deciphering the biosynthesis of aspartate-derived amino acids in plants. *Mol. Plant.* **3**, 54–65.

Joshi, V., Joung, J.-G., Fei, Z. and Jander, G. (2010) Interdependence of threonine, methionine and isoleucine metabolism in plants: accumulation and transcriptional regulation under abiotic stress. *Amino Acids*, **39**, 933–947.

Kawaura, K., Mochida, K., Enju, A., Totoki, Y., Toyoda, A., Sakaki, Y., Kai, C., Kawai, J., Hayashizaki, Y., Seki, M., Shinozaki, K. and Ogihara, Y. (2009) Assessment of adaptive evolution between wheat and rice as deduced from full-length common wheat cDNA sequence data and expression patterns. *BMC Genom.* **10**, 271.

Khakimov, B., Bak, S. and Engelsen, S.B. (2013) High-throughput cereal metabolomics: current analytical technologies, challenges and perspectives. *J. Cereal Sci.* **59**, 393–418.

Lee, D.P., Alexander, D. and Jonnalgadda, S.S. (2013) Diversity of nutrient content in grains – a pilot metabolomics analysis. *J. Nutr. Food Sci.* **3**, 2.

Martínez-Barajas, E., Delatte, T., Schluepmann, H., de Jong, G.J., Somsen, G.W., Nunes, C., Primavesi, L.F., Coello, P., Mitchell, R.A.C. and Paul, M.J. (2011) Wheat grain development is characterized by remarkable trehalose accumulation trehalose 6-phosphate accumulation pregrain filling: tissue distribution and relationship to SNF1-related protein kinase1 activation. *Plant Physiol.* **156**, 373–381.

Matthews, S.B., Santra, M., Mensack, M.M., Wolfe, P., Byrne, P.F. and Thompson, H.J. (2012) Metabolite profiling of a diverse collection of wheat lines using ultraperformance liquid chromatography coupled with time-of-flight mass spectrometry. *PLoS ONE*, **7**, e44179.

Müller, J., Aeschbacher, R.A., Wingler, A., Boller, T. and Wiemken, A. (2001) Trehalose and trehalase in Arabidopsis. *Plant Physiol.* **125**, 1086–1093.

Ortiz-Lopez, A., Chang, H.-C. and Bush, D.R. (2000) Amino acid transporters in plants. *Biochim. Biophys. Acta*, **1465**, 275–280.

Ouyang, S., Zhu, W., Hamilton, J., Lin, H., Campbell, M., Childs, K., Thibaud-Nissen, F., Malek, R.L., Lee, Y., Zheng, L., Orvis, J., Haas, B., Wortman, J. and Buell, C.R. (2007) The TIGR rice genome annotation resource: improvement and new features. *Nucleic Acids Res.* **35**, D883–D887.

Paul, M.J., Jhurreea, D., Zhang, Y., Primavesi, L.F., Delatte, T., Schluepmann, H. and Wingler, A. (2010) Upregulation of biosynthetic processes associated with growth by trehalose 6-phosphate. *Plant Signal. Behav.* **5**, 386–392.

Penna, S. (2003) Building stress tolerance through over-producing trehalose in transgenic plants. *Trends Plant Sci.* **8**, 355–357.

Pfeifer, M., Kugler, K.G., Sandve, S.R., Zhan, B., Rudi, H., Hvidsten, T.R., International Wheat Genome Sequencing Consortium, Mayer, K.F.X. and Olsen, O.-A. (2014) Genome interplay in the grain transcriptome of hexaploid bread wheat. *Science*, **345**, 1250091.

Saito, K. and Matsuda, F. (2010) Metabolomics for functional genomics, systems biology and biotechnology. *Ann. Rev. Plant Biol.* **61**, 463–489.

Saito, K., Hirai, M.Y. and Yonekura-Sakakibara, K. (2008) Decoding genes with co-expression networks and metabolomics- 'majority report by precogs'. *Trends Plant Sci.* **13**, 36–43.

Schauer, N. and Fernie, A.R. (2006) Plant metabolomics: towards biological function and mechanism. *Trends Plant Sci.* **11**, 508–516.

The International Wheat Genome Consortium (IWGSC). (2014) A chromosome-based draft sequence of the hexaploid bread wheat (*Triticum aestivum*) genome. *Science*, **345**, 1251788.

The Rice Full-Length cDNA Consortium. (2013) Collection, mapping and annotation of over 28,000 cDNA clones from *japonica* rice. *Science*, **301**, 376–379.

Ufaz, S. and Galili, G. (2008) Improving the content of essential amino acids in crop plants: goals and opportunities. *Plant Physiol.* **147**, 954–961.

Ward, J.L., Harris, C., Lewis, J. and Beale, M.H. (2003) Assessment of ^{1}H NMR spectroscopy and multivariate analysis as a technique for metabolite fingerprinting of *Arabidopsis thaliana*. *Phytochemistry*, **62**, 949–957.

Wishart, D.S., Knox, C., Guo, A.C., Eisner, R., Young, N., Gautam, B., Hau, D.D., Psychogios, N., Dong, E., Bouatra, S., Mandal, R., Sinelnikov, I., Xia, J., Jia, L., Cruz, J.A., Lim, E., Sobsey, C.A., Shrivastava, S., Huang, P., Liu, P., Fang, L., Peng, J., Fradette, R., Cheng, D., Tzur, D., Clements, M., Lewis, A., De Souza, A., Zuniga, A., Dawe, M., Xiong, Y., Clive, D., Greiner, R., Nazyrova, A., Shaykhutdinov, R., Li, L., Vogel, H.J. and Forsythe, I. (2009) HMDB: a knowledgebase for the human metabolome. *Nucleic Acids Res.* **37**, D603–D610.

Yuan, J.S., Galbraith, D.W., Dai, S.Y., Griffin, P. and Stewart, C.N. (2008) Plant systems biology comes of age. *Trends Plant Sci.* **13**, 165–171.

Zhang, Y., Primavesi, L.F., Jhurreea, D., Andralocj, P.J., Mitchell, R.A.C., Powers, S.J., Schluepmann, H., Delatte, T., Wingler, A. and Paul, M.J. (2009) Inhibition of SNF1-related protein kinase1 activity and regulation of metabolic pathways by trehalose 6-phosphate. *Plant Physiol.* **149**, 1860–1871.

Zhou, S., Wei, F., Nguyen, J., Bechner, M., Potamousis, K., Goldstein, S., Pape, L., Mehan, M.R., Churas, C., Pasternak, S., Forrest, D.K., Wise, R., Ware, D., Wing, R.A., Waterman, M.S., Livny, M. and Schwatrz, D.C. (2009) A single molecule scaffold for the maize genome. *PLoS Genet.* **5**, e1000711.

13

Transcriptional programs regulating seed dormancy and its release by after-ripening in common wheat (*Triticum aestivum* L.)

Feng Gao[1], Mark C. Jordan[2] and Belay T. Ayele[1]*

[1]*Department of Plant Science, University of Manitoba, Winnipeg, MB, Canada*
[2]*Cereal Research Centre, Agriculture and Agri-Food Canada, Winnipeg, MB, Canada*

Summary

Seed dormancy is an important agronomic trait in wheat (*Trticum aestivum*). Seeds can be released from a physiologically dormant state by after-ripening. To understand the molecular mechanisms underlying the role of after-ripening in conferring developmental switches from dormancy to germination in wheat seeds, we performed comparative transcriptomic analyses between dormant (D) and after-ripened (AR) seeds in both dry and imbibed states. Transcriptional activation of genes represented by a core of 22 and 435 probesets was evident in the dry and imbibed states of D seeds, respectively. Furthermore, two-way ANOVA analysis identified 36 probesets as specifically regulated by dormancy. These data suggest that biological functions associated with these genes are involved in the maintenance of seed dormancy. Expression of genes encoding protein synthesis/activity inhibitors was significantly repressed during after-ripening, leading to dormancy decay. Imbibing AR seeds led to transcriptional activation of distinct biological processes, including those related to DNA replication, nitrogen metabolism, cytoplasmic membrane-bound vesicle, jasmonate biosynthesis and cell wall modification. These after-ripening-mediated transcriptional programs appear to be regulated by epigenetic mechanisms. Clustering of our microarray data produced 16 gene clusters; dormancy-specific probesets and abscisic acid (ABA)-responsive elements were significantly overrepresented in two clusters, indicating the linkage of dormancy in wheat with that of seed sensitivity to ABA. The role of ABA signalling in regulating wheat seed dormancy was further supported by the down-regulation of ABA response-related probesets in AR seeds and absence of differential expression of ABA metabolic genes between D and AR seeds.

*Correspondence

email b_ayele@umanitoba.ca

Keywords: wheat, gene expression, seed dormancy, after-ripening, seed imbibition, transcriptomics.

Introduction

Seed dormancy is one of the most important adaptive traits in plants that allow seeds to avoid environments unfavourable for germination. Excessive seed dormancy is an undesirable trait in crop plants, as it causes delay in germination and poor stand establishment (Derera, 1989), whereas induction of some degree of dormancy during seed development is desirable as it avoids precocious germination. Domesticated crop species such as wheat have undergone selection against dormancy to achieve quick and uniform germination (Simpson, 1990). This has resulted in modern cultivars that have low dormancy and are prone to field sprouting that causes loss in seed yield and quality. Thus, to develop wheat cultivars with moderate dormancy and thereby prevent field sprouting, it is important to understand molecular mechanisms underlying seed dormancy maintenance and release.

Dormancy is acquired during seed development and its maintenance can be regulated by complex interactions of many factors including plant hormones, light quality, temperature, nutrition and after-ripening (Finkelstein *et al.*, 2008). The role of after-ripening in breaking seed dormancy has been demonstrated in many species (Iglesias-Fernandez *et al.*, 2011), and this process requires seed moisture above a threshold value (Bair *et al.*, 2006; Foley, 2008), thus does not occur in very dry

seeds. Leubner-Metzger (2005) showed the existence of localized hydrated regions in dry after-ripened (AR) tobacco seeds that can allow transcriptional and translational activities to take place. Indeed, changes in gene expression have been observed in seeds of tobacco during storage (Bove *et al.*, 2005; Leubner-Metzger, 2005). It has also been demonstrated that seed stored rather than *de-novo* synthesized mRNAs play key roles during germination (Rajjou *et al.*, 2004). Consistently, over half of the mRNA species represented in the GeneChips of *Arabidopsis thaliana* (Arabidopsis), rice and barley were detected in their respective dry non-dormant seeds (An and Lin, 2011; Howell *et al.*, 2009; Nakabayashi *et al.*, 2005). The highly abundant mRNA species found in the dry non-dormant seeds of barley encode proteins related to nutrient reservoir, stress tolerance, protein biosynthesis, glycolysis, lipid metabolism and oxidoreduction, and are conserved with those of Arabidopsis (An and Lin, 2011). However, very little is known about molecular mechanisms underlying the role of after-ripening in triggering developmental switches from dormancy to germination in wheat seeds.

Under favourable environmental conditions, AR seeds germinate following imbibition, and differential gene expression patterns have been reported between dormant (D) and AR seeds of Arabidopsis and barley (Barrero *et al.*, 2009; Cadman *et al.*, 2006). Higher expressions of genes for cell wall modification, reserve mobilization and many protein synthesis factors were evident in imbibing AR than D seeds, and these biological processes are proposed to be linked with dormancy release rather than induction of germination. For example, genes coding for cell wall modifying enzymes such as glucan endo-1,3-β-glucosidases, xyloglucan endotransglycosylases and expansins exhibited up-regulation in the coleorhiza of AR relative to that of D seeds in barley (Barrero et al., 2009). Moreover, the germination of non-dormant seeds is associated with down-regulation of trypsin and α-amylase inhibitor proteins at both transcript and protein levels (Mak et al., 2009; Potokina et al., 2002). In contrary, D seeds are enriched with genes involved in abscisic acid (ABA) biosynthesis, gibberellin (GA) catabolism, stress response and genes with repressed translation capacity (Bassel et al., 2011; Cadman et al., 2006; Rajjou et al., 2004). Furthermore, many genes differentially expressed between D and non-dormant/AR seeds possess ABA-responsive motifs (Bassel et al., 2011; Cadman et al., 2006), providing an insight into the transcriptional regulatory mechanisms underlying seed dormancy.

Previous studies have shown the significance of epigenetic mechanisms in controlling seed dormancy and germination. Genome-wide gene expression analysis during Arabidopsis seed germination has provided an insight into the involvement of epigenetic mechanisms such as methylation in regulating seed transcriptional programs (Nakabayashi et al., 2005). Indeed, a study by Liu et al. (2007) showed that mutation in HISTONE MONOUBIQUITINATION1 (HUB1) gene, which is necessary for histone H2B monoubiquitination, results in reduced seed dormancy. Polycomb repressive complex 2 (PRC2), which plays a key role in regulating epigenetic states by catalysing histone H3 lysine 27 trimethylation (H3K27me3), is also found to repress genes controlling seed dormancy (Bouyer et al., 2011). Furthermore, histone deacetylases HDA6 and HDA19 are implicated in controlling germination by repressing embryonic properties (Tanaka et al., 2008).

Abscisic acid and GA are reported to be major regulators of seed dormancy and germination (Finch-Savage and Leubner-Metzger, 2006). Several studies have demonstrated that dormancy is associated with embryo ABA level (Gubler et al., 2005). However, in wheat no direct correlation between dormancy and embryo ABA levels could be established (Morris et al., 1989; Walker-Simmons, 1987), although increased capacity of ABA biosynthesis in the embryo is required for induction of dormancy during seed development (Garello and Le Page-Degivry, 1999). After-ripening of D seeds of wheat causes decay of dormancy mainly via loss of seed sensitivity to ABA (Corbineau et al., 2000; Walker-Simmons, 1987), and this is supported by studies that involved hexaploid wheat mutants and demonstrated a strong association between loss of embryo sensitivity to ABA and decay of seed dormancy (Kawakami et al., 1997; Noda et al., 2002; Rikiishi and Maekawa, 2010). Whereas GA promotes seed germination mainly through destabilizing the growth-repressing DELLA proteins (McGinnis et al., 2003), and its accumulation is associated with dormancy release or germination.

Other plant hormones such as jasmonate (JA) and auxin have also been implicated in regulating seed dormancy. In Arabidopsis, the expression of JA biosynthetic genes and endogenous JA and JA-Ile levels have been found to be higher in dry and imbibing non-dormant than D seeds (Preston et al., 2009). After-ripening of D barley seeds also induced the expression of two JA biosynthetic and one signalling gene (Barrero et al.,

2009). As the JA-insensitive mutants, jin4 and jar1, showed increased sensitivity to ABA (Berger et al., 1996), it has been proposed that JA stimulates dormancy decay by decreasing sensitivity of seeds to ABA. Although auxin has been suggested to complement ABA during wheat seed germination (Ramaih et al., 2003) and seeds from auxin-insensitive mutants exhibit reduced dormancy (Rousselin et al., 1992), its role in seed dormancy and germination is still unclear.

Global gene expression analysis has been used in Arabidopsis and barley as a powerful approach to identify genes controlling seed dormancy and its release by after-ripening, providing an insight into the physiological and metabolic states of D and AR seeds. To date, however, there has been no study that investigates genes involved in regulating seed dormancy and its release by after-ripening in dry and imbibing states in wheat. To this end, we studied transcriptional changes in D and AR wheat seeds before and after imbibition by using Affymetrix GeneChip Wheat Genome Array (Affymetrix, Santa Clara, CA). Moreover, available transcriptomic data of Arabidopsis and barley were used to assess unique and conserved transcriptional regulatory mechanisms controlling dormancy in wheat seeds.

Results and discussion

Effect of after-ripening on the germination of dormant seeds

The germinability of D and AR seeds of common wheat (Triticum aestivum L.) cultivar AC Domain was compared. The coleorhiza was visible in 95% of the AR seeds after 24-h imbibition at room temperature in darkness, but in none of the D seeds (Figure 1). Protrusion of seminal roots through the coleorhiza was evident in the AR seeds by 36 hours after imbibition (HAI), when there was still no coleorhiza emergence from the D seeds. We monitored the germination of D seeds through 7 days after imbibition and only 20% of the D seeds completed their germination.

Changes in transcriptomic profile during after-ripening of air-dry dormant seeds

As mRNAs stored in a seed might reflect its ability to maintain dormancy or germinate, we examined the mRNA species stored in air-dry D and AR seeds. The dry (0 HAI) D and AR seeds contained 38.0% and 38.7%, respectively, of the mRNAs represented on the GeneChip Wheat Genome Array at detectable level (Table 1), suggesting the sequestration of these mRNA species from seed developmental processes to function in dormancy maintenance/release processes. It has also been shown that non-dormant dry seeds of Arabidopsis, rice and barley contain over half of the mRNA species represented in their respective GeneChips at detectable level (An and Lin, 2011; Howell et al., 2009; Nakabayashi et al., 2005). As shown in Figure 2 and Table S1, after-ripening of the mature air-dry D seeds triggered differential expression of 58 probesets (36 up and 22 down), at cut-off values of twofold change and $P \leq 0.05$, indicating that the genes represented by these probesets play a key role in regulating seed dormancy release by after-ripening. However, the differential transcript abundance of most (54 of 58) of these probesets was subtle (only two to threefold change), reflecting the low transcriptional activity present in air-dry seeds (Bove et al., 2005; Leubner-Metzger, 2005). Only four probesets that

Figure 1 Dormant (lower-row) and AR (upper-row) seeds of common wheat cultivar AC Domain imbibed for 24 h.

Table 1 Expressed probes as a percentage of the total number of probes in wheat chip*

Sample	Hours after imbibition		
	0 (dry seed)	12	24
D	38.0 ± 0.4	44.1 ± 1.5	44.6 ± 1.1
AR	38.7 ± 1.0	45.4 ± 1.1	46.8 ± 0.2

*Probesets with a 'present' detection call in the original microarray hybridization data.

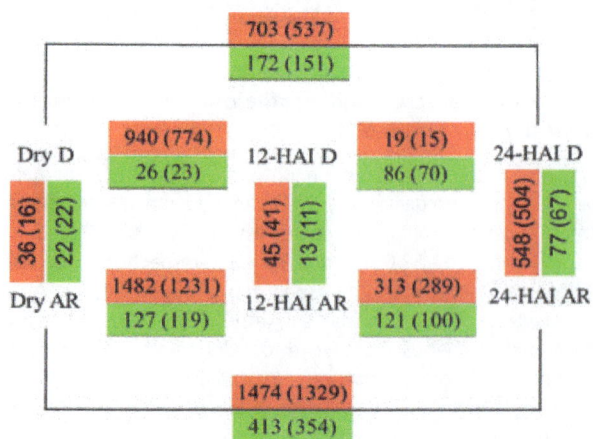

Figure 2 Total number of probesets and respective annotations (shown in bracket) significantly up (red) or down-regulated (green) at fivefold change and $P \leq 0.05$ cut-off values in the different comparisons (12-HAI D/dry D, 24-HAI D/12-HAI D, 24-HAI D/dry D, 12-HAI AR/dry AR, 24-HAI AR/12-HAI AR, 24-HAI AR/dry AR, 12-HAI AR/12-HAI D, 24-HAI AR/24-HAI D). Cut-off values of twofold change and $P \leq 0.05$ were used for dry AR/dry D comparison. Annotation of probesets was performed using HarvEST WheatChip.

correspond to three unigene sequences exhibited over three-fold differential expression (up-regulated in air-dry D relative to that in AR seeds, Table S1). As these probesets are more likely to have functional importance in regulating seed dormancy, they were annotated using HarvEST WheatChip (http://harvest.ucr.edu/) and the corresponding putative products and functions predicted based on sequence similarity with the closest homologous genes from other species. One of the probesets corresponds to the gene encoding CRUCIFE-RINA (CRA1) storage protein, a member of the downstream targets of ABA, and another probeset to a gene encoding unknown protein. It has been reported that expression of CRA1 decreases during the germination of non-dormant Arabidopsis seeds (Li et al., 2007). Thus, the higher expression of CRA1 homologue in dry D than AR wheat seeds may suggest its significance in dormancy maintenance. The other two probesets correspond to a gene encoding a protein synthesis inhibitor. Moreover, a probeset annotated in wheat as α-amylase/trypsin inhibitor (designated as CM1 because of its solubility in Chloroform: Methanol) showed a twofold higher expression in D than AR seeds (Table S1). As a decrease in the mRNA and protein levels of trypsin and α-amylase inhibitors was observed during wheat seed germination (Mak et al., 2009; Potokina et al., 2002), the transcriptional activation of these genes in D seeds highlights their importance in controlling seed dormancy in wheat. However, it should be noted that the predicted functions of wheat genes represented by the probesets may differ from the actual one, as similar sequences may have different functions across species.

Comparative transcriptomic analysis between imbibing D and AR seeds

Imbibition of the air-dry seeds for 12 h induced the expression of more genes in both D and AR samples, increasing the percentage of genes expressed at detectable level from 38.0% to 44.1% and 38.7% to 45.4% in D and AR seeds, respectively (Table 1). Although only subtle increases in the percentage of genes expressed were observed between 12 and 24 HAI in both D and AR seeds, our data analysis showed significantly more genes as being expressed in 24 HAI AR than D seeds ($P = 0.028$). To further examine changes in gene expression, comparative transcriptomic analyses were performed between imbibed and dry D, and imbibed and dry AR seeds. The total number of probesets differentially expressed in each comparison and their respective annotations were identified using a fivefold change and $P \leq 0.05$ cut-off values, and Harvest WheatChip, respectively (Figure 2). As shown in Figure 3, more probesets exhibited differential expression specific to D seeds within the first 12 HAI (194 up and 0 down) than between 12 and 24 HAI (nine up and 64 down), while a total of 325 probesets (277 up and 48 down) were differentially expressed after 24-h imbibition. Whereas 837 probesets (736 up and 101 down) exhibited differential expression specific to AR seeds within the first 12 HAI, 402 (303 up and 99 down) between 12 and 24 HAI, and a total of 1337 (1048 up and 289 down) following 24-h imbibition (Figure 3). These results indicate the role of after-ripening in initiating changes in transcriptional programs during seed imbibition. Changes in gene expression were also compared between imbibed AR and imbibed D

seeds at each time point. Only 58 probesets were differentially expressed between 12 HAI AR and D seeds (45 up and 13 down in AR seeds; Figures 2 and 3a,d), and further imbibition for an additional 12 h increased the number to 625 (548 up and 77 down; Figures 2 and 3c,f). This comparison resulted in 8 (two up and six down; Figure 3a,d) and 105 (77 up and 28 down; Figure 3c,f) probesets as specifically and differentially regulated by after-ripening after 12- and 24-h imbibitions, respectively, in addition to those probesets identified by the corresponding imbibed D/dry D and imbibed AR/dry AR comparisons. Analysis of the expression pattern of seven randomly selected differentially expressed probesets with qPCR showed a strong correlation with our microarray data (Table S2). As there are cases where several probesets on the Wheat Genome Array represent a single unique gene in the wheat EST database, the total number of differentially expressed unique genes would be less than that of the anno-

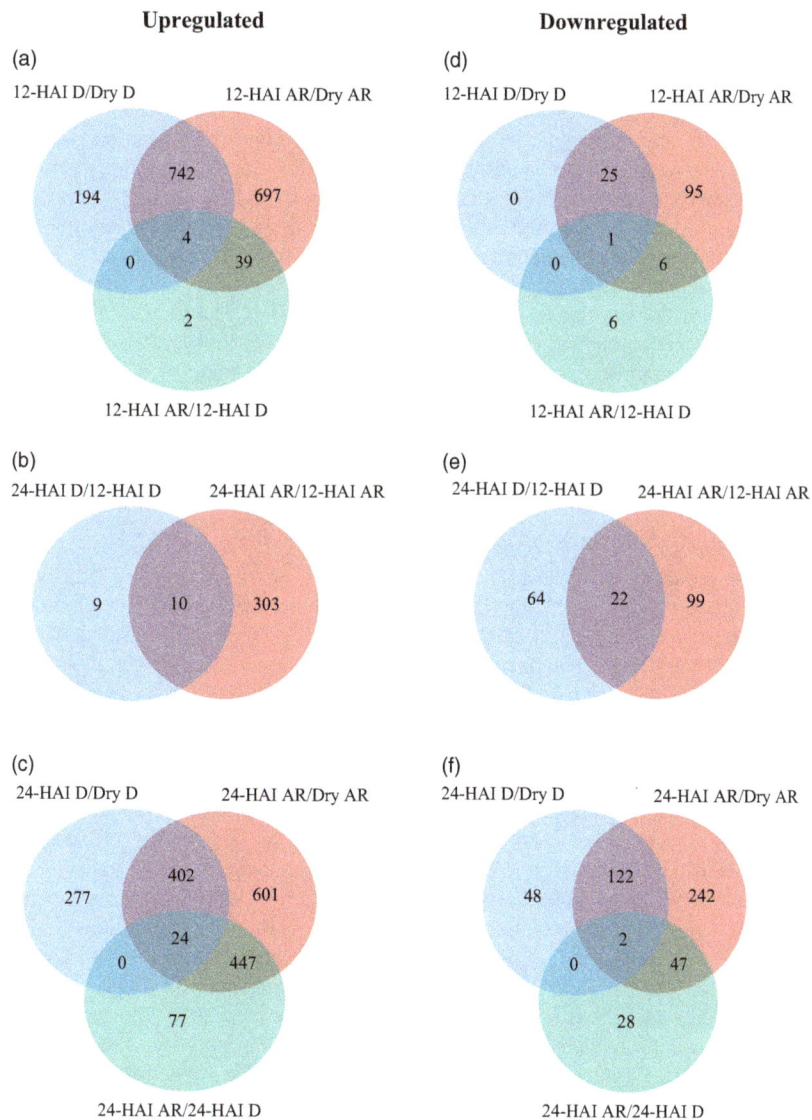

Upregulated

(a)

12-HAI D/Dry D   12-HAI AR/Dry AR

194   742   697

4

0   39

0

2

12-HAI AR/12-HAI D

Downregulated

(d)

12-HAI D/Dry D   12-HAI AR/Dry AR

0   25   95

1

0   6

6

12-HAI AR/12-HAI D

(b)

24-HAI D/12-HAI D   24-HAI AR/12-HAI AR

9   10   303

(e)

24-HAI D/12-HAI D   24-HAI AR/12-HAI AR

64   22   99

(c)

24-HAI D/Dry D   24-HAI AR/Dry AR

277   402   601

24

0   447

77

24-HAI AR/24-HAI D

(f)

24-HAI D/Dry D   24-HAI AR/Dry AR

48   122   242

2

0   47

28

24-HAI AR/24-HAI D

Figure 3 Probesets differentially expressed in imbibed/dry [12 HAI D/Dry D and 12 HAI AR/Dry AR (a,d), 24 HAI D/Dry D and 24 HAI AR/Dry AR (c,f)] and imbibed/imbibed [12 HAI AR/12 HAI D (a,d), 24 HAI D/12 HAI D and 24 HAI AR/12 HAI AR (b,e), 24 HAI AR/24 HAI D (c,f)] comparisons of D and AR samples. Venn diagram of the number of significantly upregulated (a–c) and significantly downregulated (d–f) probesets in each comparison at cutoff values 5-fold change and $P \leq 0.05$. The intersections (areas of overlap) indicate probesets regulated in common.

tated probesets (Figure 2). Overall, these data reflect the role of after-ripening in inducing imbibition-mediated repression or activation of genes responsible for biological processes associated with dormancy maintenance or release, respectively, and thereby promoting germination.

Biased gene ontology of probesets regulated by imbibition

Differential expression of 772 probesets (746 up and 26 down; Figure 3a,d) was observed within the first 12 HAI irrespective of seed dormancy status, but only 32 probesets (10 up and 22 down; Figure 3b,e) between 12 and 24 HAI. Imbibing the seeds for 24 h induced differential expression of 550 probesets (426 up and 124 down; Figure 3c,f) that are common to both D and AR samples. Thus, it is not likely that the biological processes represented by these probesets are functionally involved in the breaking of seed dormancy. Our analysis showed that those imbibition-specific up-regulated probesets are enriched in gene ontology (GO) category of DNA integration (GO: 0015074, $P \leq 4.1e{-}09$; Table S3), whereas the down-regulated probesets are enriched in nutrient reservoir activity (GO: 0045735, $P = 1.8e{-}30$; Table S3). Twelve of the nutrient reservoir activity-related probesets correspond to genes encoding gliadin, and three probesets to genes encoding glutenin. These two proteins constitute approximately 80% wheat seed storage protein (Payne et al., 1982), which upon hydrolysis will supply part of the energy required for germinating seeds. Interestingly, degradation of transcripts encoding some seed storage proteins was also observed during the germination of barley seeds (Potokina et al., 2002). Thus, it is likely that transcriptional suppression of probesets related to storage proteins is employed by imbibing seeds as a mechanism to conserve energy and nutrients to meet the post-germination demands.

Chromatin assembly is associated with seed dormancy release by after-ripening

Comparison of changes in gene expression between dry and 12 HAI seeds showed that imbibition induced up-regulation/down-regulation of 736/101 and 194/0 probesets specifically in AR and D seeds (Figure 3a,d), respectively, while only 303/99 and 9/64 between 12 and 24 HAI (Figure 3b,e). Imbibition over 24 h induced up-regulation/down-regulation of 1048/289 and 277/48 probesets specifically in AR and D seeds, respectively (Figure 3c,f). No enriched GO category was found for the probesets differentially expressed in D seeds at any imbibition period. However, those probesets up-regulated specifically in the AR seeds between 12 and 24 HAI or following 24 HAI are overrepresented in GO categories of chromatin assembly (GO: 0031497, $P \leq 2.5e{-}18$; Table S4) and DNA replication (GO: 0006260, $P \leq 6.0e{-}06$; Table S4). Over 23 chromatin assembly-related probesets showed similarity with genes encoding histone proteins H4, HTA11, HTA12, HTB11, HTB9, and FASCIATA1, a histone binding protein (Table S4). This is in agreement with up-regulation of histone genes during the late germination phase of barley seeds (An and Lin, 2011) and consistently high expression of SET family transcription factors with a role in histone methylation (Malagnac et al., 2002; Xiao et al., 2003) in germinating rice seeds (Howell et al., 2009). As histone proteins are important for nucleosome and chromatin formation, their transcriptional activation by after-ripening may indicate their role in the modulation of chromatin structure and regulation

of gene expression for seeds to switch from dormant to non-dormant status.

Evidences for the role of histone modification and chromatin remodelling in the regulation of seed dormancy and germination are emerging. For example, HUB1 activates the transcription of several dormancy-related genes, and its mutation causes reduced seed dormancy (Liu et al., 2007). Whereas PRC2 represses dormancy-related genes and its null mutant exhibits enhanced dormancy and germination defects (Bouyer et al., 2011). Based on studies that involved a chemical inhibitor and repression lines, histone deacetylases, HDA6 and HDA19, have been shown to regulate germination via repressing specific transcription factors LEC1, FUS3 and ABI3 that are responsible for embryonic properties (Tanaka et al., 2008). Consistently, genes susceptible to DNA methylation exhibit differential expression during germination (Nakabayashi et al., 2005). Probesets corresponding to genes coding for DNA and histone modification enzymes were also found to be up-regulated in AR seeds of wheat. Three probesets showing 17-fold higher expression following 24 HAI in AR than D seeds correspond to CMT3 (encoding chromomethylase 3) and MET1 (encoding methyltransferase 1), which are reported to have roles in epigenetic regulation of gene expression (Berr and Shen, 2010). Furthermore, a probeset that corresponds to a gene encoding a jmjC domain-containing transcription factor exhibited twofold higher expression in dry AR than in dry D seeds. The jmjC domain is involved in the removal of histone methyl groups (Tsukada et al., 2006), and thereby take part in epigenetic regulation of gene expression (Chen et al., 2011). The fact that this epigenetic-related probeset is differentially expressed between unimbibed AR and D seeds suggests that histone and chromatin modifications occur during after-ripening of dry dormant seeds, and possibly mediate the release of dormancy (Liu et al., 2007). However, if these modifications are the cause or effect of after-ripening is unclear.

Other biased gene ontology of probesets regulated by after-ripening

Probesets up-regulated specifically in AR seeds during the first 12 HAI are enriched in the GO class of cellular nitrogen compound metabolic process (GO: 0034641, $P = 7.2e{-}06$; Table S4). Such nitrogen metabolism genes are also activated in germinating barley seeds (Sreenivasulu et al., 2008) and have a potential role of supplying nitrogen to the developing embryo. Whereas those up-regulated between 12 and 24 HAI and following 24 HAI are also overrepresented in the GO class of cytoplasmic membrane-bound vesicle (GO: 0016023, $P \leq 6.1e{-}15$; Table S4). As vesicles are involved in multiple biological processes including storage, transport and digestion of cellular products and wastes, our data indicate the potential role of after-ripening in inducing cellular energy metabolism and transport in imbibing seeds. Interestingly, three probesets in this GO category correspond to a gene encoding GLS1_ARATH (a ferredoxin-dependent glutamate synthase 1) that is involved in oxido-reductase activity (Lea and Miflin, 2003), reflecting the role of cellular signals associated with reactive oxygen species in seed dormancy decay (El-Maarouf-Bouteau and Bailly, 2008).

Seed dormancy has been shown to be strongly associated with stress and ABA-responsive genes (Bassel et al., 2011). Consistently, probesets down-regulated specifically in AR seeds during the first 12 HAI are enriched in GO category for response to stress (GO: 0006950, $P = 2.2e{-}06$; Table S4), whereas those down-regulated between 12 and 24 HAI and following 24 HAI are enriched in the GO category for ABA response (GO: 0009737, $P \leq 7.4e{-}10$; Table S4). One of the ABA response-related probesets up-regulated following 24-h imbibition corresponds to a gene encoding G-box binding transcription factor 1 (GBF1) that can bind to an ABA-responsive CACGTG motif shown to be enriched in genes differentially expressed between D and non-dormant/AR seeds (Bassel et al., 2011; Cadman et al., 2006). Furthermore, two probesets correspond to genes encoding environmental stress response proteins, ATHVA22 and a hydrophobic protein, whereas nine other to genes encoding late embryogenesis abundant (LEA) proteins and the down-regulation of these probesets can be associated with loss of desiccation tolerance in germinating AR seeds (An and Lin, 2011). As all these probesets represent genes regulated by ABA, their transcriptional suppression may suggest that one mechanism by which after-ripening mediates dormancy release and subsequent germination is by repressing ABA activated functions. This conclusion is consistent with physiological studies that showed the association of seed dormancy with ABA sensitivity rather than embryo ABA levels (Morris et al., 1989; Walker-Simmons, 1987).

Expression of probesets involved in plant hormone metabolism

Plant hormones such as ABA, GA and JA are reported to be involved in regulating seed dormancy and germination. The expression of ABA metabolic genes has been shown to be correlated with the level of seed dormancy (Gubler et al., 2008; Millar et al., 2006; Okamoto et al., 2006). However, no significant differential expression was evident in probesets representing genes involved in ABA metabolism including those that correspond to the Arabidopsis NCED and ABA8'OH genes (Table S5), suggesting the absence of direct correlation between wheat seed dormancy and embryo ABA levels (Walker-Simmons, 1987). Although GA is not required for the germination of cereal seeds (Ueguchi-Tanaka et al., 2005), the approximately two to threefold up-regulation of two probesets that correspond to a gene encoding GA 20 oxidase (Table S5), which is involved in GA synthesis, during imbibition of AR seeds suggest the role of GA in enhancing the germination process in AR seeds.

The wheat GeneChip contains 19 probesets identified as JA biosynthetic genes (Table S5), four of which exhibited approximately 3- or more fold up-regulation within the first 12 HAI of AR relative to D seeds. These probesets correspond to genes encoding 12-oxophytodienoate reductase 1 (OPR1) and OPR2 that are involved in the early part of the JA biosynthetic pathway. Furthermore, drastic up-regulation of two other JA-related probesets was induced after 24-h imbibition of AR seeds, one corresponding to a gene encoding OPR1 (17-fold) and the other one encoding LIPOXYGENASE 5 (LOX5; 10-fold) that catalyses the first committed step of JA biosynthesis. In agreement with our data, LOX6 and OPR3 of Arabidopsis exhibited higher expression in the non-dormant Col-0 than dormant Cvi seeds (Preston et al., 2009). Consistently, the dry and imbibed seeds of Col-0 contained more JA and JA-Ile than those of Cvi. Moreover, the barley JA metabolic genes, OPR and ALLENE OXIDE SYNTHASE (AOS), are also shown to be up-regulated in coleorhiza of AR seeds (Barrero et al., 2009). As seeds from JA-insensitive mutants of Arabidopsis (jin4 and jar1) exhibit increased sensitivity to ABA-mediated inhibition of germination (Berger

et al., 1996), it has been proposed that JA stimulates the germination of AR seeds by conferring them resistance to ABA.

Probesets that represent genes in auxin biosynthesis are found in the KaPPA-view4 wheat database (Sakurai et al., 2011). Two probesets representing genes encoding tryptophan synthase and indole-3-glycerol phosphate synthase (Table S5), which are involved in the synthesis of tryptophan (a precursor for the synthesis of auxin), were up-regulated 32- to 43-fold in 24 HAI AR relative to D seeds. Although the expression profile of these genes suggests change in auxin level during imbibition of AR seeds, treatment of non-dormant wheat embryos with tryptophan and IAA, which strongly inhibited the germination of embryos from dormant seeds, did not affect their germination (Ramaih et al., 2003) indicating the significance of auxin sensitivity with respect to seed dormancy release rather than embryo auxin level. In agreement with this, several auxin-insensitive mutants of Nicotiana plumbaginifolia exhibit reduced seed dormancy (Rousselin et al., 1992).

Probesets involved in carbohydrate metabolic processes

Germination-related biological processes such as radicle protrusion and mobilization of storage reserves are mediated by a number of enzymes (Bewley and Black, 1994). Cell wall synthesis, degradation and modification play important roles during seed germination (Kucera et al., 2005). For example, a probeset corresponding to a cell wall–related gene encoding endoxyloglucan transferase A4 exhibited over fivefold higher expression in AR than D seeds within the first 12 HAI, and further imbibition for 12 h induced over fourfold higher expression of five probesets corresponding to genes encoding xyloglucan endotransglycosylase 7, basic chitinase and beta-D-xylosidases in AR than D seeds (Table S5). Similarly, genes related to cell wall modification have been shown to be up-regulated in imbibing AR barley seeds (Barrero et al., 2009) and during the germination of non-dormant barley and rice seeds (An and Lin, 2011; Howell et al., 2009). The coleorhiza of imbibing cereal seeds is proposed to act as a barrier to radicle elongation in a similar manner to the mature endosperm in dicot seeds (Barrero et al., 2009). Thus, our data suggest that after-ripening of D wheat seeds activates transcriptional programs underlying cell wall degradation and remodelling to facilitate coleorhiza weakening, and radicle cell expansion and protrusion during imbibition, reflecting the functional significance of cell wall genes in breaking seed dormancy.

Embryos must mobilize storage reserves, which is mainly starch in cereal seeds, to germinate and grow (Finkelstein et al., 2008). Forty-four probesets related to carbohydrate metabolic process (GO: 0005975, $P = 8.8e{-}07$; Table S4) such as those involved in starch and maltose degradation and glycoside hydrolysis were up-regulated specifically in 12 HAI AR seeds, and further imbibition for 12 h increased the number of such probesets to 69 (GO: 0005975, $P = 2.3e{-}07$; Table S4). For example, four probesets that correspond to genes encoding starch degrading α-amylase and maltose degrading α-glucosidase 1 showed approximately 6- to 41-fold up-regulation following 24-h imbibition in AR relative to D seeds (Table S5). Consistent with this result, a dramatic increase in α-amylase activity was observed between 18 and 33 HAI of non-dormant barley seeds (An and Lin, 2011). Furthermore, eight probesets corresponding to genes involved in glycoside hydrolysis were up-regulated approximately threefold or more mainly following 24-h imbibition in AR relative to D seeds (Table S5). Our results imply the significance of after-ripening in activating storage reserve mobilization pathways to provide energy for embryo growth. Previous studies have also shown that the germination of non-dormant seeds of barley and rice is accompanied by up-regulation of starch metabolism-related genes (An and Lin, 2011; Howell et al., 2009).

Gene clustering

Gene clustering analysis was performed by extracting the 3067 probesets that exhibited differential expression at cut-off values of fivefold change and $P \leq 0.05$ from the original microarray data of dry and imbibed D and AR seeds. Using K means/medians clustering (KMC) algorithm of Mev (Saeed et al., 2006), these probesets were grouped into 16 clusters (Figures 4 and S1). Only 58 probesets showed differential expression, although subtle, between dry AR and D seeds (Table S1). As biological processes represented by these probesets might play important roles in the after-ripening-mediated release of seed dormancy, they were considered as an independent group. Manual clustering of the 58 probesets led to their assignment mainly into the Mev generated cluster 9 (Figures 4 and 5). Furthermore, 36 probesets identified as regulated by the main factor dormancy from a Flexarray-based two-way ANOVA (main factors-dormancy and imbibition) analysis of gene expressions in the six samples (dry D, 12 and 24 HAI D, dry AR, 12 and 24 HAI AR seeds; Table S6) were also considered as an independent group. These probesets were shown to be grouped mainly into the Mev generated cluster 4 (Figures 4 and 5). Annotation of the

probesets using HarvEST WheatChip revealed that 227 probesets in cluster 4 were annotated by 130, and 281 probesets in cluster 9 by 122 rice genes. As transcriptional regulations are mainly mediated by transcription factor proteins, the 1-kb upstream region of the resulting rice genes (e ≤ 1e−30) were analysed by Osiris (http://www.bioinformatics2.wsu.edu/cgi-bin/Osiris/cgi/visualize_select.pl; Morris et al., 2008) to identify the type of transcriptional factors that interact with these genes (Table 2). MYC transcription factor binding sites CATGTG (MYCATERD1) and CACATG (MYCATRD22) are enriched in probesets grouped in cluster 4. The MYCATRD22 motif has been shown to be enriched in genes up-regulated by ABA and take part in controlling seed dormancy (Yazaki et al., 2004). Five other motifs (ABREOSRAB21, G-box-like [CACGTG], perfect-type HSE, motifA and ACGTABREMOTIFA2OSEM) are over-represented in probesets grouped in cluster 9. Consistently, promoters of genes differentially expressed between D and non-dormant/AR seeds are enriched with ACGTABREMOTIFA2OSEM and CACGTG motifs (Bassel et al., 2011; Cadman et al., 2006). The ABREOSRAB21 and ACGTABREMOTIFA2OSEM motifs are upstream ABA-responsive elements, while the CACGTG motif (which acts as a binding site for GBFs) contains the core motif ACGT that is characteristic to ABA-responsive elements, suggesting that the expression of genes in cluster 9 might be controlled by ABA and play a role in the regulation of seed dormancy. Thus, it is likely that transcription factors that bind to these motifs act as upstream regulators of seed dormancy release by after-ripening.

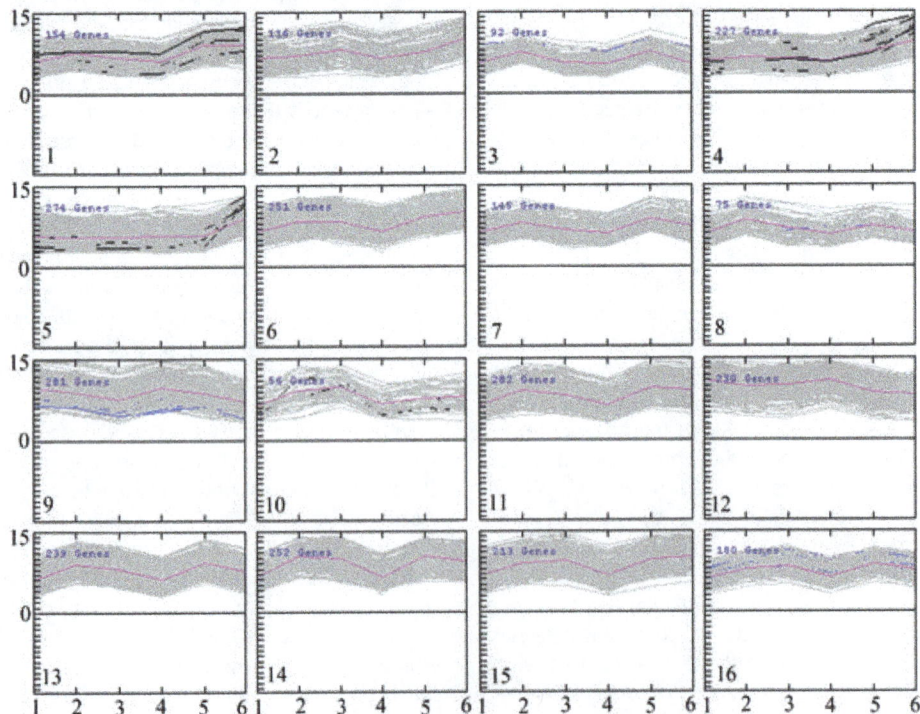

Figure 4 K means/medians clustering by expression pattern of probesets differentially regulated in the six seed samples (cut-off values fivefold change and $P \leq 0.05$). The number of probesets grouped into each cluster is indicated on the graph, which depicts their expression pattern (X axis: 1, dry D; 2, 12 HAI D; 3, 24 HAI D; 4, dry AR; 5, 12 HAI AR; 6, 24 HAI AR; Y axis: relative mRNA level with respect to the mRNA level in dry D seeds). Probesets identified as regulated preferentially by the main factor dormancy [from the two-way ANOVA analysis of probesets expressed in the six samples (main factors-dormancy and imbibition)] were grouped mainly into cluster 4, and their expression pattern is shown in black. Whereas probesets differentially expressed between dry D and dry AR seeds (twofold change and $P \leq 0.05$ cut-off value) were grouped into clusters 9, and their expression pattern is shown in blue.

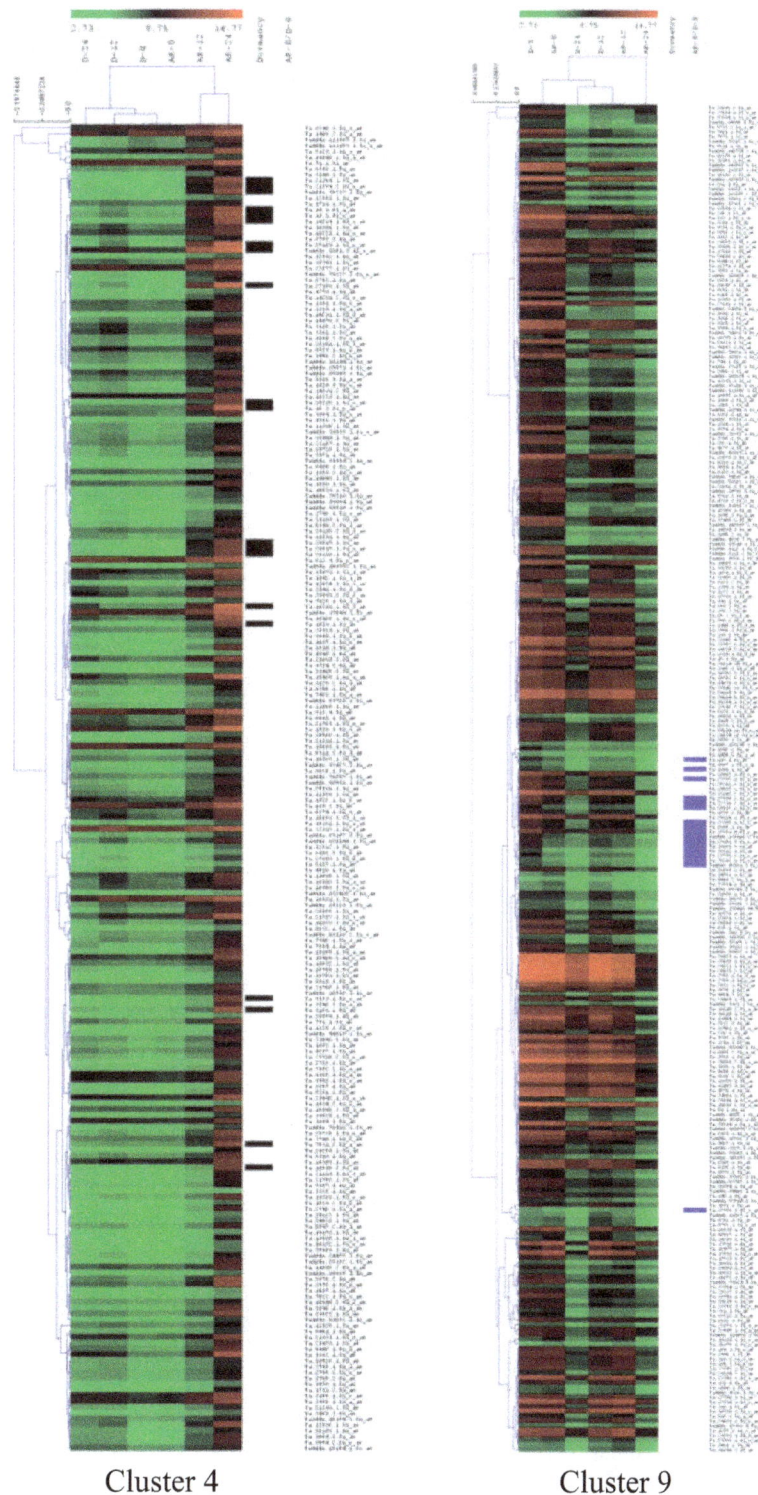

Figure 5 Hierarchical tree of clusters 4 and 9 generated based on similarities in expression profile of probesets differentially expressed in the six seed samples (at cut-off values fivefold change and $P \leq 0.05$). Probesets identified as regulated specifically by dormancy (from the two-way ANOVA analysis), and those differentially expressed between dry D and dry AR seeds (twofold change and $P \leq 0.05$ cut-off value) are shown in black and blue, respectively, and their names are indicated at right.

Comparison with transcriptomic studies of dormancy and after-ripening in other species

Transcriptomics of D and AR seeds have been studied in other species such as Arabidopsis (Cadman et al., 2006) and barley

(Barrero et al., 2009). Probesets that exhibited differential expression between imbibed AR and imbibed D seeds, at a cut-off value of fivefold change, were extracted from both barley (coleorhiza AR/coleorhiza D at 8 and 18 HAI) and Arabidopsis (DL/PD24-imbibed AR/imbibed D at 24-h imbibition). A total of

Table 2 Sequences of cis-elements overrepresented in clusters 4 and 9*

Cluster #	Motif sequence	Motif ID	P-value[†]
4	CATGTG	MYCATERD1	$<10^{-3}$
	CACATG	MYCATRD22	$<10^{-3}$
9	CACGTG	G-box-like	$<10^{-4}$
	ACGTSSSC (S = G/C)	ABREOSRAB21	$<10^{-3}$
	NTTCNNGAANNTTCN (N = A/T/C/G)	perfect-type HSE	$<10^{-4}$
	GACGTGTC	motifA	$<10^{-3}$
	ACGTGKC	ACGTABREMOTIF A2OSEM	$<10^{-3}$

*Cis-elements were identified using the OSIRIS web-based resource.
[†]Significance was determined at $P < 0.001$.

143 and 183 probesets were identified to be up-regulated by after-ripening in the coleorhiza of barley at 8 and 18 HAI, respectively, and translation of these barley probesets into their wheat equivalents by microarray platform translator (Wise et al., 2007) produced 131 and 151 unique wheat probesets, respectively. Comparison of the 131 probesets with those up-regulated in our 12 HAI AR (738 probesets; Figure 3a) and the 151 probesets with those up-regulated in the 24 HAI AR (1125 probesets; Figure 3c) wheat seeds identified 10 and 47 putative wheat-barley orthologues as up-regulated specifically by AR, respectively (Table S7). As only one orthologue is shared by these comparisons, our data indicate that after-ripening-mediated dormancy release regulatory mechanisms expressed at early and late phases of imbibition are distinct but shared between species. This result also reflects that wheat and barley, although closely related, have their own unique molecular mechanisms underlying dormancy release. In Arabidopsis, 156 genes were identified as up-regulated by after-ripening (at cut-off value of fivefold change). The 1125 probesets specifically up-regulated in our 24 HAI AR wheat seeds were annotated by 578 Arabidopsis genes. Comparison of these annotated probesets to the 156 Arabidopsis genes identified 14 wheat-Arabidopsis orthologues, which correspond to 27 probesets in the wheat Genechip, as up-regulated by after-ripening. Similarly, orthologues up-regulated in imbibing wheat embryos of dormant and non-dormant seeds have been shown to be associated with groups of Arabidopsis genes down and up-regulated, respectively, by after-ripening (Bassel et al., 2011). Although the divergence between monocots and dicots occurred 200 million years ago (Wolfe et al., 1989), these data show that specific transcriptional switches and molecular mechanisms underlying dormancy release by after-ripening are conserved between species. Furthermore, 64% of the annotated probesets up-regulated in our 24 HAI AR whole seed samples are shared by those up-regulated in the 20 HAI embryos of non-dormant wheat seeds (Bassel et al., 2011), indicating that most of the AR regulated wheat genes are expressed in the embryo. It is, however, possible that some of these genes are co-expressed in the endosperm/aleurone fraction.

In summary, transcripts of many genes synthesized during seed development are sequestered in mature D seeds, most of which likely take part in the regulation of seed dormancy. The distinct transcriptional programs activated in response to after-ripening explain the developmental switch from dormancy to germination. Both unique and conserved transcriptional cascades and molecular mechanisms underlie seed dormancy decay in wheat.

Experimental procedures

Plant materials and growth conditions

The cultivar used in this study is AC Domain, hard red spring wheat (T. aestivum L.) with high levels of pre-harvest sprouting tolerance and well adapted to the Canadian prairies (Townley-Smith and Czarnecki, 2007). To generate D seeds, wheat plants were grown in a greenhouse at 18–22 °C/14–18 °C (day/night) under a 16/8-h photoperiod until harvest. Spikes were harvested at maturity, and then seeds from the middle region of each spike were threshed by hand. Subsequently, one-half of the seeds were stored at −80 °C to retain dormancy. To obtain AR seeds, the remaining portion was first stored at room temperature for 10 months, and then at −80 °C until further use.

Germination assay

Dormant and AR seeds were surface sterilized in 5% sodium hypochlorite solution for 25 min, and rinsed five times with sterile deionized water. The surface sterilized seeds were then placed in a 9-cm sterile Petri plate (25 seeds per plate), on a sterile Whatman #1 filter paper wetted with 7 mL of sterile deionized water, and covered with a second layer of filter paper. The plates were sealed with parafilm and then placed at room temperature in darkness for 12 and 24 h. For germination analysis, seeds were scored germinated when emergence of the coleorhiza beyond the seed coat was visible (Barrero et al., 2009). Imbibed D and AR seeds were harvested in liquid nitrogen and stored at −80 °C for RNA extraction. For genome-wide profiling of mRNA species stored in air-dry seeds, RNA was extracted from unimbibed D and AR seeds.

DNA microarray analysis

Total RNA was extracted from three independent biological replicates of each sample using RNAqueous columns (Ambion, Austin, TX) as recommended by the manufacturer. The total RNA was subjected to mRNA isolation using PolyATtract kit (Promega, Madison, WI). For microarray analysis, mRNA samples were labelled and hybridized to the Affymetrix GeneChip Wheat Genome Array (Affymetrix) as described in Jordan et al. (2007). The Affymetrix GeneChip Wheat Genome Array contains 61 127 probesets representing 55 052 transcripts for all 42 chromosomes in the wheat genome. Each probeset contains 11 probe pairs of 25-mer oligonucleotides.

Data analysis

The hybridized microarrays were washed and scanned following manufacturers' instructions. Using the Affymetrix GeneChip Operating Software, the data from the 11 probe pairs were converted into a single hybridization intensity level per probeset and represented in CEL file format. Total signal intensity per chip was adjusted using the 50th percentile of all measurements. The Affymetrix Microarray Suite (MAS5) statistical algorithm was used to determine the number of probesets with a 'present' detection call for each sample. Reproducibility of the data from the three independent biological replicates was verified by scatter plot expression analysis. Following normalization of the raw data by Robust Multi-array Average (RMA) methodology, FlexArray software (http://genomequebec.mcgill.ca/FlexArray, Blazejczyk et al., 2007) was used to identify differentially expressed probesets by analysis of variance (ANOVA, $P \leq 0.05$). The resulting data were filtered to exclude genes with less than fivefold change and $P > 0.05$. Two-way ANOVA (main factors-dormancy and imbibition) was used to identify the probesets regulated by main factor dormancy in the six samples: dry D, 12- and 24-h imbibed D, dry AR and 12- and 24-h imbibed AR seeds (cut-off values of fivefold change and $P \leq 0.05$). Predicted GO for each probeset was obtained using the AgriGO analysis toolkit (http://bioinfo.cau.edu.cn/agriGO; Du et al., 2010), and candidate gene annotation was performed using HarvEST Wheat-Chip (http://harvest.ucr.edu/). MultiExperiment Viewer (MeV version 4.6; Saeed et al., 2006) software was used to generate a heat map of selected genes, and promoter sequences were analysed by Osiris (http://www.bioinformatics2.wsu.edu/cgi-bin/Osiris/cgi/visualize_select.pl; Morris et al., 2008). Microarray platform translator (http://www.plexdb.org/modules/MPT; Wise et al., 2007) software was used to translate barley probesets into their wheat equivalents.

qPCR validation of microarray results

Validation of the microarray results was performed with seven randomly selected differentially expressed genes using quantitative PCR. The same mRNA samples used for microarrays analysis were digested with DNase (DNA-free kit; Ambion) to eliminate any contamination with genomic DNA. Reverse transcription was performed using the qScript™ cDNA SuperMix Kit (Quanta Biosciences, Gaithersburg, MD). In brief, 100 ng mRNA was mixed with 4 μL of 5× qScript cDNA supermix and nuclease-free water, with a total reaction volume of 20 μL. The mixture was subjected to incubation for 5 min at 25 °C followed by 1 h at 42 °C after which the reaction was terminated by incubation at 85 °C for 5 min. Real-time PCR assays were performed on a CFX96 real-time PCR system (Bio-Rad, Hercules, CA) using a Maxima SYBR Green qPCR Master Mix (Fermentas, Glen Burnie, MD). For each 20 μL real-time PCR reaction, 4 μL of 1 : 20 diluted cDNA was mixed with 10 μL 2× SYBR Green qPCR Master Mix, 0.5 μL forward primer (10 μM; 250 nM final concentration), 0.5 μL reverse primer (10 μM; 250 nM final concentration) and 5 μL DEPC-treated water. Samples were subjected to the following thermal cycling conditions: DNA polymerase activation at 95 °C for 5 min followed by 40 cycles of denaturation at 95 °C for 15 s, annealing at 60 °C for 30 s and extension at 72 °C for 30 s in duplicate in 96-well optical reaction plates (Bio-Rad). Primers used for validation of microarray results of randomly selected probesets are listed in Table S2. Relative transcript abundance of the target genes was determined by $2^{-\Delta\Delta C_t}$ (Livak and Schmittgen, 2001) using wheat actin as the reference gene.

Acknowledgements

This work was supported by a grant from the Natural Sciences and Engineering Research Council of Canada and Western Grains Research Foundation to BTA. The authors would like to acknowledge Zhen Yao and Brenda Oosterveen for their technical assistance.

References

An, Y.Q. and Lin, L. (2011) Transcriptional regulatory programs underlying barley germination and regulatory functions of Gibberellin and abscisic acid. *BMC Plant Biol.* **11**, 105.

Bair, N.B., Meyer, S.E. and Allen, P.S. (2006) A hydrothermal after-ripening time model for seed dormancy loss in *Bromus tectorum* L. *Seed Sci. Res.* **16**, 17–28.

Barrero, J.M., Talbot, M.J., White, R.G., Jacobsen, J.V. and Gubler, F. (2009) Anatomical and transcriptomic studies of the coleorhiza reveal the importance of this tissue in regulating dormancy in barley. *Plant Physiol.* **150**, 1006–1021.

Bassel, G.W., Lan, H., Glaab, E., Gibbs, D.J., Gerjets, T., Krasnogor, N., Bonner, A.J., Holdsworth, M.J. and Provart, N.J. (2011) Genome-wide network model capturing seed germination reveals coordinated regulation of plant cellular phase transitions. *Proc. Natl Acad. Sci. USA*, **108**, 9709–9714.

Berger, S., Bell, E. and Mullet, J.E. (1996) Two methyl jasmonate-insensitive mutants show altered expression of AtVsp in response to methyl jasmonate and wounding. *Plant Physiol.* **111**, 525–531.

Berr, A. and Shen, W.H. (2010) Molecular mechanisms in epigenetic regulation of plant growth and development. *Plant Dev. Biol. Biotechnol. Perspect.* **2**, 325–344.

Bewley, J.D. and Black, M. (1994) *Seeds – Physiology of Development and Germination.* New York: Plenum Press.

Blazejczyk, M., Miron, M. and Nadon, R. (2007) FlexArray: a statistical data analysis software for gene expression microarrays. Genome Quebec, Montreal, Canada, URL http://genomequebec.mcgill.ca/FlexArray.

Bouyer, D., Roudier, F., Heese, M., Andersen, E.D., Gey, D., Nowack, M.K., Goodrich, J., Renou, J.P., Grini, P.E., Colo, V. and Schnittger, A. (2011) Polycomb repressive complex 2 controls the embryo-to-seedling phase transition. *PLoS Genet.* **7**, e1002014.

Bove, J., Lucas, P., Godin, B., Oge, L., Jullien, M. and Grappin, P. (2005) Gene expression analysis by cDNA AFLP highlights a set of new signaling networks and translational control during seed dormancy breaking in *Nicotiana plumbaginifolia*. *Plant Mol. Biol.* **57**, 593–612.

Cadman, C.S., Toorop, P.E., Hilhorst, H.W. and Finch-Savage, W.E. (2006) Gene expression profiles of Arabidopsis Cvi seeds during dormancy cycling indicate a common underlying dormancy control mechanism. *Plant J.* **46**, 805–822.

Chen, X.S., Hu, Y.F. and Zhou, D.X. (2011) Epigenetic gene regulation by plant Jumonji group of histone demethylase. *Biochim. Biophys. Acta*, **1809**, 421–426.

Corbineau, F., Benamar, A. and Come, D. (2000) Changes in sensitivity to abscisic acid of the developing and maturing embryo of two wheat cultivars with different sprouting susceptibility. *Isr. J. Plant Sci.* **48**, 189–197.

Derera, N.F. (1989) *Pre-harvest Field Sprouting in Cereals.* Boca Raton, FL: CRC Press.

Du, Z., Zhou, X., Ling, Y., Zhang, Z.H. and Su, Z. (2010) agriGO: a GO analysis toolkit for the agricultural community. *Nucleic Acids Res.* **38**, W64–W70.

El-Maarouf-Bouteau, H. and Bailly, C. (2008) Oxidative signaling in seed germination and dormancy. *Plant Signal. Behav.* **3**, 175–182.

Finch-Savage, W.E. and Leubner-Metzger, G. (2006) Seed dormancy and the control of germination. *New Phytol.* **171**, 501–523.

Finkelstein, R., Reeves, W., Ariizumi, T. and Steber, C. (2008) Molecular aspects of seed dormancy. *Annu. Rev. Plant Biol.* **59**, 387–415.

Foley, M.E. (2008) Temperature and moisture status affect afterripening of leafy spurge (*Euphorbia esula*) seeds. *Weed Sci.* **56**, 237–243.

Garello, G. and Le Page-Degivry, M.T. (1999) Evidence for the role of abscisic acid in the genetic and environmental control of dormancy in wheat (*Triticum aestivum* L.). *Seed Sci. Res.* **9**, 219–226.

Gubler, F., Millar, A.A. and Jacobsen, J.V. (2005) Dormancy release, ABA and pre-harvest sprouting. *Curr. Opin. Plant Biol.* **8**, 183–187.

Gubler, F., Hughes, T., Waterhouse, P. and Jacobsen, J. (2008) Regulation of dormancy in barley by blue light and after-Ripening: effects on abscisic acid and gibberellin metabolism. *Plant Physiol.* **147**, 886–8961.

Howell, K.A., Narsai, R., Carroll, A., Ivanova, A., Lohse, M., Usadel, B., Millar, H. and Whelan, J. (2009) Mapping metabolic and transcript temporal switches during germination in rice highlights specific transcription factors and the role of RNA instability in the germination process. *Plant Physiol.* **149**, 961–980.

Iglesias-Fernandez, R., Rodriguez-Gacio, M.C. and Matilla, A.J. (2011) Progress in research on dry after-ripening. *Seed Sci. Res.* **21**, 69–80.

Jordan, M.C., Somers, D.J. and Banks, T.W. (2007) Identifying regions of the wheat genome controlling seed development by mapping expression quantitative trait loci. *Plant Biotechnol. J.* **5**, 442–453.

Kawakami, N., Miyake, Y. and Noda, K. (1997) ABA insensitivity and low ABA levels during seed development of non-dormant wheat mutants. *J. Exp. Bot.* **48**, 1415–1421.

Kucera, B., Cohn, M.A. and Leubner-Metzger, G. (2005) Plant hormone interactions during seed dormancy release and germination. *Seed Sci. Res.* **15**, 281–307.

Lea, P.J. and Miflin, B.J. (2003) Glutamate synthase and the synthesis of glutamate in plants. *Plant Physiol. Biochem.* **41**, 555–564.

Leubner-Metzger, G. (2005) Beta-1,3-glucanase gene expression in low-hydrated seeds as a mechanism for dormancy release during tobacco after-ripening. *Plant J.* **41**, 133–145.

Li, Q., Wang, B.C., Xu, Y. and Zhu, Y.X. (2007) Systematic studies of 12S seed storage protein accumulation and degradation patterns during Arabidopsis seed maturation and early seedling germination stages. *J. Biochem. Mol. Biol.* **40**, 373–381.

Liu, Y., Koornneef, M. and Soppe, W.J. (2007) The absence of histone H2B monoubiquitination in the Arabidopsis hub1 (rdo4) mutant reveals a role for chromatin remodeling in seed dormancy. *Plant Cell* **19**, 433–444.

Livak, K.J. and Schmittgen, T.D. (2001) Analysis of relative gene expression data using real-time quantitative PCR and the 2(-Delta Delta C (T)) Method. *Methods*, **25**, 402–408.

Mak, Y.X., Willows, R.D., Roberts, T.H., Wrigley, C.W., Sharp, P.J. and Copeland, L. (2009) Germination of wheat: a functional proteomics analysis of the embryo. *Cereal Chem.* **86**, 281–289.

Malagnac, F., Bartee, L. and Bender, J. (2002) An Arabidopsis SET domain protein required for maintenance but not establishment of DNA methylation. *EMBO J.* **21**, 6842–6852.

McGinnis, K.M., Thomas, S.G., Soule, J.D., Strader, L.C., Zale, J.M., Sun, T.P. and Steber, C.M. (2003) The Arabidopsis SLEEPY1 gene encodes a putative F-box subunit of an SCF E3 ubiquitin ligase. *Plant Cell*, **15**, 1120–1130.

Millar, A.A., Jacobsen, J.V., Ross, J.J., Helliwell, C.A., Poole, A.T., Scofield, G., Reid, J.B. and Gubler, F. (2006) Seed dormancy and ABA metabolism in Arabidopsis and barley: the role of ABA 8'-hydroxylase. *Plant J.* **45**, 942–954.

Morris, C.F., Moffatt, J.M., Sears, R.G. and Paulsen, G.M. (1989) Seed dormancy and responses of caryopses, embryos, and calli to abscisic acid in wheat. *Plant Physiol.* **90**, 643–647.

Morris, R.T., O'Connor, T.R. and Wyrick, J.J. (2008) Osiris: an integrated promoter database for Oryza sativa L. *Bioinformatics*, **24**, 2915–2917.

Nakabayashi, K., Okamoto, M., Koshiba, T., Kamiya, Y. and Nambara, E. (2005) Genome-wide profiling of stored mRNA in Arabidopsis thaliana seed germination: epigenetic and genetic regulation of transcription in seed. *Plant J.* **41**, 697–709.

Noda, K., Matsuura, T., Maekawa, M. and Taketa, S. (2002) Chromosomes responsible for sensitivity of embryo to abscisic acid and dormancy in wheat. *Euphytica*, **123**, 203–209.

Okamoto, M., Kuwahara, A., Seo, M., Kushiro, T., Asami, T., Hirai, N., Kamiya, Y., Koshiba, T. and Nambara, E. (2006) CYP707A1 and CYP707A2, which encode abscisic acid 8'-hydroxylases, are indispensable for proper control of seed dormancy and germination in Arabidopsis. *Plant Physiol.* **141**, 97–107.

Payne, P.I., Holt, L.M., Lawrence, G.J. and Law, C.N. (1982) The genetics of gliadin and glutenin, the major storage proteins of the wheat endosperm. *Plant Foods Hum. Nutr.* **31**, 229–241.

Potokina, E., Sreenivasulu, N., Altschmied, L., Michalek, W. and Graner, A. (2002) Differential gene expression during seed germination in barley (*Hordeum vulgare* L.). *Funct. Integr. Genomics*, **2**, 28–39.

Preston, J., Tatematsu, K., Kanno, Y., Hobo, T., Kimura, M., Jikumaru, Y., Yano, R., Kamiya, Y. and Nambara, E. (2009) Temporal expression patterns of hormone metabolism genes during imbibition of Arabidopsis thaliana seeds: a comparative study on dormant and non-dormant accessions. *Plant Cell Physiol.* **50**, 1786–1800.

Rajjou, L., Gallardo, K., Debeaujon, I., Vandekerckhove, J., Job, C. and Job, D. (2004) The effect of a-amanitin on the Arabidopsis seed proteome highlights the distinct roles of stored and neosynthesized mRNAs during germination. *Plant Physiol.* **134**, 1598–1613.

Ramaih, S., Guedira, M. and Paulsen, G.M. (2003) Relationship of indoleacetic acid and tryptophan to dormancy and preharvest sprouting of wheat. *Funct. Plant Biol.* **30**, 939–945.

Rikiishi, K. and Maekawa, M. (2010) Characterization of a novel wheat (*Triticum aestivum* L.) mutant with reduced seed dormancy. *J. Cereal Sci.* **51**, 292–298.

Rousselin, P., Kraepiel, Y., Maldiney, R., Miginiac, E. and Caboche, M. (1992) Characterization of three hormone mutants of *Nicotiana plumbaginifolia*: evidence for a common ABA deficiency. *Theor. Appl. Genet.* **85**, 213–221.

Saeed, A.I., Bhagabati, N.K., Braisted, J.C., Liang, W., Sharov, V., Howe, E.A., Li, J.W., Thiagarajan, M., White, J.A. and Quackenbush, J. (2006) TM4 microarray software suite. *Methods Enzymol.* **411**, 134–193.

Sakurai, N., Ara, T., Ogata, Y., Sano, R., Ohno, T., Sugiyama, K., Hiruta, A., Yamazaki, K., Yano, K., Aoki, K., Aharoni, A., Hamada, K., Yokoyama, K., Kawamura, S., Otsuka, H., Tokimatsu, T., Kanehisa, M., Suzuki, H., Saito, K. and Shibata, D. (2011) KaPPA-View4: a metabolic pathway database for representation and analysis of correlation networks of gene co-expression and metabolite co-accumulation and omics data. *Nucleic Acids Res.* **39**, D677–D684.

Simpson, G.M. (1990) *Seed Dormancy in Grasses*. Cambridge, UK: Cambridge University Press.

Sreenivasulu, N., Usadel, B., Winter, A., Radchuk, V., Scholz, U., Stein, N., Weschke, W., Strickert, M., Close, T.J., Stitt, M., Graner, A. and Wobus, U. (2008) Barley grain maturation and germination: metabolic pathway and regulatory network commonalities and differences highlighted by new MapMan/PageMan profiling tools. *Plant Physiol.* **146**, 1738–1758.

Tanaka, M., Kikuchi, A. and Kamada, H. (2008) The Arabidopsis histone deacetylases HDA6 and HDA19 contribute to the repression of embryonic properties after germination. *Plant Physiol.* **146**, 149–161.

Townley-Smith, T.F. and Czarnecki, E.M. (2007) AC Domain hard red spring wheat. *Can. J. Plant Sci.* **88**, 347–350.

Tsukada, Y., Fang, J., Erdjument-Bromage, H., Warren, M.E., Borchers, C.H., Tempst, P. and Zhang, Y. (2006) Histone demethylation by a family of JmjC domain-containing proteins. *Nature*, **439**, 811–816.

Ueguchi-Tanaka, M., Ashikari, M., Nakajima, M., Itoh, H., Katoh, E., Kobayashi, M., Chow, T., Hsing, C.Y., Kitano, H., Yamaguchi, I. and Matsuoka, M. (2005) GIBBERELLIN INSENSITIVEDWARF1 encodes a soluble receptor for gibberellin. *Nature*, **437**, 693–698.

Walker-Simmons, M.K. (1987) ABA levels and sensitivity in developing wheat embryos of sprouting resistant and susceptible cultivars. *Plant Physiol.* **84**, 61–66.

Wise, R.P., Caldo, R.A., Hong, L., Shen, L., Cannon, E.K. and Dickerson, J.A. (2007) BarleyBase/PLEXdb: a unified expression profiling database for plants and plant pathogens. In *Methods in Molecular Biology. Plant Bioinformatics – Methods and Protocols*, **Vol. 406** (Edwards, D., ed.), pp. 347–363. Totowa, NJ: Humana Press.

Wolfe, K.H., Gouy, M., Yang, Y.W., Sharp, P.M. and Li, W.H. (1989) Date of the monocot-dicot divergence estimated from chloroplast DNA sequence data. *Proc. Natl Acad. Sci. USA*, **86**, 6201–6205.

Xiao, B., Wilson, J.R. and Gamblin, S.J. (2003) SET domains and histone methylation. *Curr. Opin. Struct. Biol.* **13**, 699–705.

14

A highly recombined, high-density, eight-founder wheat MAGIC map reveals extensive segregation distortion and genomic locations of introgression segments

Keith A. Gardner[1,*,†], Lukas M. Wittern[2,†] and Ian J. Mackay[1]

[1]The John Bingham Laboratory, National Institute of Agricultural Botany (NIAB), Cambridge, UK
[2]Department of Plant Sciences, University of Cambridge, Cambridge, UK

*Correspondence
email
keith.gardner@niab.com
[†]These two authors wish to be considered as equal first authors.

Keywords: Multiparent Advanced Generation Intercross (MAGIC), high-density map, wheat, recombination, segregation distortion, introgression.

Summary

Multiparent Advanced Generation Intercross (MAGIC) mapping populations offer unique opportunities and challenges for marker and QTL mapping in crop species. We have constructed the first eight-parent MAGIC genetic map for wheat, comprising 18 601 SNP markers. We validated the accuracy of our map against the wheat genome sequence and found an improvement in accuracy compared to published genetic maps. Our map shows a notable increase in precision resulting from the three generations of intercrossing required to create the population. This is most pronounced in the pericentromeric regions of the chromosomes. Sixteen percent of mapped markers exhibited segregation distortion (SD) with many occurring in long (>20 cM) blocks. Some of the longest and most distorted blocks were collinear with noncentromeric high-marker-density regions of the genome, suggesting they were candidates for introgression fragments introduced into the bread wheat gene pool from other grass species. We investigated two of these linkage blocks in detail and found strong evidence that one on chromosome 4AL, showing SD against the founder Robigus, is an interspecific introgression fragment. The completed map is available from http://www.niab.com/pages/id/326/Resources.

Introduction

Multiparent mapping populations are now being widely developed in plant species, and combine high genetic recombination with high diversity. Examples include nested association mapping populations (NAM, Yu *et al.*, 2008), the Arabidopsis multiparent RIL population (AMPRIL, Huang *et al.*, 2011) and Multiparent Advanced Generation Intercross (MAGIC) populations. The MAGIC approach was first advocated for crops in 2007 (Cavanagh *et al.*, 2008; Mackay and Powell, 2007) and MAGIC populations have subsequently been developed in a range of species, such as rice (Bandillo *et al.*, 2013), barley (Sannemann *et al.*, 2015), tomato (Pascual *et al.*, 2015) and wheat (Huang *et al.*, 2012; Mackay *et al.*, 2014; Milner *et al.*, 2015; Thepot *et al.*, 2015).

Multiparent Advanced Generation Intercross populations are multifounder equivalents of the advanced intercross introduced by Darvasi and Soller (1995) and are closely related to the heterogeneous stock and composite cross-populations used in mouse genetics (Mott *et al.*, 2000; Threadgill and Churchill, 2012). They are created by several generations of intercrossing among multiple founder lines leading to greater accumulation of recombination events and hence greater mapping precision. Multiple founders contribute more allelic and phenotypic diversity than captured in typical biparental mapping populations, raising the numbers of QTL that segregate and the types of trait and locus interactions that can be investigated. A MAGIC population founded from good representation of a breeder's gene pool also offers the opportunity to explore patterns of genomic diversity in that gene pool, such as identifying linkage blocks under fertility or viability selection and locating introgression fragments introduced from other species in the breeding process.

The NIAB eight-parent winter wheat MAGIC population (Mackay *et al.*, 2014) was developed in partnership with UK breeders to represent the diversity of UK wheat germplasm. Founder varieties were Alchemy, Brompton, Claire, Hereward, Rialto, Robigus, Soissons and Xi19. Here, we describe the creation and validation of a high-density genetic map in this population, based on the Illumina Infinium iSelect 80K SNP array (http://www.illumina.com/). This is the first publically available eight-founder MAGIC map created for wheat. We estimate the precision and accuracy of our map by reference to the wheat genome sequence and in comparison with existing wheat high-density genetic maps. Finally, we map recombination rates, blocks of segregation distortion (SD) and blocks of high marker density, and infer the genomic locations of blocks of interspecific introgression.

Results

Genotyping

A total of 643 F4 MAGIC lines (assayed as F5 progeny bulks) passed stringent quality control and were used for mapping, and 20 639 SNP markers were scorable and polymorphic, compared to 25 499 markers in a comprehensive UK wheat association mapping panel ('WAGTAIL', 520 varieties, similarly genotyped and scored by K.A. Gardner), suggesting the MAGIC population has captured >80% of the genetic diversity of UK wheat germplasm. Fifty-three markers with heterozygotes or missing data in the founder lines could not be used for mapping with R/mpMap (Huang and George, 2011). Of the remaining 20 586 markers, 18 750 (91%) were scored as codominant and 1836 (9%) as dominant; 664 of the dominant loci were nulls (3.2%) compared to 5.4% of single-locus scorable SNPs showing null alleles in Wang *et al.* (2014). For codominant markers, 2.4% residual heterozygosity was observed, compared to an expectation of 2.2%. Four PCR markers were also genotyped (Appendix S1).

Linkage map

In total, 18 601 markers were mapped across all chromosomes. The 2042 unmapped markers (Table S1) are a highly nonrandom sample: 30% are dominant (compared to 3% of mapped markers), 16% are nulls (2% mapped) and 17.5% show SD with a false discovery rate ('fdr') <0.01 (9.7% mapped). All unmapped markers were treated as traits and QTL mapped to the finished map, and their flanking markers were blasted against both wheat genome sequences (International Wheat Genome Sequencing Consortium, 2014, http://www.wheatgenome.org/, 'IWGSC'; Chapman *et al.*, 2015, 'CHAPMAP'); 1118 of the unmapped markers had matching QTL and blast hits (with ≥98% identity, Table S1). From the distribution of these unmapped markers (Table S1, 'data analysis'), both SD and translocations correlate with failure to map. About 20% of these markers were aligned to chromosome 5B or 7B, across a segregating translocation breakpoint and 12% were from chromosome 3B, which has two major SD blocks. Several large blocks of co-localized segregation distorted markers were unmappable, including a 104-marker block with SD against Robigus from the distal end of 4A.

Our NIAB MAGIC linkage map, 'NIAB2015', is dense and compact (Figure 1, data in Table S2). Summary statistics are shown in Table 1. The number of markers mapped per chromosome varied from 80 (4D) to 2327 (1B). About 37%, 50%, and 13% of markers were mapped to the A, B and D genomes, respectively. The map length totalled 5405 cM, with individual chromosomes ranging from 126 to 386 cM. A and B genome chromosomes together showed relatively low variation in length and were about 60% longer than D genome chromosomes. Altogether, the 18 601 markers were mapped to 4578 unique sites across the genome, with 41%, 47% and 12% of the unique sites mapping to the A, B and D genomes, respectively. The number of unique sites per cM is approximately one for the A and B genomes (0.94 and 1.02, respectively) but only 0.41 per cM for the D genome, reflecting the well-documented lower polymorphism of the D genome of hexaploid wheat (Wang *et al.*, 2014).

For A and B genomes, chromosome map length in cM is strongly associated with the number of unique sites (adjusted R^2 0.69, $P = 0.000132$), with a gradient of 0.53 cM/unique site (95% CI 0.32–0.74), very similar to the default minimum genetic distance of 0.5003 cM between proximal recombination fraction bins in mpMap (corresponding to a recombination fraction, rf, of 0.005). Indeed, 69% of proximal unique sites in the AB genomes are 0.5 cM apart, confirming the compactness of NIAB2015.

Internal validations

The genomewide heatmap of the recombination fraction matrix and logarithm of odds (LOD) values for NIAB2015 shows distinctly separated chromosomes and very few off-diagonal low-value (rf 0–0.1) colours (Figure 2). Two exceptions are evident. Firstly, the well-known rye introgression on chromosome 1B (Villareal *et al.*, 1991; Worland and Snape, 2001) is segregating in the population (present in founders Brompton and Rialto), causing long range linkage disequilibrium (in a similar manner to Huang *et al.*, 2012). During mapping, markers that are segregating within native wheat chromosomes in the region of the rye introgression will be 'pushed to the side' of the markers which are only segregating between rye and wheat. This is evident in our map, where we have opted to include all these markers. Secondly, there is segregation for the occurrence of a common wheat translocation between chromosomes 5BS and 7BS (Badaeva *et al.*, 2007), although with unknown parental frequencies. The largest distinct group of unmapped markers are from around the translocation breakpoint.

The heat map for 2A (Figure 2) is typical for a cleanly mapped AB genome chromosome with only the centromeric region being relatively poorly defined. The heat map for 2D (Figure 2) is typical

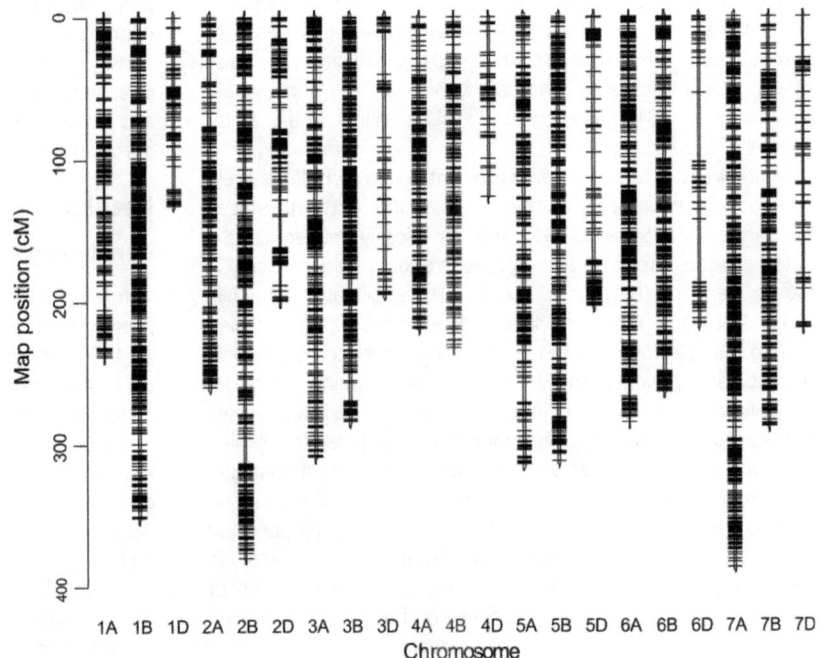

Figure 1 The NIAB2015 MAGIC genetic map. Short arm of each chromosome at top (0 cM).

1A 1B 1D 2A 2B 2D 3A 3B 3D 4A 4B 4D 5A 5B 5D 6A 6B 6D 7A 7B 7D
Chromosome

Table 1 MAGIC map summary statistics and comparison to other maps (see Table 2 for description of other maps)

| | NIAB 2015 | Other maps | | | | NIAB2015 overlap* | | |
		CM2014	9KCONS	9KMAGIC	SynOp	CM2014	9KCONS	9KMAGIC
Total length (cM)	**5405**	**11 185**	**5192**	**3722**	**3242**			
Max chromosome length	386	743	396	300	225			
Min chromosome length	126	295	99	73	67			
Average chromo length (A)[†]	287	619	284	203	156			
Average chromo length (B)	301	552	277	201	155			
Average chromo length (D)	184	427	80	128	152			
Total No markers	**18 601**	**40 267**	**7497**	**4300**	**N/A[‡]**	**15 672**	**2840**	**1663**
Max markers/chromosome	2327	3471	768	451	N/A	1906	284	162
Min markers/chromosome	80	296	38	14	N/A	67	5	4
% markers A	37	38	46	48	N/A	37	46	50
% markers B	50	46	46	45	N/A	50	46	43
% markers D	13	16	8	7	N/A	13	8	7
Total unique sites	**4578**	**5564**	**3010**	**1813**	**1446**	**2897**	**1436**	**966**
Max unique sites/chromo	452	387	265	202	114	234	119	88
Min unique sites/chromo	39	101	30	11	28	32	4	4
Average unique sites (A)	270	282	188	114	66	161	93.6	65
Average unique sites (B)	309	328	195	125	89	188	92.1	62
Average unique sites (D)	75	185	47	20	52	65	19.4	11
Unique sites/cM	**0.85**	**0.50**	**0.60**	**0.49**	**0.44**			
Unique sites/cM (A)	0.94	0.46	0.66	0.56	0.42			
Unique sites/cM (B)	1.02	0.59	0.70	0.62	0.57			
Unique sites/cM (D)	0.41	0.43	0.26	0.15	0.34			

*These comparisons only include markers variable in NIAB2015.

[†](A) refers to A genome.

[‡]>1 million contigs mapped in SynOp using GbS—see text.

of D genome chromosomes with blocks of tightly linked markers separated by tracts of low variation, although our MAGIC map has high within-block recombination. We also used principal component analysis (PCA) of the recombination fraction matrix to qualitatively assess whether chromosomes were well ordered. Figure 2 shows the first two principal components for chromosome 2B, demonstrating a horseshoe shape characteristic of a successfully mapped chromosome (Cheema and Dicks, 2009; Curtis, 1994). The curve shape of PCA plots is particularly sensitive to SD; deviations provided informative supporting evidence for our SD analysis. Figure S1 has heat maps and PCA figures for all chromosomes.

Genetic map comparisons

Table 1 compares NIAB2015 to four other genetic maps, listed in Table 2. CM2014 contains 40 267 markers (5564 unique sites) and is 11 185 cM in length. However, looking only at markers segregating in our MAGIC population, there are 1.19× as many markers but 1.58× as many unique sites as CM2014; this difference is much more pronounced for the A (1.68×) and B (1.64×) genomes than for the D genome (1.15×). The difference is probably related to the presence of the 'Synthetic' part of the SynOp cross included in CM2014; SynOp and CM2014 both have a proportionally higher number of unique sites in the D genome than NIAB2015. Our total number of unique sites is much higher than the other three maps, as expected; the 9K maps have far fewer markers, and SynOp is based on a single biparental cross and far fewer progeny lines. NIAB2015 is 51% of the length of CM2014, despite having 81% of the unique sites. In fact, for the

A and B genomes, our number of unique sites per cM (0.85) is higher than any of the other maps (range 0.44–0.60, Table 1). However, for the D genome, the unique sites/cM is roughly the same as all other maps except 9KCONS and 9KMAGIC.

Figure 3 compares NIAB2015 chromosome 3A to the CM2014, 9KCONS and SynOp maps and also to the pseudomolecule POPSEQ data from IWGSC2, derived from SynOp. Other chromosomes are in Figure S2. There is a very low gradient in the middle of most chromosomes, that is NIAB2015 is much longer in this region than CM2014 and SynOp, as a result of more recombination events. This effect is either absent or weak (Figure S2) in the 9KCONS comparisons. There is still very low recombination directly at the centromere in NIAB2015 (Figure 3f), but the surrounding nonrecombining region is smaller than other non-MAGIC maps. Figure 3b clearly demonstrates the centromeric effect in combination with the much more compact linkage map; at the distal ends of the chromosome, the markers fan outwards from NIAB2015 (left of the figure) to CM2014 (on the right) whereas in the centromeric region, the markers fan outwards from CM2014 to NIAB2015. Further examination of all chromosomes (Figure S2) reveals: (i) between NIAB2015 and CM2014, nonlocal map order in the D genome is less well conserved than in the A and B genomes; (ii) anomalous near vertical lines apparent on some chromosomes (e.g. 1B 100 cM, 4A 175 cM) reflect blocks of low recombination in the MAGIC population (possible introgression fragments); and (iii) two instances of missing chromosomal fragments. On chromosome 5A, the first 85 cM of NIAB2015 is largely absent in SynOp, whereas the first 15 cM of chromosome 5B in SynOp is largely

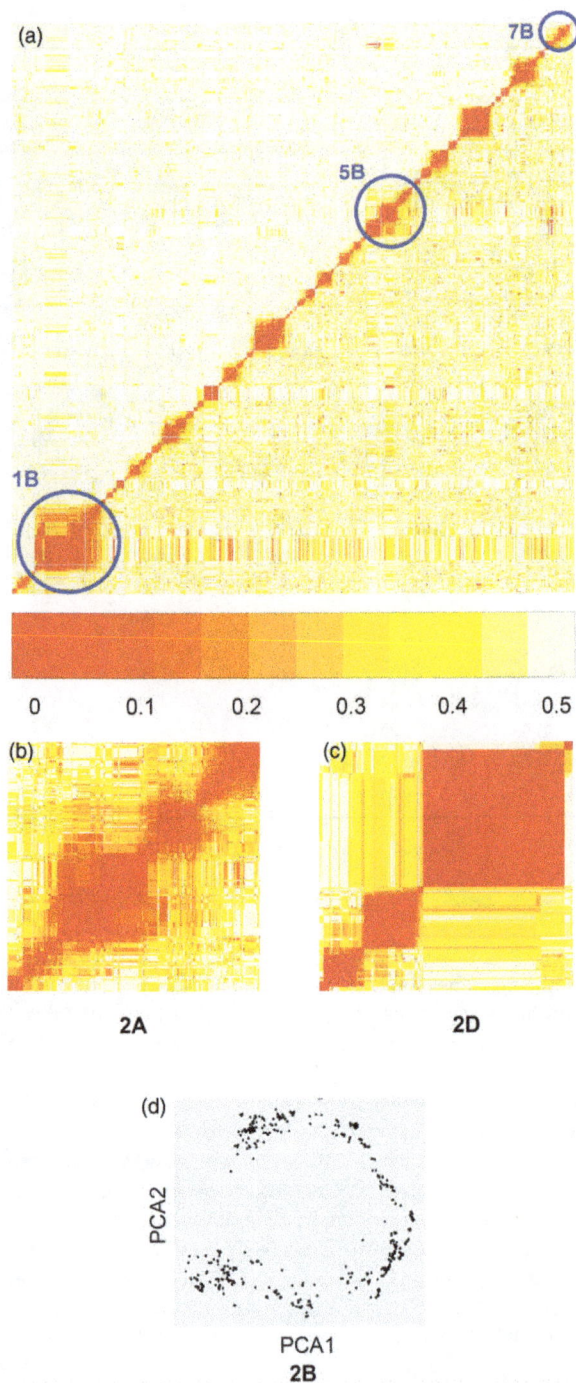

Figure 2 NIAB2015 map diagnostics. (a) Overall heatmap of recombination fraction matrix, with problematic chromosomes highlighted. Red to yellow = low (0) to high recombination (0.5). (b, c) heat maps for chromosomes 2A, 2D. (d) First two principal components (x-axis, y-axis) of recombination fraction in principal components analysis for 2B (see text).

missing in NIAB2015. The latter is the unmapped region around the translocation breakpoint of 5BS-7BS.

Figure 4 compares LD decay for chromosome 2D in NIAB2015 with 2D in CM2014, as measured by D' and R^2. For other chromosomes, see Figure S3. Graphically, it is apparent that the pattern of LD decay in the MAGIC map is considerably cleaner

Table 2 List of high-density wheat genetic maps used for comparison to NIAB2015 MAGIC map

Map	Markers	Population	Reference
SynOp	Genotyping-by-sequencing	Synthetic W7894 × Opata M85	Poland et al. (2012)
9KMAGIC	9K SNP array	4-founder MAGIC	Cavanagh et al. (2013)
9KCONS	9K SNP array	9KCONS+6 bi-parentals (inc. SynOp)	Cavanagh et al. (2013)
CM2014	80K SNP array	8 bi-parentals (inc. SynOp)	Wang et al. (2014)

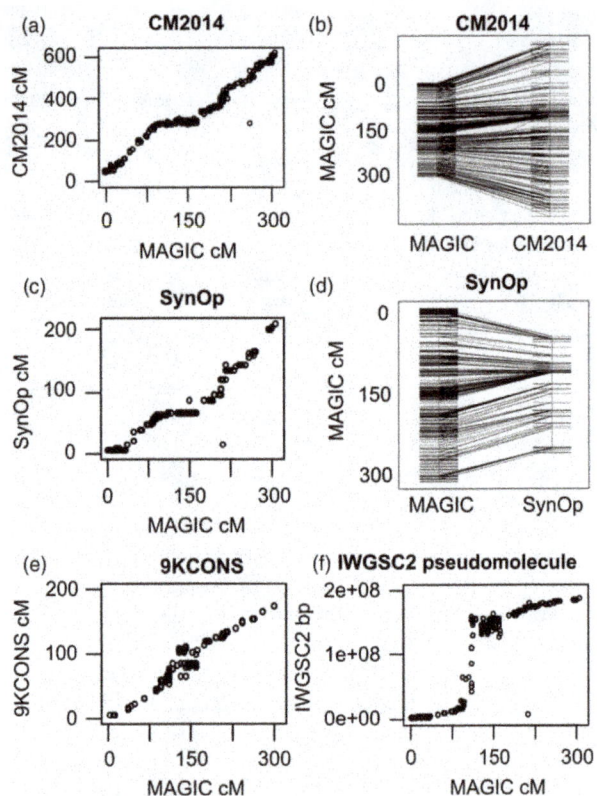

Figure 3 NIAB2015 chromosome 3A compared to four other genetic maps (a, b) CM2014 (c, d) SynOp (e) 9KCONS (f) IWGSC2 pseudomolecule. (a, c, e) cM-cM genetic map comparison, NIAB2015 on x-axis. (b, d) direct comparison between chromosome diagrams, NIAB2015 on left (f) comparison of NIAB2015 (cM, x-axis) to IWGSC2 pseudomolecule (base pairs).

than in CM2014, with many more long-distance high LD values apparent in CM2014.

Alignment to 3B reference sequence

Comparison of the alignment of CM2014 and NIAB2015 to the physical map shows differences in quality of marker alignment as well as recombination landscape (Figure 5). 2.9% of the markers in NIAB2015 which align to the 3B pseudomolecule had inconsistent orders with respect to the reference sequence (identified by a red cross in Figure 5), while in CM2014 this is 10.46%. Nineteen of the 23 misaligned markers from NIAB2015 are also found in CM2014, which showed identical

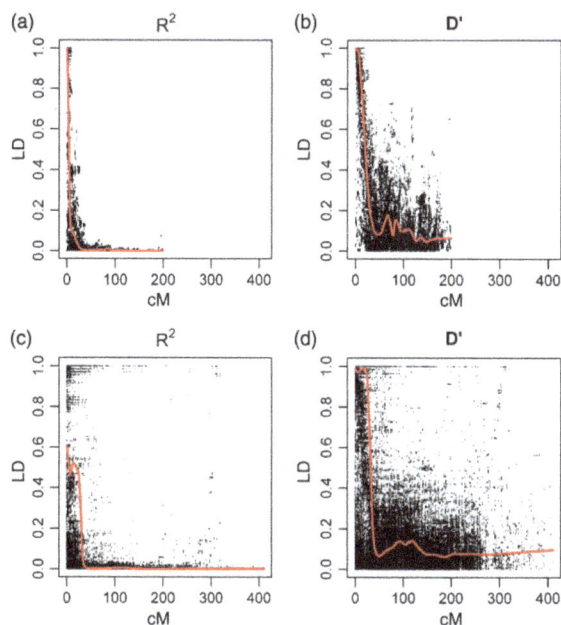

Figure 4 Linkage disequilibrium decay for chromosome 2D. (a, b) NIAB2015 (c, d) CM2014. Red line is best fit lowess curve with smoothing span parameter = 0.10.

inconsistencies. This suggests the alignment problems for these markers may lie with their physical positions within the reference sequence, rather than the genetic maps. Consistent with this explanation, nine of the 23 anomalous markers also mapped to duplicated genes with different physical positions on 3B.

Detected crossovers on chromosome 3B vary by method of analysis: haplotype analysis in R/mpMap using R/happy.hbrem detects 2655 crossovers, equalling 1.46 crossovers per line per Morgan, while haplotype analysis using R/qtl detects 2920 crossovers or 1.60 per line per Morgan. The countXO function in R/qtl detects 4117 crossovers, equalling 2.26 per line per Morgan. Choulet *et al.* (2014) report 787 crossovers in the biparental Chinese Spring × Renan population. The distribution of crossovers in NIAB2015 varies significantly over the chromosome with average chromosome-wide recombination rates of 0.30 cM/Mb and a maximum of 1.34 cM/Mb. The recombination landscape between NIAB2015 and CM2014 differs with proportionally higher recombination rates in the two distal chromosome regions (identified as 0–68 Mb and 715–774 Mb) in CM2014 (1.18 cM/Mb and 1.47 cM/Mb) compared to NIAB2015 (1 cM/Mb and 1 cM/Mb), but lower recombination rates in the large proximal regions (0.07 cM/Mb compared to 0.15 cM/Mb). Choulet *et al.* (2014) report distal recombination rates to be 0.60 and 0.96 cM/Mb and the large proximal region 0.05 cM/Mb. We observe falls in recombination rates around SD loci. SD loci associated with the founder Soissons are at 107–127 Mb and 242–414 Mb, while Robigus SD loci are at 145–150 Mb, 538–554 Mb and 641–663 Mb.

Further genome sequence comparisons

For chromosomes where no physical map is currently available, we compared NIAB2015 map locations of markers which had top BLAST hits (>99% sequence identity) to the same CHAPMAP genomic sequence contig (Table 3, Appendix S1). Sixty-six percent of markers could be assigned to a contig in CHAPMAP,

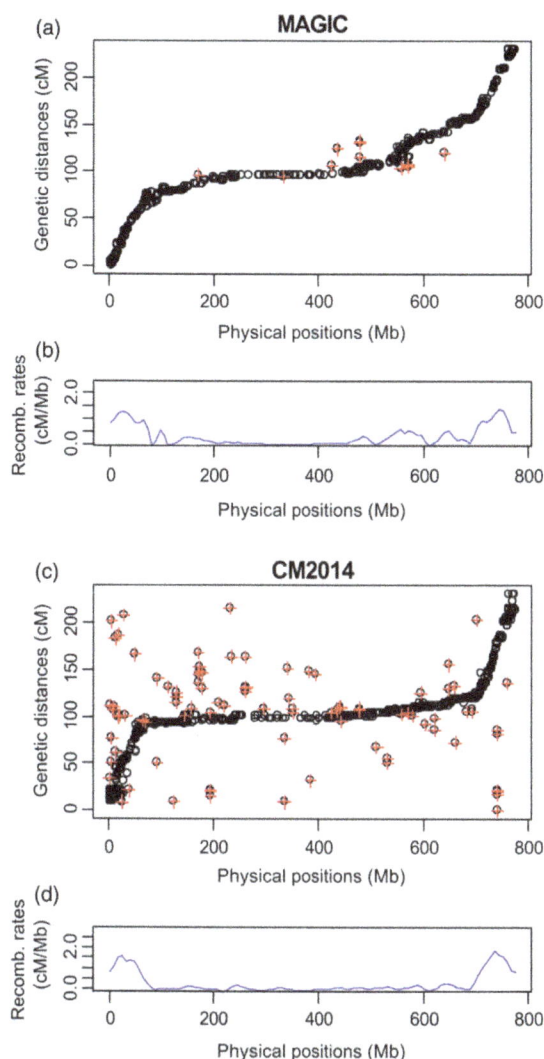

Figure 5 Chromosome 3B physical map comparison. (a, c) physical vs map distance for NIAB2015, CM2014. Note CM2014 scaled to be same length as NIAB2015. (b, d) recombination profile for NIAB2015, CM2014.

but only 39% were in shared contigs. Twenty percent (22% for CM2014) of markers in singleton contigs were mapped to a different chromosome than the BLASTn hit but for markers in shared contigs, this number was only 5% (7% for CM2014). However, 76% of singleton and 81% of shared contig disagreements mapped to homeologous chromosomes. Combined with the high percentage of no hits, this strongly suggests that the incomplete sequence coverage of the genome (estimated 10.1/17 Gb, Chapman et al., 2015) explains most of the NIAB2015-BLASTn discrepancies. Of markers in shared contigs on the same chromosome, 82% were fully consistent with our map order (same or adjacent location), a further 6% had a single nonshared unique site (usually a single marker) between the shared contig markers and the remaining 12% were further apart (median distance 6.6 cM) although 53% of these occurrences corresponded to separation by only 2–5 unique sites.

Genome diversity analysis

There has been no intentional trait selection within the MAGIC population except for the removal of double dwarf lines (lines

Table 3 Results of BLASTn analysis against Chapman *et al.* (2015)

	NIAB2015	CM2014*
Grouping		
No hit	34%	
Singleton	27%	
Same chr		80%	78%
Diff chr		20%	22%
Shared	39%	
Same chr		95%	93%
Diff chr		5%	7%
Ordering		
Same/adjacent map position		82%	
1-site gap		6%	
Multisite gap		12%	
Multisite gap median cM dist		6.6 cM	
Multisite gap ≤5 sites		53%	

*CM2014 comparison used only markers also mapped in NIAB2015.

carrying dwarfing alleles at both the *RhtB* and *RhtD* height loci). Mapping SD in the MAGIC population is thus a powerful approach to locate genomic regions causing meiotic segregation problems or under fertility or viability selection. Figure 6 shows the distribution of marker density and SD across the genome. A substantial fraction of mapped markers show statistically significant SD (Table 4): 2887 (15.5%) at the fdr<0.05 level, 1764 (9.5%) at fdr<0.01. If unmapped markers are included, these figures are higher: 16.4% fdr<0.05, 10.3% fdr<0.01. Markers exhibiting SD are nonrandomly distributed: 52% of mapped markers showing SD at fdr<0.01 map to chromosome 1B, mostly showing SD against the rye introgression (i.e. alleles from founders carrying the introgression are at lower than expected frequencies), and a further 12% of SD markers are found on each of chromosomes 3B and 6B. Otherwise, only 4A (6%) shows statistically significantly more SD loci than expected across the whole data set at the fdr <0.01 level (Table 4).

Some blocks of SD for particular founders occur over long distances along a chromosome, often >20 cM. Frequently, two single founders individually show weak SD along separate chromosome blocks but when these blocks overlap and the founders co-occur, a much stronger SD effect is seen (e.g. on 2B a weak SD against Claire block starting at 33 cM overlaps at 98.4–98.9 cM with a weak SD against Soissons block ending at 130 cM; the combined peak has stronger SD). Conversely, counteracting blocks (SD in favour of founder A but against founder B) cancel when overlapping. From these observations, we conclude that for many founder combinations, SD extends over considerable distances in the MAGIC population. We inferred 63 SD blocks at fdr < 0.01 (96 for fdr < 0.05), of which 39 consisted of at least two unique sites. This may be an underestimate: (i) it was difficult to conclude whether some widely separated SD blocks with identical founder patterns were one block or two; we treated them as one block if no contradictory founder pattern was found in between them (ii) as we can detect SD blocks only when they are adequately discriminated by the available markers, we will have by chance missed small SD blocks where the appropriate founder combination did not happen to occur (iii) a high frequency of strong SD markers are unmapped. Table 5 shows the 39 blocks with fdr < 0.01, covering more than one unique site. Blocks containing the dwarfing loci *RhtD* and *RhtB* rank 4th and 6th, with the direction of SD as expected (dwarfing allele *RhtB-1b* is found only in Robigus and Soissons, *RhtD-1b* is found in the other six founders but not in Robigus and Soissons).

Marker density peaks around the centromere in many chromosomes but noncentromeric high marker density (HMD) blocks are also visible on 1D, 2D, 3B, 4A, 5A, 5B, 6B, 7A and most notably on 1B in the centre of the region containing the rye introgression (SNP density up to 793 markers/10 cM). Twenty-seven noncontiguous blocks of markers contain more than 100 markers/10 cM. Of these, 11 are centromeric without significant SD and three are both centromeric and overlap SD blocks. Six SD blocks are associated with noncentromeric HMD blocks (Table 5, bolded). These include the blocks with the 2nd and 3rd highest chi-squared peaks: a block with SD favouring Robigus alleles on 3B and the 1B-1R rye introgression (SD against Brompton and Rialto) on 1B. Block 24, against Robigus on 4A, is under-represented in this table: 74 unmapped markers showing significant SD against Robigus were mapped as traits to this location, with lower SD *P*-values, which would make this the 2nd strongest block. Furthermore, these six SD-HMD blocks are very long: they are the six blocks with the highest number of SD markers in Table 5, 3 of the top 6 with most unique sites (all 6

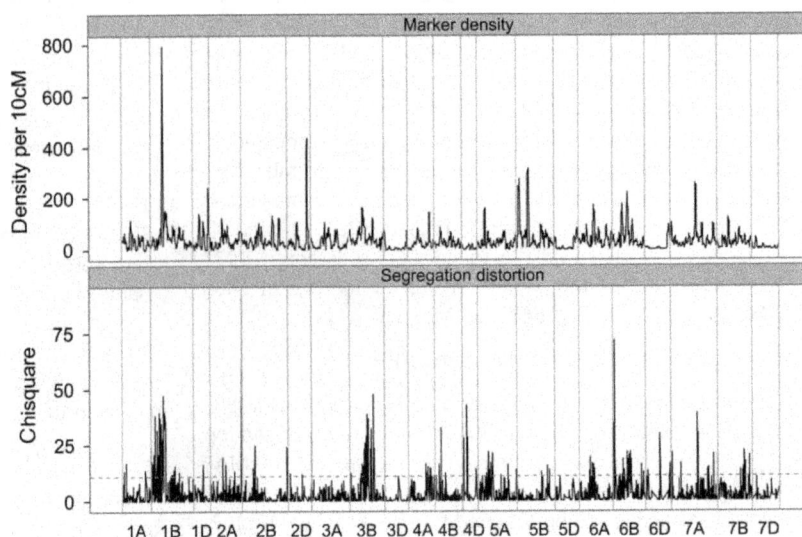

Figure 6 Genomewide patterns of marker density and segregation distortion in NIAB2015. Markers above the dotted line in the lower figure show significant SD at fdr < 0.01.

Table 4 Summary of segregation distortion results by chromosome

Chrom	Marker numbers		Percentage	
	SD, fdr < 0.05	SD, fdr < 0.01	%SD, fdr < 0.05	%SD, fdr < 0.01
ALL	2887	1764		
1A	12	3	0.42	0.17
1B	**1033**	**912**	**35.78**	**51.70**
1D	3	2	0.10	0.11
2A	37	8	1.28	0.45
2B	27	8	0.94	0.45
2D	7	5	0.24	0.28
3A	23	0	0.80	0.00
3B	**392**	**207**	**13.58**	**11.73**
3D	1	1	0.03	0.06
4A	**138**	**114**	**4.78**	**6.46**
4B	60	50	2.08	2.83
4D	14	9	0.48	0.51
5A	162	84	5.61	4.76
5B	32	4	1.11	0.23
5D	5	1	0.17	0.06
6A	122	38	4.23	2.15
6B	**502**	**213**	**17.39**	**12.07**
6D	**82**	18	**2.84**	1.02
7A	159	51	5.51	2.89
7B	60	36	2.08	2.04
7D	16	0	0.55	0.00

SD, segregation distortion; fdr, false discovery rate.

Bold = significantly more SD markers than expected for given chromosome (P < 0.01).

within top 12) and the two longest cM ranges (5 of top 10). These large SD regions posed considerable mapping challenges (chromosomes 1B, 3B and 6B, Figure S1), caused PCA plot distortions (1B, 3B, 4A, Figure S1), or were clearly visible in map comparisons (1B, 4A, Figure S2).

Robigus is thought to have emmer wheat (*Triticum dicoccoides*) in its pedigree (P. Werner, pers. comm.), but the location of any introgressions has not previously been reported. We examined the occurrence of the Robigus alleles in the two SD-HMD Robigus blocks using the Bristol University 820K SNP array database (http://www.cerealsdb.uk.net, Wilkinson *et al.*, 2012). In the 820K data set, 7202 SNPs were found in Robigus and its descendants but in no more than 15% of lines with known non-Robigus pedigree, and could be assigned an IWGSC2 location with high certainty using BLASTn (Appendix S1). Of these, 481 mapped to the MAGIC 3B SD-HMD block, most densely in the distal part of the block containing the peak SD markers. Almost all SNPs in this region had 4–6 non-Robigus pedigree lines carrying Robigus alleles: Oratorio, Moisson, Garcia and Bacanora consistently carried Robigus alleles, while Dekan and Highbury did so sporadically. In addition, Robigus alleles for most markers in the peak region are found in the variety Glasgow, also suspected to have *T. dicoccoides* in its pedigree (R. Jennaway, pers. comm.). Within the core 3B SD block, 10 'perfect match' Robigus SNPs (i.e. found in 0 non-Robigus pedigree lines) are found; however, they are not the peak markers and all have 'no call' as the Robigus allele. In contrast, 155 'perfect match' Robigus markers mapped exactly to the MAGIC 4A SD-HMD block (175–218 cM in MAGIC map), none of which had null alleles. This represents 23% of all perfect, null-free Robigus markers which

could be assigned an IWGSC location in the entire 820K data set. Furthermore, when we grouped all the perfect, null-free Robigus markers into blocks (<0.75 Mb separation between markers in IWGSC2 pseudomolecule) along chromosomes, this block was the largest, containing nearly three times as many markers as the 2nd largest block.

Discussion

Linkage map

We have constructed the first eight-parent MAGIC genetic map for wheat, comprising 18 601 mapped markers. This required strict quality control of marker calling and extensive manual curation. During map construction, we dropped 10% of markers which could not be cleanly placed along a chromosome or which could not be placed without a large increase in map distance; these were biased towards dominant, null and SD loci and included a block of loci from around the translocation breakpoint of chromosomes 5BS and 7BS. To the best of our knowledge, this small block is the only specific missing chromosomal block in our map, and 5BS-7BS was the only large-scale chromosomal rearrangement we were able to detect.

There are two major practical benefits from constructing a genetic map in an eight-founder MAGIC population. Firstly, we obtain an increase in precision from a larger number of accumulated recombination events via three rounds of inter-crossing. Secondly, while all high-density genetic maps are by necessity an average across several parental combinations, a MAGIC map produced in a single mapping experiment should have greater accuracy (lower error rate) than a map produced by merging data from several separate biparental crossing populations. Comparing our eight-founder MAGIC genetic map ('NIAB2015') to published high-density wheat genetic maps, including the current reference standard CM2014 (Wang *et al.*, 2014), we can see that NIAB2015 has either considerably more total unique sites or, compared to CM2014, more unique sites for mutually shared markers (Table 1, Figures 3 and S2). Confirming the advantage of increased recombination in multiparental populations, we have a considerably higher number of unique sites per cM in the A and B genomes than all other maps; NIAB2015 is much more compact. The most visible evidence of increased recombination in NIAB2015 is in the pericentromeric regions (Figures 3, 5 and S2) where NIAB2015 has many more unique sites than other maps. Of all map comparisons our map has most in common with the four-parent wheat MAGIC map (Cavanagh *et al.*, 2013), which has a close to linear relationship with NIAB2015 for most chromosomes, (i.e. an overall similar recombination pattern along the chromosome), confirming the expectation that MAGIC maps exhibit greater precision.

We quantified recombination rates in NIAB2015 using chromosome 3B, for which a physical map is available. Broman (2005) calculates for an eight-founder MAGIC population the expected number of informative crossovers per Morgan per line to be on average 4. Our estimates were 1.46 using R/happy.hbrem and 1.60 using R/qtl in mpMap with the mpprob function, and 2.26 using the countXO function in R/qtl. Similar to our countXO estimate, Huang *et al.* (2012) also observed about half the number of recombinations expected from simulations in a four-parent MAGIC population. In the case of using inferred haplotypes from 'mpprob' to calculate crossovers, this underestimation can partially be attributed to the extent of missing haplotype data (18.61% using R/qtl, 37.46% using R/happy.hbrem). The other

Table 5 Major segregation distortion (SD) blocks detected in the NIAB2015 population, ordered by false discovery rate (fdr) value of peak marker

Rank	PEAK SD_fdr	No SNPs	No sites	Direction	Minority founder	Ch	Start (cM)	Finish (cM)	Range (cM)	HMD	Map distort	Known Locus
1	6.48E-14	55	23	FOR	CL, SO, XI	6B	0.0	24.5	24.5		Yes	
2	**2.32E-09**	**268**	**89**	**FOR**	**RO**	**3B**	**86.6**	**198.5**	**111.9**	**HMD**	**Yes**	
3	**2.67E-09**	**931**	**80**	**AGAINST**	**BR, RI**	**1B**	**0.0**	**119.0**	**119.0**	**HMD**	**Yes**	**1BR**
4	1.69E-08	8	2	FOR	RO, SO	4D	32.2	40.1	7.9			*RhtD*
5	7.35E-08	15	6	FOR	AL, CL	7A	216.3	219.8	3.5	Edge	Yes	
6	2.78E-07	16	10	AGAINST	RO, SO	4B	10.4	52.7	42.2			*RhtB*
7	7.85E-07	17	5	AGAINST	RO, SO	7A	219.8	221.8	2.0			
8	2.63E-06	4	2	AGAINST	SO	4D	3.1	4.1	1.0			
9	1.92E-05	2	2	AGAINST	SO, XI	2B	372.4	376.0	3.5			
10	4.79E-05	33	13	AGAINST	CL	7B	194.9	226.2	31.4			
11	5.37E-05	88	27	FOR	XI	5A	58.1	118.6	60.5			
12	**5.43E-05**	**351**	**56**	**FOR**	**BR, RO, SO**	**6B**	**81.7**	**149.6**	**67.8**	**HMD**	**Yes**	
13	7.96E-05	13	7	FOR	SO, XI	7A	347.1	352.6	5.6	Edge		
14	1.71E-04	13	9	FOR	RI, SO, XI	6A	84.6	113.0	28.4			
15	7.28E-04	12	6	AGAINST	BR, SO	4A	145.5	158.7	13.2			
16	7.39E-04	70	9	AGAINST	SO	6D	192.7	204.6	11.9			
17	7.86E-04	47	17	FOR	XI	6B	227.7	252.3	24.6			
18	8.16E-04	18	5	AGAINST	XI	5A	242.5	249.2	6.6			
19	8.46E-04	4	3	FOR	RI, SO	2A	127.6	130.7	3.0			
20	1.06E-03	11	6	FOR	HE, RO, SO	1B	183.6	210.6	27.0			
21	1.12E-03	7	5	FOR	HE, SO, XI	5B	251.9	265.0	13.2			
22	1.45E-03	16	8	FOR	AL	7B	207.8	216.4	8.6			
23	1.46E-03	40	4	FOR	RI	4B	54.7	61.3	6.6			
24	**1.63E-03**	**99**	**14**	**AGAINST**	**RO**	**4A**	**170.1**	**210.2**	**40.1**	**HMD**	**Yes**	
24*	**1.35E-11**	**62**	**n/a**	**AGAINST**	**RO**	**4A**	**218.3**	**218.3**	**0.0**			
25	1.70E-03	87	36	AGAINST	AL, CL, RO	7A	240.1	310.3	70.2			
26	1.77E-03	5	3	AGAINST	CL, SO	2B	98.5	101.0	2.5			
27	2.06E-03	2	2	FOR	HE, RI	4D	104.5	106.0	1.5			
28	2.06E-03	22	5	AGAINST	RO, SO, XI	5A	310.0	313.6	3.5			
29	**2.21E-03**	**96**	**12**	**FOR**	**HE, SO, XI**	**6A**	**125.9**	**137.0**	**11.1**	**HMD**		
30	3.12E-03	7	4	AGAINST	XI	6D	11.1	19.3	8.1			
31	3.43E-03	12	7	AGAINST	BR	2A	152.5	207.6	55.1			
32	**3.73E-03**	**120**	**12**	**AGAINST**	**SO**	**3B**	**67.7**	**114.2**	**46.5**	**HMD**	**Yes**	
33	4.09E-03	22	4	FOR	CL	5B	202.4	210.0	7.6			
34	5.41E-03	18	4	FOR	XI	4A	150.1	154.7	4.6			
35	5.52E-03	55	19	AGAINST	RI, XI	1B	142.2	205.0	62.8			
36	5.81E-03	3	2	FOR	RO	6D	215.0	215.5	0.5			
37	6.00E-03	2	2	FOR	RI	2D	121.0	122.6	1.5			
38	8.26E-03	11	10	AGAINST	HE	5A	181.1	215.0	33.9			
39	8.34E-03	48	9	FOR	SO	6B	47.6	79.2	31.6	Edge		

HMD-SD ('High marker density segregation distortion') blocks in bold. Minority founder = origin of minority allele: AL Alchemy, BR Brompton, CL Claire, HE Hereward, RI Rialto, RO Robigus, SO Soissons, Xi Xi19. Ch = chromosome, HMD = overlaps with high-density block (HMD = in high-density block, edge = border of HMD-block). Map distort = visual evidence of map distortion.
*Additional unmapped markers almost certainly belong here (see text).

main reason for underestimation of crossovers is that each SNP is not fully informative in a multiparent population, making it not possible to count all recombination events directly. Nevertheless, using R/qtl (Broman *et al.*, 2003), we detected over 5× as many crossovers on Chromosome 3B (using the function countXO) as in the Chinese Spring × Renan population of Choulet *et al.* (2014).

A comparison between recombination rates on chromosome 3B using NIAB2015 and CM2014 shows that NIAB2015 has a much cleaner estimation of recombination rate as recombination rates rarely drop below 0 cM/Mb (Figure 5). NIAB2015 has a slightly higher rate of recombination in the central chromosomal region compared to CM2014, and an 18–42% lower estimated recombination rate in the distal regions. A possible explanation is that our map distances are inflated in the proximal regions, due to the increased number of unique proximal recombination fraction bins compared to CM2014. Genetic mapping errors that will lead to over- or underestimation of recombination rates, especially in the pericentromeric regions, could also be a confounding problem. Furthermore, the two SD blocks on 3B contributed to lower recombination estimates, a potential problem for all estimates of recombination rates from multifounder populations. Interestingly, Maccaferri *et al.* (2015) also report that CM2014 shows a much lower recombination rate throughout the centromeric–pericentromeric regions compared to tetraploid wheat.

Theoretically, an increase in unique sites in our map compared to others may result from erroneous ordering falsely breaking up linkage blocks, or a higher genotyping error rate, rather than from increased recombination. Several lines of evidence suggest this is not the case. For genotype calling, we explicitly aimed for high accuracy by a combination of strict QC and manual calling, and estimated error rates at several stages of the process are very low (Appendix S1). Furthermore, very few recombination bins contain only 1 marker. More generally, our detailed physical map comparison to chromosome 3B (Figure 5) shows a considerably lower rate of disagreement (2.9%) than in CM2014 (10.3%). Our marker order on 3B, especially in the pericentromeric region, matches the physical map, which would not be expected if many unique sites resulted from errors. Local inversions would also show up as negative recombination values in Figure 5 (as they do in CM2014). For other chromosomes, assessing absolute accuracy of our map, especially of grouping, is hampered by the incompleteness of the wheat genome sequences. Comparatively, grouping was better than in CM2014. For ordering, our results are in strong agreement with the 3B results. Although 18% of markers in shared contigs were not in identical or adjacent map positions, it is highly unlikely that both markers are erroneously mapped, so the best estimate of our *error* rate is 9%, of which 6% is very local (1–5 unique sites). The remaining 3% of longer-range error matches the 2.9% visible component seen in Figure 5 for chromosome 3B. For 3B, our evidence suggests that some of this error results from duplicate genes. In summary, we have some minor local ordering issues common to most high-density genetic maps, but have greatly improved longer-range error compared to CM2014. A caveat of using MAGIC populations for mapping is that map accuracy may be diminished in SD regions. Shah *et al.* (2014) developed an approach to minimize this effect for distortions involving a single founder SD locus, which we used successfully on chromosome 3B. Until a more general method-ology is developed, we believe that including and noting the introgressions, as we do here, is the optimal approach. Overall, we are confident that our map represents a significant improve-ment in both precision and map accuracy over previously published maps. The completed map is available for download from http://www.niab.com/pages/id/326/Resources.

Genome diversity

Given the lack of deliberate selection on all but one trait, plant height, we found a remarkably extensive and diverse array of linkage blocks showing SD in our MAGIC population (Figure 6, Table 5). Many of these extend for substantial distances along the chromosome (16 blocks > 20 cM), suggesting that the underly-ing cause was operating from the earliest intercrossing genera-tions. This unanticipated degree of SD may reflect negative interactions between linkage blocks from different genetic backgrounds in a multifounder population, or simply reflect lower detection ability in previous biparental and low-marker-density populations. We assume that most of these blocks either cause meiotic problems or are subject to some form of viability or fertility selection. There is evidence in wheat for meiotic problems causing SD, particularly in interspecific crosses. For example, gametocidal genes from *Aegilops* species are expressed in crosses to bread wheat (Endo, 1990) and in general crosses between even closely related species such as durum wheat and *T. dicoc-coides* show considerable SD (Avni *et al.*, 2014).

Several interspecific introgression fragments have been incor-porated into bread wheat germplasm in an attempt to broaden the genetic base and improve specific traits such as disease resistance and yield (e.g. 1B rye introgression). Often such fragments are known to exhibit SD. Such cases are likely to be characterized by a high marker density as well as significant SD for two reasons: (i) a higher frequency of markers will be polymorphic between species than within species (ii) reduced recombination between the introgressed fragment and native chromosomes. Six major SD fragments were also associated with HMD blocks in our data set, including 3 of the 4 strongest and 5 of the 10 longest SD blocks. The top 3 are the well-known rye introgression on 1B (Worland and Snape, 2001), with strong SD against the introgression, and two blocks with Robigus alone as the minority founder. Robigus is thought to have *T. dicoccoides* in its pedigree, but the locations of introgressed fragments in Robigus and its descendent lines have been a matter of speculation. One of the Robigus SD-HMD blocks at the distal end of chromosome 4AL (170–210 cM) is a perfect fit to the model of an interspecific introgressed fragment, with strong SD against Robigus in MAGIC centred on an HMD block, and restriction to Robigus and Robigus descendants only in the 820K SNP array data set from Bristol University. In the 820K data set, this block is by far the largest linkage block of Robigus-restricted markers, including 23% of all perfect match Robigus SNPs on the array. The other SD-HMD Robigus linkage block in MAGIC, on chromosome 3B, is being strongly selected *for* in the MAGIC population (allele frequency 1.8× expectation). The presence of a SD-HMD block in MAGIC, the long chromosomal range of the effect and the occurrence of Robigus alleles in Glasgow all suggest the presence of a *T. dicoccoides* introgression fragment. On the other hand, this SD block is clearly found in some older UK wheat lines and some of their more recent descendants, suggesting it is either of hexaploid wheat origin or there has been at least one more, older introgression of the fragment into the UK wheat gene pool. In the 820K array, alleles of both these Robigus fragments occur sporadically in accessions of wheat wild relatives, including a single *T. dicoccoides* accession, so definitive assignment to *T. dic-occoides* could not be confirmed from this source. Forthcoming genomic resources for *T. dicoccoides* (F. Leigh, M. Caccamo, pers. comm.) will help resolve this question.

Knowledge of the location of interspecific introgressions and other linkage blocks containing useful alleles but showing negative interactions in common genetic backgrounds is impor-tant in disentangling their desirable and undesirable effects and in tracking their inheritance during breeding. Identifying the two Robigus SD blocks is an exemplar of the potential of the MAGIC approach to mapping to aid in this process. We are presently researching the underlying genetic causes and phenotypic effects of these complex SD-HMD and translocation patterns in our MAGIC population in further detail.

Methods

Plant material

Details of the MAGIC population construction are given in Mackay *et al.* (2014) and Appendix S1. Genotyping was performed using the Illumina Infinium iSelect 80 000 SNP wheat array ('80K array', http://www.illumina.com/), described in Wang *et al.* (2014).

SNP genotype calling

The polyploid module of Genome Studio V2011.1 (Illumina, San Diego, CA) developed for wheat by Wang *et al.* (2014) is poorly suited to genotype calling in our MAGIC population, due to the presence of a low-density heterozygote cluster in most assays. We used a strategy of strict QC/error quantification followed by manual curation of all assays that failed to pass QC—about half the markers in the final data set. This data set is a larger, higher quality version of that presented in Mackay et al. (2014). See Appendix S1 for details.

Map construction

The MAGIC map was constructed in two steps using the R package mpMap version 1.25 (Huang and George, 2011) available from github (https://github.com/behuang/mpMap). For the first round of mapping, we used a subset of 18 750 SNP markers scored as co-dominant. All heterozygote calls were set to missing. These markers were then filtered for missing founder genotypes as well as SD with P-values <1e-5. Recombination fractions between all pairs of markers were calculated using the function 'mpestrf' at default values. Markers were grouped hierarchically using the 'mpgroup' function into 300 linkage groups. These linkage groups were merged using R/mpMapInteractive (https://github.com/rohan-shah/mpMapInteractive) to produce larger groups which could then be assigned chromosome names based on marker groupings in previous genetic maps (Cavanagh et al., 2013; Wang et al., 2014; Genomezipper v5 (http://wheat-urgi.versailles.inra.fr/Seq-Repository/Genes-annotations) and in the Kansas deletion lines genotyped for the 80K array by Bristol University (http://www.cerealsdb.uk.net), as well as top BLASTn (Altschul et al., 1990) hits to IWGSC contigs listed in Wang et al. (2014, Table S6).

Within linkage groups, markers were ordered using two-point ordering implemented in the function 'mporder' using default settings. Fine ordering was performed interactively using R/mpMapInteractive. Map distances were computed using the Haldane mapping function using 'computemap'. Once a draft map was built, previously excluded markers, including all those scored as dominant, were mapped as traits to the existing draft chromosomes, that is we treated the unmapped markers as phenotypes in a QTL analysis, similarly to Rostoks et al. (2006). Founder haplotype probabilities were computed with the 'mpprob' function in mpMap implemented in R/qtl (Broman et al., 2003) with a threshold of 0.6 and QTL were calculated using single-stage QTL mapping in 'mpIM'. Linkage groups were then reordered following the same steps as previously but including all the previously excluded loci which could be mapped to a chromosome with $-\log_{10} P > 16$. In the construction of this final map, manual curation was used throughout the process and loci which could not be cleanly fitted into the ordered chromosomes were dropped.

Map validation

Recombination fractions were visualized using R/mpMap, and by PCA of the recombination fraction matrix to qualitatively assess if chromosomes were well ordered. We plotted the decay of linkage disequilibrium (LD) against map distance in our MAGIC map, 'NIAB2015', and in the Wang et al. (2014) consensus map, 'CM2014', using R/popgen (Marchini, 2013) to calculate LD, then fitting a lowess curve with smoothing span parameter = 0.10. For cross-chromosome comparisons to other published maps, we used a modified version of the 'plotMap' function in R/qtl (Broman et al., 2003). We validated our map against both published wheat genome sequences (Chapman et al., 2015; International Wheat Genome Sequencing Consortium, 2014). Flanking sequences of SNP markers in our map were initially blasted against both published genomes using BLASTn (Altschul et al., 1990) with an e-value cut-off of 1e-20 and ≥99 identity. For SNP markers with multiple blast hits, we retained the top hit and all other hits which had an equal match or one mismatch worse than the top hit. Where the marker assay had only one segregating SNP and one of the multiple BLAST hits was the same as our map location, we selected this hit as our top hit. For IWGSC data, we converted IWGSC2 coordinates to IWGSC1 coordinates using EnsemblPlants (www.plants.ensembl.org).

Alignment to 3B reference sequence

Flanking marker sequences for both NIAB2105 and CM2014 were aligned to the 3B pseudomolecule (Choulet et al., 2014; https://urgi.versailles.inra.fr/download/wheat/3B/) using BLASTn. We only included alignments which were part of the final 3B pseudomolecule and had no mismatches. The strong Robigus SD locus was controlled for using the method described in Shah et al. (2014), and map distances were re-estimated using the Haldane mapping function. This 783-marker long MAGIC chromosome 3B had a reduced length of 230 cM compared to the original of 284 cM with 1408 markers. The number of crossovers per line was calculated using the function 'mpprob' in mpMap with either the options program = 'qtl" (Broman et al., 2003) or program = 'happy' (Mott et al., 2000) and a threshold of 0.5. In R/qtl the number of crossovers was calculated using the function 'calculateXO' on default settings for eight-parent RILs. Recombination per physical distance (cM/Mb) was calculated and visualized using R/MareyMap 1.3.0 (Rezvoy et al., 2007). A cubic spline with a smoothing parameter of spar 0.65 was fitted to calculate local recombination rates. Markers with inconsistent alignments were excluded from the fitting. Chromosome 3B of CM2014 was aligned to the 3B pseudomolecule as described above (with a total of 1406 markers), and its genetic distance rescaled to match the reduced 3B MAGIC map length of 230 cM.

Genome diversity analysis

Marker density was calculated in a 10-cM sliding window along each chromosome to identify HMD blocks. Segregation distortion was estimated with chi-squared tests using fdr correction for multiple comparisons. SD markers were assigned into linkage blocks by scanning along chromosomes for consistent patterns. We confirmed the SD assessment of all these blocks by manually rescoring up to 5 individual markers per SD block and recalculating SD. Flanking sequences of SNP markers in the 820K array data set were blasted against the IWGSC genome sequence in a similar manner to our map validation and then aligned to NIAB2015 via our IWGSC top hits. Details are given in Appendix S1.

Acknowledgements

Development of the NIAB MAGIC population was led by Ian Mackay and Phil Howell. Richard Horsnell, Alison Bentley, Gemma Rose, Tobias Barber, Claire Pumfrey and Nick Gosman contributed to the creation of the population. Rhian Howells and RAGT seeds UK, Ltd. provided genotyping data for mapped PCR markers. Funding for development of the population was provided by BBSRC Crop Science Initiative grant BB/E007201/1 and funding for genotyping with the 80K array was provided by the NIAB Trust. LMW was funded by BBSRC iCASE grant BB/K011790/1 awarded to Alex Webb (University of Cambridge), Matthew Hannah (Bayer CropScience) and Andy Greenland

(NIAB). Many thanks to Emma Huang and Rohan Shah for helpful discussions and comments, especially with the use of mpMap and associated software. The WAGTAIL Consortium kindly gave us permission to cite the number of scorable markers in the WAGTAIL panel for comparison to MAGIC.

References

Altschul, S.F., Gish, W., Miller, W., Myers, E.W. and Lipman, D.J. (1990) Basic local alignment search tool. *J. Mol. Biol.* **215**, 403–410.

Avni, R., Nave, M., Eilam, T., Sela, H., Alekperov, C., Peleg, Z., Dvorak, J. *et al.* (2014) Ultra-dense genetic map of durum wheat x wild emmer wheat developed using the 90K iSelect SNP genotyping assay. *Mol. Breed.* **34**, 1549–1562.

Badaeva, E.D., Dedkova, O.S., Gay, G., Pukhalskyi, V.A., Zelenin, A.V., Bernard, S. and Bernard, M. (2007) Chromosomal rearrangements in wheat: their types and distribution. *Genome*, **50**, 907–926.

Bandillo, N., Raghavan, C., Muyco, P.A., Sevilla, M.A.L., Lobina, I.T., Dilla-Ermita, C.J., Tung, C.W. *et al.* (2013) Multi-parent advanced generation inter-cross (MAGIC) populations in rice: progress and potential for genetics research and breeding. *Rice*, **6**, 15.

Broman, K.W. (2005) The genomes of recombinant inbred lines. *Genetics*, **169**, 1133–1146.

Broman, K.W., Wu, H., Sen, S. and Churchill, G.A. (2003) R/qtl: QTL mapping in experimental crosses. *Bioinformatics*, **19**, 889–890.

Cavanagh, C., Morell, M., Mackay, I. and Powell, W. (2008) From mutations to MAGIC: resources for gene discovery, validation and delivery in crop plants. *Curr. Opin. Plant Biol.* **11**, 215–221.

Cavanagh, C.R., Chao, S.M., Wang, S.C., Huang, B.E., Stephen, S., Kiani, S., Forrest, K. *et al.* (2013) Genome-wide comparative diversity uncovers multiple targets of selection for improvement in hexaploid wheat landraces and cultivars. *Proc. Natl Acad. Sci. USA*, **110**, 8057–8062.

Chapman, J.A., Mascher, M., Buluc, A., Barry, K., Georganas, E., Session, A., Strnadova, V. *et al.* (2015) A whole-genome shotgun approach for assembling and anchoring the hexaploid bread wheat genome. *Genome Biol.* **16**, 26.

Cheema, J. and Dicks, J. (2009) Computational approaches and software tools for genetic linkage map estimation in plants. *Brief. Bioinform.* **10**, 595–608.

Choulet, F., Alberti, A., Theil, S., Glover, N., Barbe, V., Daron, J., Pingault, L. *et al.* (2014) Structural and functional partitioning of bread wheat chromosome 3B. *Science*, **345**, 7.

Curtis, D. (1994) Another procedure for the preliminary ordering of loci based on 2 point LOD scores. *Ann. Hum. Genet.* **58**, 65–75.

Darvasi, A. and Soller, M. (1995) Advanced intercross lines, an experimental population for fine genetic-mapping. *Genetics*, **141**, 1199–1207.

Endo, T.R. (1990) Gametocidal chromsomes and their induction of chromosome mutations in wheat. *Jpn. J. Genet.* **65**, 135–152.

Huang, B.E. and George, A.W. (2011) R/mpMap: a computational platform for the genetic analysis of multiparent recombinant inbred lines. *Bioinformatics*, **27**, 727–729.

Huang, X.Q., Paulo, M.J., Boer, M., Effgen, S., Keizer, P., Koornneef, M. and van Eeuwijk, F.A. (2011) Analysis of natural allelic variation in Arabidopsis using a multiparent recombinant inbred line population. *Proc. Natl Acad. Sci. USA*, **108**, 4488–4493.

Huang, B.E., George, A.W., Forrest, K.L., Kilian, A., Hayden, M.J., Morell, M.K. and Cavanagh, C.R. (2012) A multiparent advanced generation inter-cross population for genetic analysis in wheat. *Plant Biotechnol. J.* **10**, 826–839.

International Wheat Genome Sequencing Consortium. (2014) A chromosome-based draft sequence of the hexaploid bread wheat (*Triticum aestivum*) genome. *Science*, **345**, 11.

Maccaferri, M., Ricci, A., Salvi, S., Milner, S.G., Noli, E., Martelli, P.L., Casadio, R. *et al.* (2015) A high-density, SNP-based consensus map of tetraploid wheat as a bridge to integrate durum and bread wheat genomics and breeding. *Plant Biotechnol. J.* **13**, 648–663.

Mackay, I. and Powell, W. (2007) Methods for linkage disequilibrium mapping in crops. *Trends Plant Sci.* **12**, 57–63.

Mackay, I.J., Bansept-Basler, P., Barber, T., Bentley, A.R., Cockram, J., Gosman, N., Greenland, A.J. *et al.* (2014) An eight-parent multiparent advanced generation inter-cross population for winter-sown wheat: creation, properties, and validation. *G3-Genes Genomes Genet.*, **4**, 1603–1610.

Marchini, J.L. (2013) *Popgen: Statistical and Population Genetics*. R package version 1.0–3. https://cran.r-project.org/web/packages/popgen/popgen.pdf.

Milner, G.M., Maccaferri, M., Huang, B.E., Mantovani, P., Massi, A., Frascaroli, E., Tuberosa, R. *et al.* (2015) A multiparental cross population for mapping QTL for agronomic traits in durum wheat (*Triticum turgidum* ssp. durum). *Plant Biotechnol. J.* **13**, 648–663.

Mott, R., Talbot, C.J., Turri, M.G., Collins, A.C. and Flint, J. (2000) A method for fine mapping quantitative trait loci in outbred animal stocks. *Proc. Natl Acad. Sci. USA*, **97**, 12649–12654.

Pascual, L., Desplat, N., Huang, B.E., Desgroux, A., Bruguier, L., Bouchet, J.P., Le, Q.H. *et al.* (2015) Potential of a tomato MAGIC population to decipher the genetic control of quantitative traits and detect causal variants in the resequencing era. *Plant Biotechnol. J.* **13**, 565–577.

Poland, J.A., Brown, P.J., Sorrells, M.E. and Jannink, J.L. (2012) Development of high-density genetic maps for barley and wheat using a novel two-enzyme genotyping-by-sequencing approach. *PLoS ONE*, **7**, 8.

Rezvoy, C., Charif, D., Gueguen, L. and Marais, G.A.B. (2007) MareyMap: an R-based tool with graphical interface for estimating recombination rates. *Bioinformatics*, **23**, 2188–2189.

Rostoks, N., Ramsay, L., Mackenzie, K., Cardle, L., Bhat, P.R., Roose, M.L., Svensson, J.T. *et al.* (2006) Recent history of artifical outcrossing facilitates whole-genome association mapping in elite inbred crop varieties. *Proc. Natl Acad. Sci. USA*, **103**, 18656–18661.

Sannemann, W., Huang, B.E., Mathew, B. and Léon, J. (2015) Multi-parent advanced generation inter-cross in barley: high-resolution quantitative trait locus mapping for flowering time as a proof of concept. *Mol. Breeding*, **35**, 86.

Shah, R., Cavanagh, C.R. and Huang, B.E. (2014) Computationally efficient map construction in the presence of segregation distortion. *Theor. Appl. Genet.* **127**, 2585–2597.

Thepot, S., Restoux, G., Goldringer, I., Hospital, F., Gouache, D., Mackay, I. and Enjalbert, J. (2015) Efficiently tracking selection in a multiparental population: the case of earliness in wheat. *Genetics*, **199**, 609–623.

Threadgill, D.W. and Churchill, G.A. (2012) Ten years of the collaborative cross. *G3-Genes Genomes Genet.*, **2**, 153–156.

Villareal, R.L., Rajaram, S., Mujeebkazi, A. and Deltoro, E. (1991) The effect of chromosome 1B/1R translocation on the yield potential of certain spring wheats (*Trtiticum aestivum*). *Plant Breed.* **106**, 77–81.

Wang, S., Wong, D., Forrest, K., Allen, A., Chao, S., Huang, B.E., Maccaferri, M. *et al.* (2014) Characterization of polyploid wheat genomic diversity using a high-density 90,000 single nucleotide polymorphism array. *Plant Biotechnol. J.* **12**, 787–796.

Wilkinson, P.A., Winfield, M.O., Barker, G.L.A., Allen, A.M., Burridge, A., Coghill, J.A., Burridge, A. *et al.* (2012) CerealsDB 2.0: an integrated resource for plant breeders and scientists. *BMC Bioinformatics*, **13**, 219.

Worland, A.J. and Snape, J.W. (2001) Genetic basis of worldwide wheat varietal improvement. In *The World Wheat Book* (Bonjean, A.P., Angus, W.J., eds), pp. 59–100. Paris, France: Lavoisier.

Yu, J.M., Holland, J.B., McMullen, M.D. and Buckler, E.S. (2008) Genetic design and statistical power of nested association mapping in maize. *Genetics*, **178**, 539–551.

15

The ERF transcription factor TaERF3 promotes tolerance to salt and drought stresses in wheat

Wei Rong[1,2,†], Lin Qi[1,†], Aiyun Wang[2,†], Xingguo Ye[1,†], Lipu Du[1], Hongxia Liang[1], Zhiyong Xin[1] and Zengyan Zhang[1,*]

[1]National Key Facility for Crop Gene Resources and Genetic Improvement/Key Laboratory of Biology and Genetic Improvement of Triticeae Crops of the Agriculture Ministry, Institute of Crop Science, Chinese Academy of Agricultural Sciences, Beijing, China
[2]Central South University of Forestry and Technology, Changsha, China

*Correspondence

email zhangzengyan@caas.cn
[†]These authors contributed equally to this work.

Keywords: Triticum aestivum, ERF protein TaERF3, TaERF3 overexpression, salt and drought tolerance, stress-related genes, TaERF3 silencing.

Summary

Salinity and drought are major limiting factors of wheat (Triticum aestivum) productivity worldwide. Here, we report the function of a wheat ERF transcription factor TaERF3 in salt and drought responses and the underlying mechanism of TaERF3 function. Upon treatment with 250 mM NaCl or 20% polyethylene glycol (PEG), transcript levels of TaERF3 were rapidly induced in wheat. Using wheat cultivar Yangmai 12 as the transformation recipient, four TaERF3-overexpressing transgenic lines were generated and functionally characterized. The seedlings of the TaERF3-overexpressing transgenic lines exhibited significantly enhanced tolerance to both salt and drought stresses as compared to untransformed wheat. In the leaves of TaERF3-overexpressing lines, accumulation levels of both proline and chlorophyll were significantly increased, whereas H_2O_2 content and stomatal conductance were significantly reduced. Conversely, TaERF3-silencing wheat plants that were generated through virus-induced gene silencing method displayed more sensitivity to salt and drought stresses compared with the control plants. Real-time quantitative RT-PCR analyses showed that transcript levels of ten stress-related genes were increased in TaERF3-overexpressing lines, but compromised in TaERF3-silencing wheat plants. Electrophoretic mobility shift assays showed that the TaERF3 protein could interact with the GCC-box cis-element present in the promoters of seven TaERF3-activated stress-related genes. These results indicate that TaERF3 positively regulates wheat adaptation responses to salt and drought stresses through the activation of stress-related genes and that TaERF3 is an attractive engineering target in applied efforts to improve abiotic stress tolerances in wheat and other cereals.

Introduction

Wheat (Triticum aestivum) is one of the most important staple crops in the world. The global demand for wheat and other foods will increase along with the constantly increasing world population. This growing demand for food is paralleled by dramatic losses of arable land due to the increasing severity of soil destruction by abiotic environmental conditions (Golldack et al., 2011). Salinity and drought are two major environmental factors that have a crucial impact on the productivity and yield of wheat and other crops. As such, it is of the utmost importance and urgency to understand the responses of wheat to salt and drought stresses, and to breed wheat cultivars with improved salt and drought tolerances.

As sessile organisms, in order to cope with abiotic stresses, plants have evolved a tuning adaptation network to perceive stress signals and modulate the expression of specific stress-responsive genes of various functions (Yamaguchi-Shinozaki and Shinozaki, 2006; Zhang et al., 2007a; Zhu, 2002). Many stress-induced genes have been identified, including those encoding key enzymes of the phytohormone abscisic acid (ABA) biosynthesis pathway, proteins involved in osmotic adaptation and tolerance to cellular dehydration (Shinozaki and Yamaguchi-Shinozaki, 1997), cellular protective enzymes (Ingram and Bartels, 1996) and signalling and regulatory proteins, including protein kinases/protein phosphatases (Hong et al., 1997) and transcription factors (TFs) (Mao et al., 2011; Yamaguchi-Shinozaki and Shinozaki, 2006). Among these stress-induced genes, TFs, such as ABA-responsive element (ABRE)-binding proteins, bZIP, zinc finger proteins, NAC, MYB and AP2/ERF proteins, are known to play pivotal roles in controlling the expression of specific stress-related genes.

The AP2/ERF TF family, being specific to plants, includes the ethylene-response factor (ERF) and dehydration-responsive element (DRE)-binding protein (DREB)/CBF subfamilies (Nakano et al., 2006). DREB/CBF members have been shown to play an important role in plant responses to cold, drought and salt stresses (Liu et al., 1998; Morran et al., 2011; Nakano et al., 2006). Earlier studies showed that the ERF subfamily members were primarily involved in biotic stresses (Berrocal-Lobo et al., 2002; Chen et al., 2008; Dong et al., 2010; Oñate-Sánchez et al., 2007). Recent investigations indicated that different ERF members may play diverse functions in plant responses to both abiotic and biotic stresses (Hattori et al., 2009; Oh et al., 2009; Schmidt et al., 2013; Xu et al., 2007; Zhang et al., 2007a, 2012a). For example, overexpression of the tomato ERF gene TSRF1 positively regulates pathogen resistance in tomato and tobacco, negatively regulates osmotic response in tobacco, but improved drought tolerance in rice (Quan et al., 2010; Zhang et al., 2007a). Distinct ERF TFs in different rice varieties participate in flooding tolerance in the escape or quiescence tolerance strategies (Hattori et al., 2009; Xu et al., 2006). Constitutive overexpression of rice ERF genes AP37 and AP59 increases the drought tolerance of rice at the vegetative stage. Interestingly, the overexpression of AP37 increased grain yield under both severe drought and normal growth conditions, whereas the overexpression of AP59 reduced grain yield (Oh *et al.*, 2009). Although there has been much progress in the identification and functional analysis of *ERF* genes in model plants, the ERF TFs in wheat have not been characterized well, specifically with respect to the involvement of these TFs in regulation of the crop's biotic and abiotic stress responses (Dong *et al.*, 2012).

Given that distinct ERF factors may have different regulation roles in response to salt and drought stresses between species, research into wheat-specific ERF genes is very important for understanding the regulatory functions of ERFs in wheat. Using *in silico* analysis based on the presence of the conserved AP2/ERF domain amino acid sequence of *Arabidopsis thaliana*, at least 47 ERF genes were identified from expressed sequence tags (ESTs) of wheat (Zhuang *et al.*, 2011), although few of these have been functionally characterized. The ectopic expression of a wheat ERF protein TaERF1, belonging to the B-2 group of the ERF subfamily, improved tolerance to salt, drought and cold stresses, as well as resistance to the bacterial pathogen *P. syringae* in transgenic *Arabidopsis* (Xu *et al.*, 2007). The overexpression of a wheat ERF transcription activator TaPIEP1, belonging to the B-3 ERF group, enhanced resistance to infection by the soilborne fungus *Bipolaris sorokiniana* in transgenic wheat (Dong *et al.*, 2010). Dong *et al.* (2012) isolated a salt-responsive ERF gene *TaERF4* from wheat, belonging to B-1 subgroup; the ectopic expression of TaERF4 in *Arabidopsis* enhanced sensitivity to salt stress. In a previous study, we demonstrated that the wheat ERF protein TaERF3 was an ERF transcription activator and that its expression was induced by pathogen infection (Zhang *et al.*, 2007b). Recently, we found the *TaERF3* gene is responsive to salt and drought stress stimuli. However, the roles of TaERF3 in wheat responses to the abiotic stresses and the mechanisms underlying the functions have not been reported previously.

In this study, we characterized the function of *TaERF3* in responses to salt and drought stresses through the generation and assessment of both overexpression and silencing wheat plants, and explored the putative molecular mechanisms underlying the observed functions. The data indicated that *TaERF3* contributes to salt and drought tolerance through the regulation of stress-responsive genes.

Results

TaERF3 expression in wheat was induced by exogenous NaCl, PEG and ABA

To evaluate whether *TaERF3* is involved in wheat responses to salt or dehydration stresses, real-time quantitative RT-PCR (qRT-PCR) was used to examine the gene transcript patterns in tolerant and sensitive wheat genotypes after treatment with either 250 mM NaCl or 20% PEG6000. PEG6000 can produce stringent water stress conditions by permeating plant roots and imitating soil dehydration (Zhang *et al.*, 2007a). As shown in Figure 1a, in the leaves of Yangmai 6 that is tolerant to salt and drought stresses (Wang *et al.*, 2012), the transcript level of *TaERF3* was markedly induced within 0.5 h of the imposition of 250 mM NaCl and reached a peak (nearly 9.5-fold over that of the control) at 2 h, then reduced, but remained higher than that of the control. Upon PEG treatment, the induction profile of *TaERF3* transcript in the leaves of Yangmai 6 was similar to the NaCl induction result;

the transcript reached a peak at 2 h (nearly 8.5-fold over that of the control) (Figure 1b). In the sensitive wheat Yangmai 12 treated with NaCl or PEG6000, transcript levels of *TaERF3* were also induced, but the degree of induction was decreased as compared to that observed for the tolerant Yangmai 6 (Figure 1a, b). The induction peak in Yangmai 12 appeared at 24 h after NaCl treatment (nearly 4-fold over that of the control) and at 12 h after PEG treatment (nearly 3-fold over that of the control). Additionally, following NaCl or PEG6000 treatment, the induction profiles of *TaERF3* transcript in roots of Yangmai 6 and Yangmai 12 were similar to those in leaves, but the expression peaks were reached more rapidly (1 h in Yangmai 6 treated by NaCl or PEG; at treatment with NaCl for 12 h or with PEG for 6 h in Yangmai 12), and the degrees of induction were higher relative to those observed in leaves (Figure S1). These results suggested that *TaERF3* may be involved in wheat early responses to salt and drought stresses.

Abscisic acid is known to play a crucial role in abiotic stress response in plants. To investigate the effect of ABA on TaERF3, the expression of *TaERF3* in wheat following treatment with ABA was analysed. As shown in Figure 1c, the transcript level of *TaERF3* in the leaves of wheat Yangmai 12 increased significantly after exogenous application of 0.1 mM ABA and reached a peak at 3 h, whereas the expression of *TaERF3* was significantly suppressed by 0.1 mM Na_2WO_4 (an ABA biosynthesis inhibitor). Furthermore, to explore whether TaERF3 is involved in salt stress response mediated by ABA, in the presence of the both Na_2WO_4 and NaCl, or in the presence of the both ABA and NaCl, the expression of *TaERF3* was investigated. As shown in Figure 1d, the transcript level of *TaERF3* was significantly increased in the leaves of Yangmai 12 that were pretreated with 0.1 mM ABA for 24 h and then subsequently treated with 250 mM NaCl solution as compared to mock+NaCl control. Conversely, following pretreatment with 0.1 mM Na_2WO_4 for 24 h, the *TaERF3* transcript in response to NaCl stimuli was significantly suppressed (Figure 1d). These data indicated that ABA positively modulated *TaERF3* expression and responses of *TaERF3* to salt and drought stresses in wheat.

Overexpression of *TaERF3* increases salt and drought tolerance in wheat

To evaluate the contribution of *TaERF3* to salinity and drought tolerance in wheat, a *TaERF3*-overexpressing transformation vector pA25-TaERF3 (Figure 2a(i)) was constructed. pA25-TaERF3 contained the *bar* gene for selection, and *TaERF3* was driven by the maize *ubiquitin* promoter. pA25-TaERF3 DNA was bombarded into 1200 Yangmai 12 immature embryos to generate *TaERF3*-overexpressing transgenic wheat plants. Positive transgenic T_0 plants were identified through phosphinothricin screening and PCR detection with transgene-specific primers (Figure 2a (i,ii)). PCR assays of the T_1–T_4 progenies (Figure 2a(ii)) showed that four stably transgenic lines, OX19, OX31, OX34 and OX78, were obtained with a transformation efficiency of 0.33% and that the introduced gene could be stably inherited. The qRT-PCR analyses indicated that compared to the wild-type (WT) Yangmai 12 or null segregants lacking the transgene, the transcript abundance of *TaERF3* in these four transgenic lines was significantly increased (Figure 2a(iii)). Furthermore, the transcript abundance of the *TaERF3* gene in the transgenic wheat lines was increased to a markedly higher level than in the WT wheat following salt or ABA treatment, but was repressed by Na_2WO_4 treatment (Figure S2).

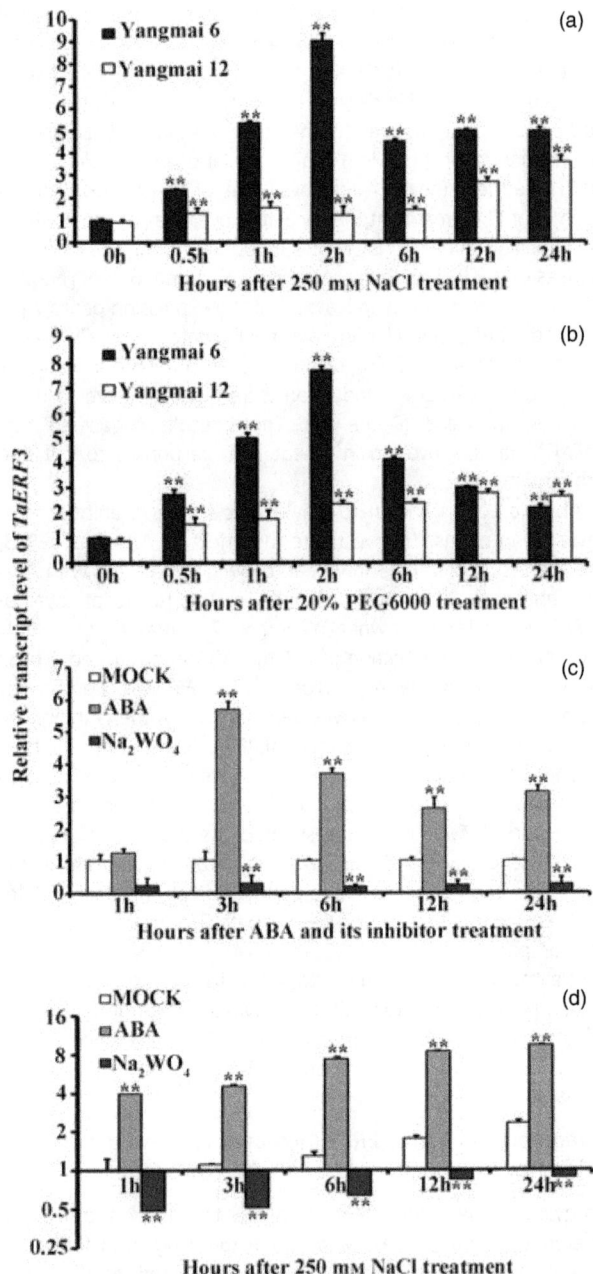

Figure 1 Expression patterns of *TaERF3* in wheat upon salt, dehydration and ABA treatment analysed by qRT-PCR. Those in leaves of tolerant wheat Yangmai 6 and sensitive wheat Yangmai 12 after treatment with (a) 250 mM NaCl and (b) 20% PEG6000. (c) Those in leaves of Yangmai 12 treated with 0.1 mM ABA, an ABA synthesis inhibitor Na_2WO_4 or mock solution (0.1% Tween 20). (d) Those in leaves of Yangmai 12 treated with 250 mM NaCl following pretreatment for 24 h with 0.1 mM ABA, Na_2WO_4 or mock solution. Relative transcript levels of *TaERF3* were quantified relative to that at 0 h (a, b) or compared with mock plants at the same time point (c, d). Three biological replicates were averaged and statistically analysed using Student's t-test (*$P < 0.05$, **$P < 0.01$). Bars indicate standard error of the mean (SE).

To test the contribution of *TaERF3* to drought stress tolerance responses in wheat, seedlings of WT Yangmai 12 and the four transgenic lines (OX19, OX31, OX34 and OX78) in T_3–T_4

generations were subjected to ~30 days of water-withholding (drought treatment). The WT Yangmai 12 plants showed leaf wilting and rolling and lost chlorophyll much earlier and more severely than did *TaERF3*-overexpressing transgenic plants. At ~30 day after water-withholding (just prior to rewatering), most WT Yangmai 12 plants showed severe wilting, whereas fewer *TaERF3*-overexpressing plants showed wilting, and the wilting that was observed in the *TaERF3*-overexpressing plants was less severe. After rewatering for 7 days, the survival rates of the four *TaERF3*-overexpressing lines were 91.58–100.00%, whereas only 15.73% of WT Yangmai 12 plants had recovered (Figure 2b,c). These results indicated that the overexpression of *TaERF3* in wheat could significantly enhance drought tolerance. Furthermore, as the water loss rate of detached leaves has been proposed as an essential parameter indicating water status in plants (Dhanda and Sethi, 1998), the water loss rates of the *TaERF3*-overexpressing wheat lines and WT Yangmai 12 were assessed at 12 time points over six hours. From 2 to 6 h, the water loss rates of *TaERF3*-overexpressing lines were much lower than that of WT Yangmai 12. At 6 h, the water loss rates of the *TaERF3*-overexpressing lines ranged from 29.50% to 35.59%, whereas that of WT Yangmai 12 was 43.99% (Figure 2d). These results indicated that *TaERF3* improved drought tolerance by enhancing plant water retention capability.

The effect of salinity on seed germination and seedling survival rates of the four transgenic lines in T_3–T_4 generations and WT Yangmai 12 was investigated. Following 200 or 300 mM NaCl treatment for 7 days, the seed germination rates of the four *TaERF3*-overexpressing transgenic lines were significantly higher than those of WT Yangmai 12 (Figure 3a,b). To further determine whether *TaERF3* overexpression affects seedling development under salt stress conditions, the seedlings at the 3-leaf stage were irrigated with 250 mM NaCl solution from 0 day and similarly irrigated every 4 days. Upon NaCl treatment for 20 days (a total of five irrigations), the majority of the *TaERF3*-overexpressing plants grew normally, while most of the WT Yangmai 12 plants displayed severe wilting (Figure 3c). Following irrigation with water for an additional 20 days, the survival rates of the four *TaERF3*-overexpressing lines were 58.00–67.35%, while that of WT Yangmai 12 was 17.78% (Figure 3d). These results indicated that *TaERF3* overexpression significantly enhanced salinity tolerance in transgenic wheat.

Physiological changes in *TaERF3*-overexpressing wheat under salt stress

Free proline content, the redox homeostasis of cells and chlorophyll content and stomatal conductance in plant leaves are important physiological indices of salinity and drought tolerance (Mao et al., 2011; Szabados and Savour, 2009; Zhang et al., 2012a). To explore the physiological mechanism of abiotic stress tolerance conferred by *TaERF3* overexpression, we investigated the proline content, chlorophyll content, H_2O_2 and stomatal conductance in *TaERF3*-overexpressing and WT Yangmai 12 wheat plants following 250 mM NaCl treatment. The free proline content in all of the *TaERF3*-overexpressing wheat lines was significantly higher than in WT Yangmai 12, suggesting that the overexpression of *TaERF3* improved the proline accumulation in leaves under stress (Figure 4a). Compared with WT Yangmai 12, *TaERF3*-overexpressing transgenic lines accumulated more chlorophyll (Figure 4b). As shown in Figure 4c, the overexpression of *TaERF3* reduced the accumulation of H_2O_2 in leaves under salt stress, suggesting that *TaERF3* enhances salt tolerance possibly by

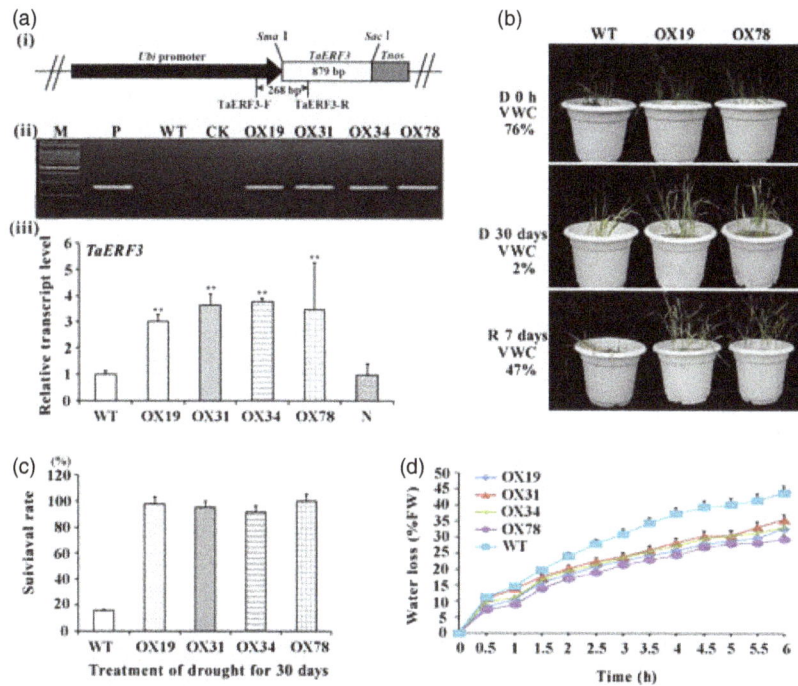

Figure 2 Molecular characterization and drought tolerance responses of *TaERF3*-overexpressing transgenic lines. (a) (i) Schema of the transformation vector pA25-TaERF3. *TaERF3* in the transformation vector was driven by the maize *ubiquitin* promoter. The arrow indicates the region amplified in the PCR assays using transgene-specific primers; (ii) PCR pattern of *TaERF3* transgenic lines and nontransgenic Yangmai 12 using transgene-specific primers. M: 100-bp DNA ladder, P: transformed vector plasmid pA25-TaERF3, WT: nontransformed Yangmai 12, N: the null segregant lacking the transgene; (iii) the transcript levels of *TaERF3* in the roots of T_4 transgenic and control wheat plants via qRT-PCR analysis. Those in transgenic lines (OX19, OX31, OX34 and OX78) and the null segregant lacking the transgene (N) were quantified relative to WT Yangmai 12. (b) Drought-tolerant phenotypes and (c) survival rates of WT Yangmai 12 and *TaERF3*-overexpressing lines of the T_4 generation. D30: water-withholding for 30 days. R7: rewatering for 7 days. (d) Water loss in detached leaves from WT Yangmai 12 and transgenic lines. Statistically significant difference was analysed using Student's t-test (**$P < 0.01$).

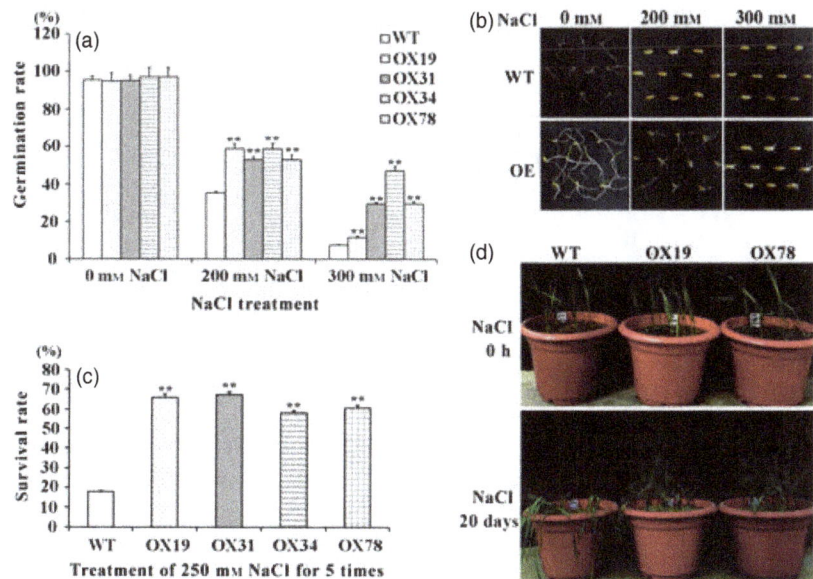

Figure 3 Salt tolerance responses of *TaERF3*-overexpressing transgenic lines. (a) Seed germination phenotypes and (b) germination rates of *TaERF3*-overexpressing lines (OX) of the T_4 generation and WT Yangmai 12 treated with 0, 200 or 300 mM NaCl solution for 7 days. (c) Salt tolerance phenotypes and (d) survival rates of WT Yangmai 12 and transgenic lines irrigated with 250 mM NaCl solution for 20 days (five times of irrigations). Statistically significant difference was analysed using Student's t-test (**$P < 0.01$).

controlling the redox homeostasis of cells during salt stress. Moreover, following NaCl treatment, *TaERF3*-overexpressing transgenic wheat maintained lower stomatal conductance than WT Yangmai 12 (Figure 4d), suggesting that *TaERF3*-overex-pressing wheat may lose less water than WT wheat. These results suggested that the overexpression of *TaERF3* resulted in physi-ological changes in transgenic wheat, which in turn enhanced the observed salt and drought tolerance.

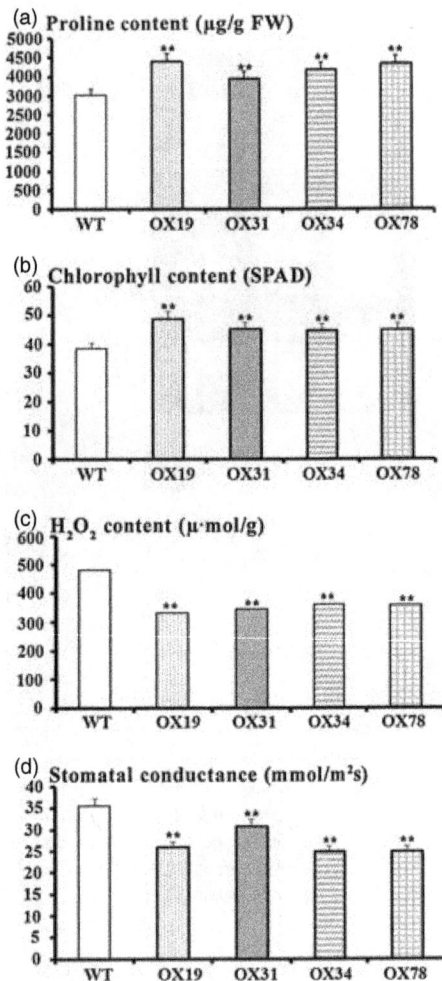

Figure 4 Physiological changes in *TaERF3*-overexpressing transgenic wheat. (a) Proline content, (b) chlorophyll content, (c) H_2O_2 content and (d) stomatal conductance in leaves of WT Yangmai 12 and *TaERF3*-overexpressing lines treated with 250 mM NaCl for 0 h or 20 days. Three replicates were averaged and statistically significant differences between OX transgenic lines with WT at the same condition were analysed using Student's *t*-test (**$P < 0.01$).

Tolerance to salt and drought was compromised by *TaERF3* silencing in wheat

Virus-induced gene silencing (VIGS) developed with barley stripe mosaic virus (BSMV) has been demonstrated to be an effective reverse genetics tool for studying the function of genes in barley and wheat (Hein *et al.*, 2005; Holzberg *et al.*, 2002; Scofield *et al.*, 2005; Zhou *et al.*, 2007). To obtain reverse genetic evidence of the regulatory role for *TaERF3* in salt and drought stress responses, a recombinant virus BSMV:TaERF3 was prepared (Figure S3) and used to silence *TaERF3* in WT Yangmai 12 and Yangmai 6. At 14 days postinoculation (dpi) with BSMV:TaERF3 virus or BSMV:GFP (green fluorescent protein) virus, yellow mosaic spots appeared in leaves of the infected wheat plants, and the expression of BSMV coat protein gene (*CP*) could be detected from these tissues (Figures 5b(i), S4a), proving that these viruses successfully infected these plants. The transcript levels of *TaERF3* were markedly decreased in Yangmai 6 and Yangmai 12 inoculated with BSMV:TaERF3 (Figures 5b,c, S4b,c),

which were henceforth referred to as *TaERF3*-silencing plants. Following 250 mM NaCl treatment, the *TaERF3*-silencing Yangmai 6 or Yangmai 12 plants displayed more severe wilting than did the BSMV:GFP or buffer-inoculated (mock) plants (Figures 5d, S4d). Following 20 days (five irrigations) of 250 mM NaCl treatment, the survival rate of the *TaERF3*-silencing Yangmai 6 plants was 21.86%, whereas those of the BSMV:GFP and the mock Yangmai 6 plants were 50.00% and 54.54%, respectively. Following 250 mM NaCl treatment for 15 days (four irrigations), the survival rate of the *TaERF3*-silencing Yangmai 12 was 8.62%, whereas those of the BSMV:GFP and the mock Yangmai 12 plants were 15.63% and 16.30%, respectively. Furthermore, following drought stimuli, the *TaERF3*-silencing wheat plants displayed more severe wilting than did the BSMV:GFP and the buffer-inoculated (mock) plants (Figure 5e). Following water-withholding for 30 days, *TaERF3*-silencing Yangmai 6 plants had a lower survival rate (25.81%) than did the BSMV:GFP (54.54%) and the mock Yangmai 6 plants (60.00%). Additionally, following water-withholding for 25 days, TaERF3-silencing Yangmai 12 plants had a lower survival rate (6.82%) than did the BSMV:GFP Yangmai 12 plants (10.31%). These results suggested that *TaERF3* silencing reduced salt and drought tolerance and that *TaERF3* was required for wheat responses to salt and drought stresses.

TaERF3 activates the expression of stress-responsive genes

Transcription factors have been implicated in stress responses through the regulation of stress-related genes. Our previous studies showed that both the wheat TaERF3 protein and its homolog TiERF1 in *Thinopyrum intermedium* contain an activation domain and were proved to be ERF transcription activators (Chen *et al.*, 2008; Zhang *et al.*, 2007b). Our study showed that the overexpression of *TiERF1* resulted in the up-regulation of many defence- and stress-related genes in transgenic wheat (Chen *et al.*, 2008; unpublished microarray data). To investigate the putative molecular genetic mechanisms of TaERF3 function, based on the integration of our microarray analysis data from the *TiERF1* study, we firstly analysed the expression of thirteen stress-related genes of interest in *TaERF3*-overexpressing and WT wheat lines. Thirteen genes investigated encode a peroxidase (*POX2*, NCBI accession number: X85228), an oxalate oxidase (*OxOx2*, BE404592), an ERF (BQ161688), 2 beta-glucosidases (*BG3*, CA608204 and *BG31*, CK163588), 2 late-embryogenesis abundant proteins (*LEA3*, AY148409 and *LEA4*, AY148492), an ABA-responsive protein (*RAB18*, CK163855), a dehydrin (*DHN*, CK163818), a short-chain dehydrogenase/reductase (*SDR*, BE604215) that catalyses the oxidation of xanthoxin to abscisic aldehyde (a regulatory step in ABA biosynthesis, Seo and Koshiba, 2002), a tonoplast intrinsic protein (*TIP2*, AY525641), a chitinase (*Chit1*, CA665158) and a glutathione s-transferase (*GST6*, CA676105). The qRT-PCR analysis showed that, in the WT and *TaERF3*-overexpressing Yangmai 12 lines, the transcript tendency of ten stress-related genes, *BG3*, *LEA3*, *DHN*, *RAB18*, *SDR*, *TIP2*, *Chit1*, *POX2*, *OxOx2* and *GST6*, was in agreement with the microarray data. Hereafter, we further analysed the transcript levels of the ten stress-related genes in *TaERF3*-overexpressing and *TaERF3*-silencing and control wheat plants. The results showed that the transcript levels of these stress-related genes were significantly higher in the *TaERF3*-overexpressing plants than in WT Yangmai 12 (Figure 6). Conversely, the transcript levels of these stress-related genes were lower in the *TaERF3*-

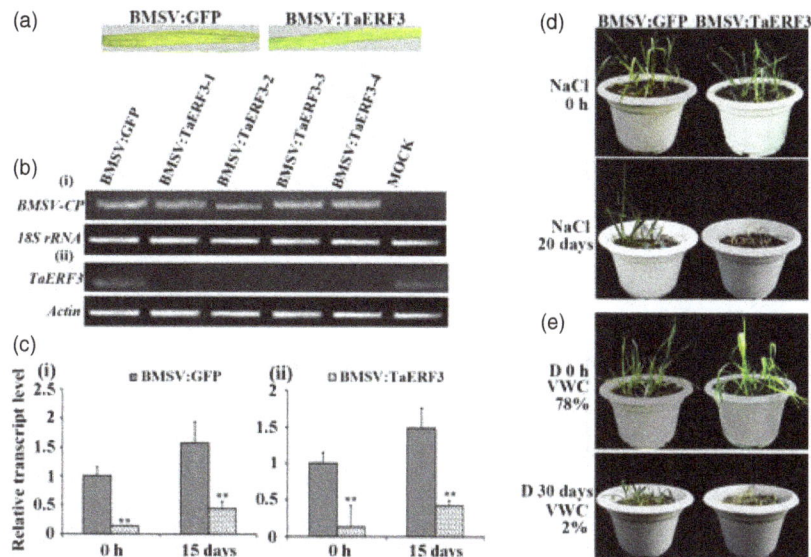

Figure 5 Molecular and stress tolerance characterizations of *TaERF3*-silencing Yangmai 6 plants. (a) BSMV-VIGS phenotypes on leaves of BSMV:GFP (control) and *TaERF3*-silencing (BSMV:TaERF3) Yangmai 6. (b) Transcript levels of *BSMV-CP* (i) and *TaERF3* (ii) in the BSMV:GFP-, BSMV:TaERF3- and mock-inoculated plants at 14 days postvirus inoculation. (c) Relative transcript levels of *TaERF3* in *TaERF3*-silencing wheat (BSMV:TaERF3) by qRT-PCR analysis following 250 mM NaCl (i) or drought (ii) treatments. Those in BSMV:TaERF3 plants were quantified relative to that in the BSMV:GFP plants at 0 h. Statistically significant differences between BSMV:TaERF3 and BSMV:GFP plants at the same time point based on three replications using Student's *t*-test (**$P < 0.01$). (d) Salt-tolerant phenotypes of the BSMV:TaERF3 and BSMV:GFP wheat following 20 days of 250 mM NaCl treatment (five irrigations). (e) Drought-tolerant phenotypes of the BSMV:TaERF3 and BSMV:GFP following water-withholding for 30 days.

silencing (BSMV:TaERF3 infected) Yangmai 6 and Yangmai 12 plants than in the BSMV:GFP plants (Figures 6, S5). These data indicated that TaERF3 functioned to activate the expression of these stress-related genes. Furthermore, these TaERF3-activated stress-related genes were also induced by drought, salt and exogenous ABA treatments, but repressed by Na_2WO_4; the transcript inductions of these genes were significantly increased in TaERF3-overexpressing wheat relative to the WT wheat (Figures 6 and 7; S6 and S7). These results suggested that TaERF3 mediated the responses of these stress-related genes to ABA, drought and salt stresses.

TaERF3 interacts with the GCC-box *cis*-element in stress-responsive genes

ERF proteins can directly or indirectly activate the expression of defence- and stress-related genes following interaction with GCC-box or non-GCC-box motifs, even through interaction with other transcription factors (Chakravarthy *et al.*, 2003; Quan *et al.*, 2010; Zhang *et al.*, 2007a, 2012a,b). In a previous study, electrophoretic mobility shift assays (EMSAs) showed that the TaERF3 protein could bind to the GCC-box motif (Zhang *et al.*, 2007b). To explore how TaERF3 activates the ten aforementioned stress-related genes, we searched *cis*-elements in 334- to 2230-bp promoter sequences upstream of ATG of all of these genes using PLACE (http://www.dna. affrc.go.jp/PLACE/) and usedEMSA to test whether TaERF3 directly interacts with the GCC- or non-GCC-boxes in these promoters. As shown in Table S1, the promoters of seven genes (*BG3*, *Chit1*, *RAB18*, *LEA3*, *TIP2*, *POX2* and *GST6*) contain several copies of GCC-boxes, whereas the GCC-box was not found in the incomplete promoters of *DHN*, *OxOx2* and *SDR*. The 10 promoters analysed also contained dehydration-responsive element (DRE) or ABA-responsive element (ABRE) or W-box (the *cis*-element interacting with WRKY TFs) or

CAAT-box sequences (Table S1). These promoter fragments, containing GCC-boxes in the above-mentioned seven genes, or DRE in *OxOx2* or ABRE in *TIP2* or CAAT-box in *GST6* or W-box in *LEA3*, were amplified or synthesized (Figure 8a), then used as probes to investigate whether TaERF3 protein binds to these *cis*-elements using EMSA. The EMSA results proved that TaERF3 bound specifically to all of the GCC-boxes in the seven stress-related genes (Figure 8b), but did not bind to the DRE, ABRE, CAAT-box or W-box *cis*-elements in the stress-related genes (Figure 8b(ii)). These results indicated that TaERF3 directly interacted with the GCC-boxes in the promoters of the stress-related genes and may thereby directly regulate the stress-related genes with the GCC-boxes in their promoters.

Discussion

Increasing evidence has shown that ERF proteins play crucial roles in regulating plant biotic and abiotic stress responses. Investigations into the roles and mechanisms of wheat ERFs in regulating stress responses are vital in efforts to understand plant adaptation to environmental stresses. TFs, including ERFs, are often up-regulated in responses to abiotic and biotic stresses (Yamaguchi-Shinozaki and Shinozaki, 2006; Zhang *et al.*, 2007a,b), and these responses are typically quite rapid (He *et al.*, 2011). In this study, qRT-PCR analyses showed that the transcriptional levels of *TaERF3* in wheat Yangmai 6 and Yangmai 12 were markedly induced in response to salt or PEG treatment. The transcript inductions of *TaERF3* in tolerant Yangmai 6 were generated more rapidly and to a higher degree than those in sensitive wheat Yangmai 12. Moreover, following irrigation with either salt or PEG solution into soil, both the degrees and peaks of *TaERF3* transcript induction were lower and slower in leaves than those observed in roots, suggesting that the responses of *TaERF3* to the stresses may

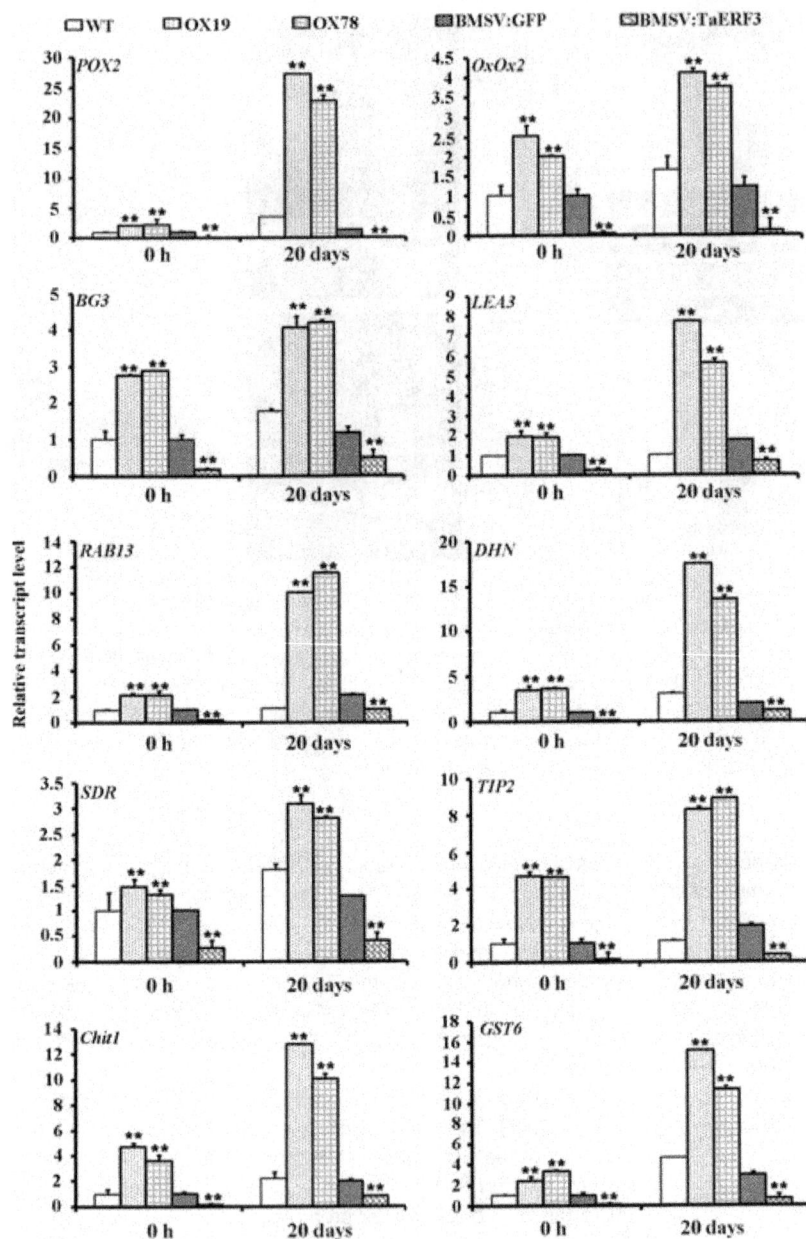

Figure 6 The transcript levels of ten stress-related genes in WT Yangmai 12, *TaERF3*-overexpressing Yangmai 12, as well *TaERF3*-silencing (BSMV:TaERF3) and BSMV:GFP Yangmai 6 treated with 250 mM NaCl for 20 days. Relative transcript abundances of the tested genes in *TaERF3*-overexpressing Yangmai 12 (OX19 and OX78), or *TaERF3*-silencing Yangmai 6, were quantified relative to that in WT Yangmai 12 at 0 h. Statistically significant differences between OX transgenic lines and WT, or between *TaERF3*-silencing and BSMV:GFP Yangmai 6 plants at the same condition, were determined based on three replications using Student's *t*-test (**$P < 0.01$).

initially occur in stress-localized tissues. These data suggested that *TaERF3* may be an important positive regulator of wheat adaptations to drought and salt stresses.

Certain phytohormones, including ABA and ethylene, play pivotal roles in abiotic stress responses in plants (Seo and Koshiba, 2002; Zhang *et al.*, 2007a, 2008). Certain ERF proteins, such as TSRF1, are molecular nodes of signal integration for the ethylene and ABA signalling pathways (Zhang *et al.*, 2008). Our previous study showed that *TaERF3* was ethylene inducible (Zhang *et al.*, 2007b). Here, the transcript abundance of *TaERF3* was also proven to be induced by exogenous ABA application, but repressed after the application of an ABA synthesis inhibitor Na_2WO_4, implying that *TaERF3* is positively responsive to ABA. To understand the molecular basis of the response of *TaERF3*, we analysed the *cis*-acting elements in the promoter of *TaERF3*. The *cis*-elements analysis (Table S1) showed that the promoter sequence of *TaERF3* contains ethylene-responsive elements (GCC-boxes) and ABA-responsive elements (ACGT and ABRE),

which may partially contribute to *TaERF3* responses to the hormone applications. Furthermore, the increased transcript abundance of *TaERF3* by salt stimuli was significantly suppressed after pretreatment with 0.1 mM Na_2WO_4, whereas that was enhanced after pretreatment with 0.1 mM ABA treatment (Figure 1d). The results indicated that in wheat, the response of *TaERF3* to salt stress was positively regulated by ABA.

In this study, to explore the roles of *TaERF3* in drought and salt stress adaptations in wheat, we generated *TaERF3*-overexpressing wheat lines via transformation and generated *TaERF3*-silencing wheat plants using BSMV-VIGS methods. The results of molecular characterization and salt and drought tolerance assays indicated that the transgene locus and the high expression levels of *TaERF3* could be inherited in these four *TaERF3*-overexpressing transgenic lines from T_1 to T_4 generations and that overexpression of *TaERF3* significantly improved drought and salt tolerance in these lines. Moreover, the reverse functional assays of *TaERF3*-silencing wheat plants indicated that silencing of *TaERF3* in wheat Yangmai

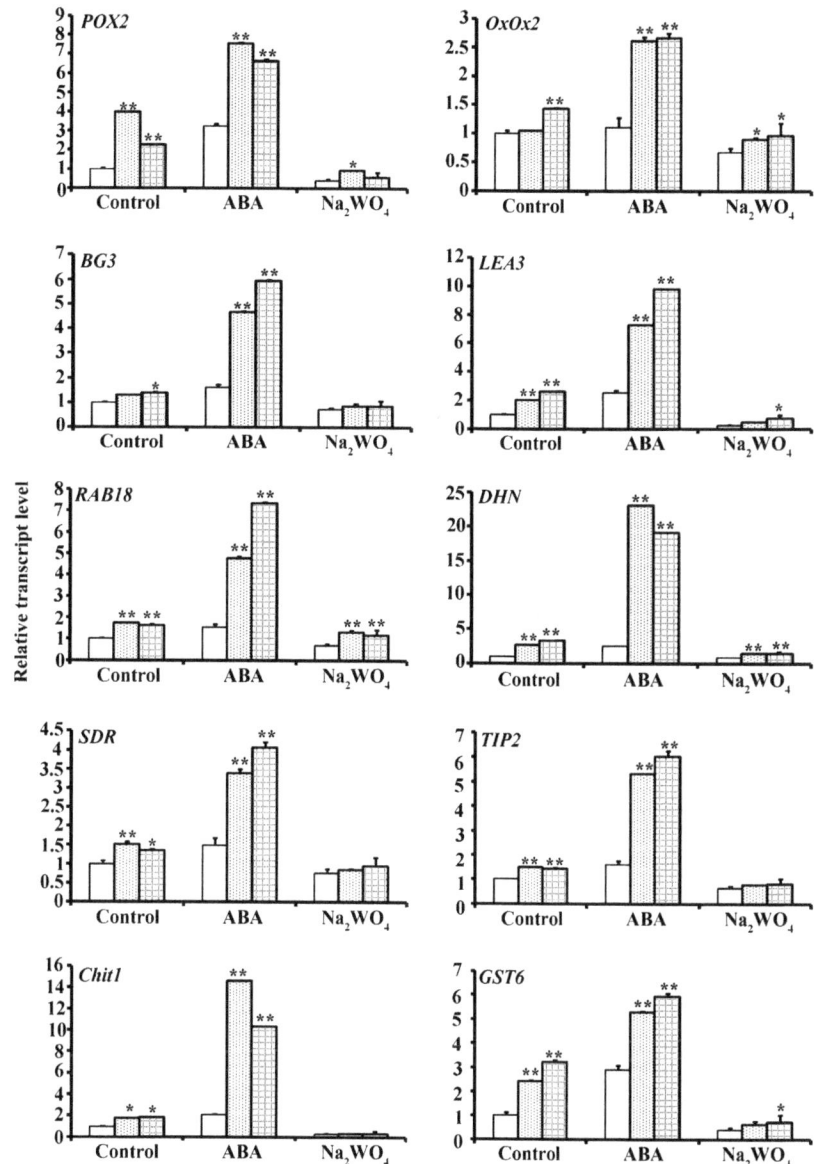

Figure 7 The transcript levels of ten stress-related genes in *TaERF3*-overexpressing and WT wheat Yangmai 12 lines treated with ABA or Na₂WO₄ for 3 h. Relative transcript abundances of the tested genes in *TaERF3*-overexpressing lines (OX19 and OX78) were quantified relative to those in the WT at 0 h. Statistically significant differences between OX transgenic lines and WT at the same condition were determined based on three replications using Student's t-test (*$P < 0.05$, **$P < 0.01$).

12 and Yangmai 6 led to suppressed tolerance to drought and salt stresses. These results indicated that TaERF3, acting as a positive regulator, plays a crucial role in wheat responses to salt and drought stresses. Interestingly, the overexpression of *TaERF3* did not affect the development and growth of the four transgenic wheat lines in the T₄ generation under normal growth conditions. Importantly, under either drought or salt stress, the seedlings of the four *TaERF3*-overexpressing transgenic wheat (Yangmai 12) lines showed a higher tolerance than did the tolerant wheat Yangmai 6. Thus, engineering and overexpression of *TaERF3* may be used for improving salt and drought tolerance of wheat.

Accumulating evidence indicates that activator-type ERF transcription factors regulate multiple stress responses through the activation of defence- and stress-related genes following interacting with various cis-acting elements including the GCC-box and even through interacting (heterodimerizing) with other transcription factor (Berrocal-Lobo et al., 2002; Buttner and Singh, 1997; Chakravarthy et al., 2003; Oñate-Sánchez et al., 2007; Zhang et al., 2007a, 2012a). Our previous study showed that the wheat ERF protein TaERF3 is an ERF transcription activator

in vivo (Zhang et al., 2007b). We supposed that TaERF3 possibly activates the expression of some stress-related genes following interaction with the GCC-box or other cis-elements. To investigate this notion, the ten aforementioned stress-related genes were subjected to qRT-PCR analysis in *TaERF3*-overexpressing and WT wheat lines, as well as in *TaERF3*-silencing and control wheat Yangmai 12 and Yangmai 6. LEA3 and DHN positively contribute to plant tolerances to cold, drought and salt stresses in other plant species (Shekhawat et al., 2011). RAB plays a crucial role in plant responses to ABA, drought and salt stresses (Sun et al., 2013). POX and GST that are ROS-scavenging enzymes are known to be involved in redox homeostasis of cells under various stress adaptive responses (Bhavanath et al., 2011; Huang et al., 2013). OxOx is involved in biotic or abiotic stress responses (Berna and Bernier, 1999). SDR is involved in responses to ABA and osmotic as well drought stresses (Quan et al., 2010; Seo and Koshiba, 2002). TIP is associated with plant tolerance to drought and salt stresses (Wang et al., 2011). BG positively contributes to drought, salt and freezing tolerance (Han et al., 2012). Chitinases participate in disease resistance and abiotic tolerance (Chen et al.,

(a)
GCC-box: 5' -AATTCATAAG<u>AGCCGCC</u>ACTCATAAG<u>AGCCGCC</u>ACTCC-3'

GCC-box (BG3): 5' - CCGAGCTGAATCCTTGATCCTGCAGCAAG<u>CCGCC</u>ATCACGATCGGCGTCTGCTACGGCGT-3'

GCC-box (Chit1): 5' - ATTCGGCACGAGGCGAACACAGCTACTA<u>GCCGCC</u>AGTACTTGGAGAAACCAACAATGGCG-3'

GCC-box (TIP2): 5' - ACCCGCTAGTGCGATTTAGG<u>CCGCC</u>TATCTAGGTGACATGATTGAACAAGTTTGTACAAT-3'

GCC-box (LEA3): 5' - TCGGAATCCGGTGTCGACCACGCCGTCC<u>GCCCGCC</u>ACGGCGTCGCGCGACTCCAAGCAC-3'

GCC-box (GST6): 5' - AGCGGCAACCTCTCGTG<u>CCGCC</u>TTGGCGAACTCAAGCAGCCTGCCCACGTCCGGCAGGGC-3'

GCC-box (POX2): 5' - CTGACCCCGACGG<u>CCGCC</u>TTGGCGGCGGGGGGAAAGACCGGCGAGGTGACGGTGTTCTGG-3'

GCC-box (RAB18): 5' - CTGACCCCGACGG<u>CCGCC</u>TTGGCGGCGGGGGGAAAGACCGGCGAGGTGACGGTGTTCTGG-3'

DRE-box: 5' -TCGACCGCAGGAGCC<u>ACCGACGG</u>CGGCGAGCGAGGCGAGG-3'

ABRE-box: 5' -GGATACTCTAGAGATTATC<u>AACGCGG</u>AACTAGCTAGTAGT-3'

W-box: 5' -GCCTAAAGCGAGT<u>GAC</u>ATTTCCCCCTTCCGAGCTAGCTCA-3'

CAAT-box: 5' -AATTACCATAAATTT<u>CAATTT</u>GTAAGCATTTTGCTGGCAG-3'

(b)

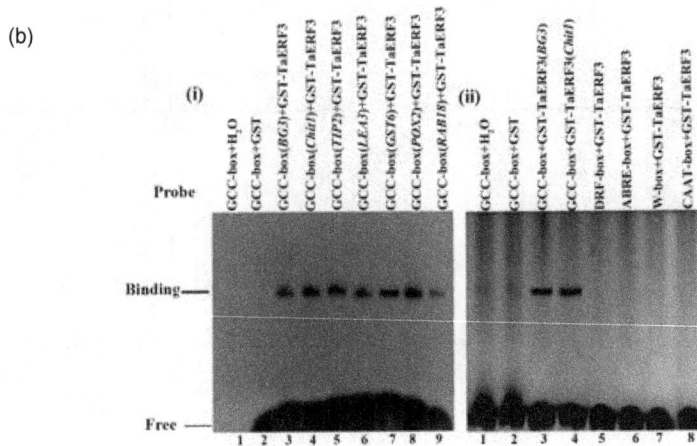

Figure 8 Assay of TaERF3 binding to the *cis*-acting element in the TaERF3-activated stress-related genes. (a) Partial sequences of different promoters with different *cis*-elements used as probes for EMSA. The core sequences of *cis*-elements are underlined. (b) EMSA indicated that TaERF3 protein (i) could and (ii) specifically bind to the GCC-boxes present in the seven stress-related genes, but failed to interact with DRE, ABRE, CAAT-box and/or W-box *cis*-elements (ii). The free probe and the binding band of recombinant GST-TaERF3 to the GCC-box are indicated.

2008; Seo and Park, 2010). In this study, the transcript levels of these stress-related genes were up-regulated in *TaERF3*-overexpressing wheat (Yangmai 12) lines relative to the WT Yangmai 12, whereas the transcript levels of those genes were suppressed in *TaERF3*-silencing (BSMV:TaERF3) Yangmai 12 and Yangmai 6 plants as compared to BSMV:GFP Yangmai 12 and Yangmai 6 (controls). These results proved that TaERF3 activated the expression of these stress-related genes. The 10 TaERF3-activated stress-related genes tested were also induced by salt, drought or exogenous ABA treatments, but repressed by Na_2WO_4. Further, the inductions were enhanced in *TaERF3*-overexpressing wheat, but compromised in *TaERF3*-silencing wheat. These results suggested that TaERF3 may function as an integrator of the responses of these stress-related genes to ABA, salt and drought stresses.

Furthermore, the *cis*-element analysis results (Table S1) showed that the promoter sequences in *BG3*, *Chit1*, *RAB18*, *LEA3*, *TIP2*, *POX2* and *GST6* contained the GCC-boxes, ABRE and/or DRE and/or W-box *cis*-acting elements. However, the GCC-box was not found in the partial promoters of *DHN*, *OxOx2* or *SDR*, perhaps due to the queried sizes of the sequences (from 334 to 918 bp), or these genes may indeed lack the GCC-box *cis*-element. Further, EMSAs showed that TaERF3 did directly bind to the GCC-boxes in the seven stress-related genes, but did not interact *in vitro* with other *cis*-elements including DRE in *OxOx2*, ABRE in *TIP2*, W-box in *LEA3* and CAAT-box in *GST6* (Figure 8). Thus, in wheat, TaERF3 may directly activate the expression of a range of stress-related genes through interacting with these GCC-boxes in their promoters or through interacting (heterodimerizing) with other TFs (Buttner and Singh, 1997; Zhang *et al.*, 2007a), and TaERF3 may indirectly regulate the expression of other stress-related genes lacking the GCC-box. Importantly, the increase in the expression of *TaERF3* and TaERF3-regulating stress-related genes results in physiological changes in cells, such as accumulation of proline and chlorophyll, redox homeostasis (H_2O_2

decrease) and stomatal closure, which may in turn increase the abiotic tolerance.

In summary, TaERF3 acts as a positive regulator, contributing to salt and drought stress tolerance in wheat possibly through the activation of a range of stress-related genes. This study provides a basis for fundamental studies on wheat regulatory responses to salt and drought stresses and identifies the perspective gene *TaERF3* for efforts to improve salt and drought stress tolerance of wheat and other cereal crops.

Experimental procedures

Plant materials and treatments

Wheat cultivars Yangmai 6 and Yangmai 12 were provided by the Lixiahe Agricultural Institute of Jiangsu, China. Yangmai 12 was used as the recipient of the *TaERF3*-overexpressing transformation and as a host for BSMV-mediated *TaERF3* silencing. Yangmai 6 was used as another host for BSMV-mediated *TaERF3* silencing.

To prepare the *TaERF3*-overexpressing transformation vector pA25-TaERF3, the *TaERF3*-coding sequence (Zhang *et al.*, 2007b) was subcloned into the *Sma*I and *Sac*I sites of the pAHC25 expression vector (Figure 2a). pA25-TaERF3 was introduced into wheat Yangmai 12 by bombardment following Chen *et al.* (2008).

To set up BSMV-VIGS, a fragment (222 bp) of *TaERF3* was subcloned into the *Nhe*I site of BSMV-γ:GFP, resulting in BSMV:TaERF3 recombinant virus (Figure S3). Following Zhou *et al.* (2007), seedlings of Yangmai 6 and Yangmai 12 at the 2-leaf stage were inoculated with BSMV:TaERF3 or BSMV:GFP and mock solution, resulting in *TaERF3*-silencing wheat plants and controls (BSMV:GFP and mock solution).

Wheat plants were cultured in soil in PVC pots in a greenhouse under a 14-h light 22 °C/10-h dark 12 °C regime. Seedlings of Yangmai 6, Yangmai 12 and *TaERF3*-overexpressing transgenics at the 3-leaf stage were watered with 250 mM NaCl or 20% PEG6000 solution, or sprayed with 0.1 mM ABA, 0.1 mM Na$_2$WO$_4$ or 0.1% Tween 20 (mock) solution, following Zhang *et al.* (2012b).

For salt tolerance comparisons, seedlings at the 3-leaf stage were watered with 250 mM NaCl solution at 0 day and then watering was repeated every 4 days (a total of five irrigations in 20 days). To assess the effect of high salinity on seed germination, seeds of *TaERF3*-overexpressing transgenics and WT Yangmai 12 were treated with 0, 200 or 300 mM NaCl solutions. For drought tolerance comparisons, water was withheld from *TaERF3*-overexpressing, *TaERF3*-silencing and control lines as in Morran *et al.* (2011) and Zhang *et al.* (2012b) until there were distinguishable differences between control lines and *TaERF3*-overexpressing or *TaERF3*-silencing lines. At 20 dpi with BSMV:TaERF3 or BSMV:GFP virus, these infected plants were water-withheld for 25–30 days, or these infected plants were irrigated with 250 mM NaCl solution at 0 day and then watering was repeated every 4 days. At least 30 plants per independent line were evaluated in each test, and all tests were repeated three times.

PCR detection of *TaERF3* transgenic wheat

Genomic DNA was extracted from leaf tissue samples using the CTAB method (Saghai-Maroof *et al.*, 1984).

The presence of the transgene was detected by PCR using specific primers (TaERF3-TF and TaERF3-TR, sequences shown in Table S2) (Figure 2a). PCR was performed in 25-μL reaction volume containing 50 ng genomic DNA, 1 × GC PCR buffer I (TaKaRa), 200 μM each dNTPs (TaKaRa), 0.4 μM each primer and 1 U rTaq polymerase (TaKaRa). The specific amplified product (268 bp) was resolved on a 1.2% agarose gel and visualized by ethidium bromide staining.

RT-PCR and qRT-PCR analyses

Total RNA was extracted from roots and leaves of wheat plants by using the TRIZOL reagent (TaKaRa, Japan) and subjected to RNase-free DNaseI (TaKaRa, Japan) for digestion and purification. Two micrograms (μg) of RNA per sample was used for synthesis of first-strand cDNA using the Superscript II First-Strand Synthesis Kit (Invitrogen).

RT-PCR was used to analyse the transcript levels of *BSMV-CP* (for the detection of BSMV infection) and *TaERF3* in the BSMV:GFP-, BSMV:TaERF3- and mock solution-inoculated Yangmai 6 and Yangmai 12 plants according to Zhang *et al.* (2012b). The wheat *18S rRNA* and *actin* genes were used as internal references to normalize *BSMV-CP* and *TaERF3*, respectively.

qRT-PCR was used to analyse transcript levels of *TaERF3* and the ten aforementioned stress-related genes following the methods described in Zhang *et al.* (2012b). The wheat *actin* was used as an internal reference to normalize all data. Three biological replications for each line were performed in each test. The relative transcript level of the target gene was calculated using the $2^{-\Delta\Delta CT}$ method (Livak and Schmittgen, 2001).

Measurement of proline, chlorophyll and H$_2$O$_2$ contents and stomatal conductance

Free proline and H$_2$O$_2$ contents in leaves were measured following Hu *et al.* (1992) and Rao *et al.* (2000), respectively. Chlorophyll content and stomatal conductance were measured following Mao *et al.* (2011). At least 15 plants per independent line were evaluated in each test, and all tests were repeated three times.

Water loss rate assay

Following the method described by Zhang *et al.* (2012b), the leaves of 15 plants per line (namely WT, OX19, OX 31, OX34 and OX78) were detached and incubated in a growth chamber under a regime of 23 °C, 45% humidity. At 30-min intervals, the leaf fresh weight was measured. Water loss was calculated by comparing the measured weights from each interval with the measurement at time zero. All tests were repeated three times.

EMSA on binding of TaERF3 to known *cis*-elements

The *TaERF3* sequence containing the AP2/ERF domain was subcloned in frame to the 3′-terminus of a *GST* gene in the pGEX-4T-1 vector (GE Amersham), resulting in GST-TaERF3 expression vector pGST-TaERF3. GST-TaERF3 recombinant protein expression and purification was performed according to Dong *et al.* (2010). The partial sequences containing the GCC-box, DRE, ABRE, CAAT or W-box *cis*-elements in the known stress-related genes (Figure 8a) were obtained through PCR or synthesized and subsequently used as probes. One μg of purified probe and 3 μg of the purified GST-TaERF3 or GST protein were mixed with binding buffer in a total volume of 15 μL. EMAS was conducted following Dong *et al.* (2010).

Primers

All primers used for vector constructions, PCR detection for the *TaERF3* transgene and qRT-PCR and RT-PCR assays are listed in Table S2.

Acknowledgements

This study was supported by the National Hi-Tech ('863') program of China (Grant no. 2012AA10A309) and NSFC program of China (Grant no. 31271799).

References

Berna, A. and Bernier, F. (1999) Regulation by biotic and abiotic stress of a wheat germin gene encoding oxalate oxidase, a H$_2$O$_2$-producing enzyme. *Plant Mol. Biol.* **39**, 539–549.

Berrocal-Lobo, M., Molina, A. and Solano, R. (2002) Constitutive expression of ETHYLENE-RESPONSE-FACTOR 1 in Arabidopsis confers resistance to several necrotrophic fungi. *Plant J.* **29**, 23–32.

Bhavanath, J., Anubha, S. and Avinash, M. (2011) Expression of *SbGSTU* (tau class glutathione S-transferase) gene isolated from *Salicornia brachiata* in tobacco for salt tolerance. *Mol. Biol. Rep.* **38**, 4823–4832.

Buttner, M. and Singh, K.B. (1997) *Arabidopsis thaliana* ethylene-responsive element binding protein (AtEBP), an ethylene-inducible, GCC box DNA-binding protein interacts with an ocs element binding protein. *Proc. Natl Acad. Sci. USA*, **94**, 5961–5966.

Chakravarthy, S., Tuori, R.P., D'Ascenzo, M.D., Fobert, P.R., Despres, C. and Martin, G.B. (2003) The tomato transcription factor *Pti4* regulates defense-related gene expression via GCC box and non-GCC box *cis* elements. *Plant Cell*, **15**, 3033–3050.

Chen, L., Zhang, Z., Liang, H., Liu, H., Du, L., Xu, H. and Xin, Z. (2008) Overexpression of *TiERF1* enhances resistance to sharp eyespot in transgenic wheat. *J. Exp. Bot.* **59**, 4195–4204.

Dhanda, S. and Sethi, G. (1998) Inheritance of excised-leaf water loss and relative water content in bread wheat (*Triticum aestivum*). *Euphytica*, **1**, 39–47.

Dong, N., Liu, X., Lu, Y., Du, L., Xu, H., Liu, H., Xin, Z. and Zhang, Z. (2010) Overexpression of *TaPIEP1*, a pathogen-induced ERF gene of wheat, confers host-enhanced resistance to fungal pathogen *Bipolaris sorokiniana*. *Funct. Integr. Genomics*, **10**, 215–226.

Dong, W., Ai, X., Xu, F., Quan, T., Liu, S. and Xia, G. (2012) Isolation and characterization of a bread wheat salinity responsive ERF transcription factor. *Gene*, **511**, 38–45.

Golldack, D., Lüking, I. and Yang, O. (2011) Plant tolerance to drought and salinity: stress regulating transcription factors and their functional significance in the cellular transcriptional network. *Plant Cell Rep.* **30**, 1383–1391.

Han, Y.J., Cho, K.C., Hwang, O.J., Choi, Y.S., Shin, A.Y., Hwang, I. and Kim, J.H. (2012) Overexpression of an *Arabidopsis* β-glucosidase gene enhances drought resistance with dwarf phenotype in creeping bentgrass. *Plant Cell Rep.* **31**, 1677–1686.

Hattori, Y., Nagai, K., Furukawa, S., Song, X.J., Kawano, R., Sakakibara, H., Wu, J., Matsumoto, T., Yoshimura, A., Kitano, H., Matsuoka, M., Mori, H. and Ashikari, M. (2009) The ethylene response factors *SNORKEL1* and *SNORKEL2* allow rice to adapt to deep water. *Nature*, **460**, 1026–1030.

He, Y., Li, W., Lv, J., Jia, Y., Wang, M. and Xia, G. (2011) Ectopic expression of a wheat MYB transcription factor gene, *TaMYB73*, improves salinity stress tolerance in *Arabidopsis thaliana*. *J. Exp. Bot.* **63**, 1511–1522.

Hein, I., Barciszewska-Pacak, M., Hrubikova, K., Williamson, S., Dinesen, M., Soenderby, I.E., Sundar, S., Jarmolowski, A., Shirasu, K. and Lacomme, C. (2005) Virus-induced gene silencing-based functional characterization of genes associated with powdery mildew resistance in barley. *Plant Physiol.* **138**, 2155–2164.

Holzberg, S., Brosio, P., Gross, C. and Pogue, G.P. (2002) Barley stripe mosaic virus-induced gene silencing in a monocot plant. *Plant J.* **30**, 315–327.

Hong, S.W., Jon, J.H., Kwak, J.M. and Nam, H.G. (1997) Identification of a receptor-like protein kinase gene rapidly induced by abscisic acid, dehydration, high salt, and cold treatments in *Arabidopsis thaliana*. *Plant Physiol.* **113**, 1203–1212.

Hu, C.A., Delauney, A.J. and Verma, D.P. (1992) A bifunctional enzyme (delta-pyrroline-5-carboxylate synthetase) catalyzes the first two steps in proline biosynthesis in plants. *Proc. Natl Acad. Sci. USA*, **89**, 9354–9358.

Huang, X.S., Wang, W., Zhang, Q. and Liu, J.H. (2013) A basic helix-loop-helix transcription factor, *PtrbHLH*, of poncirus trifoliata confers cold tolerance and modulates peroxidase-mediated scavenging of hydrogen peroxide. *Plant Physiol.* **162**, 1178–1194.

Ingram, J. and Bartels, D. (1996) The molecular basis of dehydration tolerance in plants. *Annu. Rev. Plant Physiol. Mol. Biol.* **47**, 377–403.

Liu, Q., Kasuga, M., Sakuma, Y., Abe, H., Miura, S., Yamaguchi-Shinozaki, K. and Shinozaki, K. (1998) Two transcription factors, *DREB1* and *DREB2*, with an EREBP/AP2 DNA binding domain separate two cellular signal transduction pathways in drought- and low-temperature-responsive gene expression, respectively, in *Arabidopsis*. *Plant Cell*, **10**, 1391–1406.

Livak, K.J. and Schmittgen, T.D. (2001) Analysis of relative gene expression data using real-time quantitative PCR and the $2^{-\Delta\Delta CT}$ method. *Method*, **25**, 402–408.

Mao, X., Jia, D., Li, A., Zhang, H., Tian, S., Zhang, X., Jia, J. and Jing, R. (2011) Transgenic expression of *TaMYB2A* confers enhanced tolerance to multiple abiotic stresses in *Arabidopsis*. *Funct. Integr. Genomics*, **11**, 445–465.

Morran, S., Eini, O., Pyvovarenko, T., Parent, B., Singh, R., Ismagul, A., Eliby, S., Shirley, N., Langridge, P. and Lopato, S. (2011) Improvement of stress tolerance of wheat and barley by modulation of expression of DREB/CBF factor. *Plant Biotechnol. J.* **9**, 230–249.

Nakano, T., Suzuki, K., Fujimura, T. and Shinshi, H. (2006) Genome-wide analysis of the ERF gene family in *Arabidopsis* and rice. *Plant Physiol.* **140**, 411–432.

Oh, S.J., Kim, Y.S., Kwon, C.W., Park, H.K., Jeong, J.S. and Kim, J.K. (2009) Overexpression of the transcription factor *AP37* in rice improves grain yield under drought condition. *Plant Physiol.* **150**, 1368–1379.

Oñate-Sánchez, L., Anderson, J.P., Young, J. and Singh, K.B. (2007) *AtERF14*, a member of the ERF family of transcription factors, plays a nonredundant role in plant defence. *Plant Physiol.* **143**, 400–409.

Quan, R.D., Hu, S.J., Zhang, Z.L., Zhang, H.W., Zhang, Z.J. and Huang, R.F. (2010) Overexpression of an ERF transcription factor *TSRF1* improves rice drought tolerance. *Plant Biotechnol. J.* **8**, 476–488.

Rao, M.V., Lee, H., Creelman, R.A., Mullet, J.E. and Davis, K.R. (2000) Jasmonic acid signaling modulates ozone-induced hypersensitive cell death. *Plant Cell*, **12**, 1633–1646.

Saghai-Maroof, M.A., Soliman, K.M., Soliman, K.M., Jorgensen, R.A. and Allard, R.W. (1984) Ribosomal DNA spacer-length polymorphisms in barley: mendelian inheritance, chromosomal location, and population dynamics. *Proc. Natl Acad. Sci. USA*, **81**, 8014–8019.

Schmidt, R., Mieulet, D., Hubberten, H.M., Obata, T., Hoefgen, R., Fernie, A.R., Fisahn, J., San Segundo, B., Guiderdoni, E., Schippers, J.H. and Mueller-Roeber, B. (2013) SALT-RESPONSIVE ERF1 regulates reactive oxygen species-dependent signaling during the initial response to salt stress in rice. *Plant Cell*, **25**, 2115–2131.

Scofield, S.R., Huang, L., Brandt, A.S. and Gill, B.S. (2005) Development of a virus-induced gene-silencing system for hexaploid wheat and its use in functional analysis of the *Lr21*-mediated leaf rust resistance pathway. *Plant Physiol.* **138**, 2165–2173.

Seo, M. and Koshiba, T. (2002) Complex regulation of ABA biosynthesis in plants. *Trends Plant Sci.* **7**, 41–84.

Seo, P.J. and Park, C.M. (2010) MYB96-mediated abscisic acid signals induce pathogen resistance response by promoting salicylic acid biosynthesis in *Arabidopsis*. *New Phytol.* **186**, 471–483.

Shekhawat, U.K., Srinivas, L. and Ganapathi, T.R. (2011) MusaDHN-1, a novel multiple stress-inducible SK (3)-type dehydrin gene, contributes affirmatively to drought- and salt-stress tolerance in banana. *Planta*, **234**, 915–932.

Shinozaki, K. and Yamaguchi-Shinozaki, K. (1997) Gene expression and signal transduction in water-stress response. *Plant Physiol.* **115**, 327–334.

Sun, X.L., Sun, M.Z., Luo, X., Ding, X.D., Ji, W., Cai, H., Bai, X., Liu, X.F. and Zhu, Y.M. (2013) A *Glycine soja* ABA-responsive receptor-like cytoplasmic kinase, GsRLCK, positively controls plant tolerance to salt and drought stresses. *Planta*, **237**, 1527–1545.

Szabados, L. and Savour, A. (2009) Proline: a multifunctional amino acid. *Trends Plant Sci.* **15**, 89–97.

Wang, X., Li, Y., Ji, W., Bai, X., Cai, H., Zhu, D., Sun, X., Chen, L. and Zhu, Y. (2011) A novel *Glycine soja* tonoplast intrinsic protein gene respond to abiotic stress and depressed salt and dehydration tolerance in transgenic *Arabidopsis thaliana*. *J. Plant Physiol.* **168**, 1241–1248.

Wang, M., Jiang, Q.Y., Hu, Z., Zhang, H., Fan, S., Feng, L. and Zhang, H. (2012) Evaluation for salt tolerance of wheat cultivars. *J. Plant Gene Rec.* **13**, 189–194.

Xu, K., Xu, X., Fukao, T., Canlas, P., Maghirang-Rodriguez, R., Heuer, S., Ismail, A.M., Bailey-Serres, J., Ronald, P.C. and Mackill, D.J. (2006) Sub1A is an ethylene-response-factor-like gene that confers submergence tolerance to rice. *Nature*, **442**, 705–708.

Xu, Z., Xia, L., Chen, M., Cheng, X., Zhang, R., Li, L., Zhao, Y., Lu, Y., Ni, Z., Liu, L., Qiu, Z. and Ma, Y. (2007) Isolation and characterization of the *Triticum aestivum* L. ethylene-responsive factor 1 (*TaERF1*) that increases multiple stress-tolerance. *Plant Mol. Biol.* **65**, 719–732.

Yamaguchi-Shinozaki, K. and Shinozaki, K. (2006) Transcriptional regulatory networks in cellular responses and tolerance to dehydration and cold stresses. *Annu. Rev. Plant Biol.* **57**, 781–803.

Zhang, H., Li, W., Chen, J., Yang, Y., Zhang, Z., Wang, X.C. and Huang, R. (2007a) Transcriptional activator *TSRF1* reversely regulates pathogen resistance and osmotic stress tolerance in tobacco. *Plant Mol. Biol.* **55**, 825–834.

Zhang, Z., Yao, W., Dong, N., Liang, H., Liu, H. and Huang, R. (2007b) A novel ERF transcription activator in wheat and its induction kinetics after pathogen and hormone treatments. *J. Exp. Bot.* **58**, 2993–3003.

Zhang, H., Yang, Y., Zhang, Z., Chen, J., Wang, X. and Huang, R. (2008) Expression of the ethylene response factor gene *TSRF1* enhances abscisic acid responses during seedling development in tobacco. *Planta*, **228**, 777–787.

Zhang, Z., Wang, J., Zhang, R. and Huang, R. (2012a) The ethylene response factor *AtERF98* enhances tolerance to salt through the transcriptional activation of ascorbic acid synthesis in *Arabidopsis*. *Plant J.* **71**, 273–287.

Zhang, Z., Liu, X., Wang, X., Zhou, M., Zhou, X., Ye, X. and Wei, X. (2012b) An R2R3 MYB transcription factor in wheat, TaPIMP1, mediates host resistance to *Bipolaris sorokiniana* and drought stresses through regulation of defense- and stress-related genes. *New Phytol.* **196**, 1155–1170.

Zhou, H., Li, S., Deng, Z., Wang, X., Chen, T., Zhang, J., Chen, S., Ling, H., Zhang, A., Wang, D. and Zhang, X. (2007) Molecular analysis of three new receptor-like kinase genes from hexaploid wheat and evidence for their participation in the wheat hypersensitive response to stripe rust fungus infection. *Plant J.* **52**, 420–434.

Zhu, J.K. (2002) Salt and drought stress signal transduction in plants. *Annu. Rev. Plant Biol.* **53**, 247–273.

Zhuang, J., Chen, J., Yao, Q., Xiong, F., Sun, C., Zhou, X., Zhang, J. and Xiong, A. (2011) Discovery and expression profile analysis of AP2/ERF family genes from *Triticum aestivum*. *Mol. Biol. Rep.* **38**, 745–753.

16

Discovery and development of exome-based, co-dominant single nucleotide polymorphism markers in hexaploid wheat (*Triticum aestivum* L.)

Alexandra M. Allen[1,*], Gary L. A. Barker[1], Paul Wilkinson[1], Amanda Burridge[1], Mark Winfield[1], Jane Coghill[1], Cristobal Uauy[2], Simon Griffiths[2], Peter Jack[3], Simon Berry[4], Peter Werner[5], James P. E. Melichar[6], Jane McDougall[7], Rhian Gwilliam[7], Phil Robinson[7] and Keith J. Edwards[1]

[1]*School of Biological Sciences, University of Bristol, Bristol, UK*

[2]*John Innes Centre, Norwich, UK*

[3]*RAGT, Ickleton, Essex, UK*

[4]*Limagrain, Woolpit, Suffolk, UK*

[5]*KWS, Thriplow, Hertfordshire, UK*

[6]*Syngenta Seeds Ltd, Whittlesford, Cambridge, UK*

[7]*KBioscience Unit 7, Hertfordshire, UK*

*Correspondence

email a.allen@bristol.ac.uk

Summary

Globally, wheat is the most widely grown crop and one of the three most important crops for human and livestock feed. However, the complex nature of the wheat genome has, until recently, resulted in a lack of single nucleotide polymorphism (SNP)-based molecular markers of practical use to wheat breeders. Recently, large numbers of SNP-based wheat markers have been made available via the use of next-generation sequencing combined with a variety of genotyping platforms. However, many of these markers and platforms have difficulty distinguishing between heterozygote and homozygote individuals and are therefore of limited use to wheat breeders carrying out commercial-scale breeding programmes. To identify exome-based co-dominant SNP-based assays, which are capable of distinguishing between heterozygotes and homozygotes, we have used targeted re-sequencing of the wheat exome to generate large amounts of genomic sequences from eight varieties. Using a bioinformatics approach, these sequences have been used to identify 95 266 putative single nucleotide polymorphisms, of which 10 251 were classified as being putatively co-dominant. Validation of a subset of these putative co-dominant markers confirmed that 96% were true polymorphisms and 65% were co-dominant SNP assays. The new co-dominant markers described here are capable of genotypic classification of a segregating locus in polyploid wheat and can be used on a variety of genotyping platforms; as such, they represent a powerful tool for wheat breeders. These markers and related information have been made publically available on an interactive web-based database to facilitate their use on genotyping programmes worldwide.

Keywords: wheat, next-generation sequencing, KASPar genotyping, single nucleotide polymorphism.

Introduction

Bread wheat (*Triticum aestivum*) is an allohexaploid (AABBDD) crop derived from the hybridisation of the diploid genome of *Aegilops tauschii* (DD) with the AABB tetraploid genome of *Triticum turgidum* (Dubcovsky and Dvorak, 2007). These hybridisation events, the domestication process and the inbreeding nature of wheat have together resulted in a reduced level of genetic diversity between cultivated wheat varieties, when compared with their wild ancestors (Haudry *et al.*, 2007). Wheat breeders and geneticists require tools to exploit the genetic diversity available within germplasm collections and carry out breeding programmes, which utilise this diversity to maximum effect. Molecular markers enable breeders and geneticists to carry out this process; however, in allohexaploid wheat, the development of molecular markers has, until recently, been problematic due to the presence of homoeologous and paralogous copies of the various genes (Kaur *et al.*, 2012). Recent advances in genotyping platforms have built upon the wealth of data provided by next-generation sequencing (NGS) technologies to enable, for the first time, the large-scale identification, validation and application of molecular markers in wheat breeding programmes (Berkman *et al.*, 2012; Paux *et al.*, 2011). These developments have come at a critical time, where the need for a substantial increase in yields to feed a growing global population has coincided with reduced genetic gains and increasing climatic and environmental pressures (Dixon *et al.*, 2009; Reynolds *et al.*, 2009).

Many of the recently developed genotyping platforms rely on the identification of single nucleotide polymorphisms (SNPs), which are polymorphic between different wheat varieties (Paux *et al.*, 2011). To overcome the various bottlenecks and problems associated with SNP generation, characterisation and most importantly validation in wheat, we and others have previously used NGS-based technology to identify and map relatively large numbers of gene-based SNP loci (Allen *et al.*, 2011; Akhunov *et al.*, 2009; Chao *et al.*, 2010). However, these studies used cDNA and EST sequences and were therefore subject to variation in expression of homoeologous and paralogous genes. In hexaploid wheat, this situation is further aggravated as homoeologous and paralogous genes are often silenced or can show differential spatial and/or temporal expressions (Adams and Wendel, 2005; Akhunova *et al.*, 2010; Liu *et al.*, 2009). Genomic DNA is likely to be a more reliable source of putative SNPs; however, the size of the wheat genome means that sequencing the whole genome of multiple varieties to the depths required for successful SNP identification is impractical, time consuming and costly (Biesecker *et al.*, 2011). To overcome these resource-associated problems, we have used a recently developed sequence capture targeted resequencing approach to characterise a significant proportion of the wheat exome (Winfield *et al.*, 2012). By using a reference collection of the wheat exome as the basis of our SNP collection, we have been able to sequence and compare equivalent regions of the wheat genome from several wheat varieties.

To be fully utilised in breeding programmes, putative SNPs need to be identified and converted to working assays on a high-throughput genotyping platform. Recently, several technologies have revolutionised wheat genotyping: Illumina's GoldenGate/Infinium technologies and KBioscience's KASPar (Akhunov *et al.*, 2009; Allen *et al.*, 2011). Development of these platforms has encouraged the widespread uptake of SNP-based genotyping in wheat; however, both technologies have two significant drawbacks. Firstly, they require the identification and characterisation of varietal SNPs among an excess of homoeologous and paralogous SNPs. Secondly, as both platforms were developed for diploid species, they have problems with the scoring of varietal SNPs in polyploid heterozygotes, for instance, F_2 and backcross populations. The detection of heterozygous SNPs in allohexaploid wheat is dependent on the ability of the system to accurately discriminate between different call ratios. For 'dominant' SNP assays, which amplify all three homoeologous copies, these systems are often incapable of distinguishing homozygote (having a call ratio of 4 : 2) and heterozygote lines (having a call ratio of 5 : 1) (Allen *et al.*, 2011; Paux *et al.*, 2011). In contrast, both genotyping platforms work well when the SNP is amplified from just a single homoeologous/paralogous copy. Such SNP assays are usually referred to as co-dominant SNP assays, that is, they are capable of differentiating between homozygotes (having a call ratio of 2 : 0) and heterozygotes (having a call ratio of 1 : 1). As such, co-dominant SNP assays are preferred markers compared with dominant SNP assays. Unfortunately, co-dominant SNP assays usually make up < 20% of the SNP assays generated by conventional means (Allen *et al.*, 2011). However, careful primer design can lead to the successful amplification of just one homoeolog/paralog, but this process is time consuming as the variable nature of each set of sequences demands a manual approach to primer design. To overcome this bottleneck, we have developed a SNP identification pipeline which incorporates a novel bioinformatics procedure designed to identify putative co-dominant SNP assays.

The developments described here have led to both the generation of an extensive set of putative varietal SNPs from genomic DNA and within this data set the identification of a subset of putative co-dominant SNPs. The use of an exome-based SNP discovery strategy has targeted gene discovery to genic regions. Validation of a subset of these putative co-dominant SNP assays and a comparison with dominant SNP markers has provided useful insights into their design and characteristics. Finally, the work described here has resulted in a significant increase in the number of gene-derived co-dominant SNP assays, which will be of considerable interest to wheat researchers, and in particular, the breeding community.

Results

SNP discovery

In this study, the exome of the UK varieties Alchemy, Avalon, Cadenza, Hereward, Rialto, Robigus, Savannah and Xi19 was captured using the NimbleGen capture array (NimbleGen array reference 100819_Wheat_Hall_cap_HX1) described in Winfield *et al.* (2012). This generated between 9.8 and 48.7 million reads on the Illumina GAIIx platform. Sequence data were filtered as described in the experimental procedures. Varietal SNPs were called in the filtered data where read coverage was sufficiently high that there was less than a 0.1% chance of an observed allelic difference between two varieties being due to failure to sample an allele. For example, if the varieties Avalon and Cadenza have observed calls of A(20) and A(10)G(10), respectively, we would expect half of the alleles in Avalon (10 calls) to be G under the null hypothesis that there is no real genotypic difference. Randomisation tests showed that for the data set as a whole, using a minimum expected count of 10 for null bases resulted in a false discovery rate of < 1%. Putative co-dominant SNP markers were identified as the subset of SNPs meeting the above criteria, but where every variety had only a single allele called. The SNP discovery pipeline identified 95 266 putative varietal SNPs in 26 551 distinct reference sequences (Winfield *et al.*, 2012). Examination of these SNPs suggested that as in our previous work, only 10%–20% were co-dominant (Table 1; Allen *et al.*, 2011), with 10 251 putative co-dominant SNP markers identified within 5308 contigs.

Co-dominant SNP validation

As co-dominant SNP assays are of significant interest to the wheat community, it is important that such assays have a level of polymorphism that is not significantly different to that previously shown for dominant SNP assays. In addition, it is important that the distribution of the co-dominant SNP markers across the three homoeologous genomes do not show a bias beyond those shown previously for mapped dominant SNP markers (Allen *et al.*, 2011). To assess both of these features, we selected a subset of 1337 putative co-dominant SNP assays for validation and further analysis. While the selection process used to identify this subset was essentially random, to aid further investigation, we selected those SNPs that appeared, via the sequence analysis, to be polymorphic between the parents of the UK mapping populations Avalon × Cadenza and Savannah × Rialto. Of the 1337 SNPs selected, we were able to design working KASPar assays for 1190 SNPs (89%; Data S1).

Genotyping of a panel of 47 wheat varieties using the 1190 KASPar probes resulted in 1138 probes (96%) generating data that could be scored consistently and were polymorphic in at least one of the varieties screened (Data S2). Examination of the genotypic data revealed three types of varietal SNP markers: co-dominant SNP markers (where homozygous scores were detected for all hexaploid varieties, for instance, only scores of A:A or T:T were obtained, Figure 1a); partially co-dominant SNP markers (where heterozygous and homozygous scores were detected in hexaploid varieties, that is, A:A, T:T and the mixed A:T, Figure 1b); and dominant SNP markers (where a single homozygous and heterozygous score was detected in hexaploid

Table 1 Summary of next-generation sequence data and SNPs identified for eight wheat varieties

Variety	No. of sequences (million)	No. mapped reads (million)	No. of SNPs (compared with Avalon)	No. of co-dominant SNPs (compared with Avalon)	Proportion of total SNPs that are co-dominant (%)
Avalon	27.2	10.9	N/A	N/A	N/A
Alchemy	44.4	16.9	22 092	2558	11.6
Cadenza	20.2	4.9	8909	1550	17.4
Hereward	41.0	15.2	13 379	2399	17.9
Rialto	30.4	11.8	15 141	2662	17.6
Robigus	31.4	6.3	10 823	2114	19.5
Savannah	48.7	16.8	18 648	2818	15.1
Xi19	9.8	2.9	4391	899	20.5

varieties, that is, A:A and mixed A:T only, Figure 1c). Of the 1138 validated probes, 734 (65%) were co-dominant, 194 (17%) were partially co-dominant and 210 (18%) were dominant. Dominant and co-dominant markers were used to screen an F_4 population known to contain homozygote and heterozygote individuals. Screening this population with co-dominant SNP assays resulted in three separate clusters for the various homozygote and heterozygote individuals (Figure 1d). However, screening the same population with dominant SNP assays produced a more scattered cluster where homozygous and heterozygous loci were indistinguishable (Figure 1e). Screening the F_4 population with partially co-dominant SNPs yielded the same results as described for both dominant and co-dominant SNP assays depending on the genotypes of the population parents (data not shown).

The SNP markers developed through the NimbleGen exome capture were compared with the existing database of SNP markers developed from EST and normalised cDNA sequences using the experimental procedures described in Allen *et al.*, 2011; (Table 2; Data S3). The number of co-dominant SNP assays generated was significantly higher, and dominant SNPs significantly lower, when compared with the previous data set of validated EST/cDNA-derived SNPs ($\chi^2 = 131.98$, $P < 0.001$). Comparison of the two data sets showed that the numbers of partially co-dominant SNP assays were not significantly different between the two data sets ($\chi^2 = 3.87$, $P = 0.14$). In addition, the polymorphism information content (PIC) scores and minor allele frequencies (MAF) were similar between the two data sets; they

were highest for the partially co-dominant SNP assays and lowest for the dominant SNP assays.

Characterisation of the different SNP types

To characterise the different SNP marker types identified in this study, and in particular, the co-dominant and partially co-dominant SNP assays, several analyses were performed using the contig sequences containing the SNPs. The average sizes of contigs containing different SNP types were similar (co-dominant, 692 bp; dominant, 690 bp; partially co-dominant, 666 bp). We hypothesised that co-dominant SNP assays were likely to be derived from 5′ or 3′ untranslated regions (UTRs) of genic sequences. In the absence of functional coding constraints, such regions are more likely to have diverged between homoeologs and thus represent effectively unique regions of sequence. To address this hypothesis, SNP-containing contig sequences were used to screen, via BLASTX (Altschul *et al.*, 1990), the nonredundant (nr) protein database. If a match was found (E-value 1e-5), a further analysis was then performed to identify whether the SNP was located inside or outside the coding region. This analysis showed that a higher proportion of the contig sequences used to develop co-dominant SNP assays returned no hit when subjected to a BLASTX analysis against the nr database, compared with dominant SNP assay sequences. Where a hit was identified, a higher proportion of the co-dominant SNPs were found to lie outside the coding region, compared with dominant SNPs. The number of co-dominant SNPs located within known coding

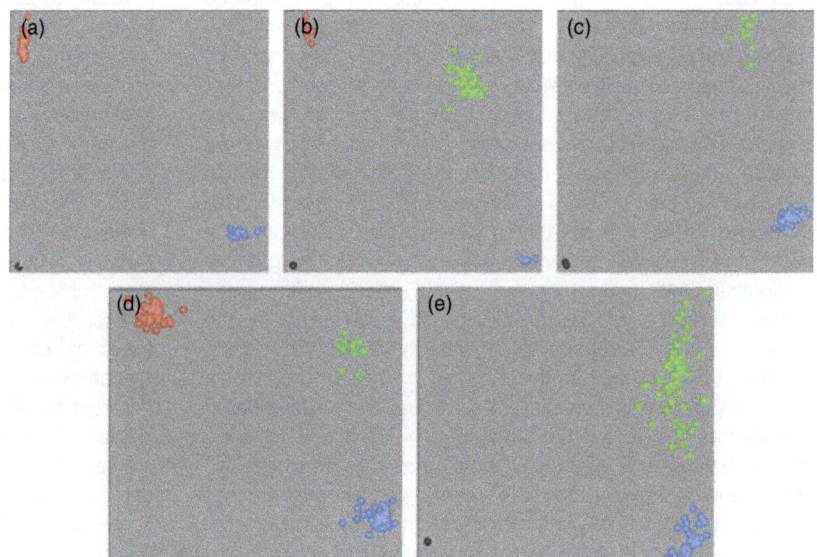

Figure 1 KASPar plots of different varietal SNP types screened against a panel of hexaploid wheat varieties with examples of (a) a co-dominant SNP assay, (b) a partially co-dominant SNP assay and (c) a dominant SNP assay. Screening an F_4 population containing heterozygotes with a co-dominant SNP assay results in a separate cluster for heterozygote individuals (d). Screening the same population with a dominant SNP assay produces a more scattered cluster where homozygote and heterozygote individuals are indistinguishable.

Table 2 Summary of validated SNPs

SNP type	Validated NimbleGen SNPs (%)	Validated EST/cDNA SNPs (%)	Total validated SNPs	Average minor allele frequency	Average PIC score
Dominant	210 (19)	1195 (58)	1407	0.249	0.273
Partially co-dominant	194 (17)	437 (21)	632	0.315	0.315
Co-dominant	734 (64)	444 (21)	1175	0.270	0.287
All SNPs	1138	2076	3214	0.270	0.286

PIC, polymorphism information content.

regions was significantly lower than would be expected to occur by chance ($\chi^2 = 7.56$, $P = 0.02$). Partially co-dominant SNPs were midway between dominant and co-dominant SNPs (Figure 2a). Across all SNP marker types, the average length of contigs returning no hit was lower (approximately 520 bp) than the average length of contigs returning BLASTX hits (approximately 770 bp), suggesting that contig length affected the likelihood of obtaining a BLASTX match. To check whether the contigs returning no hit represent genes that had not yet been annotated, the same contig sequences were subjected to a BLASTN analysis (E-value 1e-3) against the NCBI nr nucleotide database; 86% of the contigs returning no hit from the BLASTX analysis also had no match in the BLASTN nr database.

We further hypothesised that some co-dominant SNP assays may have been derived from single-copy regions of the wheat genome. Such regions may have either been unique to only one progenitor genome or alternatively one or more copies have been lost since polyploidisation. To address this second hypothesis, the same sets of sequences were screened, using BLASTN, against the 5 × Chinese Spring genomic raw reads at http://www.cerealsdb.uk.net/. BLAST hit coverage was calculated for every nucleotide position in the query sequence and averaged over the whole sequence to derive a mean contig coverage. All three SNP types peak in coverage at 15 ×, indicative of three gene copies each at 5 × coverage; however, the co-dominant SNPs and to a lesser extent the partially co-dominant SNPs had a secondary peak of coverage at fivefold coverage, indicative of single-copy number sequences. This peak is absent from the dominant SNPs (Figure 2b).

These analyses were combined by comparing the coverage of sequences containing co-dominant SNPs that had different BLASTX results. The sequences with 5 × coverage are most highly represented by those returning a 'no hit' from BLASTX analysis (Figure 2c). When contig length was plotted against median coverage for co-dominant SNPs, no relationship was observed ($r = -0.0009$). A similar result ($r = -0.0009$) was obtained by performing the same analysis using only sequences returning no BLASTX hit, suggesting that the contig length does not affect the number of hits returned from the BLASTN analysis or the estimated level of coverage.

Map location of dominant, partially co-dominant and co-dominant SNP assays

The map positions of the different SNP marker types were investigated to determine whether any bias in genetic location was introduced by using co-dominant SNP assays in two doubled-haploid mapping populations developed from UK cultivars Avalon × Cadenza (A × C) and Savannah × Rialto (S × R). Of the 3214 SNP markers developed to date (Table 2), 2109 were identified as polymorphic between Avalon and Cadenza (via

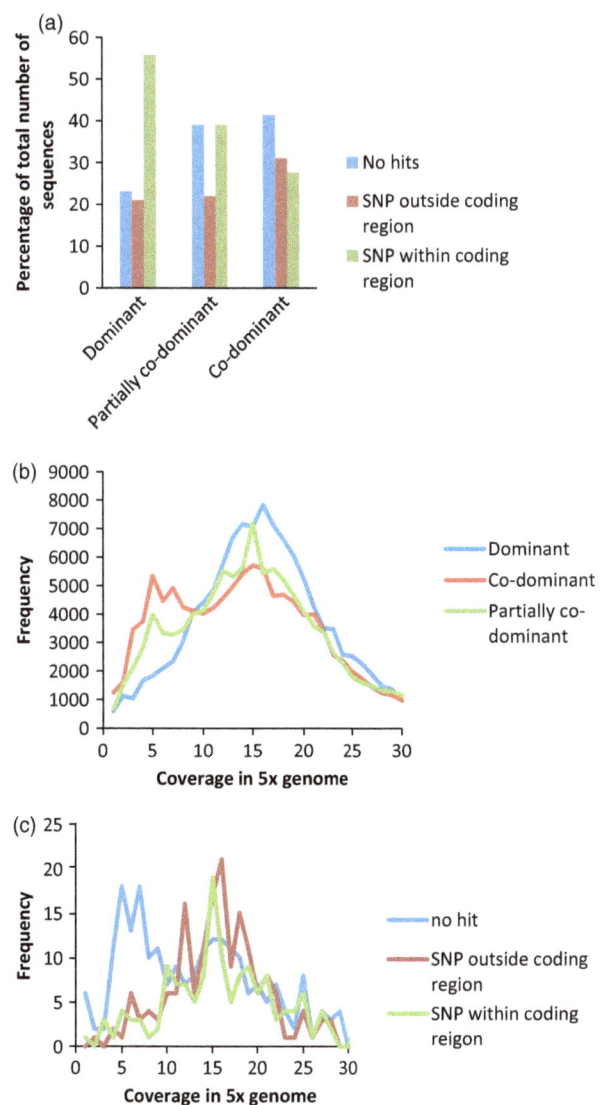

Figure 2 Characteristics of sequences containing dominant, partially dominant and co-dominant SNP types. (a) BLASTX analysis against the non-redundant (nr) protein database and (b) BLASTN against the 5 × Chinese Spring raw reads. (c) Coverage of sequences containing co-dominant SNPs against the 5 × Chinese Spring raw reads classified according to the BLASTX designation in (a).

screening of the 47 varieties above), of which 1807 were placed on the Avalon × Cadenza map. These consisted of 1152 EST/cDNA-derived markers and 655 NimbleGen-derived markers. Of the remaining 1105 SNP assays not polymorphic between Avalon and Cadenza, 562 were identified as polymorphic between Savannah and Rialto and 541 of these markers were placed on the Savannah × Rialto map. These consisted of 187 EST/cDNA-derived markers and 375 NimbleGen-derived markers. To enable comparisons between the maps, 231 evenly spaced loci from the Avalon × Cadenza map were also included on the Savannah × Rialto map (Figure 3; Data S3). For the Avalon × Cadenza map, previously mapped SSR markers were used to help assign linkage groups to chromosomes (http://www.wgin.org.uk/resources/MappingPopulation/TAmapping.php; Data S4). In total, 2350 (73%) of the validated SNP markers were mapped; these comprised of 969 dominant SNP loci, 444 partially co-dominant loci and 937 co-dominant SNP loci. In the Avalon × Cadenza map, the linkage groups ranged from 54.5 to 239.0 centiMorgans (cM) in size, with 8–214 SNP markers. The total map length was 2434.4 cM with an average spacing of 1.3 cM between SNP loci. In the Savannah × Rialto map, linkage groups ranged from 1.3 to 221.1 cM, with 2–98 SNP markers. The total Savannah × Rialto map length was 2861.8 cM with an average spacing of 3.8 cM between SNP loci (Table 3). The two linkage maps aligned well with each other, showing similar arrangements of common loci within linkage groups.

In both populations, over 97.5% of the SNP markers could be mapped unequivocally to a linkage group and assigned to a unique chromosome position. The lack of markers on the short arm of chromosome 1B in the Savannah × Rialto map can be attributed to the presence of the same 1BL.1RS rye translocation in both Savannah and Rialto, where the short arm of rye chromosome 1B has replaced the short arm of wheat chromosome 1B (Figure 3). Clustering of SNP markers was observed in both linkage maps, with 55% of A × C markers and 61% of S × R markers being completely linked (0 cM distance between them). Of the remaining markers, 81% of A × C markers are separated by < 5 cM and 90% are within 10 cM of the next marker. For the S × R map, these figures are lower (55% markers separated by < 5 cM and 74% within 10 cM), probably due to a smaller number of markers on the map. Similar levels of clustering were observed for the different SNP marker types; 60% of A × C co-dominant markers and 52% of dominant markers were completely linked. Of the remaining co-dominant markers, 77% of markers are within 5 cM of each other and 88% are within a 10 cM interval. The corresponding proportions for dominant markers are similar (84% within 5 cM and 92% within 10 cM). The different SNP types showed similar patterns of distribution between the A, B and D linkage groups in both the Avalon × Cadenza and Savannah × Rialto maps, with the only difference of a higher proportion of the partially co-dominant markers mapped to the D genome (Figure 4a). Similarly, although clustering of SNP markers was observed within the linkage groups, there was no obvious bias of different marker types (Figure 3).

Summary statistics of mapped loci

The summary statistics of mapped SNP markers were compared to assess whether different marker types had varying levels of polymorphism in the 47 varieties screened and to ensure that the co-dominant SNP markers developed in this study would be useful across a wide range of material. The mean MAF and levels of genetic diversity of SNP markers were compared between the different marker types and assigned genomes of the Avalon × Cadenza and Savannah × Rialto maps (Table 4). These summary statistics were very similar for both maps; however,

differences within the maps were observed. Partially co-dominant SNP assays had the highest average MAF and PIC scores, and dominant SNP assays had the lowest. Loci from the separate homoeologous genomes had consistent MAF and PIC measurements, although the A and D genome measurements were slightly higher than the B genome (Table 4). The different classes of SNP loci showed differences in the distribution of MAF scores. Co-dominant and partially co-dominant loci showed an increased proportion of medium and high frequency alleles compared with dominant loci (Figure 5a). Similarly, D genome loci showed a trend to have higher MAF compared with A and B genome loci (Figure 5b). The distribution of PIC scores showed that co-dominant and partially co-dominant loci types had a higher proportion of high PIC scores than dominant SNP assays (Figure 5c). A and B genome loci had a similar distribution of PIC scores, while D genome loci had a comparatively higher proportion of high PIC scores (Figure 5d).

Discussion

In this study, we present a SNP discovery pipeline capable of identifying large numbers of putative SNPs from genomic sequence obtained by targeted exome capture. This proved an efficient method to generate equivalent sequences from multiple varieties from which we were able to generate over 90 000 putative SNPs between eight elite UK cultivars. Given our results, this same approach is likely to prove highly effective at identifying SNPs across a wide range of cultivars, and in a wider range of germplasms, such as landraces, progenitors and alien species. By cataloguing SNPs using a reference collection of sequences derived from just the wheat exome, we provide a unique context for each SNP, thereby both reducing the chance of duplications within the SNP data set and allowing direct comparisons between different wheat lines. A key advantage to the SNP collection described here compared with other SNP markers such as insertion site–based polymorphisms (ISBPs; Paux et al., 2010) is that by the nature of the targeted sequencing, all the SNPs developed are associated with genes, and as such, are likely to prove useful in gene-based marker-assisted breeding.

When examined further, the SNP database was shown to contain 10 251 putative co-dominant SNPs. Validation of over 10% of the co-dominant SNP assays on the KASPar genotyping platform resulted in a significantly improved validation rate compared with our previous study with 96% being polymorphic between the varieties screened compared with 67% as described in Allen et al. (2011). This increased validation rate is probably due to the use of genomic DNA, as opposed to transcriptome-derived data, in the SNP discovery phase where the problems of expression differences and presence of intron–exon splice sites hindered effective SNP identification and primer design (Trick et al., 2012). Of the subset of putative co-dominant SNP assays, over 80% were validated as co-dominant or partially co-dominant, compared with < 20% in previous studies where random SNPs were validated (Allen et al., 2011). The occurrence of putative co-dominant SNP assays, which were dominant when validated, is likely to be due to the presence of homeologous sequences that were not represented in the sequence data but were amplified by the KASPar primers. This may be due to a feature of individual sequences that prevent them from mapping on to the assembly or a consequence of reduced sequence coverage.

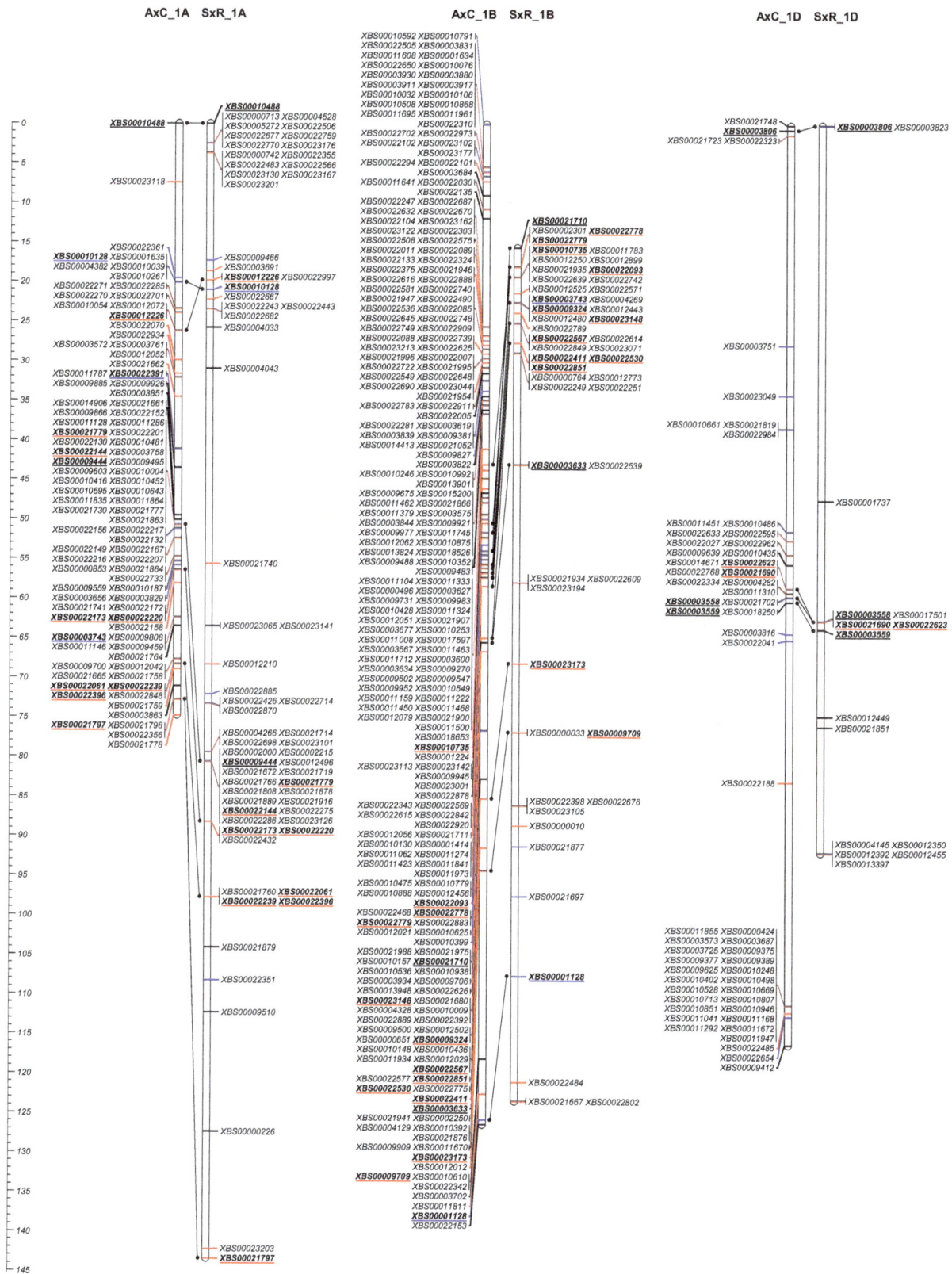

Figure 3 Genetic linkage maps of wheat derived from 190 Avalon × Cadenza doubled-haploid lines and 95 Savannah × Rialto doubled-haploid lines. Each linkage group was assigned to a chromosome indicated above the linkage group, and chromosomes are arranged with the short arm above the long arm. SNP loci mapped in this study are designated XBS and are coloured according to the SNP type: Dominant SNP loci are shown in black, co-dominant SNP loci are shown in red and partially co-dominant SNP loci are shown in blue. Common markers between the Avalon × Cadenza and Savannah × Rialto maps are underlined. Map distances, calculated using the Kosambi mapping function, are shown in centiMorgans (cM) on the ruler to the left of linkage groups.

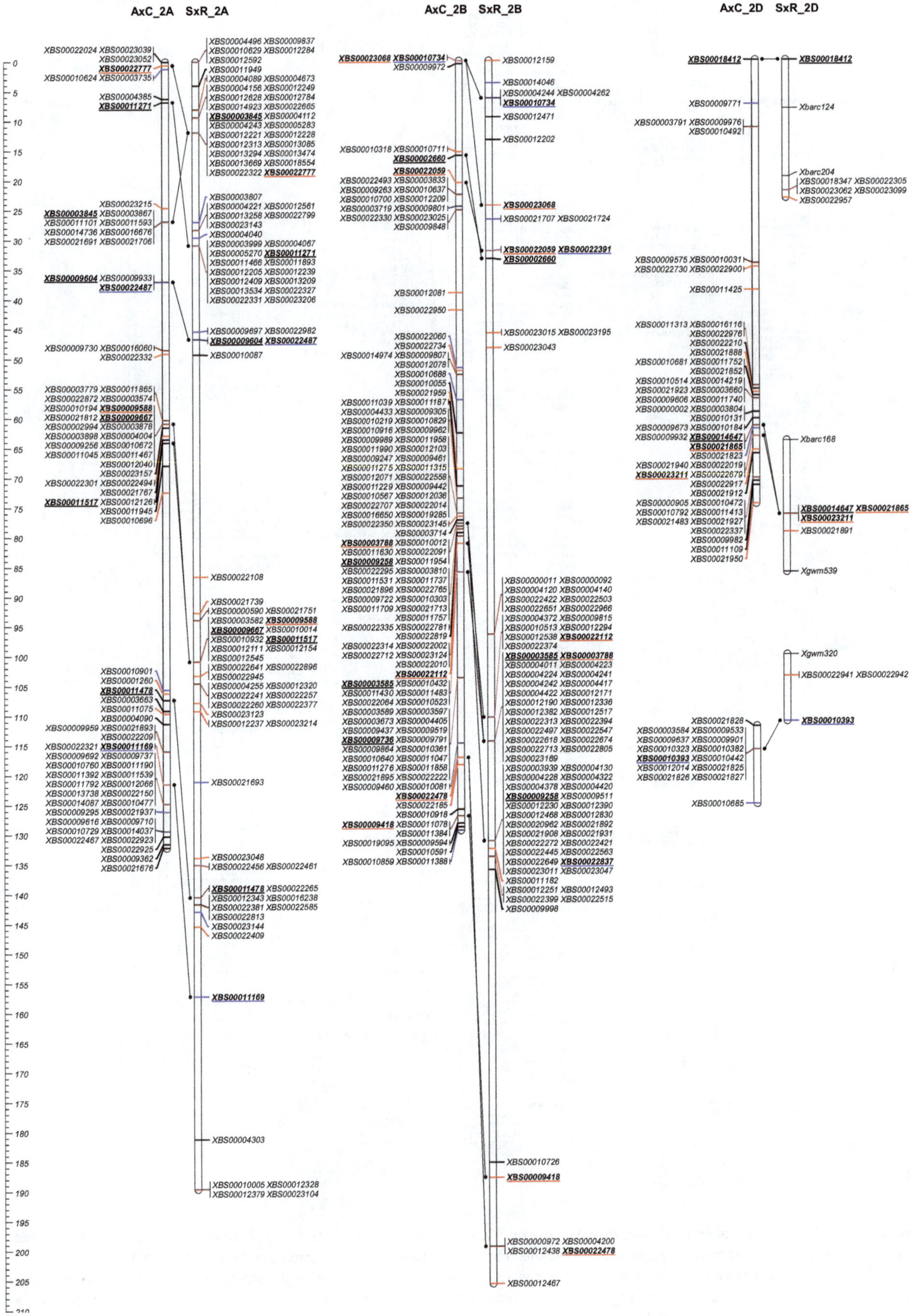

Figure 3 Continued

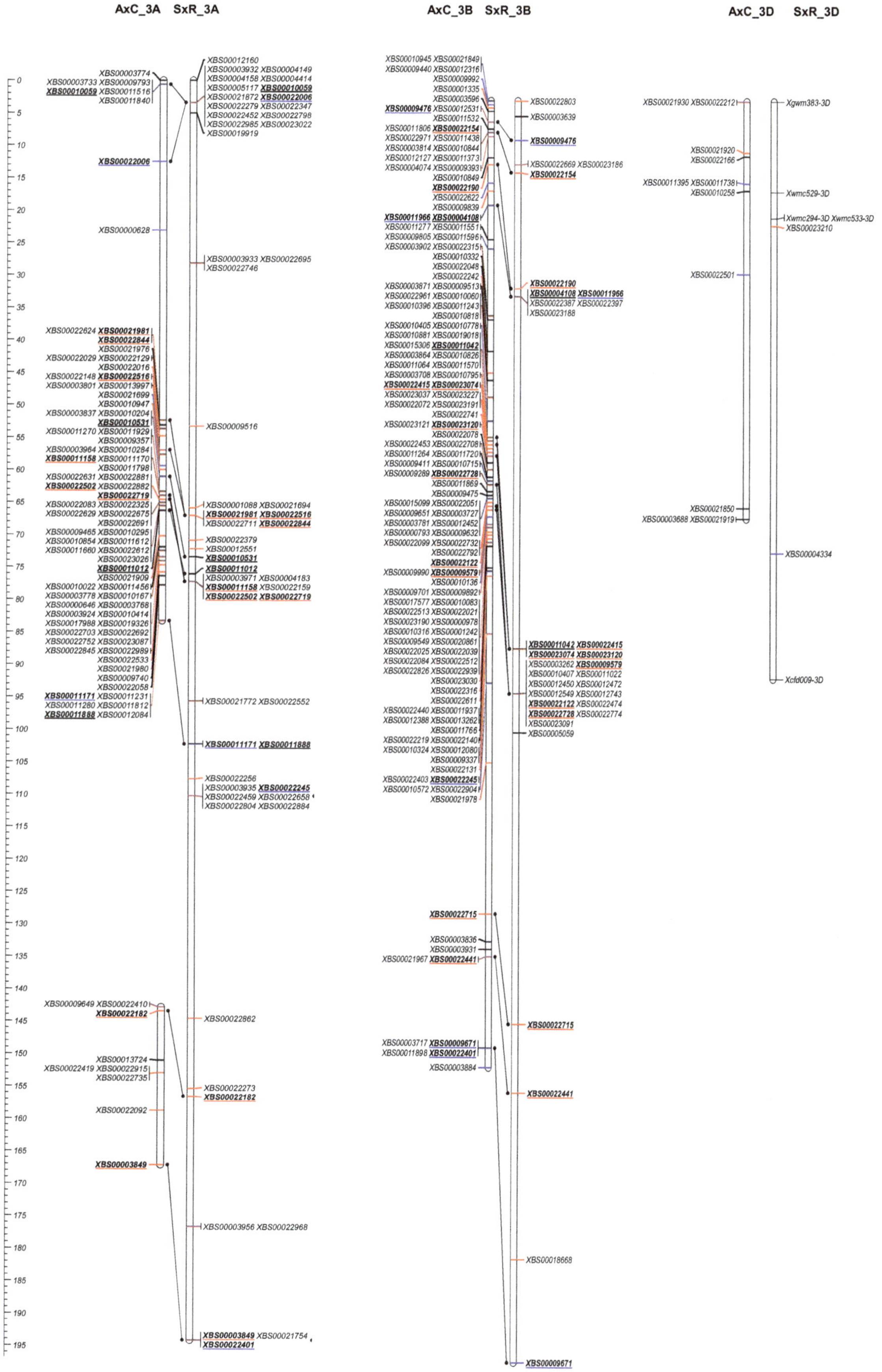

Figure 3 Continued

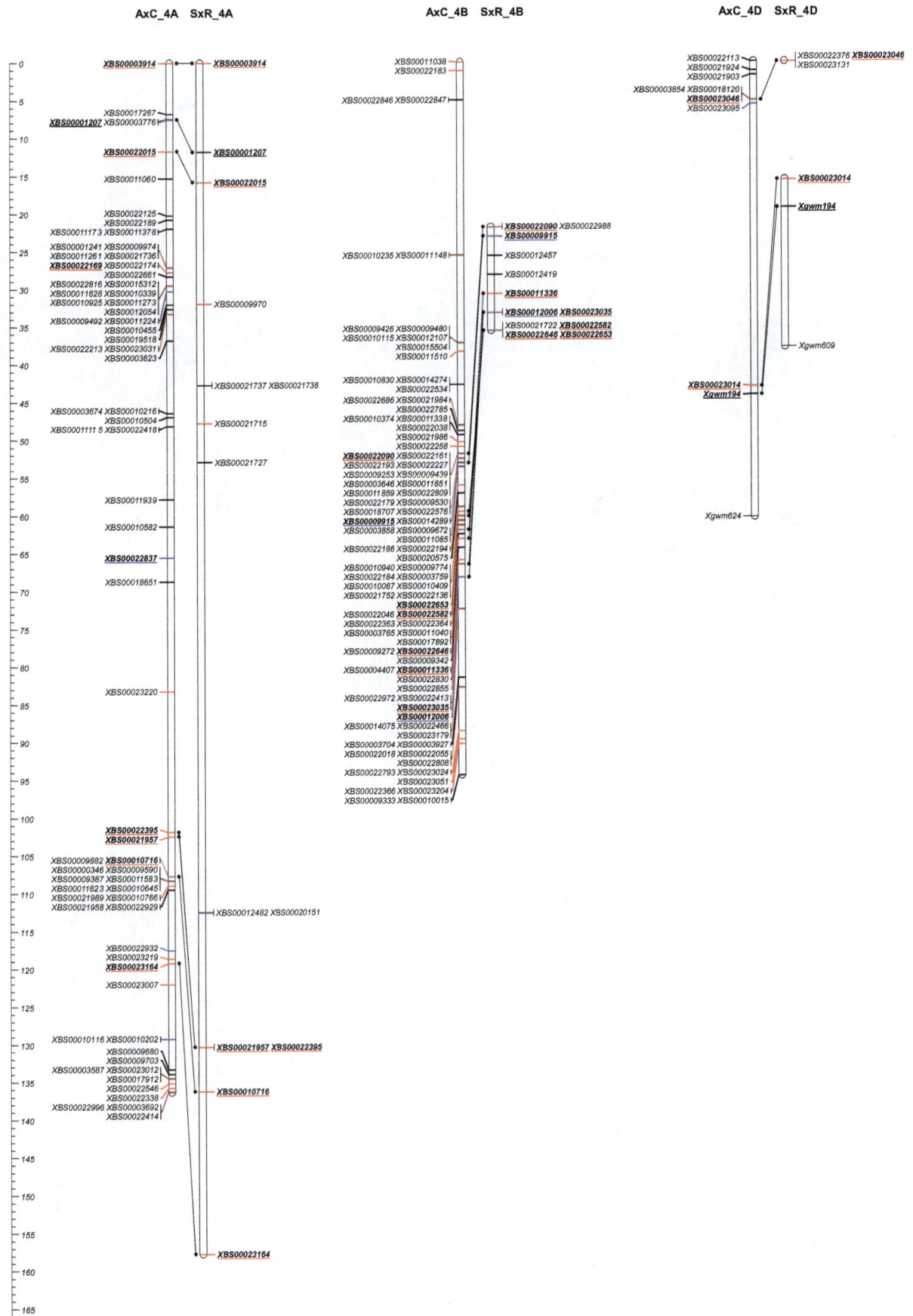

Figure 3 Continued

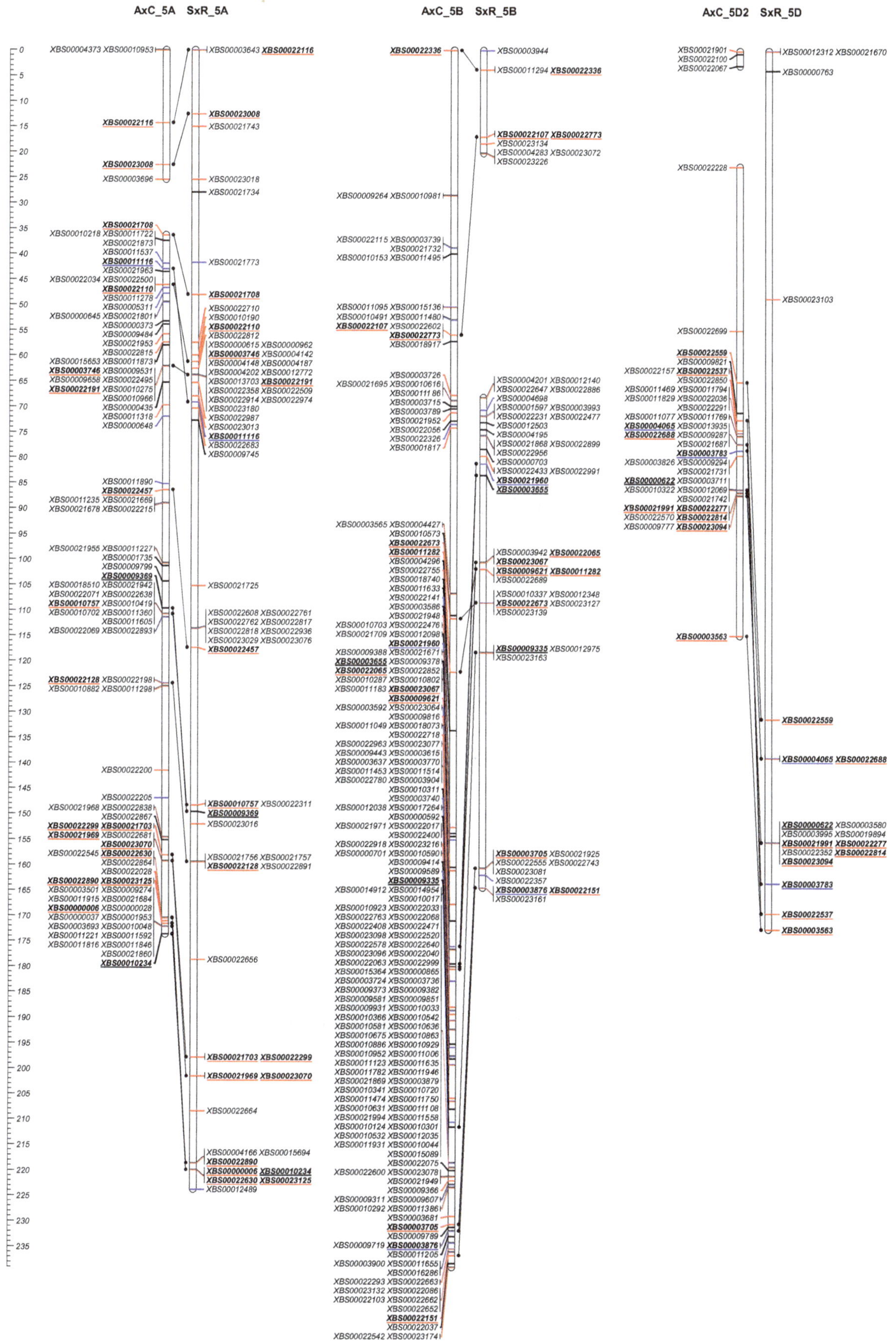

Figure 3 Continued

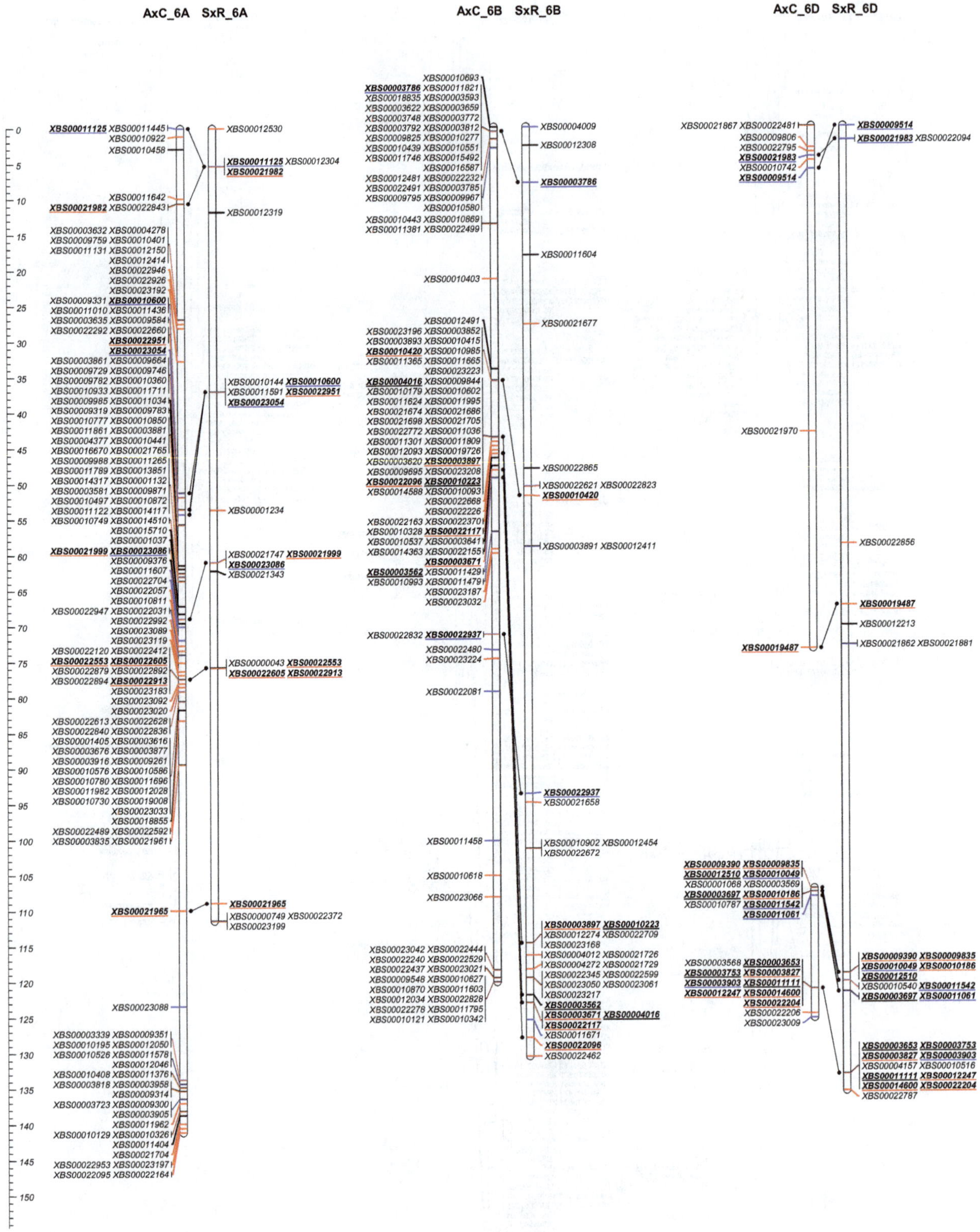

Figure 3 Continued

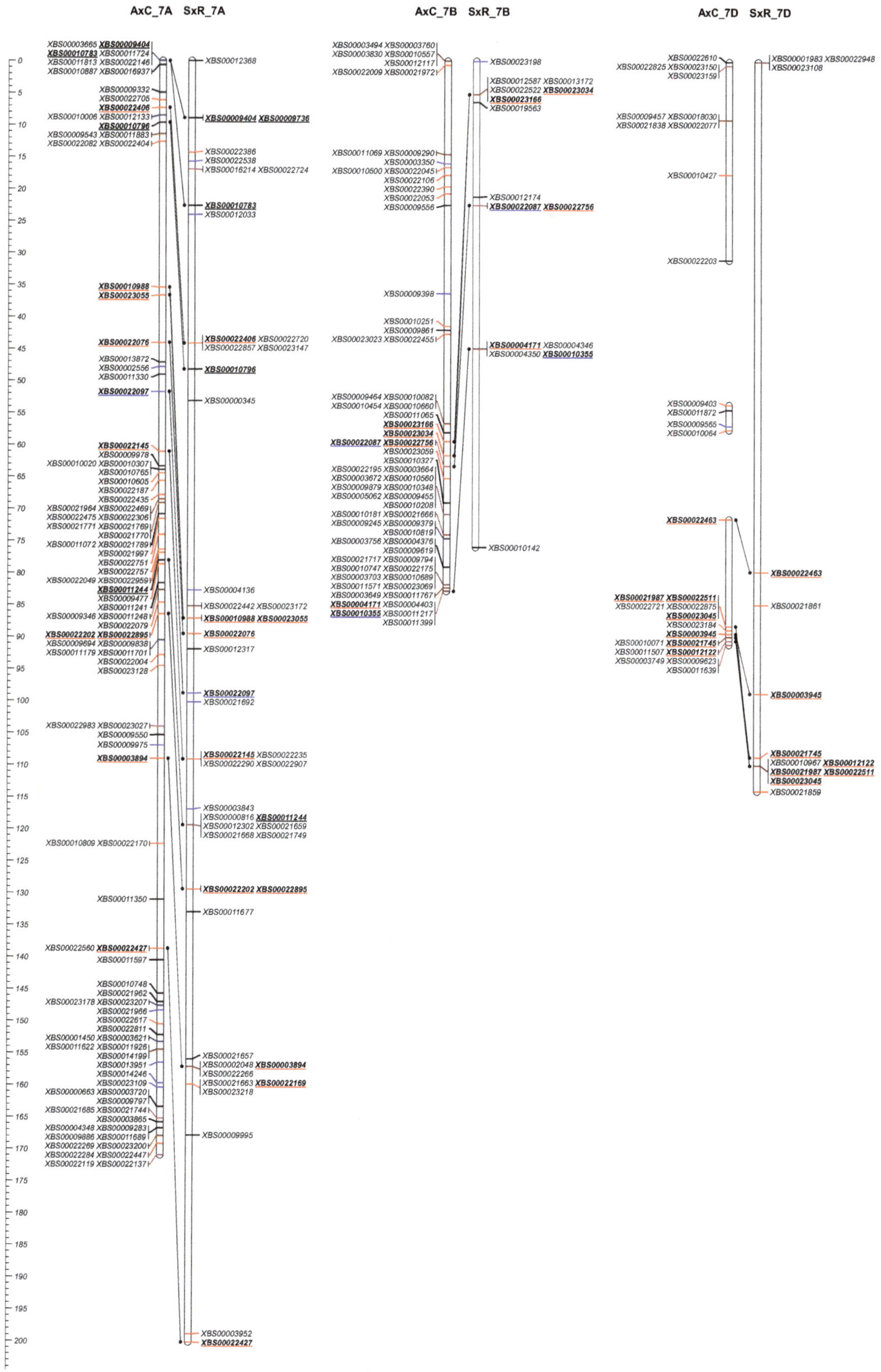

Figure 3 Continued

Table 3 Summary of linkage groups and mapped loci

Chromosome	Avalon × Cadenza map			Savannah × Rialto map		
	Number of bristol SNP loci	Size (cM)	Average spacing between loci (cM)	Number of bristol SNP loci	Size (cM)	Average spacing between loci (cM)
1A	89	74.8	0.8	68	143.4	2.1
1B	214	126.4	0.6	50	115.9	2.3
1D	57	116.2	1.6	15	104.1	6.9
2A	82	132.2	1.6	98	211.1	2.2
2B	118	129.4	1.1	91	221.7	2.4
2D	63	77.7	1.2	13	49.8	3.8
3A	86	107.6	1.3	55	194.3	3.5
3B	136	154.5	1.1	34	205.0	6.0
3D	11	66.7	6.1	2	72.3	36.2
4A	71	136.3	1.9	14	157.7	11.3
4B	87	94.4	1.1	12	23.9	2.0
4D	8	60.3	7.5	4	1.3	0.3
5A	93	162.7	1.7	64	223.8	3.5
5B	172	239.0	1.4	51	147.2	2.9
5D	38	95.0	2.5	19	194.6	10.2
6A	136	141.1	1.0	23	111.4	4.8
6B	105	120.2	1.1	37	158.6	4.3
6D	31	91.6	3.0	28	135.6	4.8
7A	103	171.0	1.7	48	200.2	4.2
7B	64	82.8	1.3	14	56.1	4.0
7D	29	54.5	1.9	13	133.8	10.3
Total	1793	2434.4	1.3	753	2861.8	3.8
A genome	660	925.7	1.4	370	1241.9	3.4
B genome	896	946.7	1.1	289	928.4	3.2
D genome	237	562	2.2	94	691.4	7.4
Group 1	360	317.4	0.8	133	363.4	2.7
Group 2	263	339.3	1.3	202	482.6	2.4
Group 3	233	328.8	1.4	91	471.6	5.2
Group 4	166	291	1.8	30	182.9	6.1
Group 5	303	496.7	1.6	134	565.6	4.2
Group 6	272	352.9	1.3	88	405.6	4.6
Group 7	193	308.3	1.6	75	390.1	5.2

Characterisation of the validated co-dominant SNP assays showed that their PIC scores and MAF were on average higher than dominant SNP assays, suggesting they are highly useful genetic markers for use on a range of materials. Analyses using the contig sequences containing the different SNP types revealed that co-dominant SNP assays were more likely to be located in contigs returning no BLAST hit to either protein or nucleotide databases, or outside coding regions in those contigs returning a BLASTX hit. Our analysis is consistent with the hypothesis that a proportion of the contigs used to develop co-dominant SNP assays represent single-copy genes. These contigs most likely represent genes that were lost before or during the domestication process as they are found as single copies in both landraces, such as Chinese Spring, and modern varieties. For those SNP contigs with 15 × Chinese Spring genomic coverage, it is quite possible that while these are represented as three homoeologs in Chinese Spring, they have undergone gene loss down to single copy in the UK germplasm we have studied. Intracultivar heterogeneity has been documented between elite inbred lines of crop species, and there are reports of intervarietal gene loss in wheat (Haun et al., 2011; Swanson-Wagner et al., 2010; Winfield et al., 2012).

In addition to the factors outlined above, the Chinese Spring reference used to map the NimbleGen-captured sequences was based upon cDNA. If only one homoeolog was sampled in the cDNA data, and this was sufficiently divergent from the other two homoeologous copies, we may have only been able to map Illumina sequence data to that single genome. This would be the case in many 3' UTR regions that are more divergent than protein-coding sequence and have diverged sufficiently during evolution to preclude their co-amplification in the KASPar PCR. This homoeolog-specific amplification could fortuitously lead to the development of co-dominant markers, yet BLAST analysis of such sequences against the Chinese Spring genome would show them to be present in three copies. In summary, investigations into the average copy number of sequences used to develop co-dominant SNP assays and the location of the SNP in the sequence suggests that these SNPs are likely to reside in single-copy genes of as yet unknown function, and/or three-copy genes which are sufficiently divergent that sequence data from one homoeolog does not map to other copies. The uncharacterised nature of these genes makes them an exciting and intriguing source of further co-dominant markers and scientific investigation.

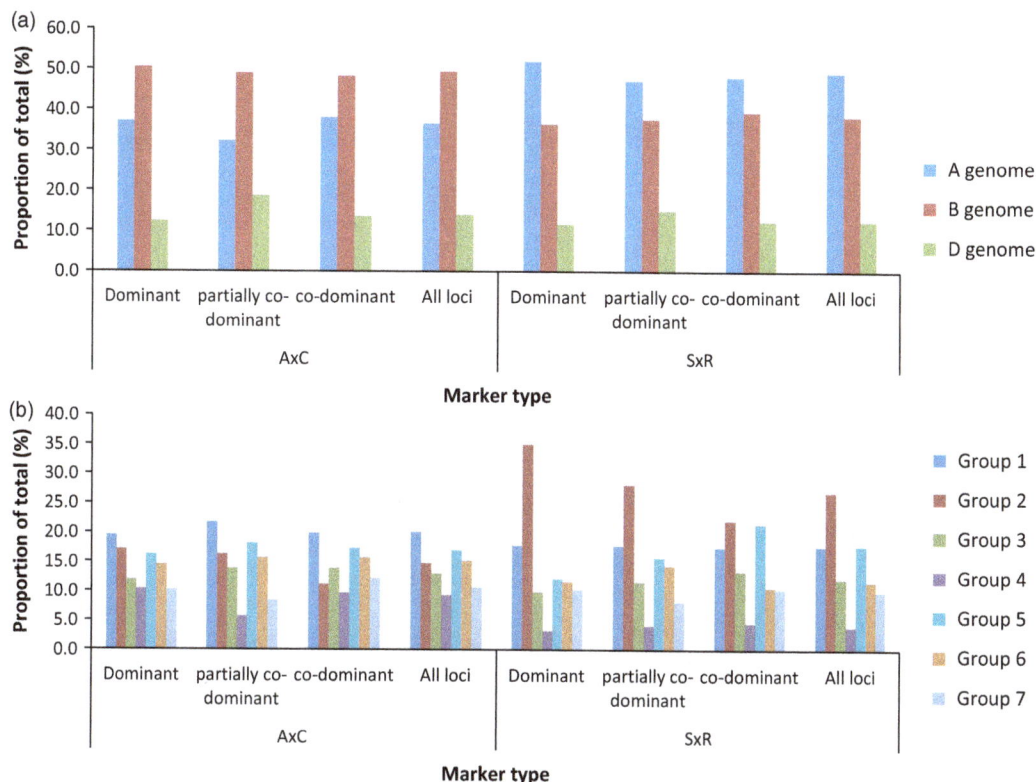

Figure 4 Distribution of the different marker types across the (a) A, B and D linkage groups and (b) homoeologous chromosome groups of the Avalon × Cadenza and Savannah × Rialto genetic maps.

Table 4 Summary statistics for mapped loci

	Number of loci	Minor allele frequency	Polymorphism information content
Avalon × Cadenza mapped loci	1793	0.264	0.284
Co-dominant loci	672	0.277	0.290
Partially co-dominant loci	332	0.308	0.313
Dominant loci	789	0.235	0.266
A genome	660	0.263	0.284
B genome	896	0.260	0.283
D genome	237	0.278	0.288
Savannah × Rialto mapped loci	753	0.284	0.299
Co-dominant loci	395	0.291	0.300
Partially co-dominant loci	213	0.319	0.323
Dominant loci	145	0.246	0.280
A genome	370	0.290	0.305
B genome	289	0.272	0.290
D genome	94	0.294	0.305

During this study, we have created two complementary genetic maps, enabling 73% of our validated SNPs to be assigned a map location. The co-dominant SNP loci had a similar pattern of distribution between linkage groups compared with dominant loci, suggesting that co-dominant SNP markers have a similar distribution to the previously used dominant markers. Analysis of the MAF and PIC scores of the different types of mapped SNPs demonstrated that the co-dominant and partially co-dominant SNP markers had higher levels of genetic diversity within the lines tested, compared with dominant SNP assays, suggesting that co-dominant SNP assays are highly suitable for use as genetic markers.

The two genetic maps aligned well with each other, with a similar assignment and order of common markers. Clustering of SNP markers was observed in both linkage maps, indicating that despite the relatively large mapping populations used, a lack of recombination events between these markers may affect map resolution. This may be overcome by mapping these markers against a larger number of individuals. Preliminary results indicate that mapping a subset of 223 evenly spaced A × C markers on 566 individuals from an extended A × C population reduced the proportion of completely linked markers from 50.4% to 44.9%, and it is likely that this figure could be further decreased by specifically targeting clustered markers. However, despite the high proportion of clustered markers, 89% of the remaining markers map to within 10 cM of the next marker, suggesting that these provide good overall coverage of the genome, with few gaps. When co-dominant and dominant markers were compared separately, similar proportions of markers were observed to map to within 10 cM of each other (88% and 92%, respectively), suggesting that both marker types are similarly distributed across the map.

Both maps had a relatively low proportion of D genome loci; this has been observed in previous studies and is likely to relate to a lower level of diversity found in the D genome due to the effects of the genetic bottleneck that accompanied the domestication of hexaploid wheat (Allen et al., 2011; Caldwell et al., 2004; Chao

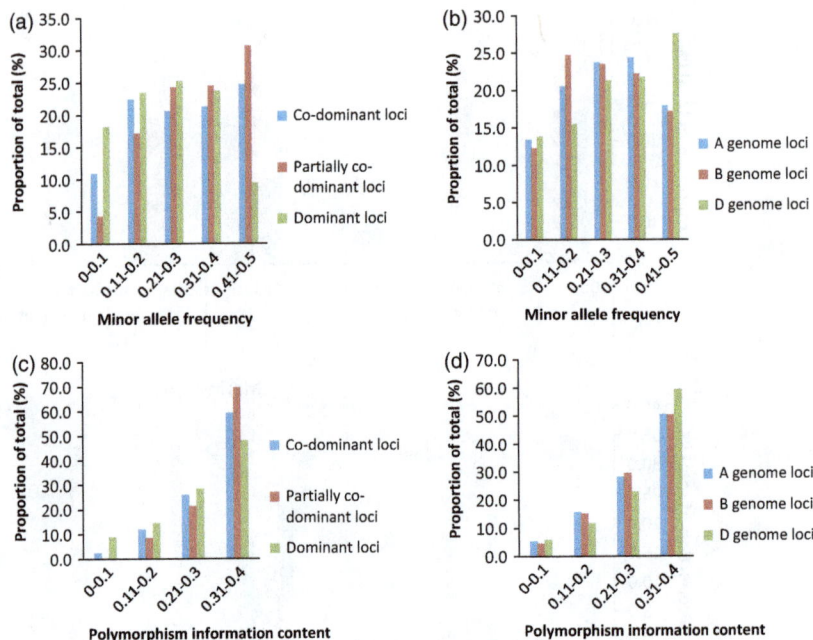

Figure 5 Distribution of minor allele frequency (MAF) and polymorphism information content (PIC) scores among the 47 wheat varieties. Loci were separated into subgroups according to (a,c) marker type and (b,d) genome.

et al., 2009). Although the mean MAF and PIC scores were similar for A, B and D genome loci, some differences were observed in the distributions of these measurements. The results for A and B genome markers were similar; however, loci assigned to the D genome had a higher proportion of high MAF and PIC scores compared with A and B genome loci. This is the opposite to what has been detected in previous studies (Akhunov et al., 2010; Chao et al., 2009) and suggests that, although the lower genetic diversity within the D genome hinders SNP discovery and marker development, the D genome SNPs identified by our pipeline are as informative and useful as loci from the A and B genome.

This study has described the design, implementation and validation of a pipeline designed to identify gene-based co-dominant SNP assays from genomic DNA sequence data. The validation results suggest that this approach is highly efficient and the resulting co-dominant SNP markers are evenly distributed across the genome with relatively high MAF and PIC scores. As such, these should prove a highly valuable resource for use in breeding programmes. The construction of two complementary genetic maps has maximised the amount of mapped SNP loci and allowed comparisons between UK breeding materials. The genotype data generated in this study for 47 widely used wheat lines, combined with genetic map locations for SNP markers, should enable wheat researchers to target their efforts to regions of interest and enable QTL studies and marker-assisted selection. The markers described in this study will be useful in linking the genetic map with the developing physical maps and so will enhance the possibility of efficient map-based cloning in hexaploid wheat. The entire data set presented in this study has been made publicly available via the provision of supplementary data sets and an interactive website (http://www.cerealsdb.uk.net/), to make this resource as accessible and useful as possible. These new co-dominant wheat SNP-based markers will be useful on a number of genotyping platforms and germplasm collections and hence should be a powerful new tool for wheat breeders and researchers alike. In addition, the pipeline developed here to identify co-dominant SNP markers should be applicable to other polyploid crops where

SNP discovery and marker development have previously been challenging (Cordeiro et al., 2006; Trick et al., 2009; Yu et al., 2012).

Experimental procedures

Plant material

Forty-seven wheat varieties were grown for DNA extraction (for details see Data S5). The Avalon × Cadenza doubled-haploid (DH) population was supplied by the John Innes Centre and was developed by Clare Ellerbrook, Liz Sayers and the late Tony Worland as part of a Defra-funded project led by ADAS. The parents were originally chosen (to contrast for canopy architecture traits) by Steve Parker (CSL), Tony Worland and Darren Lovell (Rothamsted Research). The Savannah × Rialto DH population was supplied by Limagrain UK Limited (Woolpit, Suffolk, UK). All plants were grown in pots in a peat-based soil and maintained in a glasshouse at 15–25 °C under a light regime of 16 h light and 8 h dark. Leaf tissues were harvested from 6-week-old plants and immediately frozen on liquid nitrogen and stored at −80 °C until nucleic acid extraction. Genomic DNA was prepared from leaf tissue using a phenol–chloroform extraction method (Sambrook et al., 1989).

Preparation of NimbleGen libraries

The NimbleGen capture array was designed to capture a significant proportion of the wheat exome and was developed using a gene-rich assembly of 454 titanium sequence data from normalised and non-normalised cDNA libraries of Chinese Spring line 42, publically available EST sequences and the NCBI unigene set (Winfield et al., 2012). The resulting assembly was used by NimbleGen to design an array containing 132 605 features with an average length of 426 bp (NimbleGen array reference 100819_Wheat_Hall_-cap_HX1). NimbleGen sequence libraries were prepared for eight wheat varieties (Alchemy, Avalon, Cadenza, Hereward, Rialto, Robigus, Savannah and Xi19) as described by Winfield et al. (2012). Post-capture-enriched sequencing libraries were subjected to 110 bp of paired end sequencing on a Illumina Genome

Analyser (GAIIx) using Illumina TruSeq v5 Cluster Generation (Illumina Inc., San Diego, CA) and sequencing reagents following the manufacturers preparation guides for paired end runs (Part 15019435 RevB, Oct2010 and Part 15013595 Rev C, Feb 2011, respectively).

SNP discovery

After pre-processing of reads, where adapter sequences were removed, the data were submitted to a custom pipeline (Winfield et al., 2012). NGS sequences generated from the eight varieties were mapped to the NimbleGen array reference using BWA version 0.5.9-r16 (Li and Durbin, 2009) with a seed length of 32 bases, and the resulting SAM files were used for downstream analysis. Uniquely mapped reads were analysed using a series of custom PERL scripts designed to identify only differences between varieties as opposed to those between each variety and the reference sequence. This enabled the exclusion of homoeologous SNPs (which are not useful markers), which were removed from the SNP discovery pipeline. SNPs were called where there were at least two alternative bases predicted at a reference position. An additional constraint on SNP prediction required each SNP to be represented by two or more independent reads or 2% of all reads examined (whichever was the greater). Only bases that were located at the centre of a three-base window of PHRED quality $\geq$ 20 were included in the analysis. Sequences were discarded if they displayed more than 10% sequence variation from the reference over their length or if they mapped equally well to more than one locus, as the mapping in these situations could be regarded as uncertain. In cases where multiple reads started at the same position in the reference, all but one were ignored to guard against clonal reads being sampled more than once. All NGS data generated for this study will be available at: http://www.cerealsdb.uk.net. In addition, the Illumina fastq files and associated metadata have been uploaded to NCBI Sequence Read Archive (SRA) under the study accession SRP011067. Accession numbers of fastq files for each variety are as follows: Alchemy (SRR417586.1), Avalon (SRR417587.1), Cadenza (SRR417953.1), Hereward (SRR417954.1), Rialto (SRR417955.1), Robigus (SRR418209.1), Savannah (SRR418210.1) and Xi19 (SRR418211.1).

SNP validation

For each putative varietal SNP, two allele-specific forward primers and one common reverse primer (Data S1) were designed (KBioscience, Hoddesdon, UK). Genotyping reactions were performed in a Hydrocycler (KBioscience) in a final volume of 1 µL containing 1 × KASP 1536 Reaction Mix (KBioscience), 0.07 µL assay mix (containing 12 µM each allele-specific forward primer and 30 µM reverse primer) and 10–20 ng genomic DNA. The following cycling conditions were used: 15 min at 94 °C; 10 touchdown cycles of 20 s at 94 °C, 60 s at 65–57 °C (dropping 0.8 °C per cycle); and 26–35 cycles of 20 s at 94 °C, 60 s at 57 °C. Fluorescence detection of the reactions was performed using a Omega Pherastar scanner (BMG LABTECH GmbH, Offenburg, Germany), and the data were analysed using the KlusterCaller 1.1 software (KBioscience).

Genetic map construction

The software programme MapDisto v. 1.7 (Lorieux, 2012) was used to place the SNP markers in the previously established genetic map for Avalon × Cadenza (http://www.wgin.org.uk/resources/MappingPopulation/TAmapping.php). A chi-square test was performed on all loci to test for segregation distortion from the expected 1 : 1 ratio of each allele in a DH population, and any loci showing significant distortion were removed from the data set before constructing the linkage groups. Loci were assembled into linkage groups using likelihood odds (LOD) ratios with a LOD threshold of 6.0 and a maximum recombination frequency threshold of 0.40. The linkage groups were ordered using the likelihoods of different locus-order possibilities and the iterative error removal function (maximum threshold for error probability 0.05) in MapDisto and drawn in MapChart (Voorrips, 2002). The Kosambi mapping function (Kosambi, 1944) was used to calculate map distances (cM) from recombination frequency.

SNP data analysis

Summary statistics (MAF and PIC estimates) were calculated for loci using Powermarker 3.25 software (Liu and Muse, 2005).

Acknowledgements

We are grateful to the Biotechnology and Biological Sciences Research Council, UK, and the Crop Improvement Research Club (CIRC) for providing the funding for this work (awards BB/I003207/1, BB/I017496/1). We are grateful to the Wheat Genetic Improvement Network for making the mapping data relating to the Avalon × Cadenza population public. For further details of the Avalon × Cadenza mapping population, please refer to the Wheat Genetic Improvement Network web site at: http://www.wgin.org.uk/resources/MappingPopulation/TAmapping.php. We also thank Limagrain UK limited for supplying the Savannah × Rialto mapping population and related marker data.

References

Adams, K.L. and Wendel, J.F. (2005) Novel patterns of gene expression in polyploid plants. Trends Genet. **21**, 539–543.

Akhunov, E., Nicolet, C. and Dvorak, J. (2009) Single nucleotide polymorphism genotyping in polyploidy wheat with the Illumina GoldenGate assay. Theor. Appl. Genet. **119**, 507–517.

Akhunov, E.D., Akhunova, A.R., Anderson, O.D., Anderson, J.A., Blake, N., Clegg, M.T., Coleman-Derr, D., Conley, E.J., Crossman, C.C., Deal, K.R., Dubcovsky, J., Gill, B.S., Gu, Y.Q., Hadam, J., Heo, H., Huo, N., Lazo, G.R., Luo, M.C., Ma, Y.Q., Matthews, D.E., McGuire, P.E., Morrell, P.L., Qualset, C.O., Renfro, J., Tabanao, D., Talbert, L.E., Tian, C., Toleno, D.M., Warburton, M.L., You, F.M., Zhang, W. and Dvorak, J. (2010) Nucleotide diversity maps reveal variation in diversity among wheat genomes and chromosomes. BMC Genomics, **11**, 702.

Akhunova, A.R., Matniyazov, R.T., Liang, H. and Akhunov, E.D. (2010) Homoeolog-specific transcriptional bias in allopolyploid wheat. BMC Genomics, **11**, 505.

Allen, A.M., Barker, G.L., Berry, S.T., Coghill, J.A., Gwilliam, R., Kirby, S., Robinson, P., Brenchley, R.C., D'Amore, R., McKenzie, N., Waite, D., Hall, A., Bevan, M., Hall, N. and Edwards, K.J. (2011) Transcript-specific, single-nucleotide polymorphism discovery and linkage analysis in hexaploid bread wheat (Triticum aestivum L.). Plant Biotechnol. J. **9**, 1086–1099.

Altschul, S.F., Gish, W., Miller, W., Myers, E.W. and Lipman, D.J. (1990) Basic local alignment search tool. J. Mol. Biol. **215**, 403–410.

Berkman, P.J., Lai, K., Lorenc, M.T. and Edwards, D. (2012) Next generation sequencing applications for wheat crop improvement. Am. J. Bot. **99**, 365–371.

Biesecker, L.G., Shianna, K.V. and Mullikin, J.C. (2011) Exome sequencing: the expert view. Genome Biol. **12**, 128.

Caldwell, K.S., Dvorak, J., Lagudah, E.S., Akhunov, E., Luo, M.C., Wolters, P. and Powell, W. (2004) Sequence polymorphism in polyploid wheat and their D-genome diploid ancestor. Genetics, **167**, 941–947.

Chao, S., Zhang, W., Akhunov, E., Sherman, J., Ma, Y., Luo, M.C. and Dubcovsky, J. (2009) Analysis of gene-derived SNP marker polymorphism in US wheat (*Triticum aestivum* L.) cultivars. *Mol. Breed.* **23**, 23–33.

Chao, S., Dubcovsky, J., Dvorak, J., Luo, M.C., Baenziger, S.P., Matnyazov, R., Clark, D.R., Talbert, L.E., Anderson, J.A., Dreisigacker, S., Glover, K., Chen, J., Campbell, K., Bruckner, P.L., Rudd, J.C., Haley, S., Carver, B.F., Perry, S., Sorrells, M.E. and Akhunov, E.D. (2010) Population- and genome-specific patterns of linkage disequilibrium and SNP variation in spring and winter wheat (*Triticum aestivum* L.). *BMC Genomics*, **11**, 727.

Cordeiro, G.M., Eliott, F., McIntyre, C.L., Casu, R.E. and Henry, R.J. (2006) Characterisation of single nucleotide polymorphisms in sugarcane ESTs. *Theor. Appl. Genet.* **113**, 331–343.

Dixon, J., Braun, H.J. and Crouch, J. (2009) Transitioning wheat research to serve the future needs of the developing world. In: *Wheat Facts and Futures* (Dixon, J., Braun, H.J. and Kosina, P., eds), pp. 1–19. Mexico: CIMMYT.

Dubcovsky, J. and Dvorak, J. (2007) Genome plasticity a key factor in the success of polyploid wheat under domestication. *Science*, **316**, 1862–1866.

Haudry, A., Cenci, A., Ravel, C., Bataillon, T., Brunel, D., Poncet, C., Hochu, I., Poirier, S., Santoni, S., Glémin, S. and David, J. (2007) Grinding up wheat: a massive loss of nucleotide diversity since domestication. *Mol. Biol. Evol.* **24**, 1506–1517.

Haun, W.J., Hyten, D.L., Xu, W.W., Gerhardt, D.J., Albert, T.J., Richmond, T., Jeddeloh, J.A., Jia, G., Springer, N.M., Vance, C.P. and Stupar, R.M. (2011) The composition and origins of genomic variation among individuals of the soybean reference cultivar Williams 82. *Plant Physiol.* **155**, 645–655.

Kaur, S., Francki, M.G. and Forster, J.W. (2012) Identification, characterization and interpretation of single-nucleotide sequence variation in allopolyploid crop species. *Plant Biotechnol. J.* **10**, 125–138.

Kosambi, D.D. (1944) The estimation of map distances from recombination values. *Ann. Eugen.* **12**, 172–175.

Li, H. and Durbin, R. (2009) Fast and accurate short read alignment with Burrows-Wheeler transform. *Bioinformatics*, **25**, 1754–1760.

Liu, K. and Muse, S.V. (2005) PowerMarker: an integrated analysis environment for genetic marker analysis. *Bioinformatics*, **21**, 2128–2129.

Liu, B., Xu, C., Zhao, N., Qi, B., Kimatu, J.N., Pang, J. and Han, F. (2009) Rapid genomic changes in polyploid wheat and related species: implications for genome evolution and genetic improvement. *J. Genet. Genomics*, **36**, 519–528.

Lorieux, M. (2012) MapDisto: fast and efficient computation of genetic linkage maps. *Mol. Breeding*, **30**, 1231–1235.

Paux, E., Faure, S., Choulet, F., Roger, D., Gauthier, V., Martinant, J.P., Sourdille, P., Balfourier, F., Le Paslier, M.C., Chauveau, A., Cakir, M., Gandon, B. and Feuillet, C. (2010) Insertion site-based polymorphism markers open new perspectives for genome saturation and marker-assisted selection in wheat. *Plant Biotechnol. J.* **8**, 196–210.

Paux, E., Sourdille, P., Mackay, I. and Feuillet, C. (2011) Sequence-based marker development in wheat: advances and applications to breeding. *Biotechnol. Adv.* http://dx.doi.org/10.1016/j.bbr.2011.03.031.

Reynolds, M., Foulkes, M.J., Slafer, G.A., Berry, P., Parry, M.A.J., Snape, J.W. and Angus, W.J. (2009) Raising yield potential in wheat. *J. Exp. Bot.* **60**, 1899–1918.

Sambrook, J., Fritsch, E.F. and Maniatis, T. (1989) *Molecular Cloning: A Laboratory Manual*, 2nd edn. Cold Spring Harbor: Cold Spring Harbor Laboratory Press.

Swanson-Wagner, R.A., Eichten, S.R., Kumari, S., Tiffin, P., Stein, J.C., Ware, D. and Springer, N.M. (2010) Pervasive gene content variation and copy number variation in maize and its undomesticated progenitor. *Genome Res.* **20**, 1689–1699.

Trick, M., Long, Y., Meng, J. and Bancroft, I. (2009) Single nucleotide polymorphism (SNP) discovery in the polyploid *Brassica napus* using Solexa transcriptome sequencing. *Plant Biotechnol. J.* **7**, 334–346.

Trick, M., Adamski, N., Mugford, S.G., Jiang, C., Febrer, M. and Uauy, C. (2012) Combining SNP discovery from next-generation sequencing data with bulked segregant analysis (BSA) to fine-map genes in polyploid wheat. *BMC Plant Biol.* **12**, 1–14.

Voorrips, R.E. (2002) MapChart: software for the graphical presentation of linkage maps and QTLs. *J. Hered.* **93**, 77–78.

Winfield, M.O., Wilkinson, P.A., Allen, A.M., Barker, G.L.A., Coghill, J.A., Burridge, A., Hall, A., Brenchley, R.C., D'Amore, R., Hall, N., Bevan, M., Richmond, T., Gerhardt, D.J., Jeddeloh, J.A. and Edwards, K.J. (2012) Targeted re-sequencing of the allohexaploid wheat exome. *Plant Biotechnol. J.* **10**, 733–742.

Yu, J.Z., Kohel, R.J., Fang, D.D., Cho, J., Van Deynze, A., Ulloa, M., Hoffman, S.M., Pepper, A.E., Stelly, D.M., Jenkins, J.N., Saha, S., Kumpatla, S.P., Shah, M.R., Hugie, W.V. and Percy, R.G. (2012) A high-density simple sequence repeat and single nucleotide polymorphism genetic map of the tetraploid cotton genome. *Genes Genomes Genetics*, **2**, 43–58.

17

Sequencing chromosome 5D of *Aegilops tauschii* and comparison with its allopolyploid descendant bread wheat (*Triticum aestivum*)

Bala A. Akpinar[1], Stuart J. Lucas[2], Jan Vrána[3], Jaroslav Doležel[3] and Hikmet Budak[1,2,*]

[1]*Faculty of Engineering and Natural Sciences, Sabanci University, Tuzla, Istanbul, Turkey*
[2]*Sabanci University Nanotechnology Research and Application Centre (SUNUM), Sabanci University, Tuzla, Istanbul, Turkey*
[3]*Institute of Experimental Botany, Centre of the Region Haná for Biotechnological and Agricultural Research, Olomouc, Czech Republic*

*Correspondence

email budak@sabanciuniv.edu
Sequence accessions: EBI Sequence Read
Archive, primary accession number
PRJEB5993.

Keywords: D genome donor of
wheat, chromosome 5D, comparative
genomics.

Summary

Flow cytometric sorting of individual chromosomes and chromosome-based sequencing reduces the complexity of large, repetitive Triticeae genomes. We flow-sorted chromosome 5D of *Aegilops tauschii*, the D genome donor of bread wheat and sequenced it by Roche 454 GS FLX platform to approximately 2.2x coverage. Repetitive sequences represent 81.09% of the survey sequences of this chromosome, and Class I retroelements are the prominent type, with a particular abundance of LTR/Gypsy superfamily. Nonrepetitive sequences were assembled to cover 17.76% of the total chromosome regions. Up to 6188 nonrepetitive gene loci were predicted to be encoded by the 5D chromosome. The numbers and chromosomal distribution patterns of tRNA genes suggest abundance in tRNA^{Lys} and tRNA^{Met} species, while the nonrepetitive assembly reveals tRNA^{Ala} species as the most abundant type. A comparative analysis of the genomic sequences of bread wheat and *Aegilops* chromosome 5D indicates conservation of gene content. Orthologous unique genes, matching *Aegilops* 5D sequences, numbered 3730 in barley, 5063 in *Brachypodium*, 4872 in sorghum and 4209 in rice. In this study, we provide a chromosome-specific view into the structure and organization of the 5D chromosome of *Ae. tauschii*, the D genome ancestor of bread wheat. This study contributes to our understanding of the chromosome-level evolution of the wheat genome and presents a valuable resource in wheat genomics due to the recent hybridization of *Ae. tauschii* genome with its tetraploid ancestor.

Introduction

Bread wheat (*Triticum aestivum*) has a large hexaploid genome ($\sim$17 Gb/1C) with three homoeologous subgenomes (2n = 6x = 42, genome formula AABBDD) that are believed to originate from *Triticum urartu* (2n = 2x = 14, AA), donor of the A genome, *Aegilops tauschii* (2n = 2x = 14,DD), donor of the D genome and a species similar to *Aegilops speltoides* (2n = 2x = 14, SS) as donor of the B genome (Gill *et al.*, 2004; Peng *et al.*, 2011; Tanaka *et al.*, 2014). These three diploid genomes were combined by two successive genome hybridization events; the older between A and B genomes to create AABB tetraploid (*Triticum turgidum*), to which *Ae. tauschii* contributed the D genome only about 8000 years ago (Jia *et al.*, 1996), giving rise to allohexaploid bread wheat.

Ae. tauschii retains considerable homology with the *T. aestivum* D genome; however, while the D genome of bread wheat is the least polymorphic of the three subgenomes, its progenitor *Ae. tauschii* shows much greater variation (Dvorak *et al.*, 1998). This reflects a limited pool of *Ae. tauschii* genotypes being involved in the hybridization event that formed *T. aestivum* (Dubcovsky and Dvorak, 2007). Therefore, the *Ae. tauschii* genome is of considerable interest as a source of genes and alleles for wheat improvement. The D genome harbours a number of genes that are known to be important for end-use quality of bread wheat, such as high molecular weight glutenin subunits (HMW-GS) that confer increased dough strength (Blechl and Anderson, 1996). In particular, chromosome 5D contains the *Ha* (hardness) locus that determines grain texture; the puroindoline genes that confer this trait have been lost from the A and B genomes (Chantret *et al.*, 2005). Similarly, the 5D allele of transcription factor *WAP2* is believed to be particularly important in the appearance of free-threshing wheat (Ning *et al.*, 2009) and two vernalization loci on this chromosome influence flowering time (Yoshida *et al.*, 2010; Zhang *et al.*, 2012). Quantitative trait loci (QTLs) for improved yield have been found to coincide with one of these genes, *VrnD1* (Quarrie *et al.*, 2005). Chromosome 5D also carries important stress response genes such as the leaf rust resistance gene *Lr1* (Cloutier *et al.*, 2007) and loci conferring salt and drought tolerance (Semikhodskii, 1997) and frost resistance (Galaeva *et al.*, 2013). Furthermore, chromosome 5D of *Ae. tauschii* has been shown to be useful for introducing desirable traits into bread wheat, such as powdery mildew resistance gene *Pm35* (Miranda *et al.*, 2007). *Ae. tauschii* biodiversity is being explored and exploited by the creation of synthetic hexaploid wheats, generated by artificial hybridization of *T. turgidum* with *Ae. tauschii*, which can act as a bridge to introduce *Ae. tauschii* genetic material into bread wheat (Mujeeb-Kazi *et al.*, 1996). Additionally, synthetic hexaploid lines carrying strong grain and stripe rust

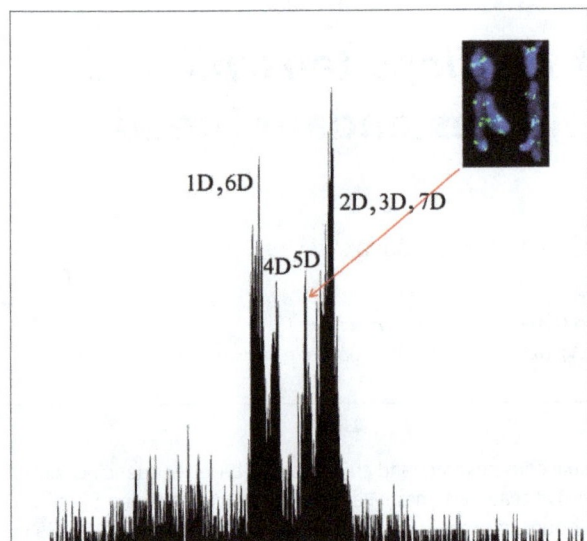

Figure 1 Histogram of relative fluorescence (flow karyotype) obtained after flow cytometric analysis of DAPI-stained mitotic metaphase chromosomes of *Aegilops tauschii*. Chromosome 5D was identified after FISH with a probe for Afa repeats (yellow–green) which give a specific banding pattern (inset). *X*-axis: relative DAPI fluorescence intensity; *Y*-axis: number of particles.

sequencing of isolated chromosomes and chromosome arms (Berkman *et al.*, 2011, 2012; Hernandez *et al.*, 2012; Vitulo *et al.*, 2011) and construction of chromosome-specific BAC libraries (Hernandez *et al.*, 2012; Safár *et al.*, 2010). In this study, we flow-sorted chromosome 5D of *Ae. tauschii*, for which individualized sequence data had not been available, and utilized Roche 454 GS-FLX Titanium technology to sequence the chromosome. The sequences were analysed to characterize the genomic composition of chromosome 5D, including repeat family identification, gene identification and annotation, chromosome-specific tRNA identification and comparative genomics with other related grasses such as barley, rice, *Brachypodium* and sorghum. These data will be useful for developing new 5D-specific molecular markers, and in combination with our recent study of *T. aestivum* chromosome 5D will help to shed light on the evolutionary and functional aspects of the wheat genome.

Results and discussion

Flow sorting, sequencing and assembly

Flow cytometric analysis of fluorescence of isolated *Ae. tauschii* chromosomes that had been stained by DAPI resulted in a flow karyotype with two composite peaks representing groups of chromosomes and two peaks representing chromosomes 4D and 5D, respectively (Figure 1). The flow karyotype was slightly different from that obtained by Molnár *et al.* (2014), who did not observe a separate peak of 4D. The difference could be due to the different genotypes used in each study. In this study, the peak of 5D was well discriminated, and microscopic observation of sorted factions after FISH with a probe for Afa repeat (Figure 1, inset) identified 91% chromosome 5D, on average. The contaminating particles comprised a random mix of other chromosomes and fragments of chromosomes and chromatids. DNA from the three batches of sorted 5D was amplified separately by multiple displacement amplification (MDA) giving a total yield of 7.4-µg DNA.

The DNA of chromosome 5D was fragmented and used to prepare three different shotgun libraries. Each of libraries was then quantitated, amplified and sequenced, giving a total of 1.24 gigabases of good quality sequence reads (Table 1). This corre-

resistance traits have outperformed commercial cultivars in some field trials (Yang *et al.*, 2009). Moreover, introgressions of D genome chromosomes of group 1 and group 6 into triticale improved the end-use quality characteristics of triticale (Mahmood *et al.*, 2004). Therefore, characterizing the genome of *Ae. tauschii* is expected to identify many genes that can be used for wheat improvement. A first draft of whole genome shotgun sequence derived from *Ae. tauschii* has recently been published (Jia *et al.*, 2013).

Flow sorting of mitotic metaphase chromosomes provides a powerful approach to dissect complex and large genomes and simplify the task of characterizing them (Doležel *et al.*, 2014). The use of this approach in bread wheat has included shotgun

Table 1 Summary of sequencing and assembly statistics for *Ae. tauschii* chromosome 5D

Library	No.of reads N	Mean read length L (bp)	Total read length (Mb)	Sequencing coverage*	P†
Aet5D-1	1 477 789	311.0	459.5		
Aet5D-2	1 020 199	373.6	381.2		
Aet5D-3	1 072 417	369.2	395.9		
Combined	3 570 405	346.3	1236.6	2.14x	0.858

Assembly statistics	No. of reads/contigs	Mean length (bp)	Total length (Mb)	Length (% of chromosome)	N50 contig size (bp)
Filtered reads	523 612	316	165.6	28.7	
LCN assembly					
Large contigs	13 719	954	13.1	2.27	1006
All contigs	28 668	601	17.2	2.99	
Singletons	278 027	307	85.2	14.77	

*Sequencing coverage was calculated using a chromosome size estimate of 577 Mb (Luo *et al.*, 2013).

†The probability of representation of any position in the data set was calculated as follows: $P = [1 - (1 - L/S)^{N \times Purity}]$, where S is the chromosome size and L and N are as listed in the table.

sponds to 2.14x coverage based on a chromosome size estimate of 577 Mb (Luo *et al.*, 2013). At this level of coverage, the probability of any given position from 5D being present at least once in the data set is estimated to be 85.8%.

The high repetitive element content of Triticeae genomes (Smith and Flavell, 1975) makes assembling reads difficult. Therefore, reads coding for repetitive elements along with rRNA coding sequences and reads derived from mitochondrial and chloroplast DNA contamination were excluded, and the remaining filtered reads used to construct an assembly of nonrepetitive regions, which should include the majority of gene coding sequences (referred to as low-copy number (LCN) assembly; Table 1). The peak assembly depth of the contigs was 2.1, very similar to the sequencing depth of 2.14x, and indicating a good quality assembly with few incorrectly merged sequences. Only 43% of the sequenced reads were assembled into contigs; this was expected as at this sequencing depth many regions of the chromosome may be covered by only a single read or by reads with insufficient overlap to be assembled into contigs (which hence remained as singletons). Therefore, both the contigs and singletons were used for the following analyses. The total length of the combined contigs and singletons was 111.3 Mb, very close to the estimated length of the nonrepetitive component of the chromosome (109 Mb; see the following section). The slightly larger size of the assembly may be explained by the presence of a small number of contaminating reads from other chromosomes among the singletons.

The low-copy number (LCN) assembly generated from *Aegilops* 5D sequence reads consisted of 28 668 contigs and 278 027 singletons, after the removal of repetitive sequences, cp/mtDNA and rRNA contaminations (Table 1). The LCN assembly was compared with the Illumina shotgun sequences obtained for 5A, 5B and 5D chromosomes of *Triticum aestivum* by the International Wheat Genome Sequencing Consortium (IWGSC) at high stringency (1E-30, 100 nt alignment length, 96% identity). A total of 26 459 contigs and 122 660 singletons (92% and 44% of all contigs and singletons, respectively) retrieved significant matches in the IWGSC data set for chromosome 5D. The comparison of *Aegilops* LCN assembly with IWGSC 5A and 5B sequences of *T. aestivum* at the same stringency yielded 4758 contig/15 707 singleton matches and 6281 contig/21 633 singleton matches, respectively (17% and 6% of all contigs and singletons for 5A; 21% and 7% of all contigs and singletons for 5B). Additionally, the LCN assembly covers all six genes, deposited in NCBI, that are mapped to *Aegilops* 5D chromosome at the same stringency level. These genes are APETALA2-like AP2 [GenBank: ABY53104.1], AetVIL1, Gsp-1, Pina-D1, Pinb-D1 and Vrn-D1 (Koyama *et al.*, 2012; Turnbull *et al.*, 2003; Yan *et al.*, 2003). At 98% and 99% identity levels, five and four of these genes were covered by the LCN assembly, respectively. These results indicate that LCN assembly provides a good coverage of the 5D chromosome representing majority of all genes encoded and, in comparison with 5A and 5B, exhibits considerably more homology to *T. aestivum* 5D chromosome.

The repeat element landscape of *Aegilops* 5D

The repetitive elements found were grouped by superfamily (Figure 2a). Over two-thirds of the repetitive space was made up of LTR (long terminal repeat) retroelements, with the Gypsy superfamily being the most abundant. There was also a high incidence of CACTA DNA transposons, contributing over a

quarter of the repetitive space. All repeat families identified are listed, along with their relative abundance, in Table S1.

At an estimated 748 Mb in size, *T. aestivum* chromosome 5D is almost 30% larger than that of its progenitor, *Ae. tauschii*. To assess the role of transposable elements in expansion of this chromosome, our previously generated *T. aestivum* 5D 454 survey sequences (Lucas *et al.*; manuscript in preparation) were re-analysed for repeat content using the same parameters as *Ae. tauschii* 5D. The proportional repeat content by length of *T. aestivum* 5D sequence reads was 82.78%, only slightly higher than for *Ae. tauschii*; however, it was observed that LTR elements of the Gypsy and Copia superfamilies made up a higher proportion of the repetitive space, with the contribution of DNA transposons correspondingly reduced (Figure 2a). This observation is brought into sharper relief by considering the total sequence length of the 20 most abundant repeat families (Figure 2b). The CACTA-type Jorge family of DNA transposons is the most highly represented in both species, but there is essentially no expansion of this family in *T. aestivum* compared with *Ae. tauschii*. In contrast, the highly promiscuous Gypsy elements, Sabrina, Wilma and Sakura, all show >75% growth in *T. aestivum*, with these three families alone supplying an estimated 40 Mb of chromosome expansion. Some growth was observed in many, but not all LTR element families; for example Cereba and Erika, although comprising over 2% of all sequences in *Ae. tauschii* 5D do not appear to have grown at all in *T. aestivum*. These findings suggest that a major component of D genome expansion since its incorporation into *T. aestivum* is due to growth of specific LTR retroelements, rather than a universal increase in TE activity. Transposable element (TE) families have been shown to vary in composition and abundance among related grass species, expanding in some, and yet, shrinking in others (Middleton *et al.*, 2012). The Class I LTR retroelements, Gypsy and to a lesser extent Copia superfamilies, might have expanded in *T. aestivum* following the polyploidization event giving rise to its hexaploid genome. Nested insertions in retroelements have been implicated to contribute to the large genomes of wheat which may not interfere with the proliferation of these retroelements (Li *et al.*, 2004). Nested insertions may be tolerated more in the hexaploid *T. aestivum* genome than in diploid *Ae. tauschii*, which could explain the relative expansion of retroelements in the D genome of bread wheat in spite of its recent origin.

Gene conservation and content of *Aegilops* 5D

A total of 6713 contigs and 32 400 singletons were found to be gene-associated as indicated by matches in related grass proteomes and UniGene/UniProt sequences (Table S2). Over half of these gene-related sequences had significant matches only in UniGene or UniProt databases (3576 contigs and 17 256 singletons), some of which may correspond to Aegilops/Triticum lineage-specific genes and thus termed as 'nonconserved' sequences. The remaining gene-associated sequences (3138 contigs and 15 144 singletons) were related to orthologous genes at least in one of the four related grass species, *Brachypodium*, rice, sorghum and barley and were termed as 'conserved' sequences. Conserved and nonconserved singletons matching to orthologous genes that were not covered by any additional contigs or singletons were further eliminated as these sequences may have arisen from the impure fraction of the sorted chromosome. This elimination left 14 191 and 11 743 contigs and singletons as candidate 'conserved' and 'noncon-

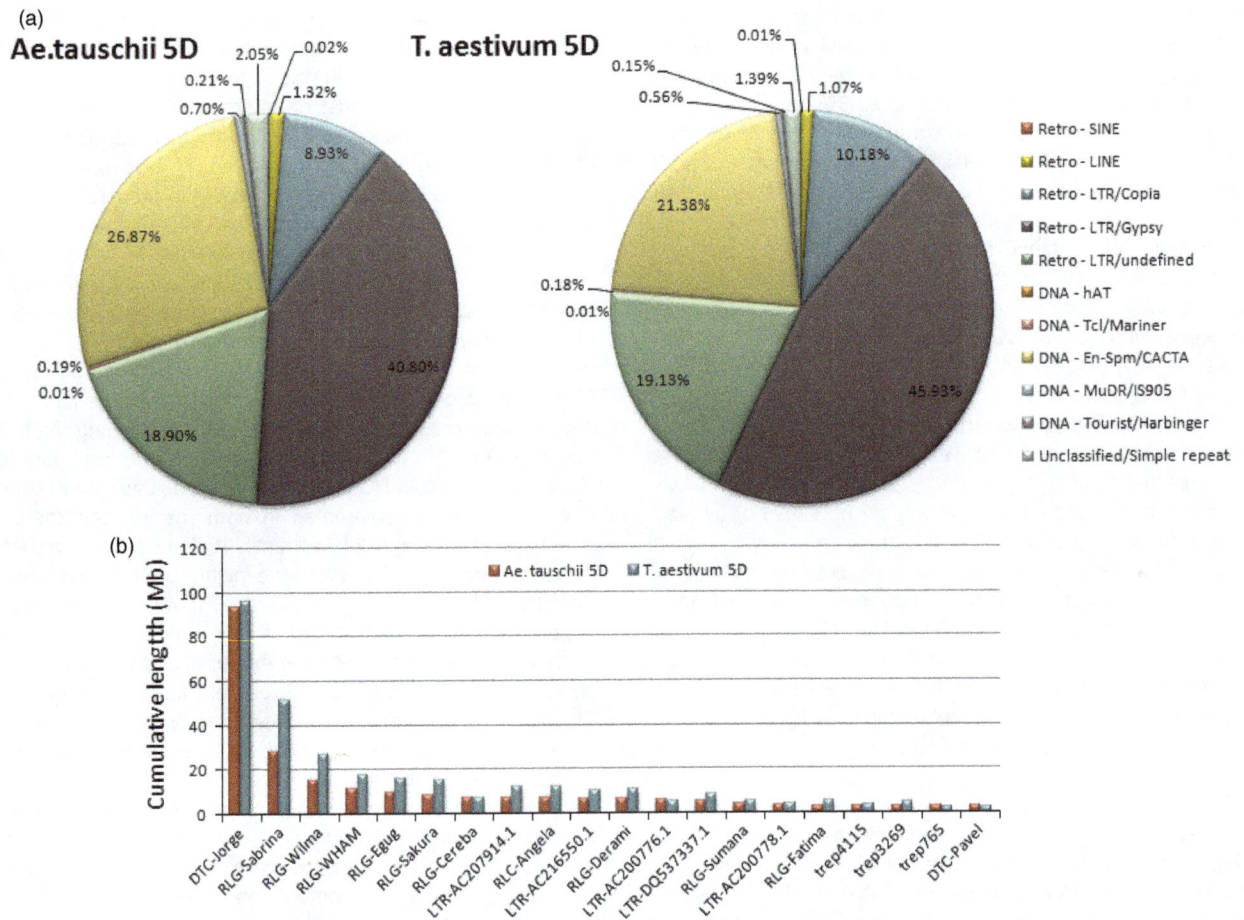

Figure 2 Composition of the repetitive space of chromosome 5D by repeat family. (a) The relative abundance of repeat superfamilies in 454 sequence reads from *Ae. tauschii* and *T. aestivum* 5D. Abundance is expressed as the percentage (by cumulative length) of all repeat sequences. (b) The physical size (in Mbp) of the 20 most abundant repeat families was estimated by multiplying the fraction of all sequences it comprised in each species, by the chromosome size estimate of 577 Mb (*Ae. tauschii*) or 748 Mb (*T. aestivum*). DTC = DNA transposon, CACTA; RLG = retroelement, LTR, Gypsy; RLC = retroelement, LTR, Copia.

served' genic sequences, respectively. These sequences were concluded to correspond to 6188 unique conserved genes, which have orthologs in at least one of the related grass species, and 4951 unique nonconserved genes (Table S2). The number of putative conserved genes, which have orthologs in related grass genomes, corresponded to a genic fraction of 2.1% on the 577 Mb long 5D chromosome (Luo *et al.*, 2013), assuming an average coding sequence length of 2000 bases (Vitulo *et al.*, 2011). The total number of genes, estimated from a gene fraction of 2.1%, for the entire 4-Gigabase long *Aegilops* genome makes roughly up to 42 000. However, assuming an average gene length of 2772 bases, as indicated by the recent physical map of *Ae. tauschii* (Luo *et al.*, 2013), gene content of chromosome 5D rises up to 2.9%, which is dramatically higher than *T. aestivum* 5D and 5A chromosomes where gene contents are ~1.15% and ~1.23%, respectively (Vitulo *et al.*, 2011). It is possible that hexaploid wheat might have lost one or more gene copies from many loci during its evolution, which might have been exacerbated by domestication. Consequently, the entire genome of *Ae. tauschii* was estimated to contain 42 000–58 000 genes, as concluded from the sequence data generated in this study. A very recent study estimated the total number of genes encoded by *Ae. tauschii* as 43 150 (Jia *et al.*, 2013), using RNA-

Seq data and *de novo* gene prediction, which falls in the range estimated by this study. The completely annotated genomes of *Brachypodium*, rice and sorghum that are close relatives of the Triticeae tribe contain over 26 000, 39 000 and 34 000 gene coding loci, respectively (www.phytozome.net); while barley is attributed to code for 30 400 genes in its ~5 Gb genome (Mayer *et al.*, 2012). *Ae. tauschii* and barley have genomes 5–15 times larger than the fully sequenced grass species, so the fact that they contain a similar total number of genes indicates that most of the genome expansion in the Triticeae line occurred in noncoding regions. The relatively smaller gene number in cultivated barley may be a sign of gene loss in modern varieties due to domestication. Additionally, 4951 putative nonconserved genes with no orthologs in any of the four grass genomes were concluded to be encoded by 5D, as evidenced by significant matches by UniGene and/or UniProt sequences from *T. aestivum* and *Ae. tauschii*, respectively. While we cannot exclude the possibility that a fraction of these putative nonconserved genes is in fact gene fragments, duplicated genes or pseudogenes, it is likely that some of these sequences belong to Triticeae-specific genes. *Aegilops* 5D sequence reads matching annotated genes from the four grass species and UniGene/UniProt sequences that were not matched by any other sequence read were considered

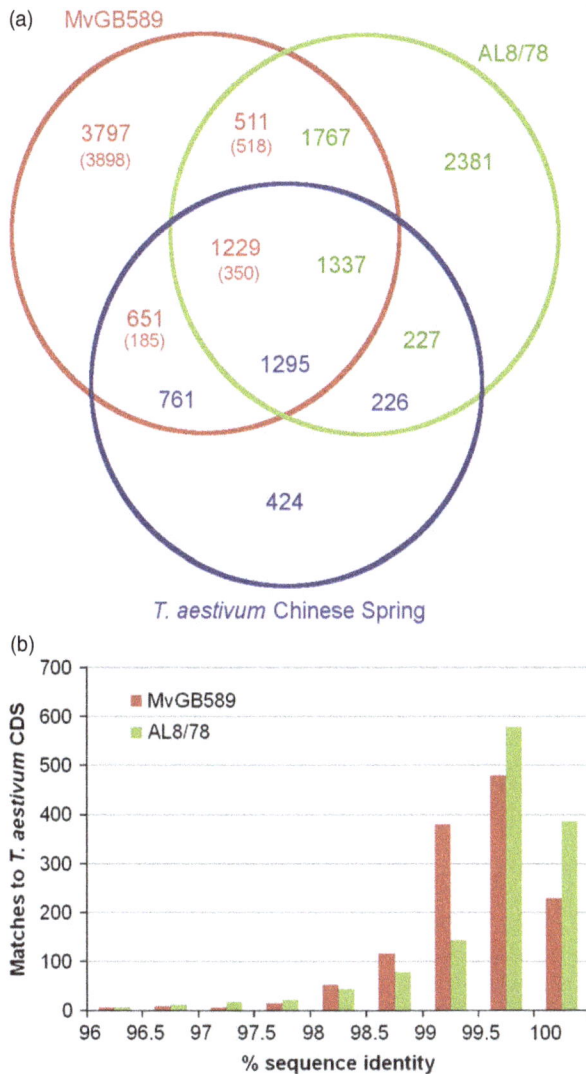

Figure 3 Comparison of predicted gene models with CDS mapped to chromosome 5D from *Ae. tauschii* var. AL8/78 and *T. aestivum* var. Chinese Spring. (a) Venn diagram showing the number of gene models/CDS matching at ≥96% identity over ≥100 nt between each data set. For MvGB589, the number of 'conserved' gene models is given, followed by the number of 'nonconserved' models in brackets. (b) For the gene models common to all three data sets, histogram showing the distribution of matches to *T. aestivum* 5D CDS from each *Ae. tauschii* data set by percentage sequence identity, in bins of 0.5%.

as pseudogenes or contaminants from genes found elsewhere in the *Aegilops* genome, as indicated by the 91% purity of the sorted 5D chromosome. In fact, considering conserved gene hits, almost 60% of these reads matched genes from only one of the four related grasses, of which half (30% overall) lacked any matches to UniGene/UniProt sequences (Table S2). In contrast, among the sequence reads matching 'conserved' genes 40% had significant matches in only one of the four grasses, while the remaining 60% of reads matched to multiple annotated genes. Similarly, only 18% of this group of sequence reads lacked matches to UniGene/UniProt sequences. Such sequences, matching only one of the four related grass proteins that were not covered by any other sequences were deemed as pseudogenes (Wicker *et al.*, 2011), while those sequences having multiple hits

on grass genomes (that were not covered by any other sequences) were indicative of contaminating chromosome fragments.

In the recently published draft genome sequence assembly of *Ae. tauschii* var. AL8/78, Illumina sequence scaffolds containing 5712 gene models predicted using RNA-seq data were assigned to chromosome 5D by genetic mapping (Jia *et al.*, 2013). Similarly, 2706 high-confidence gene models were assembled from flow-sorted chromosome 5D in the recent draft sequence of *T. aestivum* (International Wheat Genome Sequencing Consortium, 2014). To assess gene content conservation between MvGB589 and these two data sets, all gene-associated contigs and singletons from the LCN assembly were grouped into putative gene models based on their best match in other grass species as described above, and sequence similarity searches (blastn, ≥96% identity over 100 nt) were used to detect orthologs among the CDS from the aforementioned draft genome sequences. A large number of gene models from both *Ae. tauschii* varieties did not match either of the other data sets at this stringency (Figure 3a), suggesting that neither data set gives a complete representation of the gene content of 5D. This is unsurprising as both are draft-quality sequences and incomplete; in AL8/78, the total length of scaffolds mapped to 5D was approximately half the size of the chromosome (Jia *et al.*, 2013). Meanwhile, some putative gene models unique to this study are likely to be untranscribed pseudogenes or gene fragments, especially the 'nonconserved' gene models. Even so, both *Ae. tauschii* data sets give a much higher estimate of gene number on chromosome 5D than *T. aestivum*, indicating that hexaploid wheat has much less genetic variety than its diploid ancestor on this chromosome. Differences in sequencing technology and assembly methods also lead to contrasting gene models, for example 2 CDS from AL8/78 matching different parts of the same gene model from MvGB589 and vice versa. This is particularly seen in the genes unique to *Ae. tauschii* that are shared between the two varieties, where 1767 CDS from AL8/78 matched to 1029 gene models from MvGB589 (Figure 3a). These differences emphasize the need for a reference-quality genome sequence. On the other hand, the majority of *Ae. tauschii* 5D gene models that matched *T. aestivum* 5D were consistent in both sequence

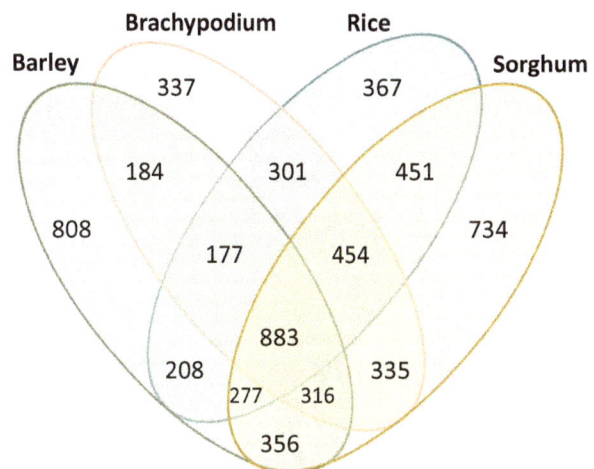

Figure 4 Venn diagram exhibiting the number of significant hits revealed by the comparison of *Aegilops* LCN assembly against the fully annotated proteomes of *Brachypodium*, rice, sorghum and against high-confidence proteins of barley.

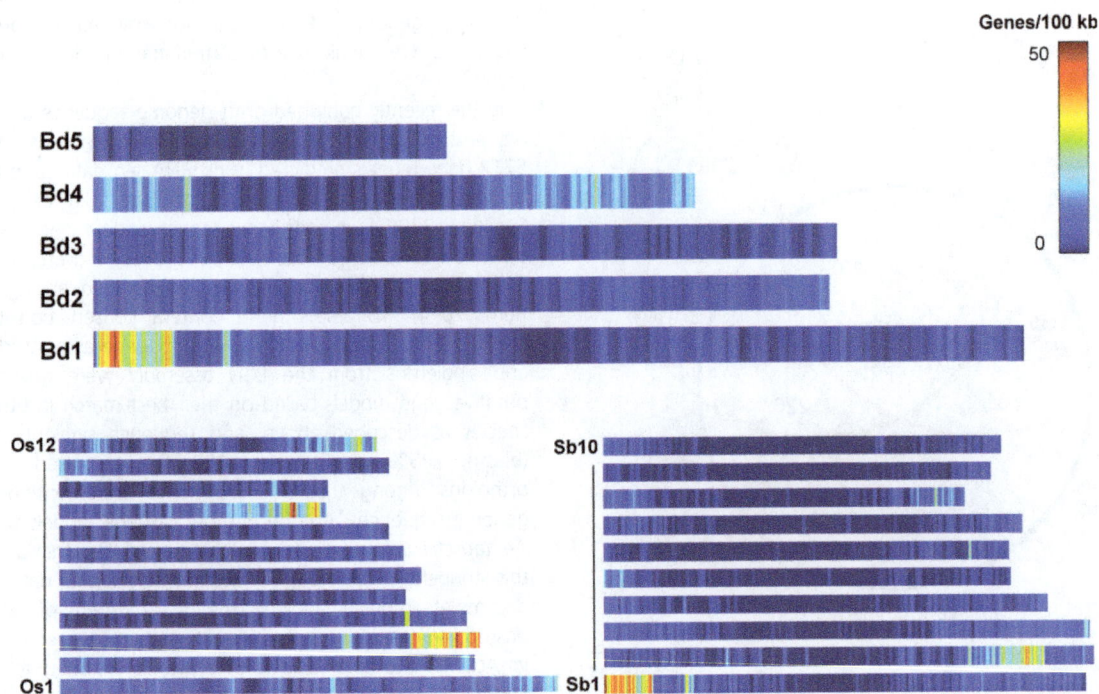

Figure 5 Heatmaps for *Brachypodium*, rice and sorghum genomes. *Aegilops* 5D sequence read matches on each genome were counted on a sliding window of 50 kb step size. *Brachypodium distachyon*, *Oryza sativa* and *Sorghum bicolor* chromosomes are abbreviated as Bd1-Bd5, Os1-Os12 and Sb1-Sb10, respectively. The colour bar indicating the level of synteny on heatmaps is given on top right.

and structure. The gene models that were common to all three data sets were analysed for sequence identity between *Ae. tauschii* and *T. aestivum* (Figure 3b). For both *Ae. tauschii* data sets, the great majority were 99%–100% identical to their wheat orthologs, but slightly higher identities were seen for AL8/78 than MvGB589, which may suggest that the former variety is genetically closer to the D genome of *T. aestivum*.

Syntenic relationships of *Aegilops* 5D with model grasses

The number of contigs or singletons with a significant match in only barley, sorghum, rice and *Brachypodium* were 808, 734, 367 and 334, respectively (Figure 4). The number of gene-related sequences having a match in all four grass species was 883, indicating highly conserved genes. Interestingly, 734 gene-related sequences matching only sorghum far exceeded 367 and 334 gene-related sequences uniquely matching rice and *Brachypodium*, which are closer relatives of *Aegilops* (International Brachypodium Initiative, 2010). Approximately, 33% of all gene-related sequences lacked a significant match in UniGene or UniProt databases. These results indicate that *Aegilops* 5D chromosome contains several lineage-specific genes, along with a considerable number of conserved genes.

In addition to conserved genes, the comparison of the LCN assembly derived from *Aegilops* 5D chromosome against the annotated genomes of *Brachypodium*, rice, sorghum and barley demonstrated significant syntenic relationships between these related grasses (Figures 5 and 6). As shown in Figure 5, one major syntenic block on the proximal end of *Brachypodium* chromosome 1 was detected along with minor syntenic blocks on the proximal and distal ends of chromosome 4. Highly syntenic regions were also identified on the distal ends of rice chromosomes 3 and 9, and, proximal and distal ends of sorghum chromosomes 1 and 2, respectively. Additionally, regions of

minor synteny were observed along rice chromosomes 1 and 12 and sorghum chromosome 8. The comparative analysis of *T. aestivum* 5D chromosome, assembled from our previously generated survey sequences, displayed a very similar pattern of synteny which is consistent with the recent hybridization of *Ae. tauschii* D genome with the AABB progenitor. The major and minor syntenic blocks observed along the model grass genomes were highly similar to those obtained with *T. aestivum* 5D sequence reads, which were assembled using the same procedure (Figure S1). The observed synteny between *Ae. tauschii* 5D chromosome and related genomes is consistent with previously established large scale conservation of genes among grasses (International Brachypodium Initiative, 2010), indicating that *Aegilops* 5D survey sequences provide a general overview of the entire chromosome even at this coverage.

The syntenic relationships and gene conservation between model grass genomes were also analysed in terms of *Aegilops* 5D survey sequences, where each sequence read matching two of the three model genomes was used as a 'link' between the respective genomes (Figure 6). Major syntenic blocks were observed on *Brachypodium* chromosomes 1 and 4, sorghum chromosomes 1 and 2 and rice chromosomes 3 and 9, similar to the heatmaps (Figure 5). Along with these major syntenic blocks, a few smaller blocks of conserved genes were also evident on *Brachypodium* chromosome 2, rice chromosomes 1, 4, 7 and 12 and sorghum chromosomes 3, 6 and 8, suggesting small-scale genome rearrangements. Strikingly, a number of *Aegilops* 5D singletons or contigs matched with genes on *Brachypodium* chromosome 2, rice chromosome 1 and sorghum chromosome 3 (Figure 6), which are known to fall into the same syntenic group (International Brachypodium Initiative, 2010). From an evolutionary perspective, a gene, found in colinear positions on rice and sorghum genomes, but not in *Brachyp-*

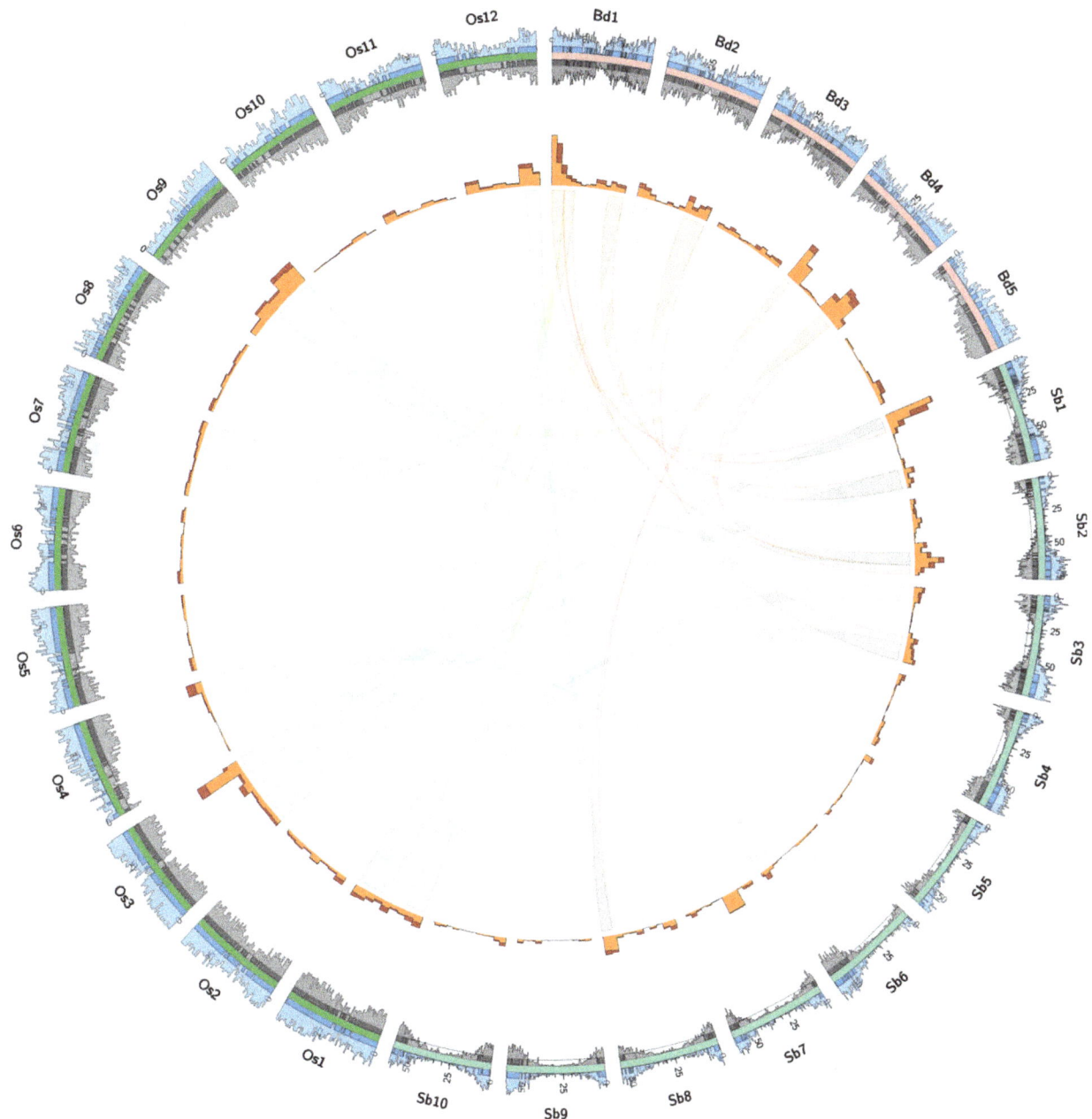

Figure 6 Circular representation of synteny among three model grass genomes in terms of *Aegilops* 5D sequence reads. Red bars correspond to histograms of the links; orange bars indicate the presence of a barley ortholog. Light blue and light grey bars surrounding chromosomes indicate gene densities, where dark blue and dark grey bars indicate a gene density of >50 genes/Mb.

odium, can be considered as a gene that was potentially 'moved' in *Brachypodium* genome (Wicker *et al.*, 2010). Similarly, these *Aegilops* 5D singletons or contigs may correspond to genes or genic sequences that were 'moved' or underwent genome-specific rearrangements after wheat and *Brachypodium* lineages were diverged from their last common ancestor.

The conservation of genes among model grass genomes determined by *Aegilops* 5D sequence reads were also supported by the close relative of wheat, barley, as well (orange bars, Figure 6). In several cases, an *Aegilops* 5D sequence read or contig matching two of the three model grass genomes (depicted as red bars) was also supported with a barley gene (depicted as orange bars). A significant number of *Aegilops* 5D sequences that

matched with barley orthologs exhibited no similarities in any of the three model grass genomes, pointing out barley-/wheat-specific genes and thereby emphasizing the importance of collaborative efforts to sequence these highly complex genomes. Additionally, conserved genes closely followed the gene density trends along the chromosomes of model genomes, as shown by light blue and light grey bars for genes on '+' and '−' strands, respectively. Similar to the positive gradient for gene density towards the telomeres, syntenic and nonsyntenic blocks of conserved genes are mostly found on proximal and distal ends of the chromosomes. This trend was particularly evident on the regions where gene densities were highest (dark blue and dark grey bars).

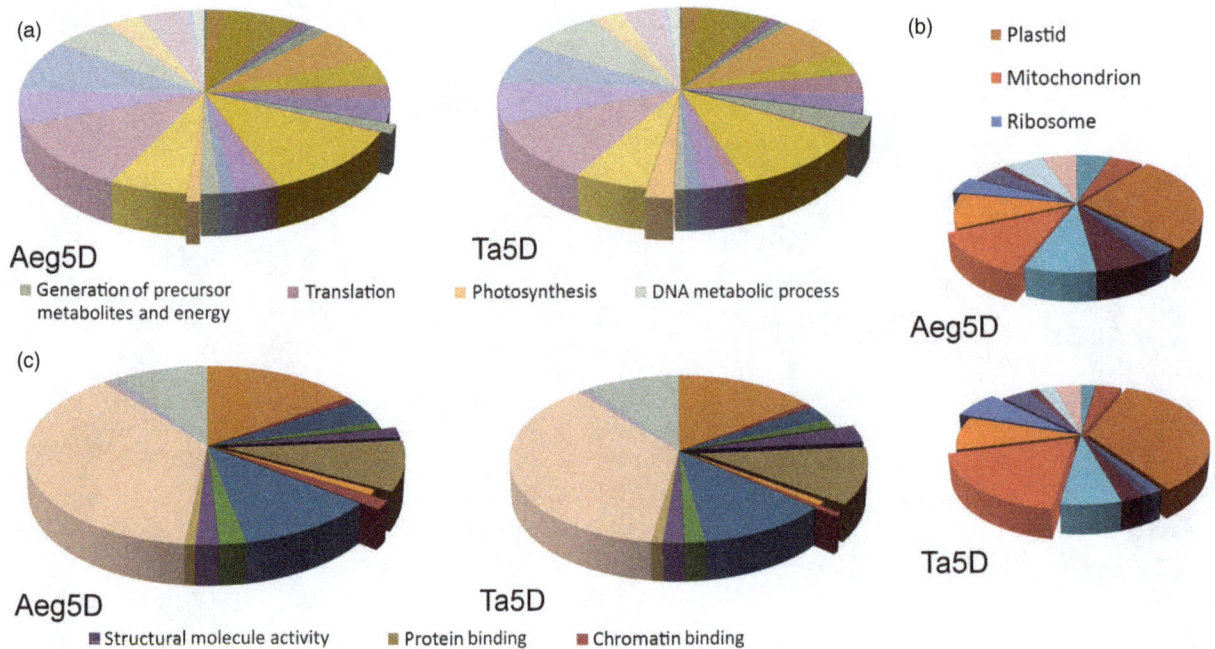

Figure 7 Gene-ontology annotations of *Aegilops* 5D sequences in terms of, (a) Biological process, (b) Cellular component, (c) Molecular function. Only GO terms with differential enrichments are emphasized.

Annotation of *Aegilops* and *T. aestivum* 5D sequences

Annotation of gene-associated sequences from *Ae. tauschii* chromosome 5D was performed to provide an insight into the molecular events involving 5D-encoded genes using a total number of 39 109 sequences. The annotation procedure was also performed for 36 535 gene-associated sequences from *T. aestivum* 5D chromosome, assembled using the same parameters. For both data sets, ~3000 sequences could be annotated, although ~22 000 sequences could be matched with a significant hit in *Viridiplantae* nonredundant protein database. Annotations were used to generate multilevel combined graphs for biological process (BP), cellular component (CC) and molecular function (MF) terms. Compared to *Aegilops* 5D, *T. aestivum* 5D chromosome was observed to be enriched for translation, generation of precursor metabolites and energy, photosynthesis, and DNA metabolic process (Figure 7a). The enrichment was particularly prominent for photosynthesis and generation of secondary metabolites and energy, suggesting that *T. aestivum* 5D chromosome may contain more sequences devoted into these processes. CC terms exhibited a pattern in accordance with the BP terms; *T. aestivum* 5D chromosome was found to be enriched for plastid, mitochondrial and ribosomal sequences. Interestingly, *Aegilops* 5D chromosome contained more sequences associated with nucleus, cytosol and extracellular region than that of *T. aestivum* 5D (Figure 7b). In terms of MF, while subtle variations were observed for a number of functions, *T. aestivum* 5D chromosome was particularly enriched for sequences with protein-binding function and structural molecule activity, while *Aegilops* 5D exhibited enrichment for chromatin-binding activity

Figure 8 Comparative map of *H. vulgare* 5H, *Ae. tauschii* 5D and *T. aestivum* 5D chromosomes.

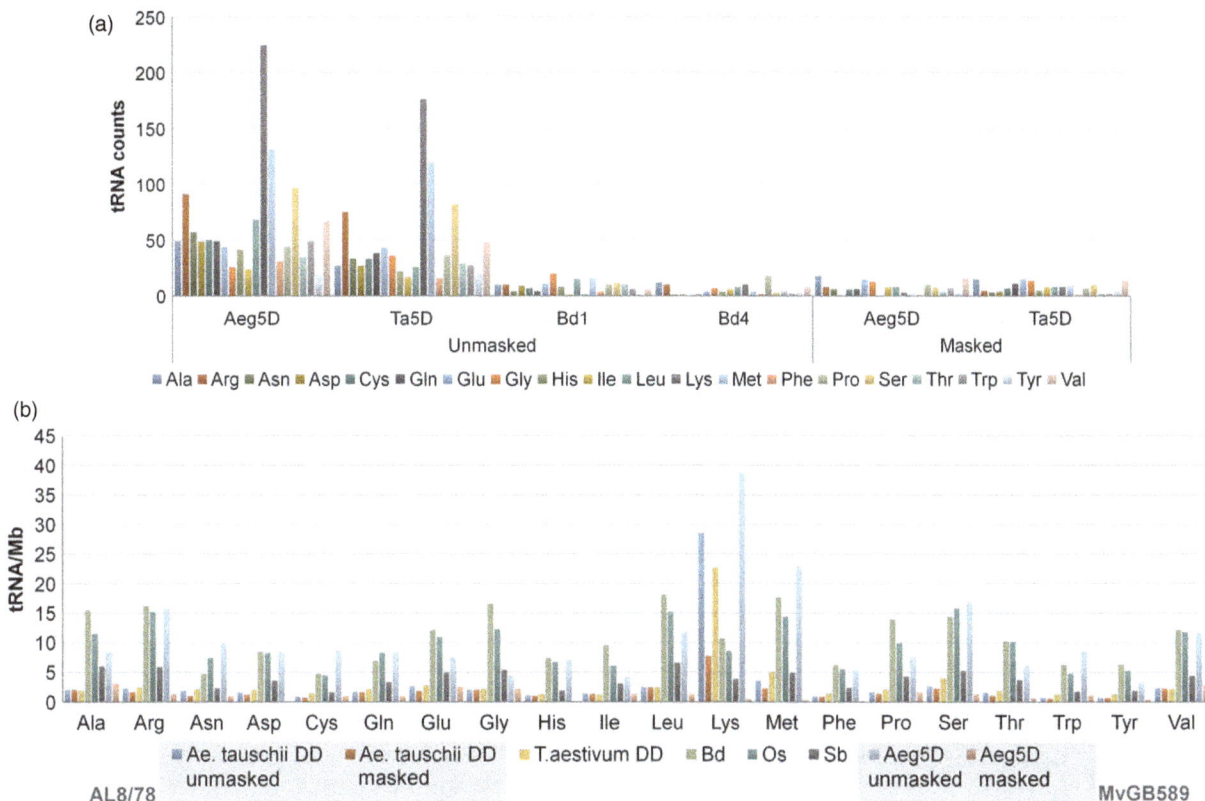

Figure 9 Putative tRNA genes predictions from *Ae. tauschii* and *T.* aestivum 5D chromosomes, D genomes and model grass genomes. (a) tRNA counts predicted from the masked and unmasked *Aegilops* 5D and *T.* aestivum 5D sequence reads, along with the orthologous *Brachypodium* chromosomes 1 and 4. (b) tRNA counts per 100 MB of the genomes of *Ae. tauschii*, *T. aestivum* and model grasses *Brachypodium* (Bd), rice (Os) and sorghum (Sb). tRNA counts per 100 MB of masked and unmasked 5D sequences of *Ae. tauschii* are also included. Aeg5D: *Ae. tauschii* var MvGB589 5D chromosome; Ta5D: *T. aestivum* var. Chinese Spring 5D chromosome; Bd1: *Brachypodium distachyon* chromosome 1; Bd4: *Brachypodium distachyon* chromosome 4; Bd: *Brachypodium distachyon*; Os: *Oryza sativa*; Sb: *Sorghum bicolor*.

(Figure 7c). Together, these results suggest that *T. aestivum* 5D chromosome encodes a wider variety of genes related to the photosynthetic machinery, energy metabolism and stress response, than that of *Aegilops* 5D chromosome. Following polyploidization, gene duplications in the hexaploid genome of bread wheat may have allowed an expansion in these genes, some of which may actually correspond to pseudogenes. Complete annotations for BP, CC and MF terms are given in Figure S2.

Virtual Gene Order and Chromosome Organization of 5D

A virtual order of the genes was determined based on the genetically mapped marker data published in a recent study (Luo *et al.*, 2013) and syntenic relationships. Of the 1029 markers genetically mapped to *Aegilops* 5D chromosome, 777 could be matched to 1693 contigs and singletons from the LCN assembly (Data S1). Two hundred thirty-four markers exhibiting homology to syntenic *Brachypodium* genes were further used to build a GenomeZipper, where additional syntenic *Brachypodium* gene blocks lacking genetic position information were added preserving their orders on *Brachypodium* genome. This virtual order of the genes exhibited several small-scale rearrangements along with conserved syntenic blocks; though it should be noted that gene orders determined solely by syntenic relationships may not reflect the actual order of genes along *Aegilops* 5D chromosome. Additionally, bin-

mapped *T. aestivum* 5D EST markers were retrieved from GrainGenes and compared against *Aegilops* 5D virtual gene order. Intriguingly, EST markers genetically mapped to the deletion bins of the short arm of chromosome 5D in *T. aestivum* were found in a reversed order in *Aegilops* 5D chromosome. In addition, one EST marker (BF201243) mapped to the centromeric region of *T. aestivum* 5D chromosome appeared among the EST markers mapped to the proximal deletion bin of the long arm, suggesting that the centromeric region may have undergone rearrangements after the divergence of *Aegilops* and *Triticum* lineages. Another EST marker (BE497595) was found to be relocated to a more distal position on *Aegilops* 5D chromosome, indicating another small-scale rearrangement. While such rearrangements can be expected in rapidly evolving grass genomes, rearrangements involving the centromere (BF201243) and the proximal region of the long arm (BE497595) are not backed up by orthologous loci in other grass genomes (Data S1), suggesting that these may correspond to pseudogenes or gene fragments.

The gene order along *Aegilops* 5D chromosome was also compared against the Genome Zippers of *T. aestivum* 5D and *H. vulgare* 5H (Mayer *et al.*, 2011). The comparative map (Figure 8) demonstrated that the gene orders are largely preserved between *Aegilops*, bread wheat and barley, with few perturbations (magenta and cyan links). The perturbations observed between *Aegilops* and *T. aestivum* chromosomes were located mostly on the telomeric region of the short arms. The relative positioning of genes along 5D and 5H chromosomes of *Aegilops* and barley revealed more disagreements, compared to wheat. In fact, a small inversion block was evident, delineated by Bradi1g08830-Bradi1g08580 (Figure 8, bundled cyan links). This inversion was not observed between *Aegilops* and *T. aestivum* 5D chromosomes; in fact, gene orders were in complete agreement along this region in these two species. This suggests that the inversion observed on 5H chromosome may have occurred in the barley lineage after its divergence from wheat. Considering evolutionary distances, rearrangements are more likely to occur between barley and *Aegilops*, compared to wheat.

Putative tRNA genes encoded by *Aegilops* chromosome 5D

Unmasked sequences of *Ae. tauschii* were found to encode 1246 putative tRNA genes, with a marked abundance of tRNA^{Lys} and tRNA^{Met} species. The same pattern was also observed among putative tRNA genes predicted from unmasked *T. aestivum* 5D survey sequences (Figure 9a) and reported for *T. aestivum* chromosome 6B (Tanaka *et al.*, 2014). However, tRNA predictions from the LCN assembly of *Ae. tauschii* 5D, which exclude the majority of known repeats, found only 142 putative tRNA genes with no bias towards any tRNA species, as was the case for an LCN assembly of *T. aestivum* 5D and the orthologous *Brachypodium* chromosomes 1 and 4 (Figure 9a). These observations support the conclusion that the unusual abundance of tRNA^{Lys} and tRNA^{Met} genes predicted from wheat chromosome 6B survey sequences and also observed here in chromosome 5D might correspond to an expansion through repetitive elements (Tanaka *et al.*, 2014). As wheat chromosomes 5D and 6B are not thought to originate from the same ancestral grass chromosomes (A12, A9 and A2, respectively), such an expansion shared between these chromosomes suggests that it is a genome-wide feature of the Triticeae (Salse *et al.*, 2008).

Genome expansion in Triticeae species is largely attributed to repetitive elements. To investigate whether the tRNA gene families are involved in genome expansion, putative tRNA genes per 100MB were compared between the draft genome sequence of *Ae. tauschii* var. AL8/78 (Jia *et al.*, 2013), D genome chromosomes of *T. aestivum* var. Chinese Spring (International Wheat Genome Sequencing Consortium, 2014) and the genomes of model grasses, *Brachypodium*, rice and sorghum (Figure 9b). The three model grass genomes, which contain much fewer repetitive elements, were observed to encode comparable numbers of putative tRNA genes, despite the differences between the genome sizes. This is reflected in higher tRNA counts per 100 MB, for which the number of putative tRNA genes were diluted by the huge genome sizes of *Ae. tauschii* and *T. aestivum* (Figure 9b). This observation suggests that, in general, tRNA gene families do not participate in the genome expansion of Triticeae genomes. However, nonrepetitive sequences of *Ae. tauschii* and *T. aestivum* D genome were also observed to retain high levels of putative tRNA^{Lys} genes. While the abundance of tRNA^{Lys} species can be largely explained by the expansion of repetitive elements that captured some putative tRNA genes, a fraction of these genes may also have been expanded in Triticeae genomes. On the chromosome level, masked and unmasked *Ae. tauschii* 5D sequences had higher tRNA counts per 100 MB than the AL8/78 genome assembly, particularly for unmasked sequences. It is possible that the different cultivars which the sequences derive

from may account for this difference; however, it is also likely that the short reads of the Illumina technology may have erroneously collapsed some putative tRNA genes into single contigs in the variety AL8/78 (Figure 9b).

Additionally, predicted proteins beginning with a Met residue were found to be followed by an Ala residue in most cases in both *Ae. tauschii* and *T. aestivum* proteomes. This trend is consistent with the tRNA gene abundances predicted from *Ae. tauschii* and *T. aestivum* LCN assemblies where the most abundant tRNA species was tRNA^{Ala}. All putative tRNA gene predictions are given in Table S3.

Experimental procedures

Flow sorting and sequencing of *Aegilops* 5D

Flow sorting was performed as described by Vrána *et al.* (2000) and Molnár *et al.* (2014) using genotype, MvGB589. Dot plots of FSC/DAPI-A (forward scatter vs. DAPI fluorescence pulse area) and DAPI-W/DAPI-A were used to discriminate 5D from other chromosomes, chromosome clumps, debris and chromatids. The level of purity of the sorted chromosomes was determined by fluorescence *in situ* hybridization (FISH) as described previously (Janda *et al.*, 2006). Three batches of 1000 chromosomes were sorted onto microscopic slide into 10-μL drop of PRINS buffer supplemented with 2.5% sucrose. After air-drying, sorted chromosomes were identified using FISH with a probe for Afa repeat (Kubaláková *et al.*, 2002).

For sequencing, chromosome 5D was sorted into 40-μL sterile-deionized water in three batches of 33 000 copies each and chromosomal DNA was purified, and subsequently amplified by multiple displacement amplification (MDA) using the illustraGenomiPhi DNA Amplification kit (GE Healthcare, Chalfont St. Giles, UK) according to Simková *et al.* (2008). Three shotgun libraries were prepared and three sequencing runs were separately performed using the Roche/454 Genome Sequencer FLX Titanium Platform, essentially following the manufacturer's protocols. All reads were combined and submitted to the EBI Sequence Read Archive under the primary accession number PRJEB5993.

Characterization and assembly of *Aegilops* 5D

Repetitive elements were identified via RepeatMasker software (http://www.repeatmasker.org/), using the MIPS Repeat Element Database (v. 9.3) for *Poaceae* as the repetitive element library (ftp://ftpmips.helmholtz-muenchen.de/plants/REdat/). Additionally, sequences not masked by the RepeatMasker software were blasted against the *Ae. tauschii* chloroplast genome (GenBank: JQ754651.1) and the *T. aestivum* mitochondrial genome (NCBI: NC_007579.1) (1E-15, -dust 'no') and all *Triticum* rRNA sequences (1E-5, -dust 'no') to identify organelle-associated sequences.

Assembly of the sequences was performed using gsAssembler tool of the Newbler Software v. 2.6 of the GS FLX Platform (Roche/454 Life Sciences, Branford, CT, USA), with the 'large and complex genome', 'heterozygotic genome', and 'extend low-depth overlaps' options. A minimum overlap identity of 98% was empirically determined to give the optimal assembly, and all other parameters were used at their default values. Repetitive sequences, along with organelle-associated sequences, were excluded from the assembly. All contigs and singletons generated by the assembly are referred to as low-copy number (LCN) assembly hereafter.

Identification and functional annotation of gene-associated reads of *Aegilops* 5D

Gene-associated reads were identified by blast searches against the model grass proteomes, *Brachypodium distachyon* (v. 1.2, http://mips.helmholtz-muenchen.de/plant/brachypodium; Nussbaumer *et al.*, 2013), *Oryza sativa* (v. 1.0, http://rapdb.dna.affrc.go.jp/download/irgsp1.html; Tanaka *et al.*, 2008) and *Sorghum bicolor* (v. 1.4, http://mips.helmholtz-muenchen.de/plant/sorghum/; Paterson *et al.*, 2009) and against the high-confidence proteins of the close relative, *Hordeum vulgare* (http://mips.helmholtz-muenchen.de/plant/barley/). Additionally, UniProt sequences of *Ae. tauschii* (http://www.uniprot.org/, a total of 34.639 sequences, last accessed on 15.09.2014) and UniGene sequences from *T. aestivum* (Build #63) were used to identify putative wheat-specific genes.

Similarity searches against the protein sequences were performed using blastx (1E-6). Matches with an alignment length of at least 30 amino acids, and 75% amino acid similarity were retained (Vitulo *et al.*, 2011); except for barley and *Ae. tauschii* UniProt sequences, where the amino acid similarity cut-off was set to 90% and 98%, respectively, due to closer evolutionary distances. In addition, tblastn searches against the LCN assembly were also performed with the same criteria using the protein databases as queries. Only reciprocal best hits (for blastx and tblastn searches) were retained for further analyses. Similarity searches against the UniGene sequences from *T. aestivum* were performed using blastn (1E-30), and the best hits with at least 96% identity over 100 nucleotides were retained. For all blast searches, of the contigs and singletons covering the exact same region on the same protein or gene match, only one is kept to eliminate amplification bias. All results were combined using in-house Perl and Matlab scripts. All blast searches were carried out using the BLAST+ stand-alone toolkit, version 2.2.25 (Camacho *et al.*, 2009).

Functional annotation of the gene-associated sequences of the LCN assembly was performed with Blast2GO (Conesa and Götz, 2008) (www.blast2go.com). Initial blast step was run locally against the nonredundant *Viridiplantae* (taxid: 33090) database (1E-6), while other steps were carried out at default parameters.

All CDS reported as mapped to chromosome 5D from *Ae. tauschii* var. AL8/78 (Jia *et al.*, 2013) and all high-confidence CDS assembled from flow-sorted *T. aestivum* chromosome 5D (International Wheat Genome Sequencing Consortium, 2014) were compared to gene-associated contigs and singletons from the LCN assembly using blastn with an e-value cut-off of $1e^{-30}$ and sequence identity ≥96% over ≥100 nucleotides. Using custom Perl scripts, multiple sequence matches between sections of one CDS (putative exons) with different parts of the same gene-associated contig/singleton were combined to give a single sequence match, retaining only the best hit for each exon.

Genome conservation and syntenic analyses

To investigate genome conservation, positions of genomic elements of the annotated model grass genomes were retrieved from MIPS database of plants (http://mips.helmholtz-muenchen.de/plant/genomes.jsp). Heatmaps were drawn on MATLAB R2010b with a sliding window approach of 50 kb step size. Circle plots were generated via Circos software (Krzywinski *et al.*, 2009). Gene densities (light blue & light grey) were counted on 500 kb intervals along chromosomes. Virtual gene order was constructed using genetically mapped markers retrieved from a recent study (Luo *et al.*, 2013). The marker sequences were blasted against the LCN assembly contigs and singletons (1E-10), and significant matches were ordered using the genetic map positions of the markers. Genome Zipper was constructed using syntenic *Brachypodium* orthologs. Comparative map of *H. vulgare* 5H, *Ae. tauschii* 5D and *T. aestivum* 5D chromosomes was visualized using MATLAB R2010b.

Mining tRNA genes in *Aegilops tauschii*

The tRNA genes were predicted from unmasked sequences and LCN assembly of *Ae. tauschii* and *T. aestivum* 5D chromosomes using tRNAscan-SE 1.21 (Lowe and Eddy, 1997). Additionally, tRNA gene prediction was performed on orthologous *Brachypodium* chromosomes 1 and 4, from unmasked genome annotation v1.2 and on the model grass genomes of *Brachypodium distachyon*, *Oryza sativa* and *Sorghum bicolor*. The program was run locally with the default parameters for eukaryotic genomes. Pseudogenes and other undetermined annotations were excluded from the analysis.

Acknowledgements

We thank Dr. István Molnár (Agricultural Institute, Centre for Agricultural Research, Martonvásár, Hungary) for providing seeds of *Ae. tauschii* and Dr. Mingcheng Luo for providing marker data for *Ae. tauschii*. We thank graduate students Zaeema Khan, Kuaybe Kurtoglu and Deniz Adali, and postdoctoral fellows Dr. Meral Yuce and Dr. Raghu Mokkapati for providing practical assistance with setting up individual GS FLX sequencing runs, and Dr. Marie Kubaláková, Jarmila Číhalíková, Romana Šperková and Zdeňka Dubská for assistance with chromosome sorting and DNA amplification. This work was supported by Sabanci University. We also thank TUBITAK-BIDEB for providing scholarship with BAA. JD and JV were supported by the Czech Science Foundation (award no. P501/12/G090) and by the grant LO1204 from the National Program of Sustainability I.

References

Berkman, P.J., Skarshewski, A., Lorenc, M.T., Lai, K., Duran, C., Ling, E.Y.S., Stiller, J., Smits, L., Imelfort, M., Manoli, S., McKenzie, M., Kubaláková, M., Šimková, H., Batley, J., Fleury, D., Doležel, J. and Edwards, D. (2011) Sequencing and assembly of low copy and genic regions of isolated *Triticum aestivum* chromosome arm 7DS. *Plant Biotechnol. J.* **9**, 768–775.

Berkman, P.J., Skarshewski, A., Manoli, S., Lorenc, M.T., Stiller, J., Smits, L., Lai, K., Campbell, E., Kubaláková, M., Simková, H., Batley, J., Doležel, J., Hernandez, P. and Edwards, D. (2012) Sequencing wheat chromosome arm 7BS delimits the 7BS/4AL translocation and reveals homoeologous gene conservation. *Theor. Appl. Genet.* **124**, 423–432.

Blechl, A.E. and Anderson, O.D. (1996) Expression of a novel high-molecular-weight glutenin subunit gene in transgenic wheat. *Nat. Biotechnol.* **14**, 875–879.

Camacho, C., Coulouris, G., Avagyan, V., Ma, N., Papadopoulos, J., Bealer, K. and Madden, T.L. (2009) BLAST+: architecture and applications. *BMC Bioinformatics*, **10**, 421.

Chantret, N., Salse, J., Sabot, F., Rahman, S., Bellec, A., Laubin, B., Dubois, I., Dossat, C., Sourdille, P., Joudrier, P., Gautier, M.-F., Cattolico, L., Beckert, M., Aubourg, S., Weissenbach, J., Caboche, M., Bernard, M., Leroy, P. and

Chalhoub, B. (2005) Molecular basis of evolutionary events that shaped the hardness locus in diploid and polyploid wheat species (*Triticum* and *Aegilops*). *Plant Cell*, **17**, 1033–1045.

Cloutier, S., McCallum, B.D., Loutre, C., Banks, T.W., Wicker, T., Feuillet, C., Keller, B. and Jordan, M.C. (2007) Leaf rust resistance gene Lr1, isolated from bread wheat (*Triticum aestivum* L.) is a member of the large psr567 gene family. *Plant Mol. Biol.* **65**, 93–106.

Conesa, A. and Götz, S. (2008) Blast2GO: a comprehensive suite for functional analysis in plant genomics. *Int. J. Plant Genomics*, **2008**, 619832.

Doležel, J., Vrána, J., Cápal, P., Kubaláková, M., Burešová, V. and Simková, H. (2014) Advances in plant chromosome genomics. *Biotechnol. Adv.* **32**, 122–136.

Dubcovsky, J. and Dvorak, J. (2007) Genome plasticity a key factor in the success of polyploid wheat under domestication. *Science*, **316**, 1862–1866.

Dvorak, J., Luo, M.-C., Yang, Z.-L. and Zhang, H.-B. (1998) The structure of the Aegilops tauschii genepool and the evolution of hexaploid wheat. *Theor. Appl. Genet.* **97**, 657–670.

Galaeva, M.V., Fayt, V.I., Chebotar, S.V., Galaev, A.V. and Sivolap, Y.M. (2013) Association of microsatellite loci alleles of the group-5 of chromosomes and the frost resistance of winter wheat. *Cytol. Genet.* **47**, 261–267.

Gill, B.S., Appels, R., Botha-Oberholster, A.-M., Buell, C.R., Bennetzen, J.L., Chalhoub, B., Chumley, F., Dvorák, J., Iwanaga, M., Keller, B., Li, W., McCombie, W.R., Ogihara, Y., Quetier, F. and Sasaki, T. (2004) A workshop report on wheat genome sequencing: International Genome Research on Wheat Consortium. *Genetics*, **168**, 1087–1096.

Hernandez, P., Martis, M., Dorado, G., Pfeifer, M., Gálvez, S., Schaaf, S., Jouve, N., Šimková, H., Valárik, M., Doležel, J. and Mayer, K.F.X. (2012) Next-generation sequencing and syntenic integration of flow-sorted arms of wheat chromosome 4A exposes the chromosome structure and gene content. *Plant J.* **69**, 377–386.

International Brachypodium Initiative (2010) Genome sequencing and analysis of the model grass *Brachypodium distachyon*. *Nature*, **463**, 763–768.

Janda, J., Safár, J., Kubaláková, M., Bartos, J., Kovárová, P., Suchánková, P., Pateyron, S., Cíhalíková, J., Sourdille, P., Simková, H., Faivre-Rampant, P., Hribová, E., Bernard, M., Lukaszewski, A., Dolezel, J. and Chalhoub, B. (2006) Advanced resources for plant genomics: a BAC library specific for the short arm of wheat chromosome 1B. *Plant J.* **47**, 977–986.

Jia, J., Devos, K.M., Chao, S., Miller, T.E., Reader, S.M. and Gale, M.D. (1996) RFLP-based maps of the homoeologous group-6 chromosomes of wheat and their application in the tagging of Pm12, a powdery mildew resistance gene transferred from Aegilops speltoides to wheat. *Theor. Appl. Genet.* **92**, 559–565.

Jia, J., Zhao, S., Kong, X., Li, Y., Zhao, G., He, W., Appels, R., Pfeifer, M., Tao, Y., Zhang, X., Jing, R., Zhang, C., Ma, Y., Gao, L., Gao, C., Spannagl, M., Mayer, K.F.X., Li, D., Pan, S., Zheng, F., Hu, Q., Xia, X., Li, J., Liang, Q., Chen, J., Wicker, T., Gou, C., Kuang, H., He, G., Luo, Y., Keller, B., Xia, Q., Lu, P., Wang, J., Zou, H., Zhang, R., Xu, J., Gao, J., Middleton, C., Quan, Z., Liu, G., Wang, J., International Wheat Genome Sequencing Consortium, Yang, H., Liu, X., He, Z., Mao, L. and Wang, J. (2013) Aegilops tauschii draft genome sequence reveals a gene repertoire for wheat adaptation. *Nature*, **496**, 91–95.

Koyama, K., Hatano, H., Nakamura, J. and Takumi, S. (2012) Characterization of three VERNALIZATION INSENSITIVE3-like (VIL) homologs in wild wheat, *Aegilops tauschii* Coss. *Hereditas*, **149**, 62–71.

Krzywinski, M., Schein, J., Birol, I., Connors, J., Gascoyne, R., Horsman, D., Jones, S.J. and Marra, M.A. (2009) Circos: an information aesthetic for comparative genomics. *Genome Res.* **19**, 1639–1645.

Kubaláková, M., Vrána, J., Cíhalíková, J., Simková, H. and Dolezel, J. (2002) Flow karyotyping and chromosome sorting in bread wheat (*Triticum aestivum* L.). *Theor. Appl. Genet.* **104**, 1362–1372.

Li, W., Zhang, P., Fellers, J.P., Friebe, B. and Gill, B.S. (2004) Sequence composition, organization, and evolution of the core Triticeae genome. *Plant J.* **40**, 500–511.

Lowe, T.M. and Eddy, S.R. (1997) tRNAscan-SE: a program for improved detection of transfer RNA genes in genomic sequence. *Nucleic Acids Res.* **25**, 955–964.

Luo, M.-C., Gu, Y.Q., You, F.M., Deal, K.R., Ma, Y., Hu, Y., Huo, N., Wang, Y., Wang, J., Chen, S., Jorgensen, C.M., Zhang, Y., McGuire, P.E., Pasternak, S., Stein, J.C., Ware, D., Kramer, M., McCombie, W.R., Kianian, S.F., Martis,

M.M., Mayer, K.F.X., Sehgal, S.K., Li, W., Gill, B.S., Bevan, M.W., Simková, H., Dolezel, J., Weining, S., Lazo, G.R., Anderson, O.D. and Dvorak, J. (2013) A 4-gigabase physical map unlocks the structure and evolution of the complex genome of Aegilops tauschii, the wheat D-genome progenitor. *Proc. Natl Acad. Sci. USA*, **110**, 7940–7945.

Mahmood, A., Baenziger, P.S., Budak, H., Gill, K.S. and Dweikat, I. (2004) The use of microsatellite markers for the detection of genetic similarity among winter bread wheat lines for chromosome 3A. *Theor. Appl. Genet.* **109**, 1494–1503.

Mayer, K.F.X., Martis, M., Hedley, P.E., Simková, H., Liu, H., Morris, J.A., Steuernagel, B., Taudien, S., Roessner, S., Gundlach, H., Kubaláková, M., Suchánková, P., Murat, F., Felder, M., Nussbaumer, T., Graner, A., Salse, J., Endo, T., Sakai, H., Tanaka, T., Itoh, T., Sato, K., Platzer, M., Matsumoto, T., Scholz, U., Dolezel, J., Waugh, R. and Stein, N. (2011) Unlocking the barley genome by chromosomal and comparative genomics. *Plant Cell*, **23**, 1249–1263.

Mayer, K.F.X., Waugh, R., Brown, J.W.S., Schulman, A., Langridge, P., Platzer, M., Fincher, G.B., Muehlbauer, G.J., Sato, K., Close, T.J., Wise, R.P. and Stein, N. (2012) A physical, genetic and functional sequence assembly of the barley genome. *Nature*, **491**, 711–716.

International Wheat Genome Sequencing Consortium (2014) A chromosome-based draft sequence of the hexaploid bread wheat (*Triticum aestivum*) genome. *Science*, **345**(6194), 1251788.

Middleton, C.P., Stein, N., Keller, B., Kilian, B. and Wicker, T. (2012) Comparative analysis of genome composition in Triticeae reveals strong variation in transposable element dynamics and nucleotide diversity. *Plant J.* Available at: http://www.ncbi.nlm.nih.gov/pubmed/23057663 [Accessed February 3, 2014].

Miranda, L.M., Murphy, J.P., Marshall, D., Cowger, C. and Leath, S. (2007) Chromosomal location of Pm35, a novel Aegilops tauschii derived powdery mildew resistance gene introgressed into common wheat (*Triticum aestivum* L.). *Theor. Appl. Genet.* **114**, 1451–1456.

Molnár, I., Kubaláková, M., Simková, H., Farkas, A., Cseh, A., Megyeri, M., Vrána, J., Molnár-Láng, M. and Doležel, J. (2014) Flow cytometric chromosome sorting from diploid progenitors of bread wheat, T. urartu, Ae. speltoides and Ae. tauschii. *Theor. Appl. Genet.* Available at: http://www.ncbi.nlm.nih.gov/pubmed/24553964 [Accessed April 3, 2014].

Mujeeb-Kazi, A., Rosas, V. and Roldan, S. (1996) Conservation of the genetic variation of Triticum tauschii (Coss.) Schmalh. (Aegilops squarrosa auct. non L.) in synthetic hexaploid wheats (*T. turgidum* L. s.lat. x T. tauschii; 2n=6x=42, AABBDD) and its potential utilization for wheat improvement. *Genet. Resour. Crop Evol.* **43**, 129–134.

Ning, S.-Z., Chen, Q.-J., Yuan, Z.-W., Zhang, L.-Q., Yan, Z.-H., Zheng, Y.-L. and Liu, D.-C. (2009) Characterization of WAP2 gene in Aegilops tauschii and comparison with homoeologous loci in wheat. *J. Syst. Evol.* **47**, 543–551.

Nussbaumer, T., Martis, M.M., Roessner, S.K., Pfeifer, M., Bader, K.C., Sharma, S., Gundlach, H. and Spannagl, M. (2013) MIPS PlantsDB: a database framework for comparative plant genome research. *Nucleic Acids Res.* **41**, D1144–D1151.

Paterson, A.H., Bowers, J.E., Bruggmann, R., Dubchak, I., Grimwood, J., Gundlach, H., Haberer, G., Hellsten, U., Mitros, T., Poliakov, A., Schmutz, J., Spannagl, M., Tang, H., Wang, X., Wicker, T., Bharti, A.K., Chapman, J., Feltus, F.A., Gowik, U., Grigoriev, I.V., Lyons, E., Maher, C.A., Martis, M., Narechania, A., Otillar, R.P., Penning, B.W., Salamov, A.A., Wang, Y., Zhang, L., Carpita, N.C., Freeling, M., Gingle, A.R., Hash, C.T., Keller, B., Klein, P., Kresovich, S., McCann, M.C., Ming, R., Peterson, D.G., Mehboob, R., Ware, D., Westhoff, P., Mayer, K.F., Messing, J. and Rokhsar, D.S. (2009) The Sorghum bicolor genome and the diversification of grasses. *Nature*, **457**, 551–556.

Peng, J.H., Sun, D. and Nevo, E. (2011) Domestication evolution, genetics and genomics in wheat. *Mol. Breed.* **28**, 281–301.

Quarrie, S.A., Steed, A., Calestani, C., Semikhodskii, A., Lebreton, C., Chinoy, C., Steele, N., Pljevljakusic, D., Waterman, E., Weyen, J., Schondelmaier, J., Habash, D.Z., Farmer, P., Saker, L., Clarkson, D.T., Abugalieva, A., Yessimbekova, M., Turuspekov, Y., Abugalieva, S., Tuberosa, R., Sanguineti, M.-C., Hollington, P.A., Aragués, R., Royo, A. and Dodig, D. (2005) A high-density genetic map of hexaploid wheat (*Triticum aestivum* L.) from the cross Chinese Spring x SQ1 and its use to compare QTLs for grain yield across a range of environments. *Theor. Appl. Genet.* **110**, 865–880.

Safář, J., Simková, H., Kubaláková, M., Cíhalíková, J., Suchánková, P., Bartos, J. and Dolezel, J. (2010) Development of chromosome-specific BAC resources for genomics of bread wheat. *Cytogenet. Genome Res.* **129**, 211–223.

Salse, J., Bolot, S., Throude, M., Jouffe, V., Piegu, B., Quraishi, U.M., Calcagno, T., Cooke, R., Delseny, M. and Feuillet, C. (2008) Identification and characterization of shared duplications between rice and wheat provide new insight into grass genome evolution. *Plant Cell,* **20**, 11–24.

Semikhodskii, A.G. (1997) *Mapping Quantitative Traits for Salinity Responses in Wheat.* University of East Anglia. Available at: http://www.amazon.co.uk/quantitative-salinity-responses-Triticum-aestivum/dp/B001NVOB6E [Accessed March 4, 2014].

Simková, H., Svensson, J.T., Condamine, P., Hribová, E., Suchánková, P., Bhat, P.R., Bartos, J., Safář, J., Close, T.J. and Dolezel, J. (2008) Coupling amplified DNA from flow-sorted chromosomes to high-density SNP mapping in barley. *BMC Genom.* **9**, 294.

Smith, D.B. and Flavell, R.B. (1975) Characterisation of the wheat genome by renaturation kinetics. *Chromosoma,* **50**. Available at: http://link.springer.com/10.1007/BF00283468 [Accessed March 22, 2014].

Tanaka, T., Antonio, B.A. and Kikuchi, S., Matsumoto, T., Nagamura, Y., Numa, H., Sakai, H., Wu, J., Itoh, T., Sasaki, T., Aono, R., Fujii, Y., Habara, T., Harada, E., Kanno, M., Kawahara, Y., Kawashima, H., Kubooka, H., Matsuya, A., Nakaoka, H., Saichi, N., Sanbonmatsu, R., Sato, Y., Shinso, Y., Suzuki, M., Takeda, J., Tanino, M., Todokoro, F., Yamaguchi, K., Yamamoto, N., Yamasaki, C., Imanishi, T., Okido, T., Tada, M., Ikeo, K., Tateno, Y., Gojobori, T., Lin, Y.C., Wei, F.J., Hsing, Y.I., Zhao, Q., Han, B., Kramer, M.R., McCombie, R.W., Lonsdale, D., O'Donovan, C.C., Whitfield, E.J., Apweiler, R., Koyanagi, K.O., Khurana, J.P., Raghuvanshi, S., Singh, N.K., Tyagi, A.K., Haberer, G., Fujisawa, M., Hosokawa, S., Ito, Y., Ikawa, H., Shibata, M., Yamamoto, M., Bruskiewich, R.M., Hoen, D.R., Bureau, T.E., Namiki, N., Ohyanagi, H., Sakai, Y., Nobushima, S., Sakata, K., Barrero, R.A., Sato, Y., Souvorov, A., Smith-White, B., Tatusova, T., An, S., An, G., OOta, S., Fuks, G., Messing, J., Christie, K.R., Lieberherr, D., Kim, H., Zuccolo, A., Wing, R.A., Nobuta, K., Green, P.J., Lu, C., Meyers, B.C., Chaparro, C., Piegu, B., Panaud, O. and Echeverria, M. (2008) The Rice Annotation Project Database (RAP-DB): 2008 update. *Nucleic Acids Res.* Available at: http://researchrepository.murdoch.edu.au/4954/1/rice_annotation_project.pdf [Accessed January 26, 2014].

Tanaka, T., Kobayashi, F., Joshi, G.P., Onuki, R., Sakai, H., Kanamori, H., Wu, J., Simkova, H., Nasuda, S., Endo, T.R., Hayakawa, K., Doležel, J., Ogihara, Y., Itoh, T., Matsumoto, T. and Handa, H. (2014) Next-generation survey sequencing and the molecular organization of wheat chromosome 6B. *DNA Res.* **21**, 103–114.

Turnbull, K.M., Turner, M., Mukai, Y., Yamamoto, M., Morell, M.K., Appels, R. and Rahman, S. (2003) The organization of genes tightly linked to the Ha locus in *Aegilops tauschii,* the D-genome donor to wheat. *Genome,* **46**, 330–338.

Vitulo, N., Albiero, A., Forcato, C., Campagna, D., Dal-Pero, F., Bagnaresi, P., Colaiacovo, M., Faccioli, P., Lamontanara, A., Šimková, H., Kubaláková, M., Perrotta, G., Facella, P., Lopez, L., Pietrella, M., Gianese, G., Doležel, J., Giuliano, G., Cattivelli, L., Valle, G. and Stanca, A.M. (2011) First survey of the wheat chromosome 5A composition through a next generation sequencing approach. E. Newbigin, ed. *PLoS ONE,* **6**, e26421.

Vrána, J., Kubaláková, M., Simková, H., Cíhalíková, J., Lysák, M.A. and Dolezel, J. (2000) Flow sorting of mitotic chromosomes in common wheat (*Triticum aestivum* L.). *Genetics,* **156**, 2033–2041.

Wicker, T., Buchmann, J.P. and Keller, B. (2010) Patching gaps in plant genomes results in gene movement and erosion of colinearity. *Genome Res.* **20**, 1229–1237.

Wicker, T., Mayer, K.F.X., Gundlach, H., Martis, M., Steuernagel, B., Scholz, U., Simkova, H., Kubalakova, M., Choulet, F., Taudien, S., Platzer, M., Feuillet, C., Fahima, T., Budak, H., Dolezel, J., Keller, B. and Stein, N. (2011) Frequent gene movement and pseudogene evolution is common to the large and complex genomes of wheat, barley, and their relatives. *Plant Cell,* **23**, 1706–1718.

Yan, L., Loukoianov, A., Tranquilli, G., Helguera, M., Fahima, T. and Dubcovsky, J. (2003) Positional cloning of the wheat vernalization gene VRN1. *Proc. Natl Acad. Sci. USA,* **100**, 6263–6268.

Yang, W., Liu, D., Li, J., Zhang, L., Wei, H., Hu, X., Zheng, Y., He, Z. and Zou, Y. (2009) Synthetic hexaploid wheat and its utilization for wheat genetic improvement in China. *J. Genet. Genomics,* **36**, 539–546.

Yoshida, T., Nishida, H., Zhu, J., Nitcher, R., Distelfeld, A., Akashi, Y., Kato, K. and Dubcovsky, J. (2010) Vrn-D4 is a vernalization gene located on the centromeric region of chromosome 5D in hexaploid wheat. *Theor. Appl. Genet.* **120**, 543–552.

Zhang, J., Wang, Y., Wu, S., Yang, J., Liu, H. and Zhou, Y. (2012) A single nucleotide polymorphism at the Vrn-D1 promoter region in common wheat is associated with vernalization response. *Theor. Appl. Genet.* **125**, 1697–1704.

mlo-based powdery mildew resistance in hexaploid bread wheat generated by a non-transgenic TILLING approach

Johanna Acevedo-Garcia[1], David Spencer[1], Hannah Thieron[1], Anja Reinstädler[1], Kim Hammond-Kosack[2], Andrew L. Phillips[2] and Ralph Panstruga[1]*

[1]*Unit of Plant Molecular Cell Biology, Institute for Biology I, RWTH Aachen University, Aachen, Germany*
[2]*Department of Plant Biology and Crop Science, Rothamsted Research, West Common, Harpenden, Hertfordshire, AL5 2JQ, UK*

Summary

*Correspondence
email:
panstruga@bio1.rwth-aachen.de*

Wheat is one of the most widely grown cereal crops in the world and is an important food grain source for humans. However, wheat yields can be reduced by many abiotic and biotic stress factors, including powdery mildew disease caused by *Blumeria graminis* f.sp. *tritici* (*Bgt*). Generating resistant varieties is thus a major effort in plant breeding. Here, we took advantage of the non-transgenic Targeting Induced Lesions IN Genomes (TILLING) technology to select partial loss-of-function alleles of *TaMlo*, the orthologue of the barley *Mlo* (*Mildew resistance locus o*) gene. Natural and induced loss-of-function alleles (*mlo*) of barley *Mlo* are known to confer durable broad-spectrum powdery mildew resistance, typically at the expense of pleiotropic phenotypes such as premature leaf senescence. We identified 16 missense mutations in the three wheat *TaMlo* homoeologues, *TaMlo-A1*, *TaMlo-B1* and *TaMlo-D1* that each lead to single amino acid exchanges. Using transient gene expression assays in barley single cells, we functionally analysed the different missense mutants and identified the most promising candidates affecting powdery mildew susceptibility. By stacking of selected mutant alleles we generated four independent lines with non-conservative mutations in each of the three *TaMlo* homoeologues. Homozygous triple mutant lines and surprisingly also some of the homozygous double mutant lines showed enhanced, yet incomplete, *Bgt* resistance without the occurrence of discernible pleiotropic phenotypes. These lines thus represent an important step towards the production of commercial non-transgenic, powdery mildew-resistant bread wheat varieties.

Keywords: Targeting Induced Local Lesions in Genomes, powdery mildew, *Mlo*, hexaploid bread wheat, *Blumeria graminis*, plant disease resistance.

Introduction

Bread wheat (*Triticum aestivum*) is the third largest cultivated crop after maize and rice and the second in terms of dietary intakes. By 2013, global wheat grain production reached 712 million tonnes (http://faostat.fao.org), and by 2050 an estimated increase of a further 60% will be required to meet the demands of our growing population (http://wheatinitiative.org).

Allohexaploid wheat has a genome of 17 Gb in size, of which more than 80% is composed of repetitive transposable elements. Genetically, it has the structure of three independent genomes in one species (*AABBDD* genome), since meiotic pairing of homoeologous chromosomes is prevented through the action of *Ph* (*pairing homoeologous*) genes (IWGSC, 2014).

Yield in wheat can be reduced by abiotic factors, pests or diseases caused by pathogens. A global disease threat is powdery mildew caused by *Blumeria graminis* f.sp. *tritici* (*Bgt*) (Conner *et al.*, 2003; Dean *et al.*, 2012), an obligate biotrophic fungus (ascomycete) of the order Erysiphales (Glawe, 2008). However, more recently, *B. graminis* f.sp. *triticale* has also been reported to reproduce on wheat (Menardo *et al.*, 2016). Control of the powdery mildew disease caused by *Bgt* in wheat is performed mainly by using fungicides and varieties containing *R* (resistance) genes, the latter typically conferring isolate-specific protection (Dean *et al.*, 2012). Going forward, the most cost effective solution to control this disease is by exploiting genetic resources to provide enhanced, durable resistance.

In barley, natural and induced loss-of-function mutations of the *Mildew resistance locus o* (*Mlo*) gene confer broad-spectrum

resistance against most *B. graminis* f.sp. *hordei* (*Bgh*) isolates, an effect that has been shown to last in the field for more than 30 years (Jørgensen, 1992; Lyngkjær *et al.*, 2000). *Mlo* is a member of an ancient eukaryotic gene family that is conserved throughout the plant kingdom (Kusch *et al.*, 2016), and although its role in powdery mildew resistance has been well studied in various species, the biochemical function of Mlo proteins is still unknown (Acevedo-Garcia *et al.*, 2014).

On susceptible host plants, once sporelings of the pathogen land on the leaf or stem surface, these germinate and form an appressorium within two hours. The appressorium attempts to penetrate the epidermal cell layer by generating a penetration peg. If the pathogen successfully enters the host cell, in the following hours of infection the penetration peg enlarges to develop the feeding structure known as haustorium. Thereafter, the pathogen will complete its asexual life cycle on the leaf surface with the development of epiphytic hyphae, the production of conidiophores and the release of new spores (reviewed in Glawe, 2008). In the case of resistant *mlo* plants, a near-complete arrest of pathogen growth occurs at the penetration stage where the germinating spore is not able to develop a haustorium (Jørgensen and Mortensen, 1977). However, in barley *mlo* lines resistance is typically associated with pleiotropic phenotypes such as the spontaneous deposition of callose-containing cell wall appositions, early chlorophyll decay and spontaneous mesophyll cell death, which together lead to chlorotic and necrotic leaf flecking and have been interpreted as signs of premature leaf senescence (Peterhänsel *et al.*, 1997; Piffanelli *et al.*, 2002; Schwarzbach, 1976; Wolter *et al.*, 1993).

In wheat, the three orthologues of barley *Mlo*, *TaMlo-A1*, *-B1* and *-D1* (Konishi and Sasanuma, 2010), are located on chromosomes 5AL, 4BL and 4DL (Elliott *et al.*, 2002). In contrast to barley, occurrence of natural wheat *mlo* mutants has not been reported. This is likely due to its hexaploid nature, which may require mutation of all six gene copies to generate a resistance phenotype that would be detectable within a breeding programme. Previously, Elliott and co-workers showed that one of the wheat orthologues of barley *Mlo*, *TaMlo-B1*, can complement powdery mildew-resistant barley *mlo* mutants at the single-cell level (Elliott *et al.*, 2002). Furthermore, Várallyay and colleagues used virus-induced gene silencing (VIGS) to demonstrate that RNAi silencing of the *TaMlo* homoeologues in wheat results in powdery mildew resistance (Várallyay *et al.*, 2012). Most recently, the teams of Caixia Gao and Jin-Long Qiu took advantage of the TALEN (transcription activator-like effector nuclease) genome editing technology to generate transgenic winter wheat plants containing simultaneous knockout lesions in the three *TaMlo* homoeologues. Compared to the respective parental line, these plants were fully resistant against *Bgt* infection (Wang *et al.*, 2014).

Targeting Induced Local Lesions IN Genomes (TILLING), first introduced 16 years ago by McCallum and collaborators (McCallum *et al.*, 2000), is a powerful approach that integrates chemical mutagenesis with a high-throughput detection method to identify single-nucleotide mutations in a specific region of a gene of interest. In principle, TILLING can be used in different plant species independent of their ploidy. Interestingly, polyploid species can tolerate higher mutation densities than diploids (Slade *et al.*, 2005; Uauy *et al.*, 2009; Wang *et al.*, 2012). This feature translates into a far smaller population size that needs to be screened to reach saturated mutagenesis in a polyploid species (Kurowska *et al.*, 2011). Chemical mutagenesis combined with TILLING provides an ample spectrum of mutations, where a pool of allelic variations can result in a range of weak to strong phenotypes (Slade *et al.*, 2005). To date there are several TILLING resources available for various crop species such as hexaploid and durum wheat, barley, rice, tomato, maize, sorghum, soybean and potato (reviewed in Chen *et al.*, 2014b). Although until recently no TILLING-derived crop variety has been released commercially, they represent a great advantage for plant breeding (especially in Europe) since these varieties will be considered non-transgenic (Chen *et al.*, 2014b).

In wheat (diploid, tetraploid or hexaploid), several genes involved in starch synthesis have been targeted by TILLING, for example, *Waxy* (Rawat *et al.*, 2012; Slade *et al.*, 2005), *Starch Branching Enzyme II* (*SBEII*) (Botticella *et al.*, 2011; Sestili *et al.*, 2015; Slade *et al.*, 2012; Uauy *et al.*, 2009) and *Starch Synthase II* (*Sgp-1/SSII* (Dong *et al.*, 2009; Sestili *et al.*, 2009). Additionally, genes involved in other processes such as carotenoid content (Colasuonno *et al.*, 2016), grain width (Simmonds *et al.*, 2016), flowering (Chen *et al.*, 2014a) or vernalization (Chen and Dubcovsky, 2012), have been also targeted by TILLING.

Currently, wheat breeders have a great challenge to select for varieties that on the one hand have a minimal disturbance in growth, development and fertility, thereby maintaining good grain quality and enhancing grain yield, and on the other hand exhibit improved resistance to pests and pathogens. At least in Europe, the deployment of transgenic plants (including those resulting from genome editing approaches) is still socially and politically contentious in plant breeding, agriculture as well as the food, feed and drinks industries. Therefore, methods that are indisputably non-transgenic are favoured for the development of new varieties.

In this study, we generated hexaploid bread wheat lines with enhanced resistance to the common powdery mildew disease. Our approach made use of the non-transgenic TILLING technology to select four different lines containing missense mutations in the *TaMlo-A1*, *TaMlo-B1* and *TaMlo-D1* homoeologues. Triple and some double mutant lines showed enhanced powdery mildew resistance compared to respective wild-type (WT) plants. So far these mutant plants did not show discernible abnormalities in growth and development.

Results

Cloning of genomic sequences of the wheat *TaMlo-A1*, *TaMlo-B1* and *TaMlo-D1* homoeologues

An essential requirement to perform a TILLING screening is the development of primers for gene-specific polymerase chain reaction (PCR) amplification. A limiting factor in hexaploid wheat in this respect is the very high nucleotide similarity between the coding sequences of its homoeologous genes, which often constrains primer design. This is also the case for *TaMlo*, which exhibits 95%, 96% and 97% nucleotide sequence identity between the coding sequences of the A and B, A and D, and B and D genomes, respectively (File S1). To overcome this limitation, homoeologue-specific PCR primers are frequently designed on the basis of intron sequences, which typically show a higher degree of nucleotide sequence polymorphisms than the exonic coding sequences. We thus first opted to obtain genomic DNA sequence information for the three *TaMlo* homoeologues. Based on the available known cDNA sequences in NCBI: (i) *TaMlo-A1*: AF361933 and AX063298; (ii) *TaMlo-B1*: AF361932, AX063294, AF384145; and (iii) *TaMlo-D1*: AX063296, we designed common oligonucleotide primers, predicted to equally bind to all three *TaMlo* homoeologues, and used them to amplify and clone several overlapping genomic *TaMlo* fragments from DNA obtained from the hexaploid spring wheat cultivar (cv.) Cadenza. Following *in silico* assembly of the amplicons and also taking information from the known cDNA sequences into account, we obtained almost complete genomic assemblies representing the three *TaMlo* homoeologues. These were reconciled with the manually annotated sequences generated on the basis of the available wheat genome survey data to finally produce tentative genomic *TaMlo* consensus sequences (cv. Cadenza and cv. Chinese Spring hybrid sequences; File S2). The determined exon/intron structure of the *TaMlo* homoeologues is similar to the one of barley *Mlo* (Büschges *et al.*, 1997) and comprises 11 exons each (Figure 1). The deduced *TaMlo* protein sequences are each one amino acid longer (534 amino acids) than barley Mlo (533 amino acids); consequently, as a result of small gaps in the alignment between wheat and barley Mlo proteins, the amino acid numbering after position 115 is shifted by +1 in the former compared to the latter.

TILLING target and primer design

Previously, we identified the second and third cytoplasmic loop of the barley Mlo protein as relevant regions for its powdery mildew susceptibility-conferring activity. Mutations in these parts of the Mlo protein that lead to a single amino acid exchange (missense mutations) often result in a loss-of-function of the protein and thus resistance against the powdery mildew

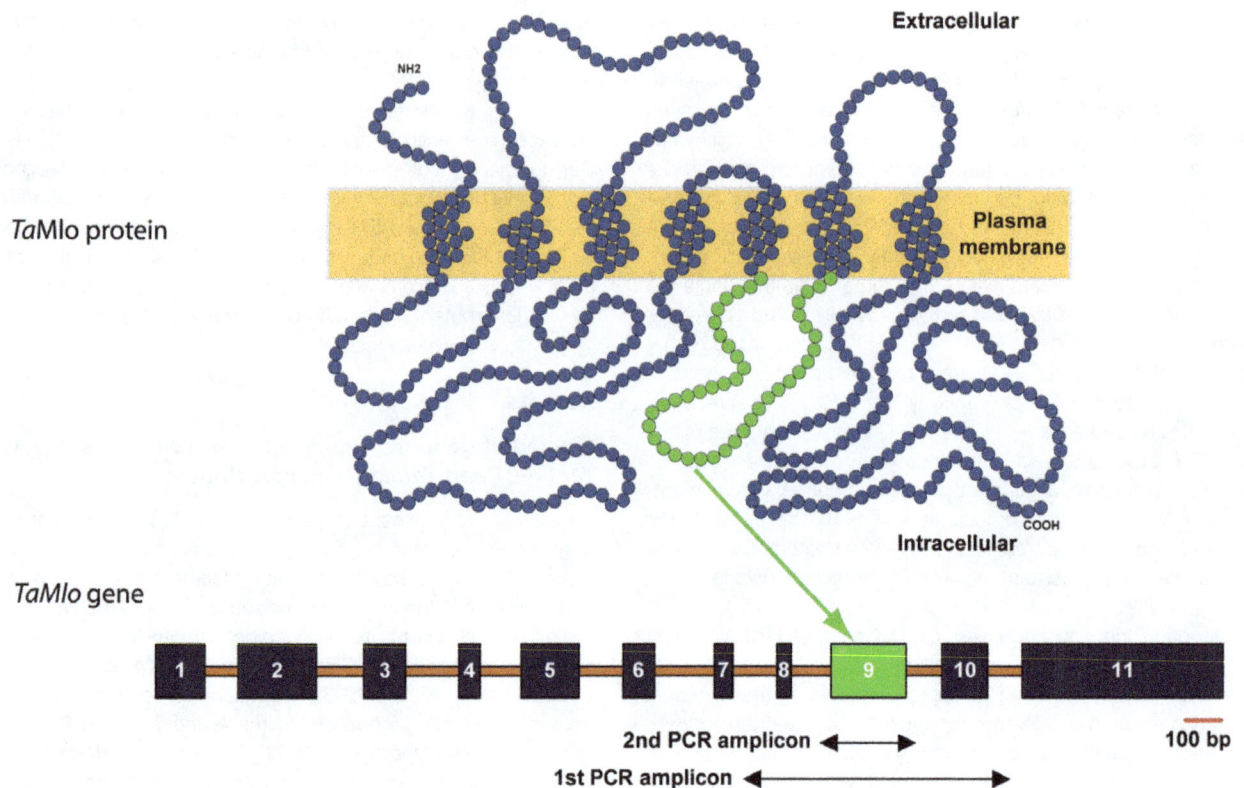

Figure 1 Exon 9 of *TaMlo* is the target for the TILLING screen. The scheme illustrates the predicted topology of the 7 transmembrane domain *Ta*Mlo proteins (top) and the experimentally determined common exon-intron structure of the corresponding *TaMlo* homoeologues (bottom). The 11 exons are indicated by dark boxes, introns by orange lines. The third cytoplasmic loop of *Ta*Mlo, depicted in green and encoded by exon 9, was the target for TILLING. Arrows below the *TaMlo* gene model specify the amplicons of the 1st and 2nd round PCR. A scale bar, shown in red, is given below exon 11 (100 bp).

pathogen (Reinstädler *et al.*, 2010). Hence, we selected the third cytoplasmic loop, encoded by exon 9 of the *TaMlo* homoeologues, as the target region for the TILLING screening in wheat (Figure 1). We developed homoeologue-specific primers for a 1st round PCR to amplify a product (between 750 and 1700 bp in size, depending on the *TaMlo* homoeologue) that contains the target exon (Figure 1 and Table S1). In addition, as High-Resolution Melt (HRM) analysis is most accurate on DNA fragments up to ~300 bp in size (Reed and Wittwer, 2004), homoeologue-specific primers were designed for a 2nd round PCR with a maximum amplicon size of 310 bp (Figure 1 and Table S1). Primer specificity was confirmed by amplification from genomic template DNA of cv. Cadenza followed by direct amplicon sequencing.

Identification of *Tamlo* mutant candidates by TILLING

The ethyl methane sulfonate (EMS)-mutagenized population of spring bread wheat cv. Cadenza has been previously described (Rakszegi *et al.*, 2010). Genomic DNA samples derived from ~2020 M_2 individuals were pooled twofold and employed as templates for 1st round PCR amplification in the target region using the validated homoeologue-specific primer pairs. The primary PCR product was used as template for a 2nd round of PCR and the resulting ~300 bp amplicons were subjected to HRM analysis to detect mutations in the target exon of the *TaMlo* homoeologues.

After screening the entire population, a total of 76 candidate mutants were identified (28, 27 and 21 in *TaMlo-A1*, *TaMlo-B1*

and *TaMlo-D1*, respectively, Table 1). Twenty of these candidate mutations could not be confirmed in a secondary round of analysis; these were considered false positives and discarded from further study. Among the remaining 56 mutants, 14 mutations were located in intron regions flanking exon 9, but did not affect the consensus GT/AG splice junctions, and 26 were silent (synonymous mutations, Table 1), resulting in an unaltered *Ta*Mlo amino acid sequence. The remaining 16 mutants each represented missense mutations (non-synonymous mutations, Table 2); nonsense mutations were not found in this screen. At the nucleotide level, the identified missense mutations were all transitions of the type C → T or G → A, as expected for EMS mutagenesis (Table 2). Furthermore, the mutations covered a broad spectrum of amino acid exchanges, ranging from substitutions of amino acids with similar biophysical properties (e.g. S315N, V323I, A354V) to exchanges with dramatic alterations in biophysical properties (e.g. R313W, P335L, G296E). Three of the

Table 1 Mutations in the region of exon 9 of *TaMlo* identified by TILLING

Target	Total candidates	False positives	Intron	Silent	Missense	Non sense
TaMlo-A1	28	9	7	10	2	0
TaMlo-B1	27	4	5	11	7	0
TaMlo-D1	21	7	2	5	7	0

Table 2 *TaMlo* exon 9 missense mutants identified by TILLING

Gene	EMS line	SNP		Amino acid exchange	Zygosity[†]	SIFT score[‡]	Previously described in barley
		WT	Mutant				
TaMlo-A1	CAD1-29-A8	C	T	P325L	Het	0.20	*mlo* P324A Reinstädler *et al.* (2010)
TaMlo-A1	CAD-22-D12	C	T	A354V	Hom	0.00*	No
TaMlo-B1	CAD1-2-H9	G	A	G296E	Het	0.00*	No
TaMlo-B1	CAD2-18-H1	C	T	T297I	Het	0.04	No
TaMlo-B1	CAD2-22-E11	C	T	R313W	Het	0.01*	No
TaMlo-B1	CAD2-1-A3	G	A	S315N	Het	0.49	No
TaMlo-B1	CAD2-21-D7	G	A	G319R	Hom	0.00*	*mlo*-38 Lundqvist *et al.* (1991), Reinstädler *et al.* (2010)
TaMlo-B1	CAD1-4-H10	G	A	A320T	Het	0.97	No
TaMlo-B1	CAD1-2-G9	C	T	T345M	Het	0.09	No
TaMlo-D1	CAD2-1-A2	G	A	A314T	Het	0.65	No
TaMlo-D1	CAD2-19-C9	G	A	G319R	Het	0.01*	*mlo*-38 Lundqvist *et al.* (1991), Reinstädler *et al.* (2010)
TaMlo-D1	CAD2-23-A2	G	A	A320T	Het	1.00	No
TaMlo-D1	CAD2-24-C8	C	T	P321S	Het	0.25	No
TaMlo-D1	CAD2-19-H4	G	A	V323I	Hom	0.13	No
TaMlo-D1	CAD2-3-B6	C	T	P335L	Hom	0.00*	*mlo*-29 Piffanelli *et al.* (2002), Müller *et al.* (2005)
TaMlo-D1	CAD2-29-A1	C	T	T345M	Hom	0.09	No

[†]Hom, Homozygous; Het, Heterozygous.

[‡]SIFT scores ≤0.05 indicate mutations that are predicted to affect protein functions (indicated by an asterisk), while SIFT scores >0.05 indicate mutations that are predicted to be tolerated.

mutations (G319R, A320T and T345M) were found in the *TaMlo-B1* as well as in the *TaMlo-D1* genome (Table 2). Interestingly, mutations G319R (*TaMlo-B1* and *TaMlo-D1*) and P335L (*TaMlo-D1*) were previously identified as sequence variants with the characteristic resistance phenotype in barley plants (*mlo*-38 and *mlo*-29, respectively; Lundqvist *et al.*, 1991; Piffanelli *et al.*, 2002; Müller *et al.*, 2005; Reinstädler *et al.*, 2010). Furthermore, a similar mutation to *Tamlo-A1* (P325L) was studied by transient gene expression in barley (P324A), showing a partial resistance phenotype (Reinstädler *et al.*, 2010). Analysis with the online tool SIFT (Sorting Intolerant From Tolerant; Kumar *et al.*, 2009) revealed in each genome at least one mutation predicted to affect protein function (Table 2). Taken together, we identified 16 missense mutations, distributed between all three homoeologues that would potentially affect *Ta*Mlo function. Since nonsense mutations were not identified (Table 1), we further functionally characterized the missense mutations (Table 2) to select the most suitable candidates for stacking of the alleles in single wheat lines.

Validation of mutant candidates by transient gene expression

Previously, we demonstrated that *TaMlo-B1* is able to complement a powdery mildew-resistant barley *mlo* mutant at the single-cell level in leaf epidermal tissue (Elliott *et al.*, 2002). Heterologous complementation is possible because of the high level of sequence identity between the wheat and barley Mlo proteins (~89%). We took advantage of this particle bombardment-based barley transient expression assay to experimentally evaluate the potential of the *Ta*Mlo variants with amino acid substitutions to alter the powdery mildew infection phenotype.

Full-length cDNAs of *TaMlo-A1*, *TaMlo-B1* and *TaMlo-D1* were cloned as expression constructs driven by the maize *Ubiquitin1* promoter (pUbiGATE) and used as a template to recreate each of

the 16 missense mutations by PCR-based, site-directed mutagenesis. In addition, WT barley *Mlo* cDNA (also driven by the *Ubiquitin1* promoter) was used to generate the mutant variants P324A, *mlo*-29 (P334L) and *mlo*-38 (G318R), which were included as additional controls.

Detached leaves of the barley *mlo*-3 null mutant were co-bombarded with the various *TaMlo* or *Mlo* constructs and a plasmid expressing the *β-glucuronidase* (GUS) reporter gene. As a negative control we employed the *GUS* plasmid only, and as a positive control we used barley *Mlo*. Twenty-four hours after bombardment the treated leaves were inoculated with the barley powdery mildew pathogen *Bgh*. All three WT wheat *TaMlo* genes showed a complementation efficiency similar to barley *Mlo* (~80% host cell entry, median value, Figure 2). Transient expression of mutants *Tamlo-A1* (P325L), its barley counterpart *mlo* (P324A), *Tamlo-B1* (R313W) and *Tamlo-D1* (A314T), (P321S) and (T345M) resulted in ~40%–60% host cell entry (median value, Figure 2). Expression of mutants *Tamlo-B1* (G296E) and *Tamlo-B1*, *Tamlo-D1* (G319R) led to an even lower host cell entry rate of ~15%–20% (Figure 2). Similar results were obtained with the corresponding G318R barley mutant (*mlo*-38, Figure 2). Interestingly, the entry rates obtained by expression of *Tamlo-D1* (P335L) and the corresponding barley mutant variant (*mlo*-29) were close to background levels (*GUS* only, Figure 2), suggesting a near-complete loss-of-function in the case of these amino acid substitutions. Expression of the remaining seven wheat mutant variants showed host cell entry rates not considerably different from the respective *TaMlo* WT versions (Figure 2). Overall there was a good correlation between the results of the *in silico* SIFT analysis and the experimental data. Exceptions were mutations *Tamlo-A1* (A354V), *Tamlo-B1* (T297I) and *Tamlo-B1* (R313W), which were predicted to have a significant impact on protein function but were only moderately affected regarding host cell entry rates (Figure 2 and Table 2).

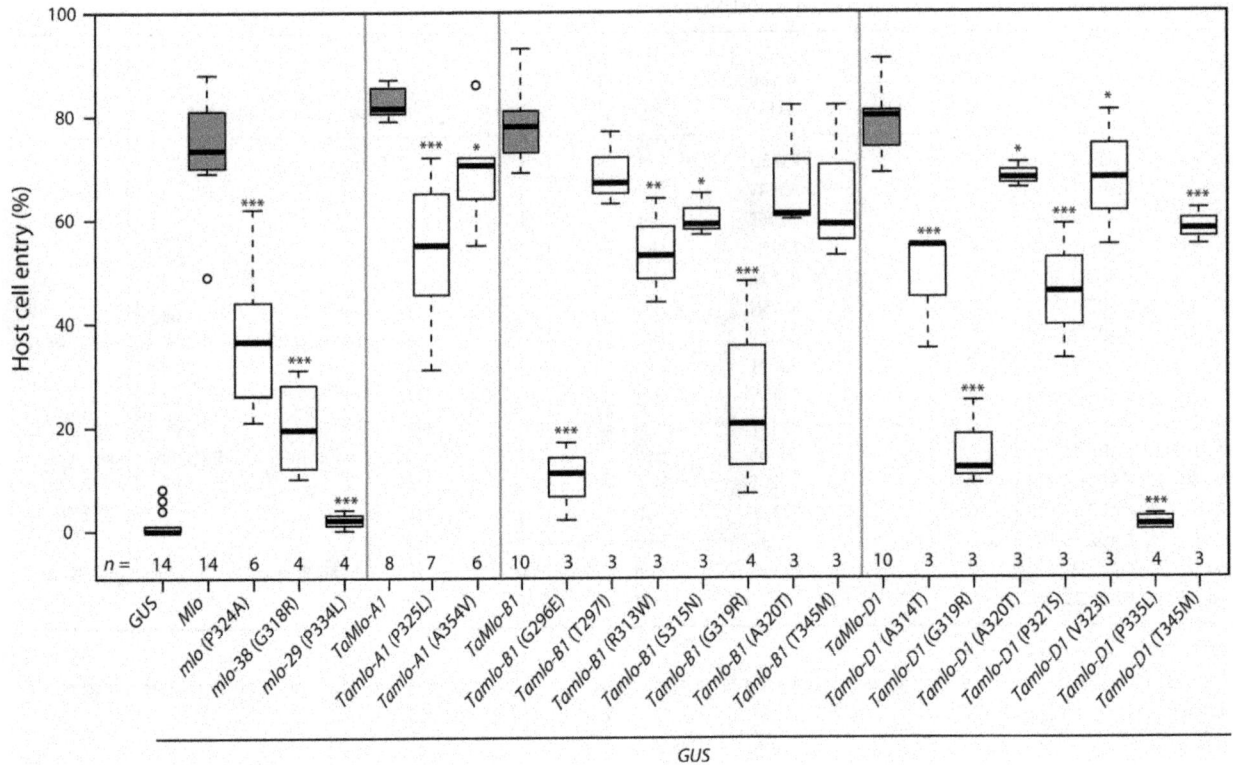

Figure 2 TILLING-derived variants of *Tamlo* homoeologues exhibit different levels of functionality in a transient gene expression assay. Detached 8-day-old leaves of the powdery mildew-resistant barley *mlo-3* mutant were co-bombarded with a *GUS* reporter plasmid and a plasmid encoding the indicated *Ta*Mlo protein variant (WT or mutant version under transcriptional control of the maize *Ubiquitin1* promoter). Expression of *GUS* alone and *GUS* plus WT barley *Mlo*, driven by the maize *Ubiquitin1* promoter, were used as negative and positive controls, respectively, for restoration of *Bgh* susceptibility. Host cell entry was scored at 48 hours post inoculation (h p.i.) in GUS-stained cells attacked by powdery mildew sporelings and results visualized as box plots. Centre lines show the medians; upper and lower box limits indicate the 25th and 75th percentiles respectively; upper and lower whiskers extend 1.5 times the interquartile range from the 25th and 75th percentiles, respectively, and outliers are represented by dots. Numbers at the bottom of the boxplots indicate the number of biological replicates per sample (*n*). One biological replicate was typically composed of six leaves with 150 scored cells. Asterisks indicate a statistically significant difference to the respective WT (barley Mlo or *Ta*Mlo) with ***$P < 0.001$, **$P < 0.01$ and *$P < 0.5$ as determined by a Generalized Linear Model (GLM) test. Statistics were performed and boxplots generated with R software.

Barley mutants *mlo*-29 and *mlo*-38 as a case study to compare results from transient gene expression assays and resistance *in planta*

Since we found wheat equivalents of barley mutants *mlo*-29 (P335L) and *mlo*-38 (G319R) in our wheat TILLING screening, we compared the powdery mildew host cell entry rate *in planta* (which has not been assessed before; Figure 3) with the one obtained in transient expression experiments (Figure 2). The respective WT barley parental lines cv. Bonus, Kristina and Sultan showed high susceptibility to *Bgh* (Figure 3a) with ~85% host cell entry (Figure 3b). By contrast, the mutants *mlo*-38 (two lines: SR59, background cv. Bonus and SR65, background cv. Kristina) and *mlo*-29 (background cv. Sultan) were highly resistant to the pathogen (Figure 3a) with a host cell entry rate of less than 6% (Figure 3b). At least in the case of *mlo*-38, this value is substantially lower than the one obtained upon transient expression of the *mlo*-38 variant (~20% entry rate, median value; Figure 2). These results suggest that overexpression of *mlo* mutant variants may result in an overestimation of the residual protein function (*i.e.* artificially high protein levels may partly compensate for the defect in the protein). Consequently, for a given *mlo* mutant allele the resulting level of resistance *in planta* can be higher (*i.e.* respective

plants are more resistant) than indicated by the results of the single-cell transient gene expression assays.

Generation of triple homozygous *Tamlo* lines

Based on the data obtained with transient gene expression experiments (Figure 2), mutant candidates that showed reduced host cell entry compared to the respective *Ta*Mlo WT genes were selected for allele stacking (Table 3). While several suitable mutants with moderately or strongly reduced *Ta*Mlo function were available for the wheat genomes B and D, only two mutants were available for the A genome, which conferred moderately (P325L) and slightly (A354V) reduced host cell entry in the transient gene expression assay (Figure 2 and Table 3). Due to the lack of alternatives for genome A, mutant P325L was used in all crossings. With the aim of balancing the degree of powdery mildew resistance and potential pleiotropic phenotypes associated with the loss of *Mlo* function (Hückelhoven *et al.*, 2013), we generated four wheat lines containing different mutant allele combinations that ranged from predicted weak to strong effects on powdery mildew susceptibility (Table 3). Depending on the zygosity of the available single mutants identified in the TILLING screen we followed a crossing scheme similar to the one shown in Figure S1 to obtain ideally all possible mutant combinations. To

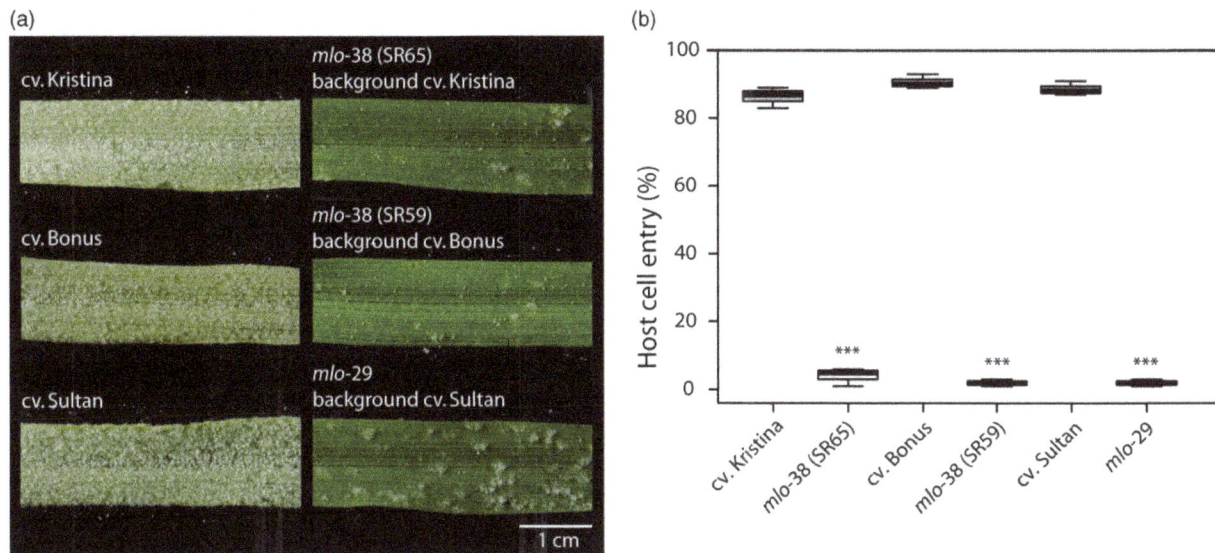

Figure 3 *Bgh* infection phenotypes of barley mutants *mlo*-38 and *mlo*-29 and their respective parental lines. Seven-day-old leaves were inoculated with *Bgh* isolate K1. (a) The macroscopic phenotype was recorded at 7 days post inoculation (d.p.i.). A scale bar shown in white is given in the lower right corner (1 cm). (b) Host cell entry was scored at 48 h p.i. Centre lines show the medians; upper and lower box limits indicate the 25th and 75th percentiles respectively; upper and lower whiskers extend 1.5 times the interquartile range from the 25th and 75th percentiles, respectively. Shown are data of $n = 3$ independent biological replicates with 3–4 leaves (typically 200 scored interaction sites per leaf) per replicate. Asterisks indicate a statistically significant difference to the respective parental background with ***$P < 0.001$ as determined by a GLM test. Statistics were performed and boxplots established by R software.

Table 3 Triple mutant wheat lines obtained by stacking of *Tamlo* mutant alleles

Line No.	Tamlo-A1		Tamlo-B1		Tamlo-D1		Current status
	Mutation	Effect*	Mutation	Effect*	Mutation	Effect*	
1	P325L	Medium	G319R	Strong	P335L	Very strong	aabbdd
2	P325L	Medium	G319R	Strong	G319R	Strong	aabbdd
3	P325L	Medium	G296E	Strong	P321S	Medium	AaBbDd
4	P325L	Medium	T297I	Weak	P321S	Medium	AaBbDd

*Effect on powdery mildew susceptibility based on data from transient gene expression assays (Figure 2).

identify the respective mutations in the progeny of the crossings we developed cleaved amplified polymorphic sequence (CAPS) markers (Table S2). Genotype confirmation was subsequently performed by DNA sequencing of the 2nd round PCR products of the particular *TaMlo* homoeologues.

Currently, we have produced four different *Tamlo* lines that bear mutations in all three *TaMlo* homoeologues. Lines 1 and 2 (Table 3) are already homozygous and are referred as *Tamlo-aabbdd*. We also obtained two of the three possible double mutant combinations (*Tamlo-aabbDD* and *Tamlo-AAbbdd*). Lines 3 and 4 (Table 3) are presently triple heterozygous plants in self-fertilization stage.

Mutations in *TaMlo* homoeologues confer enhanced *Bgt* resistance

We next aimed to test the powdery mildew infection phenotype with the set of available single, double and triple mutants. We inoculated 10-day-old leaves with a *Bgt* field isolate and estimated microscopically the host cell entry rate at 72 h p.i. (hours post inoculation; Figure 4a). Furthermore, we evaluated the macroscopic phenotype of these mutants at 6 h p.i. (hours post inoculation; Figure 4b).

The cv. Cadenza exhibited a *Bgt* penetration rate of ~85%. The various single homozygous mutants showed a slight reduction in

Figure 4 *Bgt* infection phenotypes of wheat WT (cv. Cadenza), single, double and triple *Tamlo* mutants. Ten-day-old leaves were inoculated with *Bgt* isolate JA82. (a) Host cell entry was scored at 72 h p.i. Centre lines show the medians; upper and lower box limits indicate the 25th and 75th percentiles respectively; upper and lower whiskers extend 1.5 times the interquartile range from the 25th and 75th percentiles, respectively, and outliers are represented by dots. Numbers at the bottom of the boxplots indicate the number of biological replicates per sample (*n*). One biological replicate was typically composed of four leaves with 200 scored cells. Letters indicate genotypes whose data are significantly ($P < 0.05$) different from genotypes labelled with other letters, as determined by pair-wise testing with a Games–Howell *post hoc* test. Statistics were performed and boxplots generated with R software. (b) First leaves of germinated seedlings were fixed with surgical tape to a polycarbonate platform and inoculated with *Bgt* conidiospores. The macroscopic phenotype was recorded at 6 d.p.i. A scale bar shown in white is given in the lower right corner (1 cm).

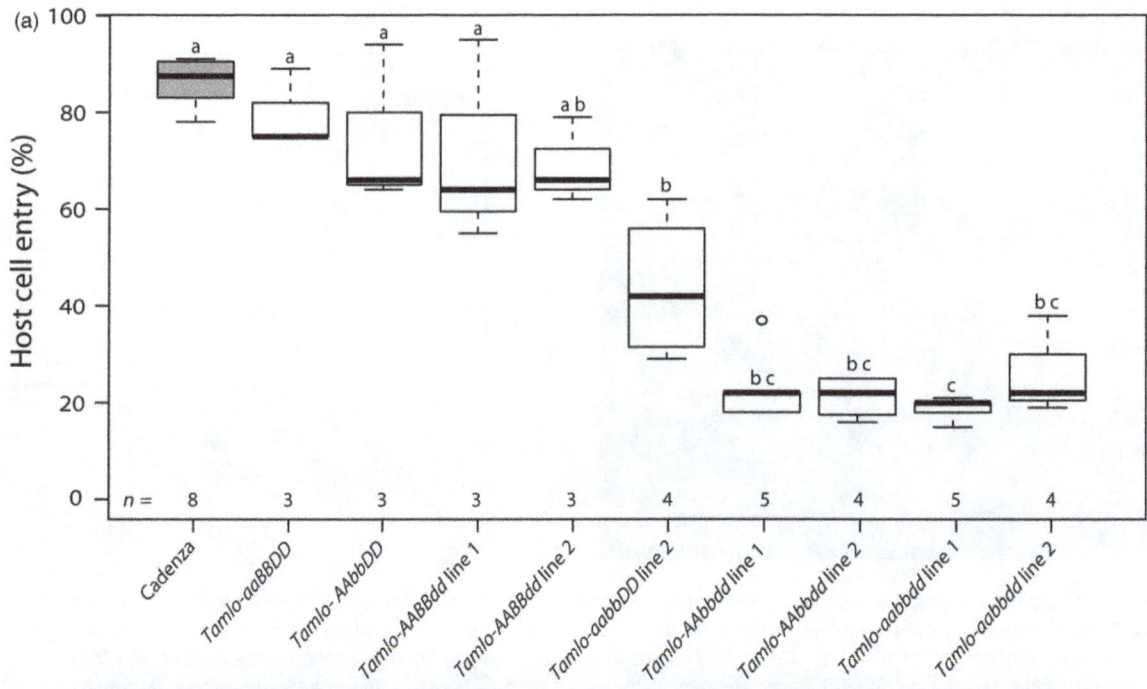

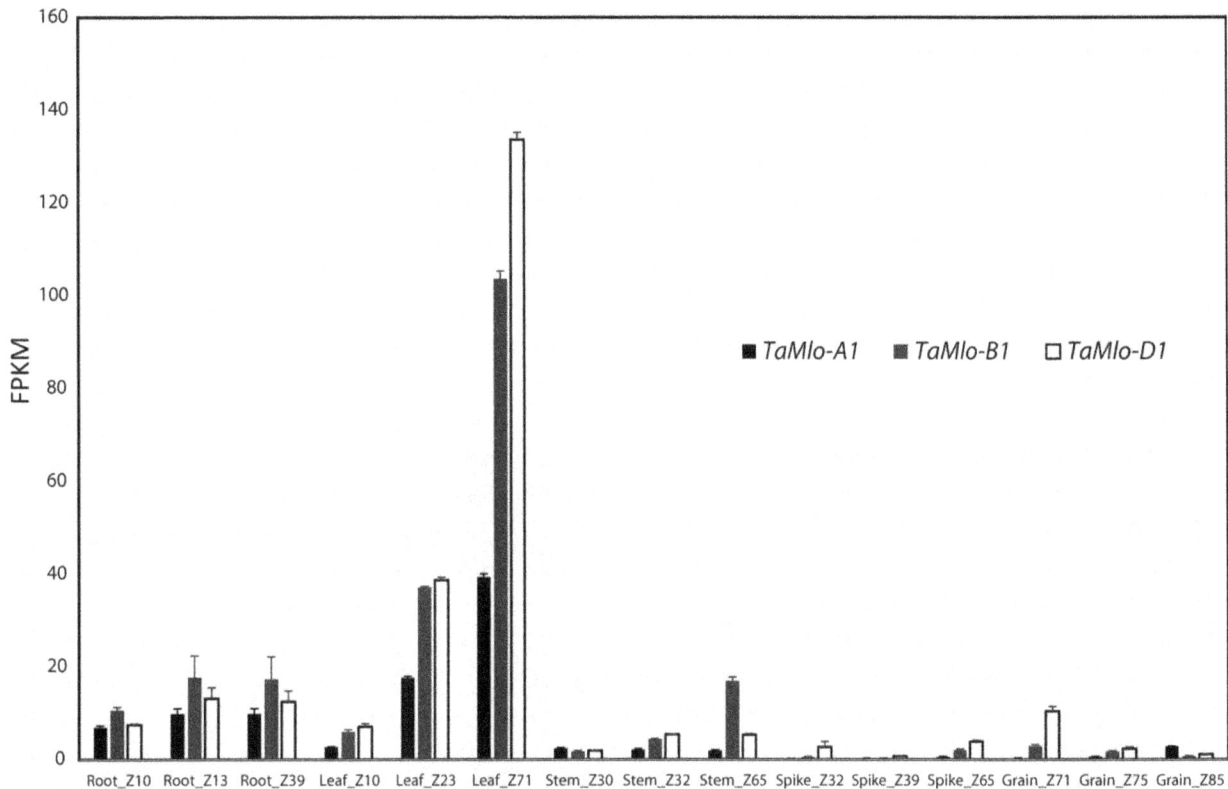

Figure 5 Transcript accumulation of *TaMlo* homoeologues in five bread wheat tissues at different developmental stages. Transcript levels of each homoeologue were summed up and then averaged across replicates, ± standard errors of the mean. FPKM: Fragments per kb per million mapped reads; Z: Zadoks developmental scale (Zadoks *et al.*, 1974).

entry rate (~65%–75%, not significantly different from cv. Cadenza), with the tendency of *Tamlo-AAbbDD* and *Tamlo-AABBdd* to be lower than *Tamlo-aaBBDD;* Figure 4a). Interestingly, both *Tamlo-AAbbdd* double mutant lines showed a lower penetration rate than the *Tamlo-aabbDD* double mutant line (between 14% and 22% for *Tamlo-AAbbdd versus* 42% for *Tamlo-aabbDD*, Figure 4a and S2a; note that *Tamlo-aabbDD* was obtained from segregating progeny in line 2, but is based on the same allele combination as in line 1). Furthermore, both *Tamlo-aabbdd* triple mutant lines exhibited enhanced resistance compared with WT, with host cell entry values of ~20%, similar to the *Tamlo-AAbbdd* double mutant (Figures 4a and S2a). This result suggests that the contribution of the *Tamlo-A1* allele to resistance in the triple homozygous line is lower than the contribution of the *Tamlo-B1* and *-D1* alleles, as was predicted by the transient expression data (Figure 2).

The evaluation of the macroscopic phenotype upon *Bgt* inoculation was largely consistent with the microscopic assessment (Figures 4b and S2b). The four single mutants exhibited an infection phenotype similar to WT, while the *Tamlo-aabbDD* and *Tamlo-AAbbdd* double mutant lines and the two *Tamlo-aabbdd* triple mutant lines showed enhanced resistance compared to the WT (Figures 4b and S2b). However, in contrast to the host cell entry results, at the level of the macroscopic infection phenotype the triple *Tamlo-aabbdd* mutants seem to be more resistant than the *Tamlo-AAbbdd* and *Tamlo-aabbDD* double mutants, as exemplified by fewer colonies on the leaf surface (Figures 4b and S2b). This observation suggests that the *Tamlo-aabbdd* lines might have an additional post-penetration effect in powdery mildew infection. Nevertheless, resistance of each mutant was clearly weaker than

seen for the triple knockout mutant obtained by TALENs (Wang *et al.*, 2014; Figures 4b and S2b).

Although resistance of the TILLING mutants in the field has not yet been scored, plants were naturally and spontaneously infected by *Bgt* in the greenhouse. Throughout the entire wheat growth cycle we observed considerably fewer powdery mildew symptoms on the *Tamlo-aabbdd* lines compared to the WT or heterozygous individuals within segregating populations (Figure S3).

Differential transcript accumulation of the *TaMlo1* homoeologues

Since we observed a higher reduction in the *Bgt* penetration rate and less sporulation on the leaf surface of the *Tamlo-AAbbdd* double mutant compared to the *Tamlo-aabbDD* double mutant (Figure 4) we reasoned that this difference could be based (i) on the particular mutant allele combinations; (ii) on potential unequal functional redundancy of the *TaMlo* homoeologues; or (iii) on differential expression of the three *TaMlo* homoeologues. To explore the latter possibility, we made use of a data set of RNA-seq samples derived from several tissues (root, leaf, stem, spike and grain) at different developmental stages of bread wheat cv. Chinese Spring (Choulet *et al.*, 2014). The paired-end RNA-seq reads were mapped to a transcriptome reference containing the coding sequences of *TaMlo* obtained in this study (File S1) together with non-redundant cDNA sequences from The International Wheat Genome Sequencing Consortium (IWGSC, 2014). The respective mean fragments per kilobase (kb) per million mapped reads (FPKM) values revealed that the *TaMlo* homoeologues are highly expressed in leaves compared to the other tissues, especially at Zadoks stage 71 (Zadoks et al., 1974; Figure 5 and Table S3). Furthermore, the FPKM values suggest that in leaves the three homoeologues have differential expression levels, with TaMlo-D1 showing the highest level of expression, followed by TaMlo-B1, while the level of TaMlo-A1 transcripts seems strikingly lower (Figure 5).

To further test the notion that TaMlo-A1 shows lower transcript levels in leaf tissue compared to TaMlo-B1 and -D1, we followed an entirely independent experimental route. We PCR-amplified and cloned a TaMlo cDNA fragment using oligonucleotide primers that have a complete match in all three TaMlo homoeologues. This cDNA fragment was further chosen to contain several homoeologue-specific single-nucleotide polymorphisms (SNPs) that distinguish them unequivocally (Figure S4). We randomly picked and sequenced 28 clones representing the various TaMlo cDNA fragments. Among these, we found 14, 10 and 4 clones corresponding to TaMlo-B1, TaMlo-D1 and TaMlo-A1, respectively. Together the results of these two approaches suggest that TaMlo-A1 is the homoeologue with the lowest transcript levels in leaves.

TILLING-derived Tamlo triple mutant lines lack obvious pleiotropic phenotypes

In our conditions, the Tamlo-aabbdd lines did not show any obvious and macroscopically discernible differences regarding overall growth habit, height, the number of spikes or seed number compared to cv. Cadenza WT plants. We also did not notice any signs of early leaf chlorosis in the TILLING mutants compared to the WT. This is in stark contrast to the pronounced leaf chlorosis observed under our growth conditions in case of the TALEN-derived Tamlo-aabbdd line, which represents a full Tamlo knockout mutant (Wang et al., 2014, Figure S5).

Discussion

In this study, we used the non-transgenic TILLING technology as a strategy to identify mutant candidates in the TaMlo genes from an EMS-mutagenized population of spring wheat cv. Cadenza. TILLING has been previously suggested as a means to create mutants in 'susceptibility genes' such as Mlo in a targeted manner (Hückelhoven et al., 2013; van Schie and Takken, 2014). By using transient gene expression assays in barley and an adapted powdery mildew species for the bioassay we were able to test the obtained missense mutant wheat variants and to select candidates that confer different degrees of reduction in powdery mildew host cell entry (Figure 2). This experimental assessment of TaMlo protein functionality broadly agreed with the results of in silico prediction via SIFT analysis (Figure 2 and Table 2). However, the observed exceptions [Tamlo-A1 (A354V), Tamlo-B1 (T297I) and Tamlo-B1 (R313W)] highlight the need for validation of mutant variants by experimentation. Stacking of the selected mutant alleles resulted in four different wheat lines carrying a missense mutation in each of the three TaMlo homoeologues (Table 3). The two triple homozygous Tamlo-aabbdd mutant lines (1 and 2) showed enhanced powdery mildew disease resistance, exemplified by a considerable reduction in Bgt host cell entry [from ~85% in WT to ~20% (median values) in the triple mutant] and a substantial decrease in macroscopically visible Bgt colonies (Figures 4 and S2). Notably, the Tamlo-AAbbdd double mutants showed a similar level of enhanced Bgt penetration resistance.

In our TILLING screening we targeted TaMlo exon 9, which encodes the third cytoplasmic loop of the TaMlo protein.

Excluding the false positives, we found a total of 56 mutants within an approximate population of 2020 individuals (Table 1), which represents an average density of 1 mutation per 32 kb screened. These results are comparable to the ones obtained by Botticella and co-workers after screening the same TILLING population for mutants in the Sb␣␣ genes (Botticella et al., 2011). These authors identified an average of 60 candidate mutants per exon and a density of approximately 1 mutation per 40 kb screened. As in our screening, the authors did not find nonsense mutations for each exon assessed. However, in contrast to our results, the number of identified missense mutations was higher than the silent ones. The lack of identification of nonsense mutations in our study is somewhat surprising because exon 9 has six glutamine (Q; CAG triplet) or tryptophan (W; TGG triplet) codons that could potentially have been mutated to stop codons.

Using transient gene expression experiments in single barley epidermal cells we demonstrated that cDNAs from each of the three TaMlo homoeologues are able to complement the barley mlo resistance phenotype and can restore host cell entry to WT-like levels (Figure 2). Of the 16 identified missense mutations, the non-conservative changes had a more dramatic effect on susceptibility to powdery mildew, especially G296E, G319R and P335L (Figure 2 and Table 3). These data are consistent with previous site-directed mutagenesis studies performed in barley, further reinforcing the relevance of the third cytoplasmic loop of Mlo as an important domain in the susceptibility-conferring activity of the protein (Reinstädler et al., 2010). Therefore, the application of transient gene expression assays was an efficient tool to select suitable mutant alleles for subsequent stacking.

Surprisingly, not only the Tamlo-aabbdd triple mutants but also the double mutants Tamlo-aabbDD and Tamlo-AAbbdd exhibited significantly enhanced Bgt resistance compared to the WT (Figures 4 and S2). These results are in stark contrast with the ones obtained by Wang and co-workers, where only the triple knockout mutant exhibited a notable difference in the powdery mildew infection phenotype (Wang et al., 2014). However, the Tamlo-AAbbdd double mutant line was missing in the set of tested TALEN plants (Wang et al., 2014); therefore, a comparison between this double mutant and ours generated by TILLING is not possible. It is conceivable that some of the discrepancies between the two data sets are due to the type of evaluation, since we scored host cell entry while Wang and co-workers evaluated the number of microcolonies per germinated spores (Wang et al., 2014). Additionally, second-site mutations present in our TILLING lines, which are derived from EMS mutagenesis, might lead to differences in the powdery mildew infection phenotype. Furthermore, the TILLING lines are based on a spring wheat cultivar (cv. Cadenza), while the TALEN lines are derived from a winter wheat. Together, our results challenge the hypothesis of strict functional redundancy between wheat TaMlo homoeologues proposed by Borrill et al. (2015) and suggest instead that the powdery mildew infection phenotype can be the result of gene dosage effects or differential expression of the respective TaMlo homoeologues. Evidence to support this notion is provided not only by the differences in host cell entry and macroscopic infection phenotypes observed between the double mutants Tamlo-aabbDD and Tamlo-AAbbdd compared to WT (Figure 4), but also the seemingly lower transcript levels of TaMlo-A1 compared to TaMlo-B1 and -D1 (Figure 5). Tamlo double mutants with enhanced powdery mildew resistance might be advantageous in future breeding programmes, since only two mutant alleles must be followed in segregating populations.

The incomplete *Bgt* resistance found in our *Tamlo-aabbdd* triple mutants is likely based on the partial loss-of-function alleles used in these lines (Figure 2; Table 3). Deployment of yet to be identified stronger EMS-induced alleles (e.g. nonsense mutations) may, similar to the TALEN-derived line (Wang *et al.*, 2014), result in a TILLING-based non-transgenic triple mutant with full *Bgt* resistance. However, the incomplete resistance phenotype in our generated *Tamlo-aabbdd* TILLING lines might in fact represent an advantage in several different ways, namely, reduced pleiotropic effects, more durable resistance and an opportunity for breeders since products of induced mutagenesis are free of government regulation.

Pleiotropic effects such as early leaf senescence associated with *mlo* mutants have been better studied in barley where lines carrying weak resistance alleles, permitting residual fungal growth, exhibit fewer pleiotropic phenotypes than *mlo* mutants harbouring strong alleles, which confer complete immunity (Hentrich, 1979; Piffanelli *et al.*, 2002). TILLING has thus been suggested as a means to find partial loss-of-function mutants of 'susceptibility genes' that show mild pleiotropy (Hückelhoven *et al.*, 2013). During the first cycle of propagation we did not observe differences in growth, development and fertility between our *Tamlo-aabbdd* TILLING lines (partial loss-of-function alleles) and the WT, nor did we observe enhanced or premature leaf senescence (Figure S5). These results are in contrast to the noticeable leaf chlorosis in TALEN-based mutants (complete loss-of-function alleles) in comparison to its parental line KN199 (Figure S5). Nevertheless, this is only a first qualitative assessment of these plants and quantitative measurements of photosynthetic performance and chlorophyll levels under various growth conditions are needed to substantiate these observations in future experiments.

The most common allele introduced in the spring barley varieties cultivated in Europe is *mlo*-11 (Brown and Rant, 2013; Kokina *et al.*, 2007; Piffanelli *et al.*, 2004). Barley plants carrying this natural partial loss-of-function allele are not fully resistant and allow the growth of powdery mildew colonies at a low level (Piffanelli *et al.*, 2004). The barley *mlo*-11 plants nevertheless confer broad-spectrum resistance to the pathogen, which has been durable in the field for more than 30 years (reviewed in Acevedo-Garcia *et al.*, 2014). One may speculate that the lower selection pressure exerted on powdery mildew populations by partial resistance might have contributed to this durability under agricultural conditions. Our *Tamlo-aabbdd* lines are also partially resistant, allowing a low level of *Bgt* sporulation (Figure 4). This feature may contribute to the robustness of the trait once *mlo*-based resistance is used in commercial wheat crops. Finally, our non-transgenic wheat *Tamlo-aabbdd* plants represent an excellent alternative for breeders and farmers in regions where growth of genetically-modified organisms (GMOs) is banned and where regulations for the cultivation of genome-edited crops are still under discussion (Chen *et al.*, 2014b; Huang *et al.*, 2016).

This study reports the first step towards the generation of new commercial bread wheat varieties with an enhanced resistance phenotype to the globally important powdery mildew disease, generated by TILLING technology targeting the *TaMlo* gene. Currently, our *Tamlo-aabbdd* plants are being backcrossed to remove excess of EMS mutations that may affect additional traits. It will be interesting to determine the performance of the backcrossed plants in the field at the agronomic level and their response to other pathogens and pests that attack wheat crops.

Experimental procedures

Plant and fungal material

The EMS-mutagenized population of the spring wheat (*Triticum aestivum*) cv. Cadenza has been previously described (Rakszegi *et al.*, 2010). Barley (*Hordeum vulgare*) seeds NGB14661.2 (cv. Kristina), NGB14682.1 (cv. Bonus) NGB119178.1 (SR59, *mlo*-38, background cv. Bonus), NGB119185.1 (SR65, *mlo*-38, background cv. Kristina) were originally reported in Lundqvist *et al.* (1991) and obtained from NordGen (Nordic Genetic Resource Centre) in Sweden. The *mlo*-29 mutant and its parent cv. Sultan were described previously (Piffanelli *et al.*, 2002). The TALEN-derived *Tamlo-aabbdd* line and its parental line KN199 (Wang *et al.*, 2014) were kindly provided by Prof. Caixia Gao and Prof. Jin-Long Qiu from the State Key Laboratories (Chinese Academy of Sciences, Beijing, China). The *Bgt* isolate, termed JA82, was obtained from plant material spontaneously infected in the greenhouse, latitude: 50.7784, longitude: 6.04863. It was propagated weekly on 10-day-old seedlings comprising a mixture of summer wheat cv. Munk and commercially available seeds of unknown identity. *Bgh* isolate K1 was propagated weekly on 7-day-old barley seedlings cv. Margret.

TILLING screening

M_2 wheat genomic DNA samples (~2020) of the TILLING library developed at Rothamsted Research were twofold pooled in 96-well plates at 25 ng/µL. The amplicons for HRM were produced by a nested PCR strategy. The 1st round PCR was carried out in 10 µL final volume using 1 µL of pooled template DNA, 5 µL of HotShot Diamond™ Master Mix (Clent Life Sciences, Stourbridge, UK) and 0.5 µM primers. The PCR program was: 98 °C, 5 min; (97 °C, 30 s; 66–67 °C, 30 s; 72 °C 1 min) × 35 cycles; 72 °C, 5 min; 10 °C until end.

For HRM, the 1st round PCR reaction was diluted 100-fold and 2 µL were used as template for the 2nd round PCR. The 2nd round PCR was prepared in FrameStar 96-well skirted plates, black frame with white wells (4titude, Surrey, UK). The reaction contained 2 µL of diluted DNA template, 0.5 µM of each primer, 5 µL HotShot Diamond Mastermix, and 1 µL of LCGreen Plus (Clent Life Sciences, Stourbridge, UK) in a total volume of 10 µL; 10 µL of mineral oil (M5904, Sigma, Deisenhofen, Germany) was added to prevent evaporation. The PCR program was similar to the primary PCR, with annealing at 66 °C and after final extension, an additional denaturation step at 98 °C, 3 min, followed by cooling to 10 °C at 0.2 °C/s.

High-Resolution Melting analysis

The 2nd round PCR plates were used for HRM analysis using the LightScanner instrument (Idaho Technology Inc., Utah) and with the following temperatures: start, 82 °C; end 98 °C and a hold of 78 °C. The analysis was performed with the software provided with the scanner and as described in Botticella *et al.* (2011). Negative samples were repeated and positive candidates were selected from the independent plate. A twofold pool of DNA from candidate lines together with cv. Cadenza DNA was generated for the 2nd round PCR and screening was repeated. The amplicon from putative positive candidates was re-amplified and sent for sequencing. DNA sequencing was performed by Eurofins MWG Operon (Ebersberg, Germany).

Transient gene expression in barley leaf epidermal cells

Ballistic transformation of 8-day-old detached barley *mlo*-3 (background cv. Ingrid) leaves was performed as previously described (Reinstädler *et al.*, 2010). Inoculation with *Bgh* isolate K1 was carried out at 24 h after bombardment and histochemical staining with X-Gluc (5-Bromo-4-chloro-3-indolyl-ß-D-glucoronide; Carl Roth GmbH, Karlsruhe, Germany) for β-glucoronidase (GUS) activity was performed at 48 h p.i. Fungal structures were stained with Coomassie Brilliant Blue R-250 (Carl Roth GmbH, Karlsruhe, Germany). The host cell entry rate was calculated as: (number of transformed, *Bgh*-attacked cells with haustoria/transformed and *Bgh*-attacked cells)*100. At least six leaves were evaluated per mutant with a minimum of 150–200 cells per experiment and at least three independent biological replicates.

For further experimental details, see Files S3 and S4.

Acknowledgements

We are thankful to Prof. Caixia Gao and Prof. Jin-Long Qiu from the State Key Laboratories (Chinese Academy of Sciences, Beijing, China) for providing seeds of the transgenic TALEN-derived *Tamlo-aabbdd* mutant and its parental line KN199, and to Prof. Hans Thordal-Christensen (University of Copenhagen, Denmark) for facilitating the delivery of the seeds. We acknowledge Stefan Kusch for his advice regarding statistical analysis. This work was funded by a shared grant of the German Federal Ministry of Food and Agriculture (BMEL; project management *via* the Specialist Agency for Renewable Resources; grant number 22030411) and the Germany Society for the Advancement of Plant Innovation (GFPi; grant number G 135/12 NR) respectively. AP and KHK are funded by the 20:20 Wheat Institute Strategic Programme Grant to Rothamsted Research from the Biotechnology and Biological Sciences Research Council of the U.K (grant number BB/J/00426X/1).

References

Acevedo-Garcia, J., Kusch, S. and Panstruga, R. (2014) Magical mystery tour: MLO proteins in plant immunity and beyond. *New Phytol.* **204**, 273–281.

Borrill, P., Adamski, N. and Uauy, C. (2015) Genomics as the key to unlocking the polyploid potential of wheat. *New Phytol.* **208**, 1008–1022.

Botticella, E., Sestili, F., Hernandez-Lopez, A., Phillips, A. and Lafiandra, D. (2011) High resolution melting analysis for the detection of EMS induced mutations in wheat SBEIIa genes. *BMC Plant Biol.* **11**, 156.

Brown, J.K.M. and Rant, J.C. (2013) Fitness costs and trade-offs of disease resistance and their consequences for breeding arable crops. *Plant. Pathol.* **62**, 83–95.

Büschges, R., Hollricher, K., Panstruga, R., Simons, G., Wolter, M., Frijters, A., van Daelen, R. *et al.* (1997) The barley *Mlo* gene: a novel control element of plant pathogen resistance. *Cell*, **88**, 695–705.

Chen, A. and Dubcovsky, J. (2012) Wheat TILLING mutants show that the vernalization gene *VRN1* down-regulates the flowering repressor *VRN2* in leaves but is not essential for flowering. *PLoS Genet.* **8**, e1003134.

Chen, A., Li, C., Hu, W., Lau, M.Y., Lin, H., Rockwell, N.C., Martin, S.S. *et al.* (2014a) PHYTOCHROME C plays a major role in the acceleration of wheat flowering under long-day photoperiod. *Proc. Natl Acad. Sci. USA*, **111**, 10037–10044.

Chen, L., Hao, L., Parry, M.A.J., Phillips, A.L. and Hu, Y.-G. (2014b) Progress in TILLING as a tool for functional genomics and improvement of crops. *J. Integr. Plant Biol.* **56**, 425–443.

Choulet, F., Alberti, A., Theil, S., Glover, N., Barbe, V., Daron, J., Pingault, L. *et al.* (2014) Structural and functional partitioning of bread wheat chromosome 3B. *Science*, **345**, 1249721.

Colasuonno, P., Incerti, O., Lozito, M.L., Simeone, R., Gadaleta, A. and Blanco, A. (2016) DHPLC technology for high-throughput detection of mutations in a durum wheat TILLING population. *BMC Genet.* **17**, 43.

Conner, R.L., Kuzyk, A.D. and Su, H. (2003) Impact of powdery mildew on the yield of soft white spring wheat cultivars. *Can. J. Plant Sci.* **83**, 725–728.

Dean, R., van Kan, J.A.L., Pretorius, Z.A., Hammond-Kosack, K., Pietro, A.Di, Spanu, P.D., Rudd, J.J. *et al.* (2012) The Top 10 fungal pathogens in molecular plant pathology. *Mol. Plant Pathol.* **13**, 414–430.

Dong, C., Vincent, K. and Sharp, P. (2009) Simultaneous mutation detection of three homoeologous genes in wheat by High Resolution Melting analysis and Mutation Surveyor®. *BMC Plant Biol.* **9**, 143.

Elliott, C., Zhou, F., Spielmeyer, W., Panstruga, R. and Schulze-Lefert, P. (2002) Functional conservation of wheat and rice *Mlo* orthologs in defense modulation to the powdery mildew fungus. *Mol. Plant Microbe Interact.* **15**, 1069–1077.

Glawe, D.A. (2008) The powdery mildews: a review of the world's most familiar (yet poorly known) plant pathogens. *Ann. Rev. Phytopathol.* **46**, 27–51.

Hentrich, W. (1979) Multiple Allelie, Pleiotropie und züchterische Nutzung mehltauresistenter Mutanten des mlo-Locus der Gerste. *Tag-Ber. Akad. Landwirtsch-Wiss.* **175**, 191–202.

Huang, S., Weigel, D., Beachy, R.N. and Li, J. (2016) A proposed regulatory framework for genome-edited crops. *Nat. Genet.* **48**, 109–111.

Hückelhoven, R., Eichmann, R., Weis, C., Hoefle, C. and Proels, R.K. (2013) Genetic loss of susceptibility: a costly route to disease resistance? *Plant. Pathol.* **62**, 56–62.

IWGSC. (2014) A chromosome-based draft sequence of the hexaploid bread wheat (*Triticum aestivum*) genome. *Science*, **345**, 1251788.

Jørgensen, J.H. (1992) Discovery, characterization and exploitation of *Mlo* powdery mildew resistance in barley. *Euphytica*, **63**, 141–152.

Jørgensen, J.H. and Mortensen, K. (1977) Primary infection by *Erysiphe graminis* f. sp. *hordei* of barley mutants with resistance genes in the ml-o locus. *Phytopathology*, **5**, 678–685.

Kokina, A., Legzdiņa, L., Bērziņa, I., Bleidere, M., Rashal, I. and Rostoks, N. (2007) Molecular marker-based characterization of barley powdery mildew *Mlo* resistance locus in European varieties and breeding lines. *Lat. J. Agron.* **11**, 77–83.

Konishi, S. and Sasanuma, T. (2010) Identification of novel Mlo family members in wheat and their genetic characterization. *Genes Genet. Syst.* **85**, 167–175.

Kumar, P., Henikoff, S. and Ng, P.C. (2009) Predicting the effects of coding non-synonymous variants on protein function using the SIFT algorithm. *Nat. Protoc.* **4**, 1073–1081.

Kurowska, M., Daszkowska-Golec, A., Gruszka, D., Marzec, M., Szurman, M., Szarejko, I. and Maluszynski, M. (2011) TILLING – a shortcut in functional genomics. *J. App. Genet.* **52**, 371–390.

Kusch, S., Pesch, L. and Panstruga, R. (2016) Comprehensive phylogenetic analysis sheds light on the diversity and origin of the MLO family of integral membrane proteins. *Genome Biol. Evol.* **8**, 878–895.

Lundqvist, U., Meyer, J. and Lundqvist, A. (1991) Mutagen specificity for 71 lines resistant to barley powdery mildew race D1 and isolated in four highbred barley varieties. *Hereditas*, **115**, 227–239.

Lyngkjær, M.F., Newton, A.C., Atzema, J.L. and Baker, S.J. (2000) The barley *mlo*-gene: an important powdery mildew resistance source. *Agronomie*, **20**, 745–756.

McCallum, C.M., Comai, L., Greene, E.A. and Henikoff, S. (2000) Targeting induced local lesions IN genomes (TILLING) for plant functional genomics. *Plant. Physiol.* **123**, 439–442.

Menardo, F., Praz, C., Wyder, S., BenDavid, R., Bourras, S.A., Matsumae, H., McNally, K.E. *et al.* (2016) Hybridization of powdery mildew strains gives raise to pathogens on novel agricultural crop species. *Nat. Genet.* **48**, 201–205.

Müller, J., Piffanelli, P., Devoto, A., Miklis, M., Elliott, C., Ortmann, B., Schulze-Lefert, P. *et al.* (2005) Conserved ERAD-like quality control of a plant polytopic membrane protein. *Plant. Cell*, **17**, 149–163.

Peterhänsel, C., Freialdenhoven, A., Kurth, J., Kolsch, R. and Schulze-Lefert, P. (1997) Interaction analyses of genes required for resistance responses to powdery mildew in barley reveal distinct pathways leading to leaf cell death. *Plant. Cell*, **9**, 1397–1409.

Piffanelli, P., Zhou, F., Casais, C., Orme, J., Jarosch, B., Schaffrath, U., Collins, N.C. *et al.* (2002) The barley MLO modulator of defense and cell death is responsive to biotic and abiotic stress stimuli. *Plant. Physiol.* **129**, 1076–1085.

Piffanelli, P., Ramsay, L., Waugh, R., Benabdelmouna, A., D'Hont, A., Hollricher, K., Jørgensen, J.H. *et al.* (2004) A barley cultivation-associated polymorphism conveys resistance to powdery mildew. *Nature*, **430**, 887–891.

Rakszegi, M., Kisgyörgy, B.N., Tearall, K., Shewry, P.R., Láng, L., Phillips, A. and Bedö, Z. (2010) Diversity of agronomic and morphological traits in a mutant population of bread wheat studied in the Healthgrain program. *Euphytica*, **174**, 409–421.

Rawat, N., Sehgal, S.K., Joshi, A., Rothe, N., Wilson, D.L., McGraw, N., Vadlani, P.V. *et al.* (2012) A diploid wheat TILLING resource for wheat functional genomics. *BMC Plant. Biol.* **12**, 205.

Reed, G.H. and Wittwer, C.T. (2004) Sensitivity and specificity of single-nucleotide polymorphism scanning by high-resolution melting analysis. *Clin. Chem.* **50**, 1748–1754.

Reinstädler, A., Müller, J., Czembor, J.H., Piffanelli, P. and Panstruga, R. (2010) Novel induced *mlo* mutant alleles in combination with site-directed mutagenesis reveal functionally important domains in the heptahelical barley Mlo protein. *BMC Plant. Biol.* **10**, 31.

van Schie, C.C.N. and Takken, F.L.W. (2014) Susceptibility genes 101: how to be a good host. *Annu. Rev. Phytopathol.* **52**, 551–581.

Schwarzbach, E. (1976) The pleiotropic effects of the *ml-o* gene and their implications in breeding. In *Barley Genetics III* (Gaul, H., ed), pp. 440–445. Munich: Karl Thiemig Verlag.

Sestili, F., Botticella, E., Bedo, Z., Phillips, A. and Lafiandra, D. (2009) Production of novel allelic variation for genes involved in starch biosynthesis through mutagenesis. *Mol. Breed.* **25**, 145–154.

Sestili, F., Palombieri, S., Botticella, E., Mantovani, P., Bovina, R. and Lafiandra, D. (2015) TILLING mutants of durum wheat result in a high amylose phenotype and provide information on alternative splicing mechanisms. *Plant. Sci.* **9116**, 1–7.

Simmonds, J., Scott, P., Brinton, J., Mestre, T.C., Bush, M., del Blanco, A. and Dubcovsky, J., *et al.* (2016) A splice acceptor site mutation in *TaGW2-A1* increases thousand grain weight in tetraploid and hexaploid wheat through wider and longer grains. *Theor. Appl. Genet.* **129**, 1099–1112.

Slade, A.J., Fuerstenberg, S.I., Loeffler, D., Steine, M.N. and Facciotti, D. (2005) A reverse genetic, nontransgenic approach to wheat crop improvement by TILLING. *Nat. Biotechnol.* **23**, 75–81.

Slade, A.J., McGuire, C., Loeffler, D., Mullenberg, J., Skinner, W., Fazio, G., Holm, A. *et al.* (2012) Development of high amylose wheat through TILLING. *BMC Plant. Biol.* **12**, 69.

Uauy, C., Paraiso, F., Colasuonno, P., Tran, R.K., Tsai, H., Berardi, S., Comai, L. *et al.* (2009) A modified TILLING approach to detect induced mutations in tetraploid and hexaploid wheat. *BMC Plant. Biol.* **9**, 115.

Várallyay, É., Giczey, G. and Burgyán, J. (2012) Virus-induced gene silencing of *Mlo* genes induces powdery mildew resistance in *Triticum aestivum. Arch. Virol.* **157**, 1345–1350.

Wang, T.L., Uauy, C., Robson, F. and Till, B. (2012) TILLING *in extremis. Plant. Biotechnol. J.* **10**, 761–772.

Wang, Y., Cheng, X., Shan, Q., Zhang, Y., Liu, J., Gao, C. and Qiu, J.L. (2014) Simultaneous editing of three homoalleles in hexaploid bread wheat confers heritable resistance to powdery mildew. *Nat. Biotechnol.* **32**, 947–951.

Wolter, M., Hollricher, K., Salamini, F. and Schulze-Lefert, P. (1993) The *mlo* resistance alleles to powdery mildew infection in barley trigger a developmentally controlled defence mimic phenotype. *Mol. Gen. Genet.* **239**, 122–128.

Zadoks, J.C., Chang, T.T. and Konzak, C.F. (1974) A decimal code for the growth stages of cereals. *Weed Res.* **14**, 415–421.

Permissions

The contributors of this book come from diverse backgrounds, making this book a truly international effort. This book will bring forth new frontiers with its revolutionizing research information and detailed analysis of the nascent developments around the world.

We would like to thank all the contributing authors for lending their expertise to make the book truly unique. They have played a crucial role in the development of this book. Without their invaluable contributions this book wouldn't have been possible. They have made vital efforts to compile up to date information on the varied aspects of this subject to make this book a valuable addition to the collection of many professionals and students.

This book was conceptualized with the vision of imparting up-to-date information and advanced data in this field. To ensure the same, a matchless editorial board was set up. Every individual on the board went through rigorous rounds of assessment to prove their worth. After which they invested a large part of their time researching and compiling the most relevant data for our readers.

The editorial board has been involved in producing this book since its inception. They have spent rigorous hours researching and exploring the diverse topics which have resulted in the successful publishing of this book. They have passed on their knowledge of decades through this book. To expedite this challenging task, the publisher supported the team at every step. A small team of assistant editors was also appointed to further simplify the editing procedure and attain best results for the readers.

Apart from the editorial board, the designing team has also invested a significant amount of their time in understanding the subject and creating the most relevant covers. They scrutinized every image to scout for the most suitable representation of the subject and create an appropriate cover for the book.

The publishing team has been an ardent support to the editorial, designing and production team. Their endless efforts to recruit the best for this project, has resulted in the accomplishment of this book. They are a veteran in the field of academics and their pool of knowledge is as vast as their experience in printing. Their expertise and guidance has proved useful at every step. Their uncompromising quality standards have made this book an exceptional effort. Their encouragement from time to time has been an inspiration for everyone.

The publisher and the editorial board hope that this book will prove to be a valuable piece of knowledge for researchers, students, practitioners and scholars across the globe.

List of Contributors

Michael Abrouk, Barbora Balcárková, Hana Šimková, Eva Komínkova, Jaroslav Doležel, Miroslav Valárik, Elodie Rey and Jan Vrána
Institute of Experimental Botany, Centre of the Region Hana for Biotechnological and Agricultural Research, Olomouc, Czech Republic

Mihaela M. Martis
Munich Information Center for Protein Sequences/ Institute of Bioinformatics and Systems Biology, Institute for Bioinformatics and Systems Biology, Helmholtz Center Munich, Neuherberg, Germany
Division of Cell Biology, Department of Clinical and Experimental Medicine, Bioinformatics Infrastructure for Life Sciences, Link€oping University, Link€oping, Sweden

Irena Jakobson and Ljudmilla Timofejeva
Department of Gene Technology, Tallinn University of Technology, Tallinn, Estonia

Andrzej Kilian
Diversity Arrays Technology Pty Ltd, Canberra, ACT, Australia

Alexandra M. Allen, Mark O. Winfield, Amanda J. Burridge, Harriet R. Benbow, Gary L. A. Barker, Paul A. Wilkinson, Jane Coghill, Christy Waterfall and Keith J. Edwards
Life Sciences, University of Bristol, Bristol, UK

Rowena C. Downie
Life Sciences, University of Bristol, Bristol, UK
The John Bingham Laboratory, NIAB, Cambridge, UK

Alison R. Bentley
The John Bingham Laboratory, NIAB, Cambridge, UK

Alessandro Davassi, Geoff Scopes, Ali Pirani, Teresa Webster, Fiona Brew and Claire Bloor
Affymetrix UK Ltd, High Wycombe, UK

Simon Griffiths
John Innes Centre, Norwich, Norfolk, UK

Mark Alda and Peter Jack
RAGT Seeds, Ickleton, Essex, UK

Andrew L. Phillips
Plant Biology and Crop Science Department, Rothamsted Research, Harpenden, UK

Francisco Barro, Julio C. M. Iehisa, María J. Gimenez, María D. Garcıa-Molina, Carmen V. Ozuna and Javier Gil-Humanes
Departamento de Mejora Genetica, Instituto de Agricultura Sostenible (IAS), Consejo Superior de Investigaciones Cientıficas (CSIC), Cordoba, Spain

Susan A. Gillies, Agnelo Futardo and Robert J. Henry
Southern Cross Plant Science, Southern Cross University, Lismore, NSW, Australia
Queensland Alliance for Agriculture and Food Innovation, The University of Queensland, St Lucia, Qld, Australia

Paul J. Berkman
School of Agriculture and Food Sciences, University of Queensland, Brisbane, QLD, Australia
Australian Centre for Plant Functional Genomics, University of Queensland, Brisbane, QLD, Australia
CSIRO Plant Industry, St Lucia, QLD, Australia

Paul Visendi, Michał T. Lorenc, Kaitao Lai and David Edwards
School of Agriculture and Food Sciences, University of Queensland, Brisbane, QLD, Australia
Australian Centre for Plant Functional Genomics, University of Queensland, Brisbane, QLD, Australia

Jiri Stiller
School of Agriculture and Food Sciences, University of Queensland, Brisbane, QLD, Australia
CSIRO Plant Industry, St Lucia, QLD, Australia

Delphine Fleury
Australian Centre for Plant Functional Genomics, University of Adelaide, Glen Osmond, SA, Australia

Hana Šimková and Jaroslav Doležel
Centre of the Region Haná for Biotechnological and Agricultural Research, Institute of Experimental Botany, Olomouc, Czech, Republic

Marie Kubaláková and Song Weining
State Key Laboratory of Crop Stress Biology in Arid Areas, College of Agronomy and Yangling Branch of China Wheat Improvement Center, Northwest A&F University, Yangling, Shaanxi, China

Julien Bonneau, Jesse Beasley and Alexander A. T. Johnson
School of BioSciences, The University of Melbourne, Melbourne, Vic., Australia

Ute Baumann and Yuan
Australian Centre for Plant Functional Genomics, The University of Adelaide, Adelaide, SA, Australia

Edward H. Byrne, Ian Prosser, Nira Muttucumaru, Tanya Y. Curtis and Nigel G. Halford
Department of Plant Science, Rothamsted Research, Harpenden, Hertfordshire, UK

Astrid Wingler
Research Department of Genetics, Evolution and Environment, University College London, Gower Street, London, UK

Stephen Powers
Department of Biomathematics and Bioinformatics, Rothamsted Research, Harpenden, Hertfordshire, UK

Ming Chen, Youzhi Ma, Lipu Du, Xingguo Ye, Junlan Pang, Xinmei Zhang, Liancheng Li and Huijun Xu
Institute of Crop Sciences, Chinese Academy of Agricultural Sciences/National Key Facility for Crop Gene Resources and Genetic Improvement, Key Laboratory of Biology and Genetic Improvement of Triticeae Crops, Ministry of Agriculture, Beijing, China

Liying Sun, Jiong Chen, Ida B. Andika and Jianping Chen
State Key Laboratory Breeding Base for Zhejiang Sustainable Pest and Disease Control, MoA Key Laboratory for Plant Protection and Biotechnology, Zhejiang Provincial Key Laboratory of Plant Virology, Institute of Virology and Biotechnology, Zhejiang Academy of Agricultural Sciences, Hangzhou, China

Hongya Wu, Xiaoxiang Zhang, Shunhe Cheng and Boqiao Zhang
Institute of Agricultural Sciences of Lixiahe Districts, Jiangsu, China

Delia I. Corol, Michael H. Beale, Peter R. Shewry and Jane L. Ward
Department of Plant Biology and Crop Science, Rothamsted Research, Harpenden, Hertfordshire, UK

Catherine Ravel and Gilles Charmet
Inra-Ubp, Umr1095 Gdec, Clermont-Ferrand Cedex, France

Marianna Rakszegi and Zoltan Bedo
Agricultural Institute, Centre for Agricultural Research of the Hungarian Academy of Sciences, Martonvásár, Hungary

David Edwards and Michał T. Lorenc
Australian Centre for Plant Functional Genomics and University of Queensland, St. Lucia, Qld, Australia

Stephen Wilcox
Australian Genome Research Facility, The Walter and Eliza Hall Institute of Medical Research, Parkville, Vic., Australia

Roberto A. Barrero, Paula Moolhuijzen, Gabriel Keeble-Gagneère, Matthew I. Bellgard and Rudi Appels
Centre for Comparative Genomics, Murdoch University, Perth, WA, Australia

Delphine Fleury, Ute Baumann and Peter Langridge
Australian Centre for Plant Functional Genomics, University of Adelaide, Urrbrae, SA, Australia

Jennifer M. Taylor and Matthew K. Morell
CSIRO Plant Industry, Black Mountain Laboratories, Canberra, ACT, Australia

Colin R. Cavanagh
CSIRO Plant Industry, Black Mountain Laboratories, Canberra, ACT, Australia
CSIRO Food Future Flagship, Black Mountain Laboratories, Canberra, ACT, Australia

Kerrie L. Forrest and Matthew J. Hayden
Department of Primary Industries, Victorian AgriBiosciences Centre, Bundoora, Vic., Australia

Catherine A. Shang and Anna Fitzgerald
Bioplatforms Australia, Macquarie University, North Ryde, NSW, Australia

Craig C. Wood and Philip J. Larkin
CSIRO Plant Industry, Canberra, ACT, Australia

Muhammad Fahim
CSIRO Plant Industry, Canberra, ACT, Australia
Division of Plant Sciences, Research School of Biology, Australian National University, Canberra, ACT, Australia

Anthony A. Millar
Division of Plant Sciences, Research School of Biology, Australian National University, Canberra, ACT, Australia

Michael G. Francki
Department of Agriculture and Food Western Australia, Grains Industry, South Perth, WA, Australia
State Agricultural Biotechnology Centre, Murdoch University, Murdoch, WA, Australia

Sarah Hayton
Separation Science and Metabolomics Laboratory, Research and Development, Murdoch University, Murdoch, WA, Australia

Catherine Rawlinson
Separation Science and Metabolomics Laboratory, Research and Development, Murdoch University, Murdoch, WA, Australia
Metabolomics Australia, Murdoch University Node, Murdoch, WA, Australia

Joel P. A. Gummer and Robert D. Trengove
Separation Science and Metabolomics Laboratory, Research and Development, Murdoch University, Murdoch, WA, Australia
School of Veterinary and Life Sciences, Murdoch University, Murdoch, WA, Australia
Metabolomics Australia, Murdoch University Node, Murdoch, WA, Australia

Feng Gao and Belay T. Ayele
Department of Plant Science, University of Manitoba, Winnipeg, MB, Canada

Mark C. Jordan
Cereal Research Centre, Agriculture and Agri-Food Canada, Winnipeg, MB, Canada

Keith A. Gardner and Ian J. Mackay
The John Bingham Laboratory, National Institute of Agricultural Botany (NIAB), Cambridge, UK

Lukas M. Wittern
Department of Plant Sciences, University of Cambridge, Cambridge, UK

Lin Qi, Xingguo Ye, Lipu Du, Hongxia Liang, Zhiyong Xin and Zengyan Zhang
National Key Facility for Crop Gene Resources and Genetic Improvement/Key Laboratory of Biology and Genetic Improvement of Triticeae Crops of the Agriculture Ministry, Institute of Crop Science, Chinese Academy of Agricultural Sciences, Beijing, China

Wei Rong
National Key Facility for Crop Gene Resources and Genetic Improvement/Key Laboratory of Biology and Genetic Improvement of Triticeae Crops of the Agriculture Ministry, Institute of Crop Science, Chinese Academy of Agricultural Sciences, Beijing, China
Central South University of Forestry and Technology, Changsha, China

Aiyun Wang
Central South University of Forestry and Technology, Changsha, China

Alexandra M. Allen, Gary L. A. Barker, Paul Wilkinson, Amanda Burridge, Mark Winfield, Jane Coghill and Keith J. Edwards
School of Biological Sciences, University of Bristol, Bristol, UK

Cristobal Uauy and Simon Griffiths
John Innes Centre, Norwich, UK

Peter Jack
RAGT, Ickleton, Essex, UK

Simon Berry
Limagrain, Woolpit, Suffolk, UK

Peter Werner
KWS, Thriplow, Hertfordshire, UK

James P. E. Melichar
Syngenta Seeds Ltd, Whittlesford, Cambridge, UK

Jane McDougall, Rhian Gwilliam and Phil Robinson
KBioscience Unit 7, Hertfordshire, UK

Bala A. Akpinar
Faculty of Engineering and Natural Sciences, Sabanci University, Tuzla, Istanbul, Turkey

Hikmet Budak
Faculty of Engineering and Natural Sciences, Sabanci University, Tuzla, Istanbul, Turkey
Sabanci University Nanotechnology Research and Application Centre (SUNUM), Sabanci University, Tuzla, Istanbul, Turkey

Stuart J. Lucas
Sabanci University Nanotechnology Research and Application Centre (SUNUM), Sabanci University, Tuzla, Istanbul, Turkey

Jan Vrána and Jaroslav Doležel
Institute of Experimental Botany, Centre of the Region Hana for Biotechnological and Agricultural Research, Olomouc, Czech Republic

Johanna Acevedo-Garcia, David Spencer, Hannah Thieron, Anja Reinstädler and Ralph Panstruga
Unit of Plant Molecular Cell Biology, Institute for Biology I, RWTH Aachen University, Aachen, Germany

Kim Hammond-Kosack and Andrew L. Phillips
Department of Plant Biology and Crop Science, Rothamsted Research, West Common, Harpenden, Hertfordshire, AL5 2JQ, UK

Index

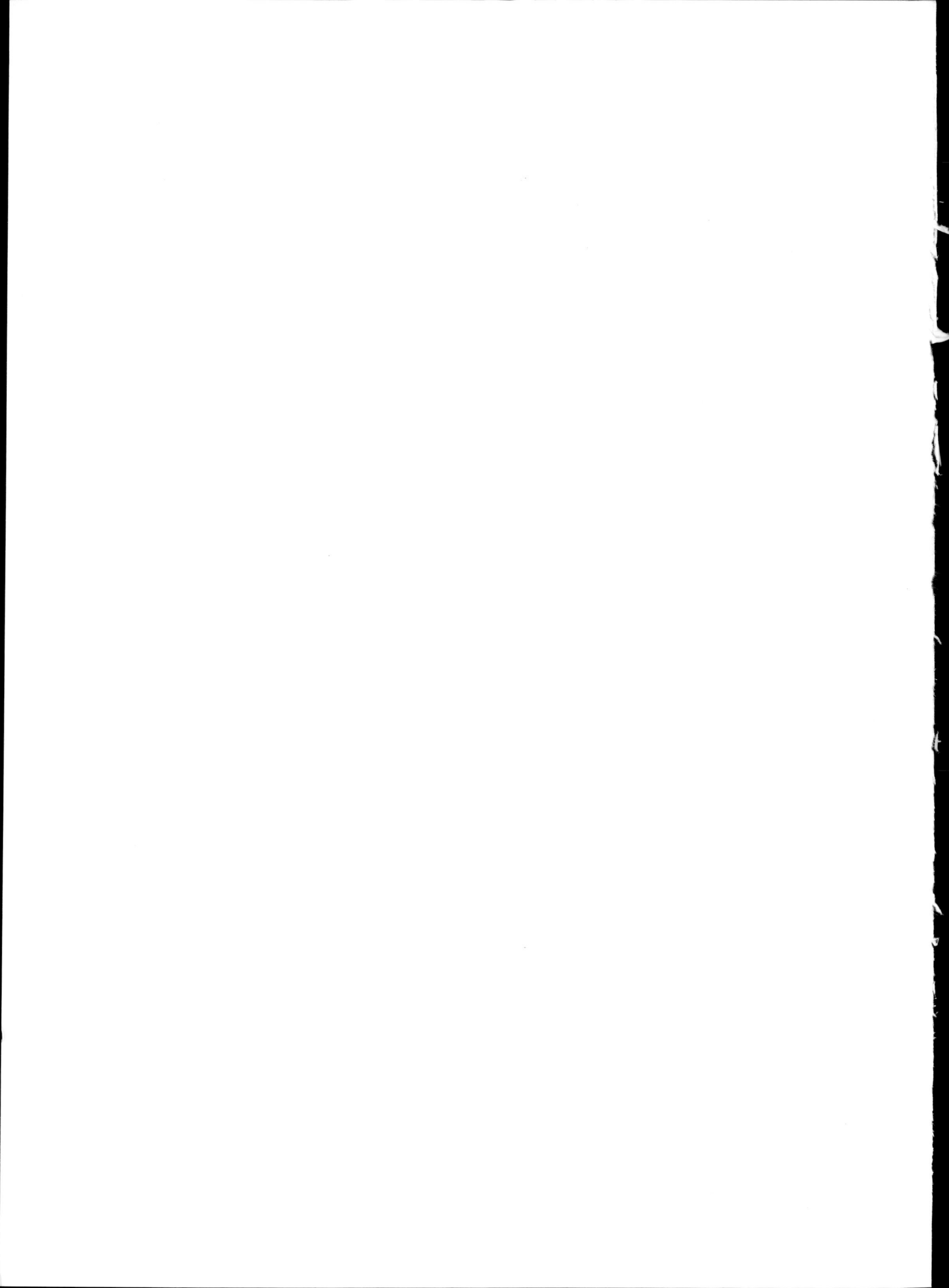